IET TELECOMMUNICATIONS SERIES 79

Satellite Communications in the 5G Era

Other volumes in this series:

Volume 9 **Phase Noise in Signal Sources** W.P. Robins
Volume 12 **Spread Spectrum in Communications** R. Skaug and J.F. Hjelmstad
Volume 13 **Advanced Signal Processing** D.J. Creasey (Editor)
Volume 19 **Telecommunications Traffic, Tariffs and Costs** R.E. Farr
Volume 20 **An Introduction to Satellite Communications** D.I. Dalgleish
Volume 26 **Common-Channel Signalling** R.J. Manterfield
Volume 28 **Very Small Aperture Terminals (VSATs)** J.L. Everett (Editor)
Volume 29 **ATM: The broadband telecommunications solution** L.G. Cuthbert and J.C. Sapanel
Volume 31 **Data Communications and Networks, 3rd Edition** R.L. Brewster (Editor)
Volume 32 **Analogue Optical Fibre Communications** B. Wilson, Z. Ghassemlooy and I.Z. Darwazeh (Editors)
Volume 33 **Modern Personal Radio Systems** R.C.V. Macario (Editor)
Volume 34 **Digital Broadcasting** P. Dambacher
Volume 35 **Principles of Performance Engineering for Telecommunication and Information Systems** M. Ghanbari, C.J. Hughes, M.C. Sinclair and J.P. Eade
Volume 36 **Telecommunication Networks, 2nd Edition** J.E. Flood (Editor)
Volume 37 **Optical Communication Receiver Design** S.B. Alexander
Volume 38 **Satellite Communication Systems, 3rd Edition** B.G. Evans (Editor)
Volume 40 **Spread Spectrum in Mobile Communication** O. Berg, T. Berg, J.F. Hjelmstad, S. Haavik and R. Skaug
Volume 41 **World Telecommunications Economics** J.J. Wheatley
Volume 43 **Telecommunications Signalling** R.J. Manterfield
Volume 44 **Digital Signal Filtering, Analysis and Restoration** J. Jan
Volume 45 **Radio Spectrum Management, 2nd Edition** D.J. Withers
Volume 46 **Intelligent Networks: Principles and applications** J.R. Anderson
Volume 47 **Local Access Network Technologies** P. France
Volume 48 **Telecommunications Quality of Service Management** A.P. Oodan (Editor)
Volume 49 **Standard Codecs: Image compression to advanced video coding** M. Ghanbari
Volume 50 **Telecommunications Regulation** J. Buckley
Volume 51 **Security for Mobility** C. Mitchell (Editor)
Volume 52 **Understanding Telecommunications Networks** A. Valdar
Volume 53 **Video Compression Systems: From first principles to concatenated codecs** A. Bock
Volume 54 **Standard Codecs: Image compression to advanced video coding, 3rd Edition** M. Ghanbari
Volume 59 **Dynamic Ad Hoc Networks** H. Rashvand and H. Chao (Editors)
Volume 60 **Understanding Telecommunications Business** A. Valdar and I. Morfett
Volume 65 **Advances in Body-Centric Wireless Communication: Applications and state-of-the-art** Q.H. Abbasi, M.U. Rehman, K. Qaraqe and A. Alomainy (Editors)
Volume 67 **Managing the Internet of Things: Architectures, theories and applications** J. Huang and K. Hua (Editors)
Volume 68 **Advanced Relay Technologies in Next Generation Wireless Communications** I. Krikidis and G. Zheng
Volume 69 **5G Wireless Technologies** Dr. Angeliki Alexiou (Editor)
Volume 70 **Cloud and Fog Computing in 5G Mobile Networks** Dr. Evangelos Markakis, Dr. George Mastorakis, Dr. Constandinos X. Mavromoustakis and Dr. Evangelos Pallis (Editors)
Volume 71 **Understanding Telecommunications Networks, 2nd Edition** A. Valdar
Volume 72 **Introduction to Digital Wireless Communications** Hong-Chuan Yang
Volume 73 **Network as a Service for Next Generation Internet** Q. Duan and S. Wang (Editors)
Volume 74 **Access, Fronthaul and Backhaul Networks for 5G and Beyond** M.A. Imran, S.A.R. Zaidi and M.Z. Shakir (Editors)
Volume 76 **Trusted Communications with Physical Layer Security for 5G and Beyond** T.Q. Duong, X. Zhou and H.V. Poor (Editors)
Volume 80 **Transceiver and System Design for Digital Communications, 5th Edition** Scott R. Bullock
Volume 905 **ISDN Applications in Education and Training** R. Mason and P.D. Bacsich

Satellite Communications in the 5G Era

Edited by
Shree Krishna Sharma, Symeon Chatzinotas
and Pantelis-Daniel Arapoglou

The Institution of Engineering and Technology

Published by The Institution of Engineering and Technology, London, United Kingdom

The Institution of Engineering and Technology is registered as a Charity in England & Wales (no. 211014) and Scotland (no. SC038698).

First published 2018

The Institution of Engineering and Technology
Michael Faraday House
Six Hills Way, Stevenage
Herts, SG1 2AY, United Kingdom

www.theiet.org

British Library Cataloguing in Publication Data
A catalogue record for this product is available from the British Library

ISBN 978-1-78561-427-9 (hardback)
ISBN 978-1-78561-428-6 (PDF)

Typeset in India by MPS Limited
Printed in the UK by CPI Group (UK) Ltd, Croydon

Contents

Preface

Satellite Communication (SatCom) has been playing a vital role in the wireless world due to its capability of broadcasting telecommunication services to wider geographical areas and delivering broadband connectivity to sparsely populated remote regions, which are typically inaccessible or under-served by the terrestrial communication infrastructures. SatCom technologies have been significantly useful in bridging the digital gap in today's information age by fostering the economic and social development of rural communities and developing countries. Although there are several advances in the terrestrial wireless world in terms of capacity and coverage enhancement, SatCom is the only viable option for delivering telecommunication services in a wide range of sectors such as aeronautical, maritime, military, rescue and disaster relief. Moreover, the demand for emerging applications such as high definition television, interactive multimedia services and broadband internet access is rapidly increasing, thus leading to the ever-increasing need of SatCom systems. More importantly, in order to meet the consumer expectation of the seamless access to any telecommunications services anytime and anywhere including the scenarios like traveling on cruise liners, planes and high-speed trains, satellite should be an important component of the upcoming fifth generation (5G) and beyond wireless architectures.

The upcoming 5G and beyond wireless communications are expected to support a massive number of smart devices, connected sensors and massive machine type communication (MTC) devices having diverse quality-of-service (QoS) requirements. In this direction, 5G wireless systems are envisioned to provide 1,000 times increased capacity, 10–100 times higher end-user data-rates, 5 times lower latency, 10 times increased energy efficiency for low-power devices and to support 10–100 times higher number of connected devices as compared to the current 4G systems. Also, various emerging wireless systems such as broadband systems, Internet of Things (IoT) and MTC systems are expected to be integrated with the legacy networks to utilize the already deployed technologies such as 2G, 3G, long-term evolution (LTE), LTE-advanced, Wi-Fi and satellite. However, there are several challenges in meeting heterogeneous service requirements in terms of achievable coverage, data rates, latency, reliability and energy consumption, and in delivering converged wireless solutions to the end-users. Mainly, future wireless networks will need to provide anywhere, anytime and any device connectivity in a wide range of emerging application scenarios including industrial automation, connected car, E-Healthcare, smart city, smart home, smart grid, communications-on-the-move and high-speed platforms such as trains, airplanes and unmanned aerial vehicles (UAVs).

In the era of 5G wireless, SatCom solutions can complement terrestrial telecommunication solutions in all geographical regions including rural/inaccessible places and urban/suburban areas in terms of providing telecommunication services to the end-users. Satellite backhaul becomes an ideal solution to deliver telecommunication services to the geographically challenging areas since it is difficult to deploy wired backhaul solutions such as copper and optical fiber due to cost and implementation issues. As compared to the terrestrial backhaul, satellite backhaul not only can reduce the infrastructure cost but also can be a backup solution to the terrestrial backhaul links in case of failure or for load balancing in the places/events with high traffic demand. Furthermore, in many applications targeted by 5G and beyond systems such as distributed IoT/MTC networks, content delivery networks (CDNs), and highly distributed small/medium size networks, satellite networks are better suited than the terrestrial only solutions. Therefore, SatComs can be considered as an important means to support the expansion of 5G ecosystem toward highly reliable and secure global networks.

The recent advances in Ku-band, Ka-band, extra high frequency (EHF) band and free space optical technologies have led to the new era of high throughput satellite (HTS) systems. These HTSs are expected to significantly reduce the communication cost of the next generation satellite systems and of the integrated satellite-terrestrial systems. However, the main challenge in emerging HTS systems and non-geostationary (NGSO) constellations is the integration of satellite and terrestrial systems from an architectural perspective so that SatCom systems become an active part of the access network rather than another transparent backhaul medium for the 5G and beyond systems. In this regard, the concepts of employing software-defined networking (SDN) and network function virtualization (NFV) toward enabling the seamless integration of satellite-terrestrial are emerging. These SDN- and NFV-based solutions are envisioned to drastically shift the existing hardware-based system design and implementation toward full softwarization, thus enabling the flexible and adaptive implementation of the 5G ecosystem toward fulfilling the diverse QoS requirements of the end-users.

Satellite systems are challenged in meeting the latency requirements for some applications such as tactile Internet, and there are other challenges such as improving reliability, efficiency, coverage and reducing costs in the dense areas. This is particularly true for geostationary (GSO) satellites, whereas NGSO satellite constellations are in much better position in terms of latency. On the other hand, terrestrial wireless can provide connectivity to indoor and ground-mobile users with low latency but is economically challenging in sparse or intermittent areas. In this direction, the convergence of mobile, fixed and broadcasting systems with the possibility of coexistence of satellite networks with the terrestrial systems is one of the promising future directions. Toward enabling this convergence, SatComs can play a key role in building heterogeneous architectures through hybrid and integrated satellite-terrestrial paradigms. Furthermore, the involvement of satellite makes the deployment of IoT involving sensors and M2M connections in wide areas feasible. Besides, in order to enable Internet of everything, providing precise positioning and context capabilities is a crucial aspect and can be achieved by the combination of satellite and cellular positioning systems. Moreover, the integration of SatComs with the cellular will lead to better availability in emergency and disaster applications. As an example, the delivery of real-time high definition video using satellite in the UAV surveillance applications can be considered. In addition, there is a growing interest for satellite delivery in the transport sector safety services and vehicle-to-vehicle applications.

The aforementioned aspects clearly highlight the need of integrating satellite in 5G and beyond wireless architectures toward enabling the increased convergence as targeted by the 5G community. One promising way of taking mutual benefits from satellite and terrestrial technologies in the 5G ecosystem is to combine them in the same platform in the form of hybrid/integrated networks. However, satellite systems have been mostly used in an overlay manner rather than an integrated form except in the S-band. Also, enhancing the spectral efficiency as well as the total system throughput has been an important concern for future SatCom systems due to continuously increasing demand for broadcast, multimedia and interactive services and the lack of usable satellite spectrum. Although SatCom systems have moved from the traditional monobeam satellites to the multibeam platform and the emerging full-frequency reuse concept can provide significant capacity gains as compared to the conventional four-color reuse method, the problem of cochannel interference needs to be addressed with the help of advanced precoding and multiuser detection schemes. Besides, as the number of cochannel satellites (GSO and NGSO) as well as other cochannel terrestrial systems increases, handling inter-system interference becomes another issue. In this regard, the investigation of suitable spectrum sharing, resource allocation and interference avoidance/mitigation techniques has become crucial toward realizing the next generation Terabit/s SatCom systems.

Motivated by the above-mentioned numerous benefits and the role of SatComs in 5G systems and the associated challenges, several academic institutions, regulators and industries are putting significant efforts in investigating novel satellite-terrestrial integrated solutions and the next generation SatComs technologies/architectures. Mainly, several enabling technologies and architectures such as traffic offloading via satellite-terrestrial hybrid backhaul, high resolution content delivery via satellite-assisted CDN networks, advanced satellite constellation networks such as low Earth orbit (LEO) mega-constellations, medium Earth orbit (MEO) constellations and multilayered LEO/MEO satellite networks, extremely HTS systems of the Terabit/s class, beamhopping satellite systems, onboard signal processing, IoT via satellite, software-defined payloads, SDN- and NFV-based satellite-terrestrial integrated networks are being investigated in the related research communities. Also, there are ongoing activities in the areas of dynamic spectrum sharing, cognitive and cooperative SatComs, resource allocation, advanced interference mitigation techniques, multi-beam joint processing, multiuser detection, advanced precoding techniques, design of smart antennas, optical intersatellite/space-ground links and the exploitation of high frequency bands (Q/V/W/optical) for the gateway connections.

Although there are some recent books in the literature discussing the aspects of 5G cellular communications, the importance of SatComs in 5G and beyond wireless systems has been neglected. In this direction, this book focuses on recent research efforts being carried out toward integrating SatCom systems in the upcoming 5G and beyond systems, and also on various novel enabling technologies for the next generation of Terabit/s SatComs. This book aims to provide significant inputs to academics, researchers, telecom engineers, industrial actors and policy makers such as 5G stakeholders, regulators and research agencies to stimulate future activities in strengthening the role of SatCom in the 5G and beyond wireless systems.

In the above context, this book discusses various emerging concepts/technologies/architectures in the domain of next generation SatComs and integrated satellite-terrestrial systems. The chapters included in this book are presented in the logical sequence of 5G SatCom scenarios and services/networking (Chapters 1–4), channel and propagation aspects (Chapters 5 and 6), physical- and system-level techniques (Chapters 7–10), optical technology-based satellite systems (Chapters 11–12), onboard processing (OBP) systems and techniques (Chapters 13 and 14), advanced collision/interference mitigation, spectrum sharing and latency reduction techniques (Chapters 15–18).

The book starts with an overview of the role of SatCom in the 5G era and the related use cases (Chapters 1 and 2), and then presents the emerging concepts related to SDN (Chapter 3) and NFV (Chapter 4) along with their applications toward the seamless integration of satellite and terrestrial networks. Then, the book analyzes the feasibility of using satellite systems in EHF bands for aeronautical broadband applications along with the characteristics of the aeronautical to satellite channel (Chapter 5). The book advances by presenting the main propagation characteristics of NGSO satellite systems along with some promising capacity enhancement techniques (Chapter 6). Subsequently, various aspects of MEO satellites such as diversity combining and handover techniques are discussed and an SDN-based cost-effective handover architecture is proposed along with some prototype-based test results (Chapter 7). Then, the book presents several advanced compensation techniques which can mitigate the effect of nonlinear distortions in emerging multicarrier satellite systems (Chapter 8). Subsequently, the book analyzes the feasibility of a software-defined radio (SDR)-based precoder for broadband multibeam satellite systems with the help of in-lab validation results (Chapter 9).

The book then proceeds by presenting emerging beamhopping technologies for the next generation satellite systems with a particular focus on the upcoming Eutelsat Quantum-class satellite (Chapter 10). In the context of emerging optical technologies, the book discusses several aspects of optical on–off keying (OOK) data links for emerging LEO downlink applications along with a detailed analysis of the laser communication channel (Chapter 11). In addition, the main elements involved in the design of optical technology-based ultra-high speed relay systems are discussed and the link budget calculation of various associated links is presented (Chapter 12). Next, the book includes two chapters related to the promising OBP paradigm in the next generation satellite systems. Mainly, various design aspects related to OBP are presented toward enabling the satellite-terrestrial integration along with an OBP example use case by employing LEO satellites (Chapter 13). And, some promising onboard interference detection and localization techniques are presented along with their performance evaluation via numerical results (Chapter 14). The book then discusses various conventional and advanced random access (RA) schemes and analyzes their performance with respect to various system constraints (Chapter 15). In the context of hybrid satellite-terrestrial mobile backhaul (MBH) systems, various interference avoidance and mitigation techniques including user-level linear precoding schemes and symbol-level precoding (SLP) schemes are discussed along with their performance analysis (Chapter 16). Moreover, toward enabling dynamic sharing of radio spectrum between satellite and terrestrial systems, various spectrum sharing techniques are discussed along with a practical coexistence example of a fixed satellite service (FSS) system and a terrestrial fixed service (FS) system (Chapter 17). Finally, the book discusses various aspects of two-way satellite relaying (TWSR) including a detailed mathematical analysis of beamforming and combining techniques in TWSR communication systems (Chapter 18). In the following paragraphs, an overview of the main contents of all the chapters is presented.

In Chapter 1, *O. Onireti and M. A. Imran* discuss several key areas where satellites can play significant roles in the 5G systems starting with the highlights on the 5G vision. The key areas discussed include providing ubiquitous connectivity to inaccessible areas such as remote locations, passengers in aircrafts/trains/vessels, emergency and critical scenarios, massive MTCs, resilience provisioning, content caching and multicasting, satellite-terrestrial integrated network (trunking and head-end feed, backhauling and communication on the move) and ultrareliable communication. Furthermore, the authors highlighted and discussed the recent advances in 5G SatComs systems including some ongoing projects on satellite-terrestrial integration [Satellite and Terrestrial Network for 5G (SAT5G), SANSA and VITAL], spectral sharing between satellite and terrestrial systems, mega-LEO constellation, OBP, gallium nitride technology, SDN, multicasting and integrated signaling. Finally, some research challenges and recommendations associated with the integrated satellite-terrestrial architecture, integrated signaling and OBP are provided.

In Chapter 2, *K. Liolis et al.* discuss various promising use cases and scenarios for 5G enhanced Mobile Broadband (eMBB) defined in the context of the European Commission H2020 5G PPP Phase 2 project SaT5G. Starting with a brief discussion on the role of satellite in the 5G ecosystem and the SaT5G project, the chapter presents four different use cases for the eMBB and provides their relevance to the key research pillars, the main 5G PPP Key Performance Indicators, the 3GPP SA1 SMARTER use case families and 5G market verticals. The main use cases included in this chapter include delivery and offloading of multimedia content to the network edges, 5G fixed backhaul to provide broadband connectivity to the places inaccessible by terrestrial communications, complementary connectivity to terrestrial networks in under-served areas and broadband connectivity to the platforms on the move. Furthermore, the chapter provides the qualitative market size assessment for the selected satellite use cases for eMBB based on the satellite operators' perspective and recent industrial developments. Moreover, the chapter describes a set of scenarios for each of the selected use cases along with their qualitative high-level description. Finally, the chapter concludes by highlighting the key aspects of the presented use cases and scenarios.

In Chapter 3, *F. Mendoza et al.* discuss the role of SDN technologies in facilitating the seamless integration and operation of integrated satellite and terrestrial networks. In particular, the realization of end-to-end (E2E) traffic engineering (TE) in a combined terrestrial-satellite network by using SDN technologies is discussed with a specific focus on an MBH network scenario. Furthermore, a system architecture for an SDN-enabled ground segment system is presented, and several candidate SDN data models and interfaces are discussed. Moreover, an integration approach for the realization of E2E SDN-based TE in satellite-terrestrial backhaul networks is presented by abstracting the satellite component as an open flow switch, and two central TE workflows are illustrated to validate the proposed integration approach with one workflow toward computing an optimal path and another workflow to overcome congestion/failures. In addition, the performance of the proposed SDN-based TE application is analyzed via numerical simulations in various scenarios including homogeneous and heterogeneous load situations, unavailability of terrestrial links and the presence of transportable base stations which exclusively rely on the satellite capacity for backhauling. Finally, the chapter provides some concluding remarks and future recommendations.

In Chapter 4, *H. Koumaras et al.* first provide a brief introduction to cloud computing and discuss the functionalities of NFVs. Subsequently, the chapter presents the promising use case scenarios for the integration of cloud networking techniques into satellite networks which are derived from the terrestrial NFV use cases, and adapted to the SatCom context. The discussed scenarios include virtual CDN as-a-service, satellite virtual network operator scenario, dynamic backhauling with edge-processing-as-a-service scenario, and customer functions virtualization scenario. For each of these scenarios, the associated benefits and implementation challenges are discussed. The chapter concludes by providing some future recommendations for the efficient implementation of NFV technology in SatCom systems.

In Chapter 5, *N. Jeannin et al.* discuss the potentialities of using EHF on satellite systems for the provision of aeronautical broadband communication. Starting with an overview of existing or planned systems dedicated to broadband communication, the authors analyze the projected commercial aviation generated traffic demand by considering current commercial aviation traffic and the forecasted data usage. Subsequently, the characteristics of the aeronautical to satellite channel at the EHF bands are presented with a particular focus on the impact of the altitude on the tropospheric impairments. Furthermore, the authors present the latest ITU-R standards which assess the impact of the troposphere on an aircraft-space link along with a discussion on the associated propagation characteristics. Moreover, the authors extrapolate the current aeronautical terminals and satellite characteristics to EHF range to demonstrate the performance improvement at the EHF bands and it is shown that the capacities provided can be enhanced by the use of conformal antennas and provide about 4–10 times capacity improvements over current Ka-band systems. Finally, the chapter concludes by demonstrating the feasibility of EHF satellite systems to meet future Aero passenger requirements.

In Chapter 6, *C. Kourogiorgas et al.* present various link characteristics and system perspectives of the next generation NGSO SatCom systems. The chapter first discusses the main propagation characteristics for the links between ground stations and NGSO satellites including local environmental effects and propagation characteristics via atmosphere. The operation of NGSO systems in lower bands (L-/S-bands) is mostly affected by the local environment, while in high RF bands and optical range, atmospheric effects become dominant and they must be considered for the system design. Regarding atmospheric propagation features, authors provide a detailed discussion on propagation characteristics for RF systems at the Ka-band along with different existing models for calculating total atmospheric attenuation and rain attenuation. Subsequently, the chapter discusses the propagation characteristics for optical NGSO systems by highlighting the effects of clouds and turbulence. Furthermore, the chapter presents some promising techniques to enhance the capacity of NGSO systems including variable and adaptive coding and modulation, and spatial diversity and multiple antenna techniques. Finally, the chapter provides a brief discussion on interference issues and the perspective of NGSO–GSO cooperation.

In Chapter 7, starting with the role of MEO satellites in 5G systems, *Nicolò Mazzali et al.* discuss the system architecture, services, applications and challenges of MEO satellites with a particular focus on the O3b satellite network. Subsequently, the chapter describes the key elements of the E2E channel of MEO satellites including the uplink and downlink radio propagation effects, payload effects and user terminal effects. Furthermore, an overview of the existing handover techniques for MEO applications is provided along with the details on the seamless handover concept. To address the shortcomings of the existing handover solutions in achieving optimal performance and zero packet loss, the chapter proposes an SDN-based cost-effective handover architecture which enables the combination of the concepts of "make-before-break" and "unidirectional switching." Also, the chapter describes a prototype built to demonstrate the handover performance of the proposed solution along with some test results. Moreover, the chapter provides a detailed review of the diversity combining techniques for MEO satellites along with their advantages, drawbacks and trade-offs, and presents the performance of three classic combining algorithms in MEO applications by considering realistic signal and channel models. Finally, the chapter concludes by providing the main insights and future roadmap.

In Chapter 8, *B. F. Beidas* first presents an analytical framework based on Volterra series representation, which characterizes the distortion among carriers suitable for multicarrier satellite applications. Subsequently, several advanced compensation techniques to be applied at the transmitter and receiver to effectively minimize the linear and nonlinear distortion in SatCom systems are presented. As one of the promising solutions at the receiver, the author describes the Turbo Volterra equalization method which iteratively exchanges soft information between equalizer and forward error correcting (FEC) decoders. Furthermore, three different types of predistortion solutions, namely, Volterra-based data predistortion, Volterra based successive signal predistortion and successive data predistortion are discussed for the transmitter side. Moreover, the application of orthogonal frequency division multiplexing (OFDM) signaling for broadband satellite transmission in the forward direction (from the gateway to terminals) is discussed and suitable countermeasure strategies are employed to mitigate the effect of nonlinear distortion in OFDM-based satellite systems. Finally, the chapter concludes by providing recommendations for the applications of the proposed non-linear distortion countermeasures in precoding-based satellite systems and cognitive satellite systems.

In Chapter 9, *Stefano et al.* demonstrate the capability of an SDR-based precoder in enabling the operation of broadband multibeam satellite systems with aggressive frequency reuse modes in the presence of practical impairments. Starting with the discussion of recent projects on precoding, the authors provide a brief review of the related works on precoding for SatComs. Subsequently, a detailed analysis of practical constraints such as instantaneous differential phase distortion, timing mis-alignment and channel state information (CSI) estimation errors for precoding is provided and possible solutions are discussed. Furthermore, the authors describe the practical implementation of precoding techniques with a particular focus on SLP. Moreover, in-lab validation of precoding techniques is presented including the experimental validation of precoded transmission in 2×2 multiple-input–multiple-output (MIMO) system, the evaluation of symbol-level optimized precoding and the uncoded bit error rate (BER) performance analysis of non-negative least squares (NNLS)-SLP. In addition, experimental results for BER performance of the NNLS-SLP and zero forcing precoding are provided and compared with that of the standard non-coded system which employs a frequency division scheme. Finally, the chapter concludes by providing a list of possible future in-lab validation scenarios.

In Chapter 10, *C. Rohde et al.* discuss promising beamhopping technologies for the next generation SatCom systems with a particular focus on the upcoming Eutelsat Quantum-Class satellite designed for the beamhopping operation. Starting with the basic concepts and the benefits of beamhopping technology, the chapter presents the application of DVB-S2X waveform for beamhopping by considering both the traditional DVB-S2X framing and the superframing. It is concluded that superframing has higher relevance for practical feasibility of beamforming systems than the conventional framing. Based on the identified waveform key requirements for applying beamhopping, the superframing specification of the already released DVB-S2X standard is reviewed and the formats 2, 3, and 4 are found to be ready to use for beam-hopping configurations. Subsequently, the technical details and implementation aspects of the upcoming Eutelsat Quantum satellite for beamhopping are presented along with the features like re-configurable beamforming and the highlights on potential applications. Also, the corresponding ground equipment is discussed by exploiting the advantages of wideband processing. Moreover, implementation feasibility of the beamhopping system is demonstrated by means of detection performance results by considering DVB-S2X Super-Frame Format 4. Finally, the chapter is concluded by providing the overall summary and some important benefits of beamhopping for addressing varying traffic demands and enhancing the usable throughput.

In Chapter 11, *Dirk Giggenbach et al.* present various aspects of optical OOK data links for emerging LEO downlink applications. Starting with an overview of previous experimental projects on optical LEO downlinks (OLEODLs), the performance and geometrical restrictions of OLEODLs are discussed along with some insights on the modes of varying data rates in an OLEODL system. Subsequently, the chapter provides a detailed analysis of the laser communication channel including propagation effects, transmission equation, link budget calculation and the process of pointing, acquisition and tracking. Also, the chapter discusses modulation formats based on OOK of the laser signal along with the effectiveness of data rate variation with a different OOK modulation scheme and presents the performance ranges for different receiver implementations and the impact of bit coding and higher layer coding and protocols. Furthermore, the system and component aspects of space hardware for an OLEODL link are described along with the comparison of pros and cons of monostatic and bistatic system designs. Moreover, the chapter discusses the details of ground hardware along with the basic block diagram of DLR's Optical Ground Station Oberpfaffenhofen. Finally, the chapter concludes by providing an outlook to ongoing and future developments.

In Chapter 12, *R. Barrios et al.* define and analyze the key elements involved in the design of future ultra-high-speed relay systems based on optical technologies. Starting with an overview of the relevant missions and demos related to optical communications in the space, the chapter describes a system architecture of a GSO-based relay system along with several physical layer forward error correction (FEC) coding termination options. Subsequently, various aspects of optical channel model including atmospheric effects, pointing errors and microvibrations, and light coupling efficiency are detailed. Furthermore, the chapter presents the relevant noise models and the calculation of link budget of various links including LEO to relay for a small and a big platform, UAV to relay and relay to the ground to provide insights on the possibilities of future ultra-high-speed data relay systems. Moreover, an overview of different FEC codes defined in the framework of the Consultative Committee for Space Data Systems for near earth and deep space communications is provided. In addition, the chapter provides a receiver sensitivity analysis based on the extrapolation of previously reported experiments and presents the comparison of layered coding and decoding on the board of a satellite. Finally, the chapter concludes by providing the main insights and a discussion on complexity constraints for the code design.

In Chapter 13, *R. Wansch et al.* discuss several aspects of OBP to enable the integration of satellite-terrestrial systems in the upcoming 5G ecosystem. Starting with a brief history of onboard processers (OBP), the chapter provides a detailed classification and the applications of OBPs. Mainly, the chapter presents three different types of satellite payload architectures, namely, bent-pipe, digital transparent, and regenerative and describes their components and advantages. Subsequently, the chapter presents a digital payload technology matrix by mapping the circuit technologies to signal architectures and reconfiguration grades along with the comparison of possible FPGA solutions and the advantages of reconfigurable OBPs. Furthermore, the chapter discusses various design aspects of Fraunhofer OBP as an example, including a payload architecture, the main building blocks, digital signal processing module, and a virtual telemetry/telecommand system. Finally, an exemplary 5G Use Case for OBP using LEO satellites is presented and future application scenarios in the 5G landscape are identified.

In Chapter 14, *C. Politis et al.* present interference detection and localization techniques to be applied onboard of a digital transparent processor satellite or in a partially regenerative satellite. Starting with the recent trends on onboard digitization, the chapter discusses the main sources of intrasystem and external interference for satellite systems and describes interference detection techniques with a particular focus on the energy detection approach. Mainly, the performance of an energy detector is analyzed with imperfect signal cancellation in the pilot and data domains by considering phase shift keying modulated signals. Subsequently, the chapter provides an overview of the current localization techniques and describes an interference localization method using the frequency of arrival approach to localize an unknown interferer while only relying on either the affected satellite or the satellite dedicated for interference localization. Furthermore, the chapter presents a localization algorithm to calculate the location of the interferer using the estimated and the calibrated frequencies at the gateway. Moreover, the chapter analyzes the performance of the presented interference detection and localization techniques via numerical results.

In Chapter 15, *K. Zidane et al.* discuss various existing and advanced RA schemes for SatCom systems. Starting with the main motivations for enhancing RA performance on the return link, the chapter describes the widely used legacy RA protocols such as ALOHA, slotted ALOHA and diversity slotted ALOHA along with their performance comparison in terms of the MAC layer analytical throughput. Subsequently, by highlighting the need for advanced RA techniques, the chapter presents several advanced RA schemes, by dividing them into two main categories, namely, synchronous RA and asynchronous RA. Among advanced synchronous RA techniques, the performance of various methods such as contention resolution diversity slotted ALOHA, irregular repetition slotted ALOHA, coded slotted ALOHA, multislot coded ALOHA, multireplica decoding using correlation-based localization and multifrequency contention resolution diversity slotted ALOHA is analyzed for the SatCom return link. Similarly, among advanced asynchronous techniques, various approaches such as enhanced spread spectrum ALOHA, enhanced contention resolution ALOHA and asynchronous contention resolution diversity ALOHA are discussed. Furthermore, the chapter provides general performance comparison metrics for a global comparison of different advanced RA techniques and discusses the application of each scheme with respect to system constraints such as power limitations, lower data rates and signaling overhead reduction. Finally, the chapter concludes by providing some open research challenges for advanced RA schemes in the 5G SatCom systems.

In Chapter 16, *K. Ntougias et al.* describe various interference avoidance and mitigation techniques for hybrid satellite-terrestrial MBH systems. Starting with a brief discussion of 5G radio access technologies and MIMO communication technologies, authors present a high-level overview of a hybrid satellite-terrestrial MBH system. Subsequently, along with the benefits of load-controlled parasitic antenna arrays (LC-PAA) technology, the chapter presents the channel-dependent precoding and low-complexity communication protocol by considering the application of LC-PAA technology in single-cell and multicell MU-MIMO setups. Furthermore, the chapter presents mathematical signal and interference models for single user-MIMO and single-cell MU-MIMO setups, the analysis of user-level linear precoding schemes and the extension to SLP. Moreover, the chapter describes an optimal transmission technique to maximize the capacity of a desired link under an interfered receiver constraint and presents an interference-constrained water-filling algorithm for the effective power allocation. In addition, the chapter proposes a load-controlled multiple-active multiple-passive design for the considered hybrid satellite-terrestrial MBH networks based on a bowtie patch antenna operating at 19.25 GHz. Finally, the chapter analyzes the performance of the presented interference scenarios and techniques via numerical results and demonstrates the feasibility of the employed precoding method and the communication protocol.

In Chapter 17, *M. Hoyhtya and S. Boumard* present various aspects of dynamic spectrum sharing in hybrid satellite-terrestrial systems. Starting with the classification of hybrid satellite-terrestrial systems from the spectrum sharing perspective, the chapter describes various techniques to enable spectrum sharing in hybrid satellite-terrestrial systems along with their applicability in different scenarios. The main dynamic sharing techniques discussed include spectrum sensing, databases, beamforming, beamhopping, and adaptive frequency and power allocation. Subsequently, the chapter presents the interference modeling and analysis of a Ka-band coexistence scenario between an FSS system and a terrestrial FS system along with some example results. Furthermore, the authors describe two promising application scenarios of hybrid satellite-terrestrial systems, namely, autonomous ships and the Citizens Broadband Radio Service (CBRS) system. Under the first application scenario, a high-level communications architecture consisting of satellite and terrestrial components for an autonomous/remote-controlled ship is presented and some related issues are identified. Also, under the CBRS scenario, the chapter presents an architecture of the implemented CBRS system in Finland along with a brief discussion on the trial environment. Finally, the chapter provides some future recommendations for different dynamic spectrum sharing techniques and implementation challenges for hybrid satellite-terrestrial systems.

In Chapter 18, *Arti M. K.* presents various aspects of TWSR including the challenges associated with channel estimation, differential modulation, and beamforming and combining schemes. Starting with a brief discussion on the advantages of two-way relaying and its importance for SatCom systems, the chapter provides a generic signal model for TWSR between two earth stations via a satellite. Subsequently, a detailed theoretical analysis of the training-based TWSR system is presented and the expressions for average BER and ergodic capacity are derived. The presented analysis is validated under different fading scenarios via simulation results. Furthermore, the chapter presents the detailed analysis of the differential modulation-based TWSR, which does not need the CSI, along with the details on the constellation rotation angle calculation. Moreover, beamforming and combining techniques in TWSR communication systems with multiantenna-equipped earth stations are discussed by categorizing them into two categories: beamforming and combining technique based on local channel information, and optimal beamforming and combining technique. For both these categories, theoretical performance analysis is presented in terms of average symbol error rate and diversity order, and the presented analysis is validated via numerical results.

About the editors

Shree Krishna Sharma holds a Ph.D. degree in wireless communications from the University of Luxembourg, the M. Res. degree in computing science from Staffordshire University, United Kingdom, and the M.Sc. degree in information and communication engineering from the Institute of Engineering, Nepal. He has more than 3 years of postdoctoral research experience at SnT, University of Luxembourg and at the University of Western Ontario, Canada in the areas of cognitive satellite communications, 5G wireless and Internet of Things (IoT). In the past, he was with Nepal Telecom for over 4 years as a Telecom Engineer in the field of information technology and telecommunication. He has published more than 80 technical papers in scholarly journals and international conferences and has over 1,200 Google Scholar citations. He is the recipient of several awards including FNR Award for Outstanding PhD Thesis 2015, CROWNCOM 2015 Best Paper Award and 2018 EURASIP Best Paper Award. He is a senior member of IEEE and has been actively serving as a reviewer and a TPC member for several international conferences including IEEE ICC, IEEE GLOBECOM, IEEE VTC and IEEE PIMRC.

Symeon Chatzinotas is the Deputy Head of the SIGCOM Research Group, Interdisciplinary Centre for Security, Reliability, and Trust (SnT), University of Luxembourg, Luxembourg and a Visiting Professor at the University of Parma, Italy. He received the M. Eng. degree in telecommunications from the Aristotle University of Thessaloniki, Thessaloniki, Greece, in 2003, and the M.Sc. and Ph.D. degrees in electronic engineering from the University of Surrey, Surrey, United Kingdom, in 2006 and 2009, respectively. He has over 250 publications, 2,800 citations and an H-Index of 26 according to Google Scholar. His research interests include multiuser information theory, co-operative/cognitive communications and wireless networks optimization. He is the co-recipient of the 2014 Distinguished Contributions to Satellite Communications Award from the Satellite and Space Communications Technical Committee, IEEE Communications Society, the CROWNCOM 2015 Best Paper Award and 2018 EURASIP Best Paper Award.

Pantelis-Daniel Arapoglou has been a communications system engineer with the European Space Agency ESA/ESTEC since September 2011 where he is supporting R&D activities and developments in the areas of satellite telecommunications, digital and optical communications, and high-data-rate telemetry for Earth observation applications. He received the Dr. Eng. Degree from the National Technical University of Athens (NTUA), Greece, in 2007 and the Diploma degree in Electrical and Computer Engineering in 2003. He has participated in the work of Study Group 3 of the ITU-R in SatNEx III and in COST Action IC0802. Currently, he is following SatNEx IV which is funded by ESA. He is also involved in the standardization work of the CCSDS optical working group.

Chapter 1

Role of satellite communications in 5G ecosystem: perspectives and challenges

Oluwakayode Onireti[1] and Muhammad Ali Imran[1]

The next generation of mobile radio communication systems – so-called 5G – will provide some major changes to those generations to date. The ability to cope with huge increases in data traffic at reduced latencies and improved quality of user experience together with a major reduction in energy usage are big challenges. In addition, future systems will need to embody connections to billions of objects – the so-called Internet of Things (IoT) which raises new challenges. Visions of 5G are now available from regions across the world and research is ongoing towards new standards. The consensus is a flatter architecture that adds a dense network of small cells operating in the millimetre wave bands and which are adaptable and software controlled. But what is the place for satellites in such a vision? The chapter examines several potential roles for satellites in 5G including coverage extension, IoT, providing resilience, content caching and multi-cast, and the integrated architecture. Furthermore, the recent advances in satellite communications together with the challenges associated with the use of satellite in the integrated satellite-terrestrial architecture are also discussed.

1.1 Introduction

Mobile cellular communication systems have evolved through a series of standards known as 'Generations' from Analogue (1G) through GSM (2G) via IMT 2000 (3G) to today's LTE (4G) systems. Satellite mobile systems have developed independently of the terrestrial systems and have largely been proprietary, e.g., the Inmarsat system. There has been a loose connection in that the latter has generally used the GSM network model, and more recently, there have been versions of GSM/GPRS and 3G adapted for satellites, e.g., the ETSI GMR series of standards. The result of this separation between the communities is that it is very difficult to integrate the two networks and thus join them so as to provide seamless services over both. Recently, we are waking up to this problem and work is ongoing to enable some integration of 4G

[1]School of Engineering, University of Glasgow, United Kingdom

between satellite and mobile. The next generation of cellular networks, i.e., 5G, is likely to come into operation around 2020. It is seen that satellites will integrate with other networks rather than be a stand-alone network to provide 5G services. Satellite systems are fundamental components to deliver reliably 5G services in all regions of the world, all the time and at an affordable cost. Thanks to their inherent characteristics, the satellite component will contribute to augment the 5G service capability and address some of the major challenges in relation to the support of multimedia traffic growth, ubiquitous coverage, machine-to-machine (M2M) communications and critical telecommunication missions whilst optimising the value for money to the end users.

In this chapter, we set out to discuss the 5G vision. Then, the historical review of the mobile satellite systems (MSSs) is presented stating the key ideas behind each generation and the main operational/proposed satellite systems. Next, the key areas where satellites can play a part in 5G are defined while also illustrating how satellite services can contribute to the 5G key performance indicators (KPIs). In particular, the key areas discussed include coverage, massive machine type communications, resilience provisioning, content caching and multi-cast, satellite-terrestrial integrated network (trunking and head-end feed, backhauling and communication on the move) and ultra-reliable communications. The recent advances in 5G satellite communications are also highlighted and discussed. The discussed topics include the terrestrial and satellite spectrum in 5G, mega-low earth orbit (LEO) constellation, on-board processing technology, gallium nitride (GaN) technology, software-defined networking (SDN) and the integrated signalling. Finally, the concluding remarks are drawn.

1.2 The 5G vision

The global consensus developing is that 5G will be the integration of a number of use cases, techniques, and use environment rather than the development and deployment of a new radio access technology (RAT). 5G aims to provide ubiquitous access to high data services, applications from any device, anywhere and anytime. 5G is expected to be based on customer experience and quality of service (QoS) with the aim of giving the customer the impression of an infinite capacity experience. In order to create such environment, there is the need to integrate various service applications, emerging from various services and access via a mix of access to different wireless and fixed networks. The vision of 5G mobile [1–3] is driven by the predictions of up to 1,000 times data requirement by 2020 and the fact that the traffic could be two-thirds video embedded. Another key driver for 5G is the emergence of Internet of Things (IoT) and the vision of Billions of objects being connected to the internet. This is the enabler to 'smart cities' and other such 'smart' environments and the emergence of what is called 'Big Data' applications where massive amounts of data can be processed to feed a plethora of new applications. For 5G, this implies being able to handle large quantities of low-data communications efficiently covering widespread sensor networks and M2M communications. There are two remaining pillars of the

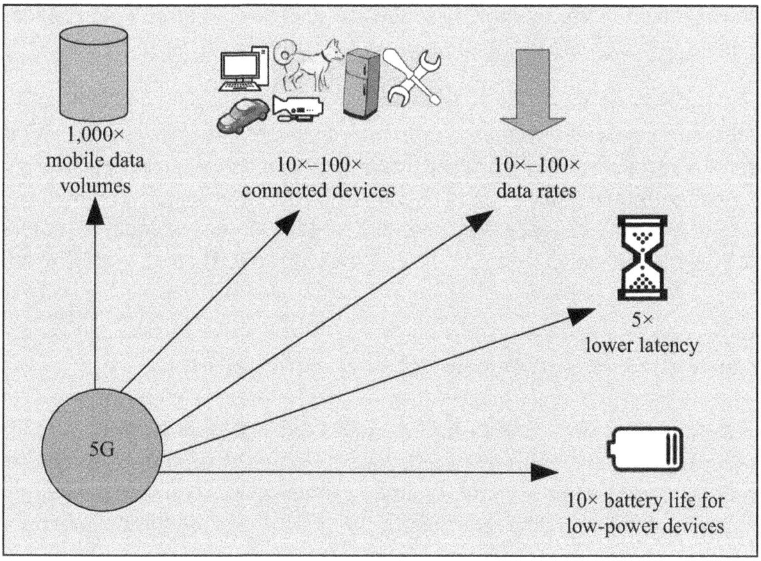

Figure 1.1 5G requirements

5G vision. The first is ensuring availability, reliability and robustness. The second and increasingly important issue is that of reducing energy. The target is a reduction by 90% of today's total energy by 2020 at no reduction in performance or increase in cost. Thus, 5G network design becomes a complex task involving link and area spectral efficiency together with energy efficiency [4]. The overall technical requirements for a 5G network as highlighted by the 5G Infrastructure Public–Private Partnership (5G PPP) can be summarised as follows [5,6]:

- 1,000 times higher mobile data volume per area,
- 10–100 times higher number of connected devices,
- 10–100 times higher typical user data rate,
- 5 times reduced end-to-end latency,
- 10 times longer battery life for low-power devices and
- Ubiquitous 5G access including in low-density areas.

Figure 1.1 shows the estimated requirement in 5G as compared with the 4G system. Of all the technical goals for 5G, the higher data rate requirement is the one that gets the most attention across the board, and this will be achieved in terrestrial systems through the combined gain from three key technologies, namely, [7]

- Increase spectral efficiency, through advance multiple-input–multiple-output (MIMO) technology, to support more bits/s/Hz per node.
- Extreme densification and offloading to improve the area spectral efficiency, i.e., more active nodes per unit area and bandwidth.

- Increase bandwidth, by moving to the millimetre wave (mmWave) spectrum and by making better use of the unlicensed spectrum in the 5-GHz band.

The combination of more Hz (bandwidth), more nodes per unit area and Hz and more bits/s/Hz will lead to many more bits/s per unit area. In general, 5G research activities are in an effort to deliver the technology that meets the ambitious KPIs of the 5G vision highlighted in the 5G-PPP. Meanwhile, the 5G research activities are mainly driven by the terrestrial operators, and hence, they do not adequately consider and evaluate the requirements from use cases which are specific to the satellite operators.

1.3 Satellites and previous cellular generations

Table 1.1 shows the evolution of the MSS and the key ideas behind them. The first major satellite operator, Inmarsat, came into existence at around the same time as the first cellular operators providing 1G analogue services. Over this period, using the L band and global beam coverage satellite, Inmarsat provided low data rate services and speech services to the maritime market of ships. In the early 1990s, Inmarsat was able to add aeronautical services to passenger aircraft with its advancement into spot beam higher power satellites. Later in 1997, worldwide spot beam operation, paging, navigation and higher rate digital to desktop terminals were introduced in the MSS. In the mid-1990s, several regional geostationary earth orbit (GEO) satellite systems such as OPTUS, AMSC, EUTELTRACS and OMNITRACS emerged focusing on land vehicles while using both the Ku and L bands. Research activities in the late 1980s and early 1990s focused on non-GEO constellations and resulted in the proposal of medium earth orbit (MEO) and LEO satellite system. Typical examples such as Globalstar and Iridium MSS came into service but were too late to compete with the spread of terrestrial GSM. A major issue with both companies was the business case as the cost of constellations was too expensive leading to their bankruptcy. Other companies such as ICO and Orbcomm have also suffered a similar fate.

In the mid-1990s, super GEO satellites were proposed with around 100–200 spots rather than the earlier generation GEOs' with 5–10 spots. Of the proposed systems, Thuraya [8] was the one that reached the market in the early 2000s, offering GPRS and GSM like services to Asia and much of Europe. Super GEO found a niche with travellers, trucks and in areas where terrestrial mobile was too expensive to deploy. Inmarsat IV a super GEO took the digital service rate from 64 to 432 kb/s from the global area network (GAN) to the broadband GAN (BGAN) [9]. Despite the move of terrestrial operators to code division multiple access in 2004–05, Inmarsat developed its proprietary time division multiple access system to deliver 3G equivalent packet services. High data rate (HDR) BGAN, which exceeds symmetric 700 kb/s, became available since 2013. HDR also supports bonding for a total bandwidth exceeding 1 Mb/s. Regarding M2M communications, Orbcomm offers data-only M2M services with a constellation of LEO in the VHF band, and in partnership with Inmarsat, they offer M2M services in the L band. Inmarsat also offers the M2M version of

Table 1.1 Mobile satellite developments

Cellular		Research ideas	Operational/Proposed systems
1G	1970s	Mobile satellite expts ATS-6	Inmarsat formed
	1980s	Non-GEO mobile cellular architecture proposed	Inmarsat operates – maritime
2G GSM	1990s	Motorola announce Iridium system LEO Orcom system proposed Teledesic announce non-GEO fixed systems Globalstar/ICO proposed Super GEO's announced Agrani/Apmt/Aces/Thuraya	Inmarsat operates – land/aero Regionals: Omnitracs, Euteltracs, Amsc, Optus Inmarsat Sats-spots Iridium operational Orbcom operational Globalstar operational World space radio
3G IMT-2000	2000s	Integrates S/T/UMTS for content proposed Satin EU project DVB-S2 standard	Iridium/Globstar/Orbcoms Thuraya operational Inmarsat IV's –100's spots and DSP processor Xm, SIRIUS, DARS MBSAT
4G	2010s	High throughput satellite	Inmarsat Global express constellation – 100's fixed spots and additional steerable spot beams Iridium-NEXT operational-features data transmission O3b satellite constellation
5G	2020s	High throughput satellite Several hundred spot beams Higher frequency bands – Q/V/W Optical for gateway connections Up to 30 m deployable antennas at L/S bands Adaptive beam hopping and forming Mobility management integration Progressive pitch technology	OneWeb satellite constellation SpaceX satellite constellation Samsung satellite constellation LeoSat constellation

BGAN called BGAN M2M, while Iridium's low bandwidth modes are also often used for M2M.

For the period 2020/25, a trend to larger and more powerful GEO satellites that will take capacities from 100's Gbps to over a Terabit/s is expected. The capacity increase will be achieved via several hundreds of spot beams and higher order frequency reuse despite the limitation in the spectrum. Furthermore, higher frequency bands such as the Q, V and W bands will be used together with optical technology for the gateway connections. Also, advances in satellite payload technology through optimised designs and new materials will enable an increase in the payload power from 20 to 30 kW and the use of up to 30 m deployable antenna. Techniques such as adaptive beam hopping and forming, and interference management will be utilised to improve connectivity and flexibility to fluctuating traffic demands and patterns. In addition, following the innovations of using different orbits by O3b, new non-GEO systems that utilises all-optical technology, i.e., between satellites and from satellite to ground, are likely to appear.

1.4 Areas where satellite can play a part in 5G

Satellite communication is becoming an important element in the 5G ecosystem, complementing wireless and fixed terrestrial communications. In the light of this, the third generation partnership project (3GPP) has identified 5G use cases wherein non-terrestrial network components and in particular, satellites have a role. The three main roles identified by 3GPP for satellites in 5G are [10–12]

- Fostering the roll-out of 5G service in un-served areas that cannot be covered by the terrestrial 5G network (e.g., isolated/remote areas, onboard aircrafts or vessels) and under-served areas (e.g., sub-urban/rural areas). Furthermore, to upgrade the performance of limited terrestrial networks in a cost-effective manner.
- Reinforcing the 5G service reliability by providing service continuity for M2M/IoT devices or for passengers on board moving platforms. Also, to ensure service availability anywhere especially for critical communications.
- Enabling 5G network scalability by providing efficient multi-cast/broadcast resource for data delivery towards the network edged.

In this section, we expatiate further on the key areas where satellites can play a part in the 5G ecosystem. The areas discussed include coverage, massive machine type communication, resilience provisioning, content caching and multi-cast, satellite-terrestrial integration (trunking and head-end feed, backhauling and tower feed, and communication on the move) and ultra-reliable communications.

1.4.1 Coverage

The overall aim of 5G is to provide ubiquitous connectivity for any kind of device and any kind of application. This can only be realised by the integration of satellites with the 5G network. Compared to the terrestrial cellular operators, satellite communications operators can provide a single global network and reduced operational and business support cost. This makes cost-effective global service and data delivery only possible via satellite technology. Hence, data and service delivery to remote locations, passengers in aircrafts, trains and vessels, difficult to reach areas (emergency and critical scenarios) as well as beyond country boarders are the leading market opportunity for the satellite network operators (SNOs). Moreover, the advantage of satellites regarding coverage are expected to further increase in the light of the following:

- A mega-constellation of LEO satellites that can offer services such as effective global transit and fine-grained geo-location ubiquitous access.

- Future deployment of cloud computing resources in space.
- Capacity increase due to new concepts such as spatial reuse of frequencies and spectral efficiency gain via new modulation codes.
- Advancement in technologies that exploit the predictive position of satellites and the geo-location of ground equipment leading to devising adaptive and more efficient schemes.

1.4.2 Massive machine-type communications

Massive machine-type communications entail the ability to support a massive number of low-cost IoT devices (connections) with very long battery life and wide coverage including the indoor environment. The exponential increase in the number of connected devices requires that new technologies towards massive data aggregation and data broadcasting which are beyond terrestrial radio must be considered. The intrinsic broadcasting capabilities of satellites which enable them to reach a very high number of devices while consuming only a limited number of resources makes them highly suitable for dispersed M2M networks. Satellite networks offer the means for massive data aggregation through the geo-observation environment as well as a means to share uplink connectivity in a very efficient manner from a very large number of connected network area. In addition, satellites already support asset tracking applications which can be scaled to support future M2M/IoT communications.

On another note, the deployment of a large number of devices poses a clear operational challenge, as the devices have to be maintained (security patches, etc.), configured and upgraded from time to time. Satellites can support/overcome the operational challenges associated with the massive deployment through the following:

- Efficient distribution of data on a massive scale and with global reach, complementing terrestrial deployments.
- Offering an on-demand backhaul capacity without the need for deploying additional terrestrial infrastructures. The on-demand nature is due to the fact that majority of the M2M services require intermittent backhaul.
- Providing a very efficient connectivity alternative for M2M communication. Satellites can also provide an alternative for remote and isolated areas as well as in dense inter-domain networks where data packets have to be passed through multiple autonomous systems to reach their destination. This represents the current market of the satellite network, where M2M is now becoming one of the important connectivity services.
- Roaming using a single satellite operator. Satellite networks can reach a wide area, crossing any type of boarders and through this ensuring the availability of connectivity through a single provider.
- Device activation and configuration via satellite for using local network infrastructure.
- Backup for continuous connectivity availability of the communication when no terrestrial network is available.

1.4.3 Resilience provisioning

Global coverage and dependability are and will remain the pivotal added value of the satellite (space)-related communication services while using the minimum amount of infrastructures on the ground. Due to these unique characteristics, satellite networks are currently used for highly reliable communications and for safety and security critical systems such as navigation information in the maritime domain. Satellites have an important potential role to play in supporting the overall resilience by complementing other communications' infrastructures. Satellites can support a resilient 5G network to mitigate the problems of overload/congestion to meet the 5G KPI 'ensure for everyone and everywhere access to a wider portfolio of services and applications at lower cost'. In order to achieve this, intelligent decisions can be made about traffic routing by placing an intelligent router functionality (IRF) at the radio access network (RAN). The IRF specifically make decisions concerning traffic routing over heterogeneous links taking into account the requirements of the applications. For example, in times of congestion on the normal terrestrial links, the IRF ensures that traffic flows over the satellite link in a seamless manner until the terrestrial links are restored. Hence, the satellite can be used to sustain ultra-high availability from the end-user perspective. Moreover, the cost can be scaled by sharing the satellite capacity over a large number of sites.

1.4.4 Content caching and multi-cast

Caching in terrestrial networks has been proven as an effective approach for improving the network performance in terms of delay and throughput. The limitation in the terrestrial storage capacity and the tendency for it not to be available in network scenarios such as for sailors on board of a ship makes the satellite option very important. Also, the larger storage capacity and the introduction of more advanced on-board processing have made satellites to become more powerful [13]. Satellites have a major role to play in content caching near the edge, i.e., bringing the content closer to the user in order to achieve the 5G KPI of zero perceived delay and 1,000 times higher wireless capacity. The benefits of using satellite for providing content multi-cast and caching include the following:

- It offers a wide coverage with low number of intermediary autonomous systems.
- It offers ultra-low content access latency.
- Offloading the cache content population from terrestrial networks.

Caching content closer to the edge using efficient multi-cast delivery will improve the end user quality of experience (QoE) and reduce backhaul traffic load. This form of content delivery can be managed using information centric network systems or other variations incorporating SDN/network functions virtualisation (NFV) with a centralised controller function that optimises delivery using satellite links when appropriate to provide immediate and on-demand content access.

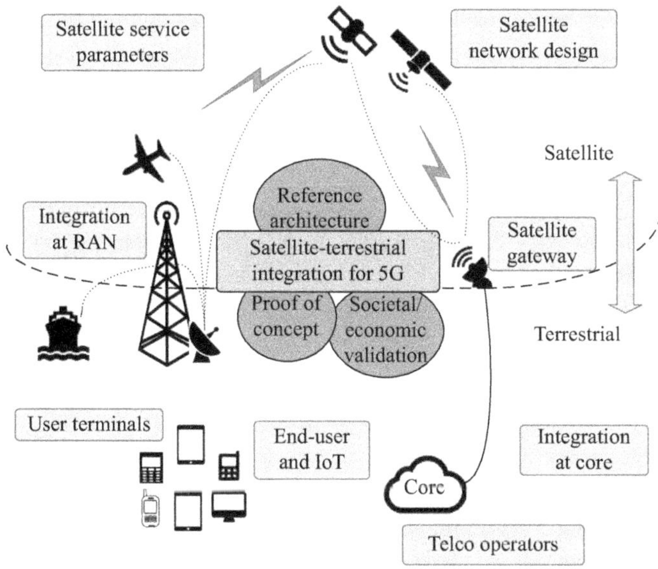

Figure 1.2 Integrated satellite-terrestrial architecture

1.4.5 Satellite-terrestrial integration in 5G

The integration of satellite communications with the terrestrial mobile communication systems has always been difficult due to the stove pipe approach of each sector [4]. Hence, massive re-engineering and cost are usually associated with such integration. For instance, the current satellite networks mainly support 2G network backhaul for fixed sites with limited connectivity and emergency scenarios, while 3G and LTE networks are now following an extensive engineering effort for standards adaptation towards the specific satellite characteristics. Meanwhile, the convergence requirement of the new 5G ecosystem offers a rare opportunity of overcoming some of the barriers associated with integration in the previous generations of terrestrial network deployments through the development of a single environment from the initial stages of development. In addition, it also enables the satellite and mobile communication industries to work together on defining and specifying a holistic 5G system. Such holistic approach will ensure that satellite communication can address some of the challenges associated with supporting the requirements envisaged for 5G networks.

An integrated satellite-terrestrial network ecosystem is shown in Figure 1.2 with integration at the RAN as well as the core network. The network model assumes the satellite network architecture consisting of satellites which connect to the satellite gateway and the satellite terminals via asymmetric links. The terrestrial RAT could include the new 5G radio, Wi-Fi and the LTE, as well as radio technologies developed for ship-to-ship and device-to-device communications. The integrated satellite-terrestrial architecture requires a holistic evaluation in terms of proof of concept for various scenarios. Key components of such evaluation include adding the satellite parameters to the 5G requirements, new satellite-based service and the end user which consist of a multi-radio terminal. The societal, economical and business validation of the integrated architecture is also very important.

Integrating satellites with the terrestrial system is perhaps the key area that enables many advantages. One of such is improving the user's QoE by intelligently routing traffic between the delivery systems and caching high capacity video for onward transmission terrestrially. This can be empowered by the inherent multi-cast/broadcast capabilities of satellite systems, while propagation latency is no longer an issue thanks to intelligent caching. Offloading traffic from the terrestrial system to save on valuable terrestrial spectrum opens up the possibility of improving resilience and security using the two networks. Three main use cases can be identified for the integration of satellite-based solution in 5G namely, trunking and head-end feed, backhauling and tower feed, and communication on the move.

1.4.5.1 Trunking and head-end feed

Satellites can provide a very high-speed direct connectivity option to remote/hard to reach locations. A very high-speed satellite link, which can be up to 1 Gbps or more, from GEO and or non-geostationary satellite will complement the existing terrestrial connectivity to enable:

- High-speed trunking of video, IoT and other data to a central site, with further terrestrial distribution to local cell sites (3G/4G/5G cellular), for instance neighbouring villages.
- Inter-cluster satellite link for remote clustering.
- Inter-cluster satellite link for edge communities.
- Inter-cluster satellite link for overflow communities.
- Remote IoT system with satellite integration.
- LEO satellite providing low latency control plane offloading.

1.4.5.2 Backhauling and tower feed

One of the major issues in 5G is seen to be the increased demands on the backhaul with very large numbers of small cells. Hence, an obvious application of satellite communication in a 5G delivery architecture is in the backhaul segment of the network. High throughput satellites (HTS) can be used here to complement terrestrial provision and provide backhaul in areas where it is difficult to do so terrestrially. HTS can provide a high-speed connectivity complement (include multi-cast content) to wireless towers, access points, and the cloud, as illustrated in Figure 1.3. In general, a very high-speed satellite link (up to 1 Gbps or more) direct to base stations, from GEO and or non-geostationary satellite will complement the existing terrestrial connectivity to enable:

- Backhaul connectivity to individual cells with the ability to multi-cast the same content (e.g., video, HD/UHD TV, as well as non-video data) across a large coverage area.
- Efficient backhauling of aggregated IoT traffic to multiple sites.

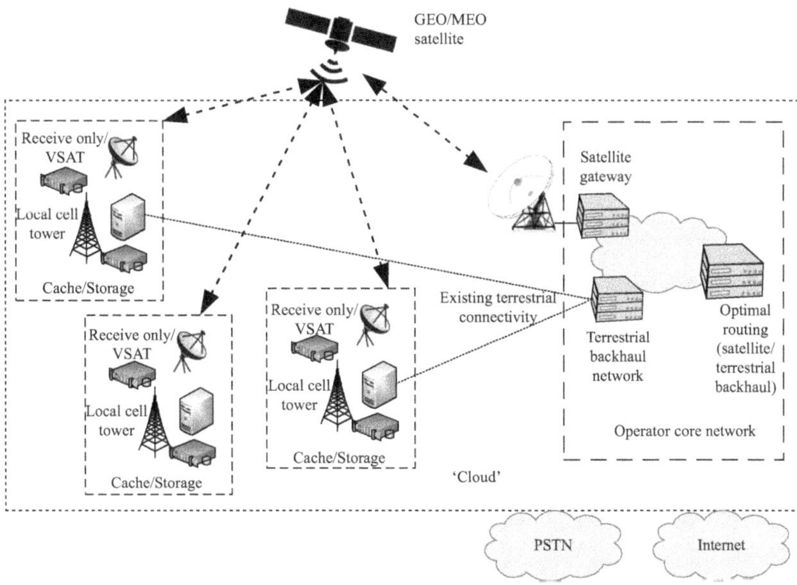

Figure 1.3 Satellite for backhauling and tower feed

In a virtualised and SDN, it might also be possible to include some of the network node functions on board the satellite and thus save on physical sites on the ground. Moreover, satellite in the backhaul can assist with populating content caches closer to the edge, deliver over-the-air configuration updates and software patched for M2M solutions and support the instantiation of network functions at the edge in mobile edge computing solutions through replication of virtual machines via broadcast.

1.4.5.3 Communication on the move

One of the 5G aims to cover mobility use cases that are beyond the reach of the current technology. This entails providing support via a global network that spans across different countries and to high-speed platforms such as airplanes, train and automotive. In such use cases, satellite networks have already proven themselves to be a viable alternative. The integrated satellite and terrestrial solution offers an efficient solution for both the relatively low-speed mobility use case via terrestrial means and through satellite communication for the high-speed mobile device while offering a smooth handover and seamless user experience [14]. Satellites provide a direct and/or complementary connection to users on the move (e.g., on airplanes, trains, automobiles and ships), as illustrated in Figure 1.4. Very high speed, multi-cast-enabled satellite link (up to 1 Gbps or more) direct to plane, train, car or vessel, from GEO and or non-geostationary satellite will enable:

- Backhaul connectivity and multi-casting of (e.g., video, HD/UHD TV and non-video data) where it may not be otherwise possible.
- Direct connectivity and/or efficient backhauling of aggregated IoT traffic.

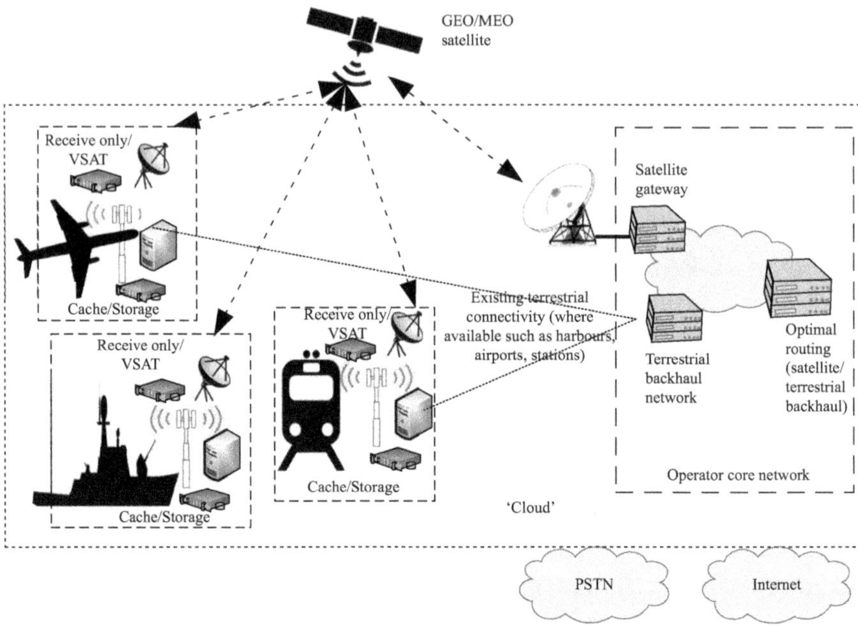

Figure 1.4 Satellite for communications on the move

- Entertainment update with satellite integration for air (connected aircraft) and sea (connected ships).
- Freight and logistics.
- Lorry monitoring and communications in a dual mode terrestrial and satellite solution.

1.4.6 Ultra-reliable communications

New applications and use cases in 5G, such as mobile healthcare and autonomous vehicles, require the support of a very low latency typically sub-1ms, and very high availability, security and reliability. Hence, a very low latency over the radio network is one of the aims of 5G. Achieving a low latency over the end-to-end service level communication is restricted by the physical limitations and is impossible without moving the functionality to the edge of the network, at a location close to the termination of the very low delay 5G radio network. Consequently, in order to meet the delay requirements, the only economical alternative is to make available compute capacity at the edge of the network and short-circuit the end-to-end network for the stringent latency services [14]. Services requiring a delay time less than 1 ms must have all their content served from a physical location very close to the user device. Possibly at the base of every cell, including the many small cells that are predicted to be fundamental to meeting the densification requirements [15]. In order to achieve the short service path, all the obligatory functionalities for the service delivery should be made available at the edge, thus making the backhaul capacity and delay characteristics beyond the edge node irrelevant to the actual service delivery delay.

The propagation latency of GEO satellite, which is about 270 ms (540 ms round trip), is acceptable in some 5G use cases. The MEO and LEO satellite network will be able to support more latency sensitive applications. The propagation latency of the connectivity service will also be managed by an adequate size and topology of the constellations, the dynamic configuration of client beams as well as delay-tolerant networking. Meanwhile, the processing latency can be managed by an adequate distribution of the execution of the virtual functions across space-and-ground-based data centres.

1.5 Recent advances in 5G satellite communications

In this section, we present the recent advances in 5G satellite communication. The recent advances covered include ongoing projects on satellite-terrestrial integration, terrestrial and satellite spectrum, mega-LEO constellation, on-board processing, GaN technology, SDN, multi-casting and integrated signalling.

1.5.1 Ongoing project works on satellite-terrestrial integration

European Commission-funded projects on the satellite-terrestrial integration under the horizon 2020 (H2020) framework include the following.

1.5.1.1 Satellite and terrestrial networks for 5G

Satellite and terrestrial networks for 5G (SAT5G) will bring satellite communication into 5G by defining optimal satellite-based backhaul and traffic offloading solutions. It will research, develop and validate key 5G technologies in order to take the best value of satellite communication capabilities and mitigate its inherent constraints such as latency. SAT5G will identify novel business models and economically viable operational collaborations that integrate the satellite and terrestrial stakeholders in a win-win situation [16].

1.5.1.2 Shared access terrestrial–satellite backhaul network enabled by smart antennas (SANSA)

The aim of SANSA project is to boost the performance of mobile wireless backhaul networks in terms of capacity and resilience while assuring an efficient use of the spectrum. SANSA project proposes a spectrum efficient self-organising hybrid terrestrial-satellite backhaul network based on three key principles.

- A seamless integration of the satellite segment into terrestrial backhaul networks.
- A terrestrial wireless network capable of reconfiguring its topology according to traffic demands.
- A shared spectrum between satellite and terrestrial segments.

It is expected that a combination of the principles will result in a flexible solution that can efficiently route the mobile traffic in terms of capacity and energy efficiency while providing resilience against link failures or congestion and easy deployment in rural areas [17].

1.5.1.3 VIrtualised hybrid satellite-terrestrial systems for resilient and flexible future networks (VITAL)

The VITAL project addresses the combination of terrestrial and satellite networks by pursuing two key innovation areas, by bringing NFV into the satellite domain and by enabling SDN-based, federated resources management in hybrid satellite communication-terrestrial networks. Enabling SDN-based, federated resource management paves way for a unified control plane that would allow operators to efficiently manage and optimise the operation of hybrid satellite communication-terrestrial networks [18].

1.5.2 Terrestrial and satellite spectrum in 5G

The use of the larger bandwidth in the mmWave band is fundamental to meeting the 5G terrestrial networks requirement. With part of the mmWave band currently allocated on a co-primary basis to a number of other applications such as the fixed satellite services (FSSs), the Federal Communications Commission (FCC) wants a more flexible framework for the use of the electromagnetic spectrum above 24 GHz. Recently, field test data were used to assess the potential interference between terrestrial mobile broadband (5G) and FSS systems sharing the 28-GHz band [19]. The aim of the work in [19] was to create service rules for the use of four spectrum bands to be shared by terrestrial and satellite systems. The bands are namely 28 GHz (27.5–28.35 GHz), 37 GHz (37–38.6 GHz) and 39 GHz (38.6–40 GHz) bands, and an unlicensed band at 64–71 GHz. These high frequencies were traditionally for satellite or fixed microwave. The field test measurement showed that the interference from existing transmit FSS earth station into 5G networks can be controlled by limiting the power flux density at 10 m above the ground level to $-77.6\ \text{dBm/m}^2/\text{MHz}$.

The feasibility of the co-existence between FSSs and mmWave terrestrial network was also investigated in [20] by evaluating the interference to noise level at the FSS and different terrestrial base station deployment and configurations. The configurations considered include multi-tier distribution of base stations and having RF beamforming at the transmitters. It was shown that by exploiting the characteristics of the mmWave scenario such as large antenna array and high pathloss, the co-existence of the mmWave terrestrial base station and FSS in the same area can be made possible. Furthermore, it was established that parameters such as the FSS elevation angle, the base station density and the protection distance are vital in the network deployment in order to guarantee the FSS functionalities.

1.5.3 Mega-LEO constellation

HTSs provide large capacity connectivity via multi-spot beam technology and frequency reuse at a reduced cost. The integration of GEO HTS with the terrestrial

Table 1.2 Planned LEO-HTS mega-constellations

Constellation	LeoSat	SpaceX	OneWeb
No. of satellites	78–108	4,000	640+
Altitude (km)	1,400	1,100	1,200
Latency (ms)	50	20–30	20–30
User speed	1.6 Gbps	1 Gbps	50 Mbps
Cost ($)	3.5B	10B	2.3B
Announced market	Enterprise, mobility, backhaul	Broadband, backhaul	Broadband, mobility

systems will provide a global large-capacity coverage. However, this comes with the challenge of a large propagation delay. Mega-LEO constellations, which are LEO systems of hundreds of satellites, can circumvent this issue and it has recently received significant attention. Mega-LEO constellation can be used to provide LTE broadband services to areas that are not connected to a terrestrial infrastructure as demonstrated in [21,22]. In [21], the authors analysed the impact of propagation delay and Doppler shift in LEO systems on the LTE PHY and MAC layer procedures. An extension of the analysis with a focus on the waveform design, random access and hybrid automatic repeat request procedure is presented in [23]. The effect of the Doppler shift in LEO systems on the waveform can be compensated by accurate location estimation. Furthermore, the impact of the propagation delay on the random access procedure can be limited by increasing the random access response timer. Table 1.2 shows some planned mega-LEO constellation and their specifications.

1.5.4 On-board processing

In on-board digital processed systems, the received waveforms are demodulated and decoded to the digital packet or bit level. This leads to increased system flexibility in terms of signal and information routing, mesh connectivity and resource management. Other gains include higher user and system throughput and higher link efficiency, which are gained from predistortion and interference mitigation, use of newer waveform and full duplexing. On-board digital processed systems are thus the future for satellite communications and this is stimulated by the following:

- An increase in the operational lifespan of the satellite. Over this period, new access characteristics may be required or the need to support a new service/user connectivity topology may arise.
- An increase in the flexibility of the payload in terms of bandwidth and agility in frequency configuration at the payload level.
- Increased configurability and reconfigurability of the payload to support cross-band inter-transponders and/or inter-beam configuration in a high spot beam coverage.

Even though, many applications only need the conventional bent-pipe delivery of bandwidth, as it remains the most efficient way of supporting services such as broadcast television. The evolution in technology and trend in service providers means an increase in the contents that are being personalised and delivered in unicast or multi-cast rather than the traditional broadcast. Hence, on-board processing will play a prominent role in the future as more and more services and content are delivered by Internet protocol connection. Meanwhile, a hybrid payload where the bent-pipe and on-board processing technologies co-exist such as the Intelsat 14 payload, reflects how the near future satellite could look like. Such hybrid deployment is expected for many years until the volume of space routers go up and the technology cost goes down. The new potential solutions for the next generation on-board processing systems must consider the following:

- reduction in the size, weight and power (SW&P) consumption at the payload level;
- reduction in the component integration scale;
- improvement in the payload reconfigurability and flexibility;
- improvement in the uplink and downlink performances.

1.5.5 GaN technology

GaN technology is a promising candidate for the next generation satellite communication subsystems [24]. Satellites in existence rely on the proven gallium arsenide (GaAs) and travelling wave tubes (TWT) technologies for most of its radio frequency (RF) front-end hardware. Moreover, the maturation of GaN technology and its commercial adoption gives way to striking advancement in the space industry. The advantages which make GaN the main candidate for space include reliability, radiation hardness and high-temperature operation, in addition to the generic advantages of high-added efficiency, high power density and high operational frequency [25,26]. The latter three, which also improves the overall efficiency in the RF chain, makes GaN technology very suitable for the 5G base station design where MIMO and mmWave technologies will be operational.

The cost advantage of GaN over TWT amplifiers (TWTAs) and GaAs solid-state power amplifiers (SSPAs) is realised by eliminating kW power supplies for TWTAs and cooling hardware for GaAs SSPAs. This leads to a reduction in size and weight which saves fuel and area on the payload. GaN technology's offer of a lightweight compact form factor is undeniable and also offers the possibility for achieving small form-factor nano- and micro-satellites where the physical size, mass, power consumption and cost pose serious restrictions. It is expected that the development of the GaN technology will continue to be by the high power RF properties [26]. The potential to achieve the whole receive front-end of the satellite with GaN technology will further create the advantage of a lower cost and improved integration [24]. To this end, several projects have been initiated to perform intensive test and analysis on GaN technology in order to exploit its potential. Such project include GaN powered Ka-band high-efficiency multi-beam transceivers for SATellites (GANSAT), GaN

Reliability Enhancement and Technology Transfer Initiative (GREAT), AlGaN and InAlN-based microwave components (AL-IN-WON).

1.5.6 Software-defined networking

SDN and NFV technologies are key enablers of a more flexible and improved integration of the terrestrial and satellite segments. SDN involves decoupling the control and user planes of the network equipment and logical centralisation of the network intelligence, i.e., the control plane [27–29]. The user plane, i.e., the underlying network infrastructure, is abstracted for external applications requesting services through the control plane. On the other hand, NFV involves decoupling the network functions from the proprietary hardware, thus making it possible to run such functions in general purpose commodity servers, switches and storage units, which can be deployed in a network's data centre. Network virtualisation enables the creation and co-existence of multiple isolated and independent virtual networks over a shared network infrastructure [27,30]. NFV provides improvement in the use of the physical resources by allowing multiple instances of the same or different virtual network functions to coexist over a common pool of compute, network and storage resource. Hence, these technologies provide the satellite network with further innovation in service and business agility via cutting-edge network resource management tools. Unlike SDN, NFV does not necessarily introduce any architectural change in the network functions. The introduction of SDN/NFV within the satellite network will contribute towards the following objectives among others [29].

- Automated customised on-demand networking with efficient and optimal sharing of the satellite network resources and infrastructure.
- Improved profit on resource and customer satisfaction via the availability of wide range of services such as on-demand QoS and on-demand bandwidth.
- Support satellite as a multi-service network with each service requiring a specific performance guarantee.
- Efficient and dynamic sharing of the satellite core network infrastructure by many SNOs and other players such as satellite virtual network operators (SVNO) and service providers.
- Simplification of the management of network services and integration via the provisioning interface for resource provisioning and invocation.
- Determining the functionalities that can run in a cloud-based environment, the right functional split between the virtualised and the non-virtualised part of the satellite.

Some of the use cases of SDN/NFV in satellite communications include (1) on-demand satellite bandwidth via SDN, (2) SVNO, (3) satellite network as a service (SatNaas) where the satellite hub functional entities are implemented as software workloads instantiated on a cloud infrastructure using the infrastructure-as-a-service and platform-as-a-service paradigms.

The key challenges to this objectives include:

- Support of SDN and NVF techniques by remote terminal and satellite gateways for different use cases.
- Support for dynamic network configuration for on-demand purpose and making the satellite network resources available when prompted.
- Using SDN/NVF techniques to enhance multi-tenancy of the satellite hub components among multiple SVNOs. This entails enabling each SVNO to have advanced control via more programmability and flexibility of the resources allocated to it by the satellite hub.

1.5.7 Multi-casting

Radio resource management (RRM) techniques for offering multimedia content in LTE-satellite networks were presented in [31]. The RRM is performed on a per-group basis, since a group of users is served by the satellite in one radio transmission. Consequently, the selection of the modulation and coding scheme must take the channel qualities of all multi-cast members into consideration. The conventional approaches such as opportunistic and conservative multi-casting scheme suffer [32] from inefficiencies relating to inadequate short-term fairness and poor spectral efficiency, respectively. A promising RRM approach in 5G satellite multi-casting environment is subgrouping. All multi-cast terminals are served in every time slot by splitting the group into different subgroups based on the experience channel qualities. It has been shown in [31] that multi-cast subgrouping overcomes the weakness of the conventional techniques and allows for the efficient delivery of multimedia content over the emerging satellite systems.

1.5.8 Integrated signalling

The 5G environment is driven by a very dense deployment of small cells delivering HDR communication services to the user equipment. A key challenge with such architecture is the limited available capacity for user data due to the increased signalling capacity. Furthermore, the base station's signalling cost contributes to the total system energy consumption, and thus, hampers energy reduction. Decoupling the control (C) plane and data (U) plane together with SDN has recently been identified as one of the promising techniques towards meeting the 5G KPI target for energy reduction. The techniques also provide an improvement in the manageability and adaptability of the 5G networks. In the split C&U plane architecture the base stations deliver data on the U plane using terrestrial link when present and route the C plane via an overlay macro-cell backhauled over a satellite link [33]. Consequently, this gives the network operator more flexibility, since the small/data cells can be activated on demand to deliver user-specific data only when and where needed. Thus, the energy consumption is improved, since the split architecture also leads to longer data cell sleep periods, due to their on-demand activation [34–36]. In the rural context, the focus is to identify C plane traffic that can be managed locally and only utilise the satellite link when required. The hybrid system with split C&U planes can achieve approximately

40% and 80% energy efficiency improvement in sparse and ultra-dense networks, respectively, as compared with the conventional LTE networks [37].

1.6 Challenges and future research recommendations

In this section, we discuss the challenges associated with the some of the recent advances in satellite communications. Furthermore, some future recommendations are also presented alongside.

1.6.1 Integrated satellite-terrestrial architecture

Focusing on multimedia distribution, significant research and development effort on the integrated architecture is required in order to satisfy the challenging requirements of future users in terms of cost, performance, QoE and QoS. Such challenges include parallel and transparent access of the user to both the broadband and broadcast networks, smart management of both the broadband and broadcast resources, and managing the user content. Also, service continuity is an essential feature of the integrated satellite-terrestrial architecture as it aims to provide seamless service delivery to 5G end-users while roaming between the terrestrial and the satellite backhauled cells. The key challenges associated with this include (1) seamless mobility support in terms of vertical handovers, (2) design of networking protocols which can cope with the different latencies, (3) design of cost-effective 5G devices which supports the satellite-terrestrial dual mode operation, (4) designing the business model for access points and addressing the service level agreement issue that could arise between the satellite and terrestrial service providers.

For the M2M application of the integrated satellite-terrestrial architecture, the key research challenges relate to designing protocols that are appropriate for the satellite M2M. Noting that significant research effort has been put into the terrestrial IoT design in terms of battery-powered M2M systems, security and integrity, energy efficient waveforms and hardware design, a similar effort is also required towards the satellite IoT design. Furthermore, routing protocol redesign is also required for IoT scenarios that involve the satellite since delay becomes more crucial in such deployment. Also, with the planned utilisation of the frequencies above 10 GHz for terrestrial deployment in 5G, there is the need to investigate various scenarios of the integrated satellite-terrestrial architecture in terms of the resource allocation (specifically, carrier, bandwidth and power) between the satellite and the terrestrial systems. The multiple antenna satellite system brings significant gain in terms of coverage and capacity. Hence, investigating its performance within terrestrial and satellite networks requires attention in a future study.

1.6.2 Integrated signalling in satellite communications

Similar to the terrestrial C/U plane split architecture, the satellite integrated split architecture must also meet the 5G engineering requirements. In addition to this, the requirements for managing ultra-dense cells must also be met in such integrated architecture. These requirements include handover and mobility management, backhauling management and data-cell discovery. User association with the data cells in the conventional split architecture are managed by the macro cells which provide control plane functionalities, whereas in the integrated architecture, satellites will handle control signalling and, hence, user-data cell association. One of the propositions in the conventional split architecture is for the macro/control cells to handle data transmission for high-mobility and low-rate users in order to reduce handover failures; the feasibility of satellites serving high mobility and low data rate has to be investigated.

Specifying the functionality of each plane and dimensioning their physical layer frames is a challenge in both the conventional and the integrated split architecture. This challenge arises from the fact that certain user activity such as handover requires several functionalities such as broadcast and synchronisation functionalities, while the frame control signal is required for more than one network functionality [34–36]. Hence, the signalling and functionalities associated with each plane must be correctly allocated. Moreover, the ability of satellites to cache certain user information and its associated latency and channel condition issues further add to the challenge experienced with the conventional split architecture.

1.6.3 On-board processing

On-board processing functionality in satellites implies having additional hardware which could lead to an increase in transponder mass and power consumption. In the light of this, the additional heat generated by the processor must be properly managed. Reliability is another key challenge with on-board processing. The backup digital signal processing (DSP) which is required in case of component failure can scale-up the cost significantly. Other challenges associated with on-board processing include the limitation to the reconfigurability of the hardware chains and the sampling capability. Low cost and reliable processing techniques are key to on-board processing in satellites.

1.7 Conclusion

This chapter has presented the key areas in which satellite can play a part in the 5G network. The examined potential areas include coverage, massive machine type communications, resilience and overspill, content multi-cast and caching, integrated network, ultra-reliable communications and spectrum utilisation. We have also highlighted the recent advances and a number of research challenges associated with the satellite-terrestrial integrated architecture. It has been emphasised that to achieve and exploit the potential of satellites in 5G and stimulate investments, the satellite community must work in close collaboration with terrestrial players in the 5G activities on areas including technology standardisation, demonstration and regulatory issues.

References

[1] Thompson J, Ge X, Wu HC, *et al.* 5G wireless communication systems: prospects and challenges. IEEE Communications Magazine. 2014 Feb; 52:62–64.

[2] Cardona N. (ed.). *Cost IC 1004 White Paper on Scientific Challenges towards 5G Mobile Communications* [online]. 2013. Available from: http://www.ic1004.org.

[3] Fallgren M., Timus B. (eds.). *Scenarios, Requirements and KPI's for 5G Mobile and Wireless Systems [ICT-317669-METIS/D1.1]* [online]. 2013. Available from: https://www.metis2020.com/wp-content/uploads/deliverables/METIS_D1.1_v1.pdf.

[4] Evans B, Onireti O, Spathopoulos T, *et al.* The role of satellites in 5G. In: 23rd European Signal Processing Conference (EUSIPCO); 2015. p. 2756–2760.

[5] *5G PPP: The 5G Infrastructure Public Private Partnership* [online]. Available from: https://5g-ppp.eu/kpis/.

[6] Popovski P, Braun V, Mayer HP, *et al.* ICT-317669-METIS/D1.1 Scenarios, Requirements and KPIs for 5G Mobile and Wireless System; 2013. Available from: https://www.metis2020.com/wp-content/uploads/deliverables/METIS_D1.1_v1.pdf.

[7] Andrews JG, Buzzi S, Choi W, *et al.* What will 5G be? IEEE Journal on Selected Areas in Communications. 2014 Jun;32(6): 1065–1082.

[8] Aston PS. Satellite telephony for field and mobile applications. In: IEEE Aerospace; 2000. p. 179–190.

[9] Franchi A, Howell A, Sengupta J. Broadband mobile via satellite: Inmarsat BGAN. In: IEE Seminar on Broadband Satellite: The Critical Success Factors – Technology, Services and Markets (Ref. No. 2000/067); 2000. p. 23/1–23/7.

[10] 3rd Generation Partnership Project (3GPP); Technical Specification Group Radio Access Network. Study on New Radio (NR) to Support Non Terrestrial Networks (Release 15). TR 38.811V0.1.0 (2017-06); 2017.

[11] 3rd Generation Partnership Project (3GPP); Technical Specification Group Radio Access Network. Propagation delay and Doppler in Non-Terrestrial Networks (Release 15). RP-171578 (2017-09).

[12] 3rd Generation Partnership Project (3GPP). 5G: Study on Scenarios and Requirements for Next Generation Access Technologies;. TR 38.913 V14.2.0; 2017.

[13] Wu H, Li J, Lu H, *et al.* A two-layer caching model for content delivery services in satellite-terrestrial networks. In: IEEE Global Communications Conference (GLOBECOM); 2016. p. 1–6.

[14] Corici M, Kapovits A, Covaci S, *et al.* Assessing Satellite-Terrestrial Integration Opportunities in the 5G Environment; 2016. Available from: https://artes.esa.int/sites/default/files/Whitepaper%20-%20Satellite_5G%20final.pdf.

[15] GSMA Intelligence. Understanding 5G: Perspectives on Future Technological Advancements in Mobile; 2014.
[16] Satellite and Terrestrial Network for 5G (SAT5G). Available from: http://sat5g-project.eu/.
[17] Shared Access Terrestrial-Satellite Backhaul Network Enabled by Smart Antennas (SANSA). Available from: http://sansa-h2020.eu/.
[18] H2020-ICT-2014-1 Project VITAL. Deliverable D2.3, System Architecture: Final Report; 2016. Available from: http://www.ict-vital.eu/.
[19] Majkowski WS. 5G US spectrum development, products and mmWave testing including measurements of fixed satellite service (FSS) earth station spillover emissions. In: IEEE International Symposium on Electromagnetic Compatibility Signal/Power Integrity (EMCSI); 2017.
[20] Guidolin F, Nekovee M, Badia L, *et al.* A study on the coexistence of fixed satellite service and cellular networks in a mmWave scenario. In: IEEE International Conference on Communications (ICC); 2015. p. 2444–2449.
[21] Guidotti A, Vanelli-Coralli A, Caus M, *et al.* Satellite-enabled LTE systems in LEO constellations. In: 2017 IEEE International Conference on Communications Workshops (ICC Workshops); 2017. p. 876–881.
[22] Guidotti A, Vanelli-Coralli A, Kodheli O, *et al.* Integration of 5G Technologies in LEO Mega-Constellations; 2017. Available from: https://arxiv.org/abs/1709.05807.
[23] Kodheli O, Guidotti A, Vanelli-Coralli A. Integration of satellites in 5G through LEO constellations. In: IEEE Global Communications Conference (GLOBECOM); 2017.
[24] Muraro JL, Nicolas G, Nhut DM, *et al.* GaN for space application: almost ready for flight. International Journal of Microwave and Wireless Technologies. 2010; Apr;2:121–133.
[25] Noh Y, Choi YH, Yom I. Ka-band GaN power amplifier MMIC chipset for satellite and 5G cellular communications. In: IEEE 4th Asia-Pacific Conference on Antennas and Propagation (APCAP); 2015. p. 453–456.
[26] Yuk K, Branner GR, Cui C. Future directions for GaN in 5G and satellite communications. In: IEEE 60th International Midwest Symposium on Circuits and Systems (MWSCAS); 2017. p. 803–806.
[27] Bertaux L, Medjiah S, Berthou P, *et al.* Software defined networking and virtualization for broadband satellite networks. IEEE Communications Magazine. 2015 Mar;53(3):54–60.
[28] Li T, Zhou H, Luo H, *et al.* Using SDN and NFV to implement satellite communication networks. In: 2016 International Conference on Networking and Network Applications (NaNA); 2016. p. 131–134.
[29] Ferrús R, Koumaras H, Sallent O, *et al.* SDN/NFV-enabled satellite communications networks: opportunities, scenarios and challenges. Physical Communication. 2016;18(Part 2):95–112. Special Issue on Radio Access Network Architectures and Resource Management for 5G. Available from: http://www.sciencedirect.com/science/article/pii/S1874490715000543.

[30] Chowdhury NMMK, Boutaba R. Network virtualization: state of the art and research challenges. IEEE Communications Magazine. 2009 Jul;47(7):20–26.
[31] Araniti G, Bisio I, Sanctis MD, *et al.* Multimedia content delivery for emerging 5G-satellite networks. IEEE Transactions on Broadcasting. 2016;62(1):10–23.
[32] Sali A, Karim HA, Acar G, *et al.* Multicast link adaptation in reliable transmission over geostationary satellite networks. Wireless Personal Communications. 2012;62(4):759–782.
[33] Spathopoulos T, Onireti O, Khan AH, *et al.* Hybrid cognitive satellite terrestrial coverage: a case study for 5G deployment strategies. In: 10th International Conference on Cognitive Radio Oriented Wireless Networks (CROWNCOM 2015); 2015.
[34] Mohamed A, Onireti O, Qi Y, *et al.* Physical layer frame in signalling-data separation architecture: overhead and performance evaluation. In: Proc. European Wireless Conference; 2014.
[35] Mohamed A, Onireti O, Imran MA, *et al.* Correlation-based adaptive pilot pattern in control/data separation architecture. In: Proc. IEEE International Conference on Communications (ICC); 2015.
[36] Mohamed A, Onireti O, Imran MA, *et al.* Control-data separation architecture for cellular radio access networks: a survey and outlook. IEEE Communications Surveys Tutorials. 2015 Firstquarter;18(1):446–465.
[37] Zhang J, Zhang X, Imran MA, *et al.* Energy efficient hybrid satellite terrestrial 5G networks with software defined features. Journal of Communications and Networks. 2017 Apr;18(2):147–161.

Chapter 2
Satellite use cases and scenarios for 5G eMBB

Konstantinos Liolis[1], Alexander Geurtz[1], Ray Sperber[1], Detlef Schulz[1], Simon Watts[2], Georgia Poziopoulou[2], Barry Evans[3], Ning Wang[3], Oriol Vidal[4], Boris Tiomela Jou[4], Michael Fitch[5], Salva Sendra Diaz[5], Pouria Sayyad Khodashenas[6], and Nicolas Chuberre[7]

This chapter presents initial results available from the European Commission H2020 5G PPP Phase 2 project SaT5G (Satellite and Terrestrial Network for 5G) [1]. It specifically elaborates on the selected use cases and scenarios for satellite communications (SatCom) positioning in the 5G usage scenario of eMBB (enhanced mobile broadband), which appears the most commercially attractive for SatCom. After a short introduction to the satellite role in the 5G ecosystem and the SaT5G project, the chapter addresses the selected satellite use cases for eMBB by presenting their relevance to the key research pillars (RPs), their relevance to key 5G PPP key performance indicators (KPIs), their relevance to the 3rd Generation Partnership Project (3GPP) SA1 New Services and Markets Technology Enablers (SMARTER) use case families, their relevance to key 5G market verticals, and their market size assessment. The chapter then continues by providing a qualitative high-level description of multiple scenarios associated to each of the four selected satellite use cases for eMBB. Useful conclusions are drawn at the end of the chapter.

2.1 Introduction

5G is the next generation of communication technology that much of the world is moving to. By supporting a world in which 'anyone and anything will be connected

[1]SES S.A., Luxembourg
[2]Avanti Communications Ltd., United Kingdom
[3]University of Surrey, Institute for Communication Systems (ICS), United Kingdom
[4]Airbus Defence and Space SAS, France
[5]British Telecommunications PLC, United Kingdom
[6]i2CAT Foundation, Spain
[7]Thales Alenia Space, France

at anytime and anywhere' [2], 5G is expected to enable new applications in various domains, including media and entertainment, health, automotive, transport, and industry. The advanced communications of 5G are expected to bring eMBB, ultra reliable low latency communications (URLLC), and massive machine type communications (mMTC), which correspond to the 5G usage scenarios defined by International Telecommunication Union – Radiocommunication Sector (ITU-R) [3] for International Mobile Telecommunications for 2020 and beyond. Practical 5G network deployments are expected to be evolutionary rather than revolutionary, simply because 4G network capabilities are progressing as well and because not all applications require all 5G features [4,5].

The next-generation network of 5G has been described as having significantly more capacity and higher user data rates than today's capabilities, so as to meet the growing demands of users. In addition, an important goal of 5G is to provide increased resilience, continuity, and much higher resource efficiency including a significant decrease in energy consumption. Finally, security and privacy will need to be ensured to protect users and the important amounts of data that will be carried across the network. The 5G KPIs are summarized in [6,7]. Note that these 5G KPIs are not expected to be met all at the same time. No single technology will meet all of these needs, and not all of these characteristics will be required for every 5G application. On the contrary, as the European Commission [8] and other governments around the world have correctly recognized, to be successful and meet user demands, the 5G infrastructure will be an ecosystem of networked networks, utilizing multiple different and complementary technologies. Thus, it is believed that the complexity of these diverse requirements offers an opportunity for satellite to support 5G roll-out and success.

To this end, many organizations, including the European Commission [8], recognize that satellite networks will be an element of 5G infrastructure. Among others, the role of satellites in 5G has been studied in the SatCom Working Group of the European Technology Platform NetWorld2020 [9] as well as in relevant R&D projects, such as SPECSI [10], MENDHOSA [11], INSTINCT [12,13], CloudSat [14], SANSA [15], VITAL [16], RIFE [17], SCORSESE [18], and High Throughput Digital Broadcasting Satellite Systems (HTS-DBS) [19]. Moreover, the EMEA (Europe, the Middle East, and Africa) Satellite Operators Association (ESOA) has published a 5G White Paper on the SatCom services' role as an integral part of the 5G ecosystem [20]. The consensus and wider agreement on what satellite brings towards achieving the 5G KPIs are:

- **Ubiquity:** Satellite provides high-speed capacity across the globe using the following enablers – capacity in-fill inside geographic gaps, overspill to satellite when terrestrial links are over capacity, general global wide coverage, backup/resilience for network fall-back, and especially communication during emergency.
- **Mobility:** Satellite is the only readily available technology capable of providing connectivity anywhere on the ground, in sea or air for moving platforms, such as airplanes, ships, and trains.

- **Broadcast (simultaneity):** Satellite can efficiently deliver rich multimedia and other content across multiple sites simultaneously using broadcast and multicast streams with information-centric networking and content caching for local distribution.
- **Security:** Satellite networks can provide efficient solutions for secure, highly reliable, rapid, and resilient deployment in challenging communication scenarios, such as emergency response and public safety communications.

In this context, the project SaT5G [1] is a European Commission H2020 5G PPP Phase 2 project, kicked off in June 2017, whose vision is to develop cost-effective 'plug and play' SatCom solutions for 5G to enable telecom operators and service providers to accelerate 5G deployment in all geographies and at the same time create new and growing market opportunities for SatCom industry stakeholders. The six principal SaT5G project objectives are to:

- leverage relevant ongoing 5G and satellite research activities to assess and define solutions integrating satellite into the 5G network architecture;
- develop the commercial value propositions for satellite-based network solutions for 5G;
- define and develop key technical enablers for the identified research challenges;
- validate key technical enablers in a lab test environment;
- demonstrate selected features and use cases with in-orbit geostationary and non-geostationary high-throughput satellite (HTS) systems; and
- contribute to the standardization at European Telecommunications Standards Institute (ETSI) and 3GPP of the features enabling the integration of SatCom solutions in 5G.

With the identified satellite strengths and based on the anticipated market needs, SaT5G focuses on the 5G usage scenario of eMBB. Based on the analysis results obtained from relevant R&D projects funded by the European Space Agency (ESA), such as SPECSI [10] and MENDHOSA [11], the broadband and broadcast services will have the highest revenue in 2025 and thus form the primary SaT5G target markets. Furthermore, from the mobile operators' viewpoint for the inclusion of satellite support in the early 5G roll-out, congested backhaul and offloading high bandwidth video download have been found to be the major drivers. These operator drivers also fall under the 5G usage scenario of eMBB. Therefore, SaT5G addresses specifically the eMBB usage scenario for 5G towards 'broadband access everywhere'. This is not to say that SatCom may not benefit other 5G usage scenarios, such as the mMTC, for instance, but only that eMBB appears the most commercially attractive for SatCom [10,11].

The remainder of this chapter is structured as follows: Section 2.2 elaborates on the satellite use cases for eMBB selected by the SaT5G project, Section 2.3 presents the specific scenarios defined for the selected satellite use cases, and Section 2.4 concludes the chapter.

2.2 Selected satellite use cases

2.2.1 Selection methodology

By definition, a 5G use case is a particular case of how the 5G system is used, whereas a satellite use case in 5G is a particular case of how the SatCom system is integrated within the 5G ecosystem. The selected satellite use cases in 5G elaborated in this section correspond to specific satellite use cases for eMBB which have been selected to be further investigated in the SaT5G project.

With focus on the eMBB usage scenario for 5G and by following the methodology illustrated in Figure 2.1, SaT5G selected four satellite use cases for eMBB to concentrate its efforts on.

Specifically, we consolidated a 'global' list of satellite use cases in 5G by review, gap analysis and taking into account satellite use cases in 5G identified at 3GPP domain (e.g. 3GPP TR 22.891 [21], 3GPP SA1 'SMARTER' technical reports (TRs) [22], 3GPP TR 22.863 [23], 3GPP TR 22.864 [24], 3GPP TR 38.811 [25]), at SatCom domain (e.g. ESOA 5G White Paper [20]), as well as by other relevant R&D projects (e.g. SPECSI [10], MENDHOSA [11], INSTINCT [12,13], CloudSat [14], SANSA [15], VITAL [16], RIFE [17], SCORSESE [18], and HTS-DBS [19]).

Furthermore, by filtering the consolidated 'global' list of satellite use cases in 5G specifically for eMBB, we came up with a subset of satellite use cases for eMBB which were analysed in detail based on their relevance to the associated core 5G PPP KPIs, the relevant 3GPP SA1 SMARTER use case families, the 5G market verticals, as well as their relevant market size. Based on this detailed analysis, which is reported in detail in [26], the satellite use cases for eMBB selected for further investigation in the SaT5G project correspond to the SaT5G use cases, which are presented in this section.

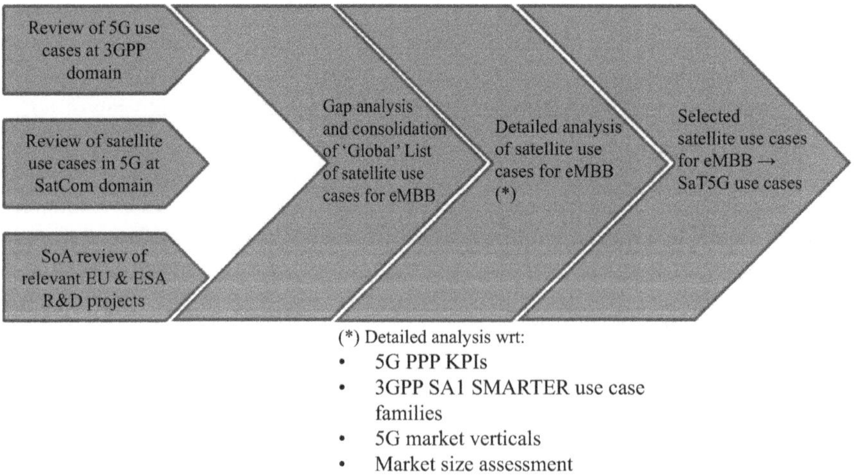

Figure 2.1 Selection methodology for SaT5G use cases

Due to space limitations, the subsequent sections hereinafter present the relevant outcomes of this analysis for the selected satellite use cases for eMBB. For further details, the interested readership is referred to [26].

2.2.2 Selected satellite use cases for eMBB

By applying the methodology described earlier, SaT5G selected four satellite use cases for eMBB to concentrate its efforts on the following ones, which are further elaborated hereinafter (see Table 2.1):

- Edge delivery and offload of multimedia content and MEC (multi-access edge computing) VNF (virtual network function) software, through multicast and caching to optimize the operation and dimensioning of the 5G network infrastructure;
- 5G fixed backhaul, to provide 5G service especially in areas where it is difficult or not possible to deploy terrestrial communications;
- 5G to premises, to provide 5G service into home/office premises in underserved areas via hybrid terrestrial-satellite broadband connections;

Table 2.1 SaT5G use cases: selected satellite use cases for eMBB

Selected satellite use case for eMBB	Description	Correspondence to satellite use case category in 5G [20]
Use case 1: Edge delivery and offload for multimedia content and MEC VNF software	Providing efficient multicast/ broadcast delivery to network edges for the contents such as live broadcasts, ad-hoc broadcast/ multicast streams, group communications, MEC VNF update distribution	Backhauling and tower feed
Use case 2: 5G fixed backhaul	Broadband connectivity where it is difficult or not (yet) possible to deploy terrestrial connections to towers, for example, coverage on lakes, islands, mountains, rural areas, isolated areas, or other areas that are best or only covered by satellites; across a wide geographic region	Trunking and head-end feed
Use case 3: 5G to premises	Connectivity complementing terrestrial networks, such as broadband connectivity to home/ office small cell in underserved areas in combination with terrestrial wireless or wireline	Hybrid multiplay
Use case 4: 5G moving platform backhaul	Broadband connectivity to platforms on the move, such as airplanes or vessels	Communications on the move

- 5G moving platform backhaul, to support 5G service on board moving platforms, such as aircraft, vessels, trains, etc.

Figure 2.2 illustrates the selected satellite use cases for eMBB and how they will be integrated into a 5G network.

2.2.3 Relevance to satellite 'sweet spots' in 5G

As can be deduced from Table 2.1, each SaT5G use case corresponds to one of the four satellite use case categories in 5G identified by ESOA [20] or else referred to as satellite 'sweet spots' in 5G (see Figure 2.3). Each satellite use case category (SUCC) in 5G has distinct connectivity characteristics, which are elaborated below.

- **Backhauling and tower feed:** This SUCC in 5G is about high-speed backhaul connectivity to individual cells, with the ability to multicast the same content (e.g. video, HD/UHD TV, as well as other non-video data) across a large coverage area (e.g. for local storage and consumption). The same capability also allows for the efficient backhauling of aggregated Internet of Things (IoT) traffic from multiple sites. A very high-speed, multicast-enabled, satellite link (up to Gbps speed), direct to the cell towers, from geostationary and/or non-geostationary satellites will complement existing terrestrial connectivity. Note that this SUCC assumes that satellite connectivity will complement existing terrestrial connectivity. Moreover, the satellite user links are either bidirectional and/or unidirectional since, depending on the case, broadband [i.e. unicast, thus VSAT (Very Small Aperture Terminal) satellite terminals] and/or broadcast/multicast (thus receive only satellite terminals) communications are supported by this category. In particular, the use of multicasting to populate edge caches is a major difference of this SUCC with respect to the next one. **Selected satellite use case 1** corresponds to this SUCC in 5G.
- **Trunking and head-end feed:** This SUCC in 5G addresses high-speed trunking of video, IoT and other data to a central site, with further terrestrial distribution to local cell sites, for instance neighbouring villages. A very-high-speed satellite link (up to Gbps speed) from geostationary and/or non-geostationary satellites will complement existing terrestrial connectivity, where available. Note that this SUCC assumes that limited or no existing terrestrial connectivity is available. Moreover, the satellite user links are bidirectional since only broadband (i.e. unicast, thus VSAT satellite terminals) communications are supported by this category (i.e. no broadcast/multicast). In particular, there is no use of multicasting to populate edge caches in this SUCC, which corresponds to a major difference with respect to the other SUCCs. **Selected satellite use case 2** corresponds to this SUCC in 5G.
- **Hybrid multiplay:** This SUCC in 5G is about high-speed connectivity including backhaul to individual homes and offices, referred to as premises, with the ability to multicast the same content (video, HD/UHD TV, as well as other non-video data) across a large coverage area (e.g. for local storage and consumption). The same capability also allows for an efficient broadband connectivity for aggregated IoT data. In-home distribution via Wi-Fi or home/office small-cell

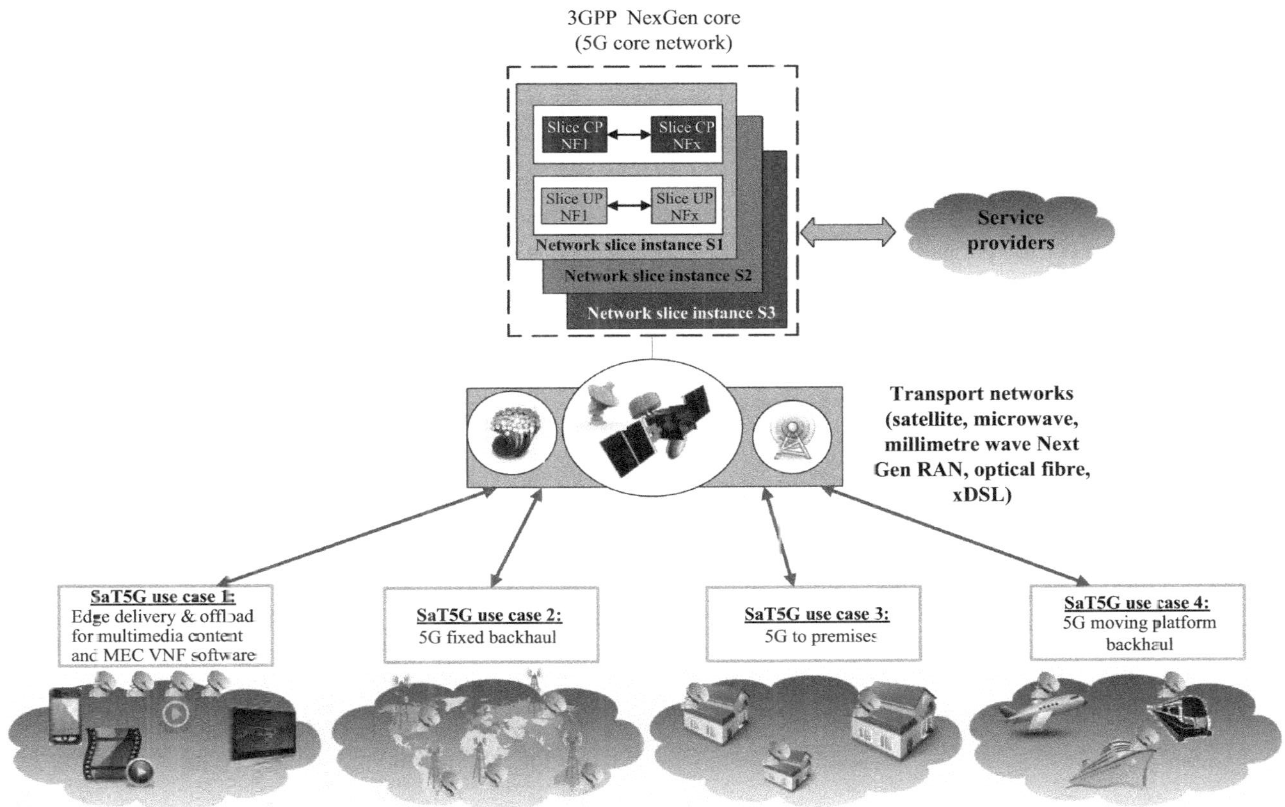

Figure 2.2 SaT5G use cases in 5G integrated satellite-terrestrial networks for eMBB

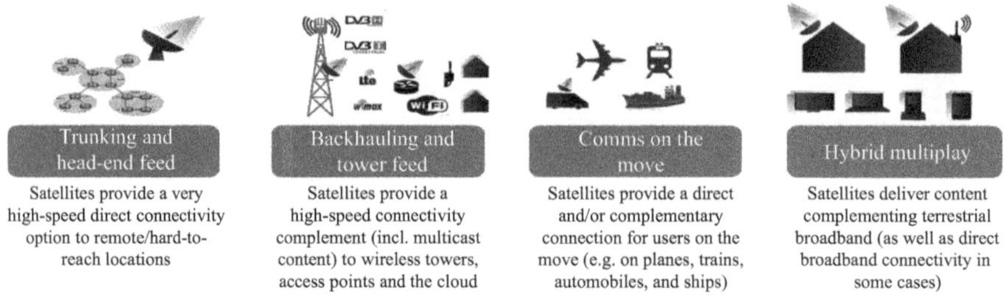

Figure 2.3 Satellite use case categories in 5G (or else referred to as satellite 'sweet spots' in 5G)

(femtocell) is considered. A very-high-speed, multicast-enabled, satellite link (up to Gbps speed), direct to the home or office, from geostationary and/or non-geostationary satellites will complement existing terrestrial connectivity. Direct-to-home (DTH) satellite TV, integrated within the home or office IP network, will further complement this use case. Note that this SUCC assumes that satellite connectivity will complement existing terrestrial connectivity. Moreover, the satellite user links are either bidirectional and/or unidirectional since, depending on the case, broadband (i.e. unicast, thus VSAT satellite terminals) and/or broadcast/multicast (thus, receive only satellite terminals) communications are supported by this category. In addition, note that in this SUCC, the local cell towers correspond to home/office small cells (femtocells). **Selected satellite use case 3** corresponds to this SUCC in 5G.

- **Communications on the move:** This SUCC in 5G is about high-speed backhaul connectivity to individual in-motion terminals on airplanes, vehicles, trains, and vessels (including cruise ships and other passenger vessels), with the ability to multicast the same content [e.g. video, HD/UHD TV, Firmware and Software Over the Air (FOTA/SOTA), as well as other non-video data] across a large coverage area (e.g. for local storage and consumption). The same capability also allows for the efficient backhauling of aggregated IoT traffic from these moving platforms. A very-high-speed, multicast-enabled, satellite link (up to Gbps speed), direct to the airplane, vehicles, train, or vessel, from geostationary and/or non-geostationary satellites will complement existing terrestrial connectivity, where available (such as, in airports, harbours, train stations, and connected cars). Moreover, the satellite user links are either bidirectional and/or unidirectional since, depending on the case, broadband (i.e. unicast, thus VSAT satellite terminals) and/or broadcast/multicast (thus receive only satellite terminals) communications are supported by this category. **Selected satellite use case 4** corresponds to this SUCC in 5G.

2.2.4 Relevance to SaT5G research pillars

The technical challenges that need to be addressed for the realization of cost-effective 'plug and play' SatCom solutions for 5G include:

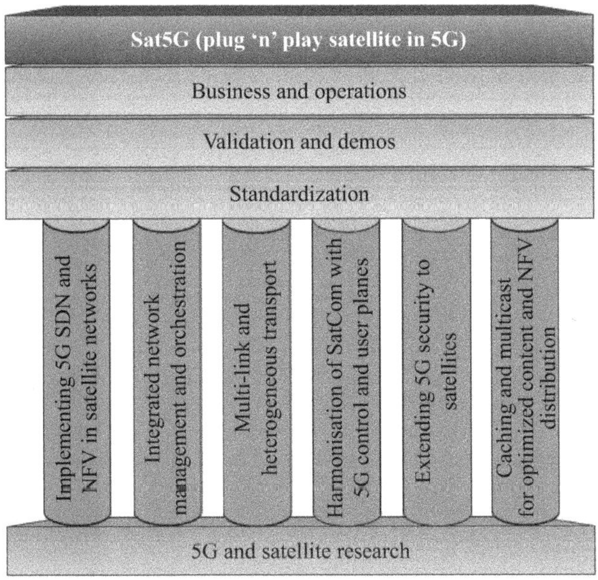

Figure 2.4 SaT5G concept

- virtualization of SatCom network functions to ensure compatibility with the 5G software-defined networking (SDN) and network functions virtualization (NFV) architecture;
- developing the enablers for an integrated 5G-SatCom virtual and physical resource orchestration and service management;
- developing link aggregation scheme for small-cell connectivity mitigating Quality of Service (QoS) and latency imbalance between satellite and cellular access;
- leveraging 5G features/technologies in SatCom;
- optimizing/harmonizing key management and authentication methods between cellular and satellite access technologies; and
- optimal integration of the multicast benefits in 5G services for both content delivery and MEC VNF distribution.

To rise to these challenges, the SaT5G concept comprises six RPs and three horizontals as shown in Figure 2.4. The horizontals address global issues cutting across the whole project whilst the RPs have been selected to address the deeper research enablers relating to the identified challenges and needed to flow into prototypes to be used in the validations and demonstrations.

The six RPs chosen along with their scope and benefits are presented in Table 2.2.

The SaT5G RPs along with the selected satellite use cases for eMBB are illustrated in Figure 2.5.

The selected satellite use cases for eMBB are mapped to the six RPs as shown in Table 2.3.

Table 2.2 Scope and benefits of the research pillars for the 5G ecosystem stakeholders

Research pillar	Scope	Benefits
RP I: Implementation of 5G SDN and NFV across satellite networks	Virtualize SatCom network functions to share the same virtualized core as cellular network functions, ensure compatibility with the SDN/NFV architecture, and support network slicing	CAPEX reduction and flexible service provisioning
RP II: Integrated network management and orchestration	Enable integrated 5G-SatCom virtual and physical resource orchestration and service management	OPEX reduction through harmonized network management between 5G and SatCom
RP III: Multi-link and heterogeneous transport	Exploit multi-link and heterogeneous links of the transport network at the backhaul level and mitigate the possible QoS and latency imbalance between the links	Improved goodput, Quality of Experience (QoE), and resiliency
RP IV: Harmonization of SatCom with 5G control and user plane	Leverage 5G features in satellite radio access network and foster the integration with other network technologies	CAPEX and OPEX reduction (especially development/maintenance effort for future SatCom)
RP V: Extending 5G security to satellite	Provide an efficient key management and authentication method and harmonize authentication and authorization between terrestrial and satellite technologies	Trust enforcement in the E2E 5G network including satellite element
RP VI: Caching and multicast for content and VNF distribution	Provide efficient delivery of multimedia content and NFV functions to mobile edge computing/caching entities through the exploitation of the intrinsic broadcast capability of SatCom	OPEX reduction through improved bandwidth efficiency

2.2.5 Relevance to 5G PPP KPIs

The 5G KPIs are summarized in [6,7]. Note that these 5G KPIs are not expected to be met all at the same time. Moreover, note that not all these 5G KPIs are relevant for SatCom.

Table 2.4 provides a mapping between the selected satellite use cases for eMBB and the core 5G PPP KPIs which are relevant to SatCom and, particularly, SaT5G.

As an illustration, particularly with regard to the 5G PPP KPI of ‘end-to-end latency of <1 ms’, the SatCom expected impact is elaborated hereinafter.

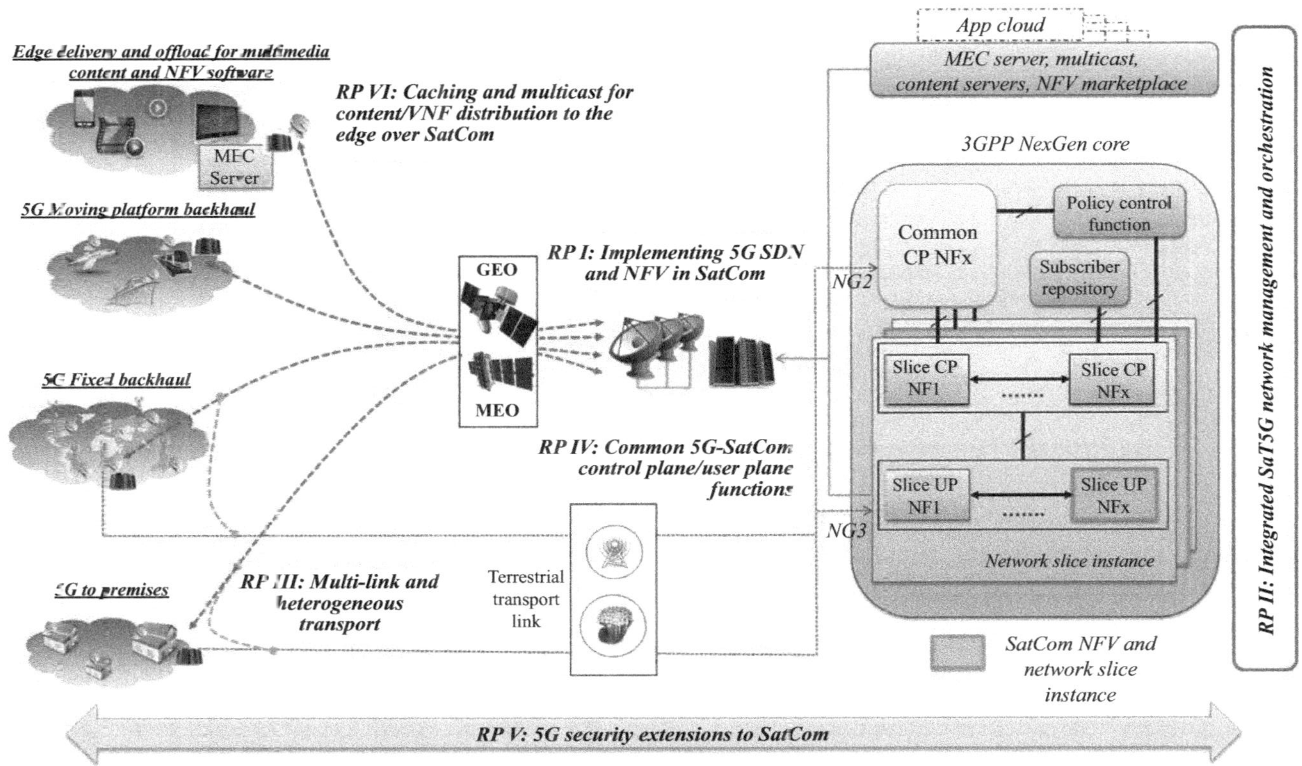

Figure 2.5 SaT5G use cases and research pillars

Table 2.3 Mapping of selected satellite use cases for eMBB to research pillars

Research pillar	Selected satellite use case 1	Selected satellite use case 2	Selected satellite use case 3	Selected satellite use case 4
RP I	Virtualization of satellite functional components and integration of the satellite transport link in the SDN/NFV architecture and support of network slicing feature			
RP II	End-to-end service life cycle management and orchestration which includes virtual and physical IT and network resources[a]	Integrated 5G-SatCom virtual and physical resource[a] orchestration and service management	Flexible integrated 5G-SatCom resource[a] orchestration and service life cycle management	Mobility aware end-to-end service life cycle management and resource[a] orchestration
RP III	Traffic splitting between multicast and unicast flows	NG2/NG3[b] protocol performance enhancement adapted to long latency link	Traffic splitting between network links with different characteristics for link aggregation	NG2/NG3[b] protocol performance enhancement adapted to long latency link
RP IV	Support of multicast traffic	Support of NG2/NG3[b] protocols	Support traffic splitting/link aggregation solutions	Support of NG2/NG3[b] dynamic relocation
RP V	Extension of the security architecture to broadcast component	Efficient key management and authentication over fixed satellite transport		Efficient key management and authentication over mobile satellite transport
RP VI	Efficient multimedia content/MEC NFV delivery over a dedicated satellite broadcast system	Efficient multimedia content/MEC NFV delivery over multicast resources of a broadband satellite transport link		

[a]In the SDN/NFV-enabled SaT5G ecosystem, network slices are composed by a wide range of resources, such as IT assets and bandwidth.
[b]NG2 corresponds to the 5G control plane interface, whereas NG3 to the 5G user plane interface.

Table 2.4 Mapping of selected satellite use cases for eMBB to relevant 5G PPP KPIs

Selected satellite use case for eMBB	1,000× capacity	10–100 Times higher typical user data rate	Better/ Increased/ Ubiquitous coverage	Service creation in minutes	End-to-end latency of <1 ms
Use case 1: 'Edge delivery & offload for multimedia content and MEC VNF software'	✓	✓	✓	✓	✓
Use case 2: '5G Fixed backhaul'	✓	✓	✓	✓	
Use case 3: '5G to premises'	✓	✓	✓	✓	✓
Use case 4: '5G Moving platform backhaul'		✓	✓	✓	✓

Sub-1 ms latency is very difficult to achieve, even for 5G mobile networks. According to GSMA Intelligence [4], 'Achieving the sub-1 ms latency rate … will likely prove to be a significant undertaking in terms of technological development and investment in infrastructure'. 'Services requiring a delay time of less than 1 ms must have all of their contents served from a physical position very close to the user's device. … possibly at the base of every cell, including the many small cells that are predicted to be fundamental to meeting densification requirements'. Moreover, according to NSR [27], 'Paradoxically, the low latency requirement for 5G network is a big ally in this vertical for satcom as many new locations for content servers will be required. In the transition to 5G, content needs to be moved to the edge and many new locations will be required, densifying content delivery networks (CDNs) and making satellite multicast a viable option'. 'The emergency and growth of applications like virtual reality, augmented reality, tactile Internet or video streaming will only accelerate this requirement to move capacities to the edge'.

Based on this justification, SatCom can indeed help 5G networks to achieve low latency by multicasting contents to caches located at individual cells, even in places without fibre. This is one of the satellite 'sweet spots'.

2.2.6 Relevance to 3GPP SA1 SMARTER use case families

Using as input the 3GPP TR 22.891 [21] and the 74, in total, 5G use cases defined therein, 3GPP SA1 has categorized these 5G use cases into four different groups:

- massive IoT (mIoT);
- critical communications;

- eMBB; and
- network operation (NEO).

This has been the result of the work conducted within 3GPP SA1, which is reported into four new TRs outlining the SMARTER for next-generation mobile telecommunications [22]. Each of the four TRs carries the title of the respective use case group:

- **3GPP TR 22.861: FS_SMARTER – mIoT:** mIoT focuses on use cases with massive number of devices (e.g. sensors and wearables). This group of use cases is particularly relevant to the new vertical services, such as smart home and city, smart utilities, e-Health, and smart wearables.
- **3GPP TR 22.862: FS_SMARTER – Critical communications:** The main areas where improvements are needed for critical communications are latency, reliability, and availability to enable, for example, industrial control applications and tactile Internet. These requirements can be met with an improved radio interface, optimized architecture, and dedicated core and radio resources.
- **3GPP TR 22.863: FS_SMARTER – eMBB:** eMBB includes a number of different use case families related to higher data rates, higher density, deployment and coverage, higher user mobility, devices with highly variable user data rates, fixed mobile convergence, and small-cell deployments.
- **3GPP TR 22.864: FS_SMARTER – NEO:** The use case group NEO addresses the functional system requirements, including aspects such as flexible functions and capabilities, new value creation, migration and interworking, optimizations and enhancements, and security.

As already stated, SaT5G puts more emphasis on the eMBB usage scenario for 5G. As such, the eMBB SMARTER use case group is of inherent relevance here. From the described categorization, the NEO SMARTER use case group is also of relevance here due to some key associated network aspects, such as network slicing, backhauling, and content delivery. Thus, the following two sections elaborate on the two 3G PPP SA1 SMARTER use case families, which are of more relevance here, that is, eMBB and NEO.

2.2.6.1 3GPP SA1 SMARTER eMBB use case families

The eMBB use case families identified by the 3GPP SA1 SMARTER study are defined in detail in 3GPP TR 22.863 [23] and overviewed below:

- **Higher data rates (3GPP TR 22.863** [23], **Section 5.1):** This use case family focuses on identifying key scenarios from which eMBB primary data rate requirements for peak, experienced, downlink, uplink, etc. data rates can be derived, as well as associated requirements pertaining to latency when applicable with user equipment (UE) relative speed to ground up to 10 km/h (pedestrian).
- **Higher density (3GPP TR 22.863** [23], **Section 5.2):** This use case family covers scenarios with system requirement for the transport of high volume of data traffic per area (traffic density) or transport of data for high number of connections

(devices density or connection density) with UE relative speed to ground up to 60 km/h (pedestrian or users moving on urban vehicle).

- **Deployment and coverage (3GPP TR 22.863** [23], **Section 5.3):** This use case family covers scenarios with system requirement considering the deployment and coverage scenario, e.g. indoor/outdoor, local area connectivity, wide area connectivity, with UE relative speed to ground up to 120 km/h.
- **Higher user mobility (3GPP TR 22.863** [23], **Section 5.4):** This use case family focuses on identifying key scenarios from which eMBB mobility requirements can be derived, with UE relative speed to ground up to 1,000 km/h. It supports eMBB for rapidly moving devices, for example the provision of Internet to road vehicles/trains/airplanes and its use for on-board entertainment and infotainment.
- **Devices with highly variable data rates (3GPP TR 22.863** [23], **Section 5.5):** This use case family focuses on identifying key scenarios from which eMBB requirements can be derived, for UE having multiple applications which exchange small amount of data and large amount of data.
- **Fixed mobile convergence (3GPP TR 22.863** [23], **Section 5.6):** This use case family focuses on identifying key scenarios for leveraging the 5G network characteristics defined in TR22.863 SMARTER eMBB (high data rates, low latency, high density, wide area coverage, and low mobility) and TR22.864 SMARTER NEO (network slicing, efficient data plane and content delivery, broadcast/multicast, policy control and charging, high availability and security) to enable combined use of fixed broadband (e.g. Fiber to the x (FTTX)/Digital Subscriber Line (xDSL)) access and next-generation radio access network (RAN).
- **Femtocell deployments (3GPP TR 22.863** [23], **Section 5.7):** This use case family focuses on identifying key scenarios for leveraging the 5G network characteristics defined in TR22.863 SMARTER eMBB (high data rates, low latency, high density, wide area coverage, and low mobility) and TR22.864 SMARTER NEO (network slicing, efficient data plane and content delivery, broadcast/multicast, policy control and charging, high availability and security) to enable use of fixed broadband (e.g. FTTx/xDSL) access and next-generation RAN for femtocell deployments. The overall aim is to deliver a seamless user experience to end users accessing operator services across any access network, including macrocellular as well as femtocell access over fixed broadband networks.

2.2.6.2 3GPP SA1 SMARTER network operation (NEO) use case families

The NEO use case families identified by the 3GPP SA1 SMARTER study are defined in detail in 3GPP TR 22.864 V14.1.0 (2016-09) [24] and are overviewed below:

- **System flexibility (3GPP TR 22.864** [24], **Section 5.1):** This family covers the use cases for building the network in a flexible manner per diverse scenario demand, e.g. slicing the network for variant market segments and verticals.
- **Scalability (3GPP TR 22.864** [24], **Section 5.2):** This family covers the use cases for enabling the operators to support an elastic and scalable network.

- **Mobility support (3GPP TR 22.864** [24], **Section 5.3):** This family covers the use cases for optimizing use of mobility management for diverse scenarios.
- **Efficient content delivery (3GPP TR 22.864** [24], **Section 5.4):** This family covers the uses cases to support efficient content delivery.
- **Self-backhauling (3GPP TR 22.864** [24], **Section 5.5):** This family covers the use cases for wireless self-backhauling.
- **Access (3GPP TR 22.864** [24], **Section 5.6):** This family covers access related use cases including selection of the most appropriate access for user traffic.
- **Migration and interworking (3GPP TR 22.864** [24], **Section 5.7):** This family covers the coexistence of the FS_SMARTER system with the legacy systems and the migration of services from early generations.
- **Security (3GPP TR 22.864** [24], **Section 5.8):** This family covers security requirements that are common to all building blocks as well as the ones that are specific to NEO.

Table 2.5 provides a mapping between the selected satellite use cases for eMBB and the relevant 3GPP SA1 SMARTER use case families (i.e. eMBB and NEO). Particularly, with respect to eMBB use case families, their associated 5G use cases and traffic scenarios are also provided in Table 2.5.

2.2.7 Relevance to 5G market verticals

In order to capture the market opportunities, an important aspect to address is the alignment of the selected satellite use cases for eMBB to the 5G vertical market segments [6]. The following verticals have been identified as of interest to SaT5G:

- **Media and entertainment** – provide broadband service with broadcast/multicast via satellite to network edges in both outdoor and direct to indoor environment.
- **Transportation and logistics, including automotive** – satellite integrate with moving platforms such as airplanes, vessels, trains, and buses to provide real-time asset monitoring, rich multimedia, and FOTA/SOTA services, etc. For many of these cases, satellite is in an augmenting role that becomes enabling when terrestrial coverage is insufficient.
- **Manufacturing** – manufacturing industries located in remote locations, such as mining, oil and gas, offshore platforms, etc.
- **Health** – improve remote access/monitoring for patients especially furthest away to health facilities.
- **Utilities (energy, water, waste)** – improve the reach for critical infrastructure monitoring, environmental monitoring, emergency prediction, etc.
- **Agro industry (agriculture, processing of food)** – as part of future food security, vast majority of industry is located outside the urban regions. Examples of relevant applications include broadband access, smart agriculture, remote monitoring, and asset tracking, etc.
- **Public safety** – satellite connectivity for public safety communications, mission critical communications, emergency communications, etc.

Table 2.5 Mapping of selected satellite use cases to relevant 3GPP SA1 SMARTER use case families

Selected satellite use case for eMBB	Relevant eMBB use case families (incl. 3GPP TR 22.863 reference section)	Relevant traffic scenarios of eMBB use case families (incl. 3GPP TR 22.863 reference section)	Relevant eMBB use cases (incl. 3GPP TR 22.891 reference section)	Relevant NEO use case families (incl. 3GPP TR 22.864 reference section)
Selected satellite use case 1	Section 5.1: Higher data rates Section 5.3: Deployment and coverage	Section 5.3.2.1: Small area connectivity Section 5.3.2.2: Wide area connectivity Section 5.3.2.4: Extreme coverage in low-density areas	Section 5.5: Mobile broadband for indoor scenario Section 5.6: Mobile broadband for hotspots scenario Section 5.10: Mobile broadband services with seamless wide-area coverage Section 5.11: Virtual presence Section 5.56: Broadcasting support Section 5.72: 5G Connectivity using satellites	Section 5.1: System flexibility Section 5.2: Scalability Section 5.4: Efficient content delivery Section 5.5: Self-backhauling Section 5.6: Access
Selected satellite use case 2	Section 5.1: Higher data rates Section 5.3: Deployment and coverage	Section 5.3.2.1: Small area connectivity Section 5.3.2.2: Wide area connectivity Section 5.3.2.4: Extreme coverage in low-density areas	Section 5.5: Mobile broadband for indoor scenario Section 5.6: Mobile broadband for hotspots scenario Section 5.10: Mobile broadband services with seamless wide-area coverage Section 5.11: Virtual presence Section 5.56: Broadcasting support Section 5.72: 5G connectivity using satellites	Section 5.1: System flexibility Section 5.2: Scalability Section 5.5: Self-backhauling Section 5.6: Access

(*Continues*)

Table 2.5 (Continued)

Selected satellite use case for eMBB	Relevant eMBB use case families (incl. 3GPP TR 22.863 reference section)	Relevant traffic scenarios of eMBB use case families (incl. 3GPP TR 22.863 reference section)	Relevant eMBB use cases (incl. 3GPP TR 22.891 reference section)	Relevant NEO use case families (incl. 3GPP TR 22.864 reference section)
Selected satellite use case 3	Section 5.1: Higher data rates Section 5.6: Fixed mobile convergence Section 5.7: Femtocell deployments	Section 5.6.2: Traffic Scenario 1 – simultaneous use of next-generation radio and fixed broadband access Section 5.6.2: Traffic Scenario 2 – 5G access as bandwidth boost Section 5.6.2: Traffic Scenario 3 – 5G access as failover Section 5.6.2: Traffic Scenario 4 – 5G access as fast provision Section 5.6.2: Traffic Scenario 5 – symmetric bandwidth Section 5.7.2: Scenario 1 – unified set of identities Section 5.7.2: Scenario 2 – consistent set of policies Section 5.7.2: Scenario 3 – access to a single set of services Section 5.7.2: Scenario 4 – access local area network services	Section 5.5: Mobile broadband for indoor scenario Section 5.6: Mobile broadband for hotspots scenario Section 5.56: Broadcasting support Section 5.64: User multi-connectivity across operators Section 5.72: 5G connectivity using satellites	Section 5.1: System flexibility Section 5.2: Scalability Section 5.4: Efficient content delivery Section 5.5: Self-backhauling Section 5.6: Access
Selected satellite use case 4	Section 5.1: Higher data rates Section 5.3: Deployment and coverage Section 5.4: Higher user mobility	Section 5.3.2.1: Small area connectivity Section 5.3.2.2: Wide area connectivity Section 5.3.2.4: Extreme coverage in low-density areas Section 5.4.2.3: Enhanced connectivity services in fast moving airplanes	Section 5.5: Mobile broadband for indoor scenario Section 5.6: Mobile broadband for hotspots scenario Section 5.10: Mobile broadband services with seamless wide-area coverage Section 5.11: Virtual presence Section 5.29: Higher user mobility Section 5.30: Connectivity everywhere Section 5.56: Broadcasting support Section 5.72: 5G connectivity using satellites	Section 5.1: System flexibility Section 5.2: Scalability Section 5.3: Mobility support Section 5.4: Efficient content delivery Section 5.5: Self-backhauling Section 5.6: Access

Table 2.6 Mapping of selected satellite use cases for eMBB to 5G market verticals

Selected satellite use case for eMBB	Media and entertainment	Transport	Manufacturing	Health	Utilities	Agriculture	Public safety
Selected satellite use case 1	✓	✓					✓
Selected satellite use case 2	✓	✓	✓	✓	✓	✓	✓
Selected satellite use case 3	✓						
Selected satellite use case 4	✓	✓		✓			✓

Table 2.7 Relevant market size of selected satellite use cases for eMBB

Selected satellite use case for eMBB	Global satellite services market size in 2030	Qualitative market size assessment
Selected satellite use case 1	>1 B€	MEC is expected [28] to make CDNs up to 40% more efficient and, as Cisco VNI [29] state that share of video traffic will grow from 60% to 78% by 2021 and also given that 70% of all traffic is already encrypted [30], a more distributed CDN system will essential to go forward
Selected satellite use case 2	>1 B€	A study [31] of rolling out 5G (min. 50 Mbps per end user) to rural areas of the United Kingdom would consume 79% of the total budget and require an alternative approach not relying on fibre. The demands for backhaul will be exponential
Selected satellite use case 3	>1 B€	Upgrade networks and deployment of FTTH on a large scale, encompassing less densely populated areas, will be cost prohibitive [32] and even the most advanced countries are far from achieving universal access [33] to fibre
Selected satellite use case 4	>1 B€	Demand generated from broadband services to mobility segments is expected to reach 480 Gbps by 2025 [34], making it a key vertical going forward

Table 2.6 provides the proposed mapping of the selected satellite use cases for eMBB to 5G market verticals.

2.2.8 Market size assessment

With regard to the market size assessment of the selected satellite use cases for eMBB, the available results from SPECSI [10] and MENDHOSA [11] projects cannot be applied as is in the case of SaT5G because they are not one-to-one applicable for the selected satellite use cases for eMBB. To this end, Table 2.7 provides a qualitative market size assessment for these selected satellite use cases for eMBB based on experts' judgement assessment from the satellite operators' viewpoint and based on the development seen in the industry.

The proposed criterion employed is 'Global Satellite Services Market Size in 2030', whereas the proposed scoring employed is the following:

- €: 1–10 M€
- €€: 10–100 M€
- €€€: 100–1,000 M€
- €€€€: >1 B€

Based on this criterion, all the selected satellite use cases for eMBB correspond to the scoring of €€€€ (>1 B€). Further details are provided in [26].

Note that the assumed timeline for the forecast is 2030 as 5G is highly unlikely to generate anywhere near the numbers mentioned as early as 2025.

2.3 Scenarios for selected satellite use cases

This section provides a qualitative high-level description of the scenarios associated to each selected SaT5G use case. By definition, the scenarios for the selected satellite use cases correspond to instantiations of the selected satellite uses cases for eMBB for the accomplishment of a specific duty. As such, a scenario for a selected satellite use case for eMBB drives the integrated network topology and the architecture design.

The scenarios for the selected satellite use cases for eMBB are summarized in Table 2.8.

Table 2.8 Scenarios for selected satellite use cases

Selected satellite use case for eMBB	Scenarios for selected satellite use case
Selected satellite use case 1	Scenario 1a: 'Offline multicasting and caching of video content and VNF software through satellite links' Scenario 1b: 'Online prefetching of video segments through satellite links'
Selected satellite use case 2	Scenario 2a: 'Satellite backhaul to groups of cell towers' Scenario 2b: 'Satellite backhaul to individual cell towers' Scenario 2c: 'Satellite backhaul to individual small cells'
Selected satellite use case 3	Scenario 3a: 'Hybrid Multiplay (satellite/xDSL) at home/office premises in underserved areas' Scenario 3b: 'Hybrid Multiplay (satellite/cellular) at home/office premises in underserved areas'
Selected satellite use case 4	Scenario 4a: 'Updating content for on-board systems and grouped media request by the moving platform company' Scenario 4b: 'Broadband access for passengers and individual media requests' Scenario 4c: 'Business and technical data transfer for the moving platform company'

2.3.1 Scenarios for selected satellite use case 1: edge delivery and offload for multimedia content and MEC VNF software

The scenarios associated to this selected satellite use case for eMBB correspond to satellite broadcast/multicast functions and the use of caching. This can be implemented via a standalone fixed terminal or via delivery to the mobile edge cache for onward delivery to UE within the 5G MNO (mobile network operator) network.

In this context, we have considered specifically the following two scenarios:

- Scenario 1a: Offline multicasting and caching of video content and VNF software through satellite links.
- Scenario 1b: Online prefetching of video segments through satellite links.

As an illustration, as estimated by Cisco [29], over three-fourths (78%) of the world's mobile data traffic will be videoed by 2021 where mobile video will increase ninefold between 2016 and 2021. For instance, more than half of YouTube views come from mobile devices [35]. Also, as can be observed from the Cisco Visual Networking Index (VNI) Global Mobile Data Traffic Forecast [29] which includes only cellular traffic and excludes traffic offloaded onto Wi-Fi and small cell from dual-mode devices, video is the highest growth application in terms of bandwidth on cellular networks. Similar growth can be assumed on offload networks (Wi-Fi, small cells). Moreover, analysts from Statista's Digital Market Outlook have revealed that the amount of time we're spending with our smartphones online has increased substantially over the last few years [36]. For instance, as reproduced by the Financial Times (2017-05-30, page 1), in the United Kingdom, the average smartphone user is spending 2 h a day with his device. Much mobile data activity takes place within users' homes. For users with fixed broadband and Wi-Fi access points at home, or for users served by operator-owned femtocells and picocells, a sizable proportion of traffic generated by mobile and portable devices is offloaded from the mobile network onto the fixed network. As estimated by Cisco [29], by 2021, 63% of all traffic from mobile-connected devices (almost 84 exabytes) will be offloaded to the fixed network by means of Wi-Fi devices and femtocells each month. Of all IP traffic (fixed and mobile) in 2021, 50% will be Wi-Fi, 30% will be wired, and 20% will be mobile. The appeal for higher definition video content is hampered by the current limited network capacity. A way to increase indefinitely the capacity to support this growth is to densify the access points or to enhance their individual capacity. However, this further increases the already predominant costs associated to backhaul. Video consumption complies with a Pareto-type law: in a standard system, 20% of content represents 80% of viewings. These figures can vary according to video services offered and users, but the principle remains: not all content has the same popularity factor with viewers. Such trends have been noticed in networks ranging from cellular, to user generated content, to Internet Protocol television (IPTV) and Video on Demand (VoD) [37]. As an illustration, for both YouTube and Daum (service in Korea), 10% of the most popular videos account for nearly 80% of views, while the remaining 90% account for total 20% of views [38]. It is a well-cited result that file popularities tend to follow a heavy-tail distribution. This means that the majority of requests occur for a relatively small fraction of the content. Observations of this characteristic have been noted in data logs for various CDNs. Such a trend is readily quantified by *Zipf's law* [37].

Typically, the effort made to create content is commensurate with the expected audience size. However, there may be exceptions with some costly content finding a small audience while cheap content finding a large audience. In any case, content delivery to a large audience is costly and not scalable. Caching and computing resources will be available in 5G network nodes. Hence, 5G networks can provide the resources upon which edge delivery node software can be deployed to facilitate a great quality of service of content delivered to end users, as well as optimizing the use of available network capacity. It may also alleviate low capacity backhaul link (e.g. in low-density populated areas) but some local congestion will still remain when feeding the most remote cache points in the network. In addition, it cannot ensure a good QoS for the delivery of live content with large audiences (e.g. news, sport events) at optimized bandwidth consumption in all the network's branches.

Adding broadcast/multicast resources in the network to be able to deliver the most popular on demand as well as live content towards the edge nodes of the network enables to offload a significant part of the traffic and/or to optimize the network infrastructure dimensioning (especially the backhaul links) in the lower density populated areas, where the cost per user is the highest.

Satellites are well suited to provide such broadcast/multicast resources over wide areas so as to aggregate the largest audience possible and hence to reduce the global delivery cost. Combining satellite broadcast/multicast resources with the terrestrial unicast resources is a powerful way to optimize the content delivery costs and improve scalability. The 5G network infrastructure selects the most appropriate resources according to the audience reached. It can convey 'video on demand' services (pull model), 'TV channels', and 'Live events' (push model) and optimize the cost in the same way. Moreover, as the audience for a TV channel varies over time, the delivery method can be adapted to optimize the network bandwidth and cost. Delivery of MEC VNF software updates can also be accommodated but would need to have greater reliability than some other services.

The service and network providers can use the geographic popularity hints to optimize the caching decision process even further and use satellite broadcast to reach the caching nodes of a popular programme region directly. This direct satellite delivery also benefits popular live content, as the time-consuming establishment of multicast trees across terrestrial networks can be avoided. Such a hybrid solution of content delivery to 5G edge nodes via (terrestrial) unicast and (satellite) broadcast/multicast resources will require adjustments to the eNB and other equipment as described in the scenarios below.

Operational scenarios embrace either the direct caching at a fixed terminal or caching at the mobile network edge for the MNO's to deliver to UE's as part of the 5G network. In the network infrastructure that is owned by a MNO, some IT resources (computing and storage) that are located at the network edge (e.g. close to eNodeBs) can be virtualized and leased by the MNO's to third-parties such as content providers. The content providers can use the virtualized storage and computing resources at the mobile edge to deploy their contents and intelligence, e.g. local caching, broadcasting, or multicasting of the contents to selected mobile edges where there are potentially large crowds of consumers on the content. We call such a virtualized mobile edge a virtual CDN node. Now, we consider the following two complementary scenarios: (1) with spatiotemporal knowledge of content popularity at different locations, selected content can be broadcasted or multicasted to the targeted mobile edge CDN nodes through satellite links a priori so that content has already been cached locally by the time the consumers make the requests; (2) a virtualized CDN node can perform *online* prefetching (through satellite links) of *just-in-time* video segments during a video session in order to ensure enhanced video quality end-to-end. This operation is useful to the video content applications where content is chunked into fixed-length segments (e.g. MPEG-DASH [Moving Picture Experts Group – Dynamic Adaptive Streaming over HTTP (Hypertext Transfer Protocol)]). In this scenario, the virtual CDN node does not need any knowledge about content popularity a priori.

2.3.1.1 Scenario 1a: offline multicasting and caching of video content and VNF software through satellite links

In case of video content, each virtual CDN node should be capable of monitoring and predicting the popularity of content objects in its local area, and making necessary decisions whether some content should be cached locally from the remote content origin a priori. In case the virtual CDN node has predicted that a specific content is expected to become popular in its region, it can issue a request to the original source to cache the content locally even before the local consumers start to make the requests. For such a purpose, satellite links can play a useful role to offload content traffic away from the terrestrial networks between the original content source and a virtual CDN node. It can be inferred that each virtual CDN node independently performs its own content popularity monitoring and prediction. As such even multiple virtual CDN nodes predict the same content object to become popular, they may issue the content requests to cache at different times. In order to maximize the benefit of broadcast and multicast, it is important that a content broadcast/multicast scheduling intelligence is in place such that (1) a (expected-to-be-popular) content object can always be delivered in time to individual virtual CDN nodes for caching before a large number of local users start to make the requests, and (2) the content traffic through the satellite links does not incur any potential congestions due to ad hoc content requests (incurred by the prediction outcome of local content popularity) from individual mobile CDN nodes. Such a concept is addressed in [39], among others.

On the other hand, such multicasting/broadcasting and caching techniques can be also applied for supporting VNF software updating at different sites. Compared to video content caching which requires popularity monitoring and prediction, the distribution task of software updates is more straightforward without complex intelligence. However, considering traffic load dynamicity over the satellite link, delay-tolerant VNF distribution operations can be scheduled during off-peak time (e.g. mid-night) when the content traffic load is expected to be on its low level.

2.3.1.2 Scenario 1b: online prefetching of video segments through satellite links

In recent years, video content providers such as YouTube, Netflix, etc. have been adopting the MPEG-DASH standard to provide streaming services [40]. In this scenario, a video content is chunked into fixed-segments which can be independently requested and adapted with multiple quality resolutions. While MPEG-DASH has many benefits such as offering flexibility through on-the-fly quality adaptation and its easy implementation over existing HTTP infrastructure, the fact that DASH uses Transmission Control Protocol (TCP) is a double-edged sword. On one hand, it means reliable content delivery and that video quality degradation caused by e.g. loss of I-frames can be avoided. On the other hand, when a wireless UE streams a video, there are two network segments on the end-to-end path that have distinctively different characteristics, which are (1) RAN that is wireless; and (2) the mobile core network and the public Internet that are typically wired or via satellite links. Specifically, the wired/satellite segment has high bandwidth-delay-product (BDP) due to the high-capacity links and long latency due to the long data transport distance across the global Internet. The worst case end-to-end latency across the terrestrial Internet is around 300 ms, and if satellite links are involved as the backhaul, this latency can be increased to 500+ ms. In contrast, the wireless access segment has much lower BDP due to limited radio resource capacity over the air interface and relatively lower latency. TCP does not perform well on end-to-end paths consisting of two segments with such different characteristics [41,42]. Even when there is no RAN resource competition, the Internet is unable to support seamless 4K video streaming in many scenarios.

Based on this observation, we propose to introduce a novel online video delivery scheme which particularly aims at providing QoE-assured 4K VoD streaming to mobile users at a global Internet scale, even with satellite links involved. The proposed scheme contains the following key operations at the mobile edge virtual CDN node: First, it realizes context awareness on network and users. For network context, it captures the RAN condition that is disseminated by the MNO through the Radio Network Information Service as specified by the ETSI MEC paradigm. Second, it performs adaptive prefetching on a per-user per-session basis, i.e. it pre-downloads video segments from the video source and maintains a progress gap ahead of the user's actual request progress. Such a gap is adaptive and is optimized based on its real-time knowledge on network and user context on-the-fly. Third, it performs DASH-based video quality adaptation on per segment basis according to its context awareness on the user side and network side. For instance, with the awareness of the UE signal strength (which is typically influenced by the user mobility), it can make appropriate decision on the current best quality that can be supported by the available throughput. Similarly, depending on the knowledge on the dynamic RAN load, the appropriate video quality can be also determined on per UE basis accordingly.

2.3.2 Scenarios for selected satellite use case 2: 5G fixed backhaul

The scenarios associated to this selected satellite use case for eMBB correspond to a wide range of scenarios. We have considered specifically the following three scenarios based on the relative geographical reach of satellite backhaul.

- Scenario 2a: satellite backhaul to groups of cell towers
- Scenario 2b: satellite backhaul to individual cell towers
- Scenario 2c: satellite backhaul to individual small cells

In all cases, the user and control plane data are interconnected to the core via the satellite gateway.

2.3.2.1 Scenario 2a: satellite backhaul to groups of cell towers

This might equally apply to an isolated town which has no practical terrestrial backhaul solution, or to a similarly isolated oceanic island.[1] In each case, a single satellite backhaul serves multiple cells – perhaps interlinked by radio [15]. A variation on this theme is that there is only one terrestrial connection and satellite is there to provide either top-up or backup capacity. In some locations, the demand for top-up capacity is seasonally driven by tourism. Such backhaul connections can be expected to carry eMBB and mMTC traffic along with any latency-tolerant URLLC applications traffic.

MIT has looked at World Bank data [43] and concluded that the number of cities with populations between 20,000 and 100,000 in sub-Saharan Africa will increase dramatically to around 2,200 from only 790 in 1990. The ITU has looked at the Information and Communication Technology (ICT) development around the world [44] and it is clear that African development lags the rest of the world.

In this context, the following representative scenario has been defined:

> ***Scenario 2a:*** *Satellite backhaul to a central node connected to five cell towers located in a rural town of 30,000 people in sub-Saharan Africa.*
>
> *The town is not one often frequented by tourists, the main foreign visitors being aid workers passing through. The satellite service is provided by a European operator and fulfilled to end users through local relationships. The predominant traffic on the cell is eMBB but there is some mMTC traffic generated by a quarry.*

2.3.2.2 Scenario 2b: satellite backhaul to individual cell towers

This would be satellite backhauls to a cell tower covering a region where there is no cost-effective terrestrial backhaul option. This includes satellite backhaul service to the following:

- rural and remote locations in developed countries;
- anywhere outside urban areas in developing countries; and
- islands, mountain regions, and other isolated areas;

Whilst there are fundamental differences in detailed implementation, there are also many similarities. The major difference between developed and developing countries is the ability to pay for service which will tend to impact on (1) the number of devices per community of a given size, (2) the amount of data an eMBB smartphone user can afford, and (3) the methods of payment.

[1]The traffic profile to a large cruise liner is also somewhat analogous though the equipment, the mMTC traffic and value chains would be somewhat different.

Undoubtedly, such a cell site would also support mMTC communications from IoT devices extending smart city like capabilities to such regions (e.g. smart villages and in particularly agri-tech functions [45]).

In this context, the following representative scenario has been defined:

> ***Scenario 2b:*** *Satellite backhaul to a single cell tower located in a rural area in the EU covering two villages about 5 km apart and a rural main road.*
>
> *The villages are home to 300 families, in summer months, an additional 50 families may be in holiday accommodation. The road can occasionally be busy with holiday traffic but is usually quiet. The predominant traffic on the cell is eMBB but there is some mMTC traffic generated by agri-tech.*

A variation on this theme is to consider such a system being used for short-term applications such as providing a new cell tower prior to the availability of fibre connectivity. Once the terrestrial connection is removed, the cell tower can then be moved to another location.

2.3.2.3 Scenario 2c: satellite backhaul to individual small cells

The most common scenario described for small cells is to increase densification in urban or other high traffic areas [46,47]. Clearly, these sites are extremely likely to have access to good terrestrial connectivity for backhaul. The instantiation of small cells in premises is considered in SaT5G use case 3 (see Section 2.3.3).

Other use cases are less often described though the Small Cell Forum provides interesting material in their release nine websites [48]. The EU has started an 'action for Smart Villages' [49] that might enable rural small cells. One advantage of small cells for remote villages over satellite is the cost-efficient coverage of small villages scattered across vast geographies, where large cell towers can be too expensive. One could also envisage one or more small cells being employed in a rural tourist location (church/temple, castle, tourist hotel/lodge, etc.) where either no signal reaches or the location's walls are too thick for the cellular radio links to penetrate.

Another interesting scenario is to provide communications when and where needed by the emergency services (e.g. [50]). This takes advantage of SatCom ability to provide and move capacity quickly from one location to another. To explore the implications of providing a service leveraging SatCom's rapid deploy/redeploy capabilities, the following representative scenario has been defined:

> ***Scenario 2c:*** *Satellite backhaul to multiple sites each with a single small cell providing the emergency services their private 5G service.*
>
> *When deployed, there will be a control room with 3 people and another 22 responders connected to this service – the cell will only carry their traffic. One such small cell will be provided per 20,000 people on average across the region or country. When analysing this, we should consider a developed country such as Belgium. All the traffic is eMBB-like traffic generated by the actions of the emergency services.*

Note that much of this analysis would apply to other related categories such as special events, humanitarian support, remote industry such as mining and even some aspects of military deployment such as for personal communications for the troops.

2.3.3 Scenarios for selected satellite use case 3: 5G to premises

The scenarios associated to this selected satellite use case for eMBB are mainly relevant to homes and small office home office (SOHO) premises located in underserved areas of developed countries, which are served with terrestrial telecommunication network infrastructure (xDSL or Cellular access) of poor bandwidth performance [e.g. users are located far from the Digital Subscriber Line Access Multiplexer (DSLAM) or far from 4G cell tower].

In such underserved areas of developed countries, the use of satellite to complement the existing terrestrial broadband access link can lead to a hybrid satellite/terrestrial multiplay scenario which can be envisaged in order to benefit from low-latency of terrestrial networks and high-bandwidth of satellite networks. In particular, complementing the existing and performance-limited terrestrial broadband link (xDSL or cellular access) by a satellite broadband link with multicast and caching capabilities is considered here.

In this context, we have considered specifically the following two scenarios:

- Scenario 3a: Hybrid multiplay (satellite/xDSL) at home/office premises in underserved areas.
- Scenario 3b: Hybrid multiplay (satellite/cellular) at home/office premises in underserved areas.

Similar hybrid satellite/xDSL scenarios have been considered in MENDHOSA [11] and other ESA studies [51,52]. This set of scenarios is particularly relevant if satellite can provide more bandwidth for premium clients, typically multiscreen and UHD, and if user experience for Internet applications is raised to a level similar than those of terrestrial networks (latency, throughput at peak hours for a large number of clients, etc.). To this end, the use of new generation HTS-DBS (or else referred to as next-generation hybrid broadband/broadcast satellites [19]) is important to maintain the satellite opportunity for broadband while bringing the communication cost significantly down but also to further boost the direct broadcast satellite services.

Cisco VNI forecast [29] for edge delivery and offload for multimedia content highlighted in Section 2.3.1 are relevant here as well. In fact, much mobile data activity take place within users' homes. For users with fixed broadband and Wi-Fi access points at home, or for users served by operator-owned femtocells and picocells, a sizable proportion of traffic generated by mobile and portable devices is offloaded from the mobile network onto the fixed network.

In this context, high-speed satellite links empowered with multicast and caching capabilities, direct to the home or office, providing the broadcast content and offloading existing terrestrial connectivity are considered here to take DTH a step beyond. The benefits of this set of scenarios are mainly twofold:

- Satellite coverage allows homogeneous service offering – anywhere.
- Multicast and caching enable bandwidth savings and improved QoS/QoE.

2.3.3.1 Scenario 3a: hybrid multiplay (satellite/xDSL) at home/office premises in underserved areas

This scenario corresponds to a multi-link network configuration with xDSL terrestrial link being augmented by the addition of a satellite broadband (bidirectional) link with broadcast/multicast and caching capabilities.

It is mainly relevant to homes and SOHO premises located in underserved areas of developed countries, which are served with xDSL links of poor bandwidth performance (e.g. users are located far from the DSLAM). As such, the xDSL link is of such poor quality that it cannot carry any multicast video.

Multiple devices are considered in the home/office environment. The advances in user devices, coupled with innovative services, drive user expectations, in particular in terms of choice, quality, availability, and affordability. There is a wide and growing choice of devices which can receive media services for viewing and listening, from stationary TV sets and home radio receivers to personal computers, tablets and smartphones, game consoles, media boxes, or even whole wall UHD TV displays. A rapidly growing number of smartphones' and tablets' displays enable high-quality video, and analysts predict that the use of video on such devices will grow substantially. Thus, DTH satellite TV integrated within the home/office IP network is considered in this scenario. Moreover, in-home/in-office distribution via Wi-Fi or home/office small-cell (femtocell) is also considered.

Such multi-link network configuration scenario requires two distinct functions as dedicated hardware or software equipment: (1) the home/office gateway located at the home/office premises and (2) Internet gateway located in the core network. Moreover, adjustments to the home/office gateway and Internet gateway in the backbone are needed to be able to split/combine the traffic over the existing terrestrial access link and the satellite link, respectively.

To this end, an intelligent user gateway which aggregates the multiple physical networks at the level of the home/office is integrated within the home/office gateway. The associated challenge here corresponds to the fact that these networks are mostly owned by different commercial entities, often competing with each other, that may have no interest in cooperating and sharing costs (e.g. on a common home/office gateway).

The home/office gateway includes caching and storage capabilities. Caching is added to each home/office premise in order to store locally the broadcasted/multicasted content. Efficient caching management algorithms push and store locally the most popular content optimizing the hit ratio and hence the bandwidth saving. Further details are provided in Section 2.3.1.

The home/office gateway also includes the satellite reception hardware. This concept is commercially appealing as it removes this functionality from the set-top-box (STB) and makes satellite delivery more independent of the main STB in-home. Satellite reception becomes a basic feature of a home and the satellite delivered services become available on any device (and not just the main TV screen STB as in the past).

It is therefore relevant to study the architecture and protocols that would be needed in order to provide a more robust satellite transmission scheme to every home/office premises and would abstract the end-devices from the actual physical delivery network.

In this context, new protocol stacks (e.g. native IP/multicast-assisted adaptive bitrate), which make the multi-device scenario significantly more attractive than it is today, should be further investigated in the subsequent work. Digital rights management and underlying protocol [e.g. DASH, HTTP Live Streaming (HLS), HTTP Smooth Streaming (HSS), etc.] will also have an impact on the home/office gateway design particularly in multiscreen environments. Other critical elements towards the realization of this scenario are caching, efficient caching management schemes, implications of chunked video, seamless blending of services, intelligent routing, network technology convergence, lowering of costs for implementing certain technologies (chipsets) because they use maximum technical commonality, standard end device functionality to provide access to all content independently on how it was delivered to that device.

2.3.3.2 Scenario 3b: hybrid multiplay (satellite/cellular) at home/office premises in underserved areas

This scenario is similar to the Scenario 3a above with the main difference being that the terrestrial xDSL is replaced by a 4G/5G cellular access link of poor bandwidth performance. Thus, it corresponds to a multi-link network configuration with 4G/5G terrestrial cellular link being augmented by the addition of a satellite broadband link with broadcast/multicast and caching capabilities.

It is mainly relevant to homes and SOHO premises located in underserved areas of developed countries, which are served with terrestrial cellular access of poor bandwidth performance (e.g. users are located far from 4G/5G cell tower).

2.3.4 Scenarios for selected satellite use case 4: 5G moving platform backhaul

The scenarios associated to this selected satellite use case for eMBB can be summarized as providing high-speed backhaul connectivity to individual moving terminals on airplanes, vehicles, trains, vessels (including cruise ships and other passenger vessels) or even future driverless cars, with the ability to multicast the same content (e.g. video, HD/UHD TV, FOTA, as well as other non-video data) across a large coverage area (e.g. for local storage and consumption) and provide efficient broadband access connectivity from/towards these moving platforms.

It should be noted that both satellite standalone backhauling and hybrid multiplay, i.e. the satellite link acting as a complement of existing terrestrial infrastructure, can be envisaged depending on the scenario and the type of targeted platform.

In this context, we have considered specifically the following three scenarios:

- Scenario 4a: updating content for on-board systems and grouped media request by the moving platform company.
- Scenario 4b: broadband access for passengers and individual media requests.
- Scenario 4c: business and technical data transfer for the moving platform company.

2.3.4.1 Scenario 4a: updating content for on-board systems and grouped media request by the moving platform company

This scenario can be resumed as a grouped request for media by the moving platform company.

This would be a way to update the content proposed by the moving platform company to passengers and subscription to live TV. The end user would use a standalone application on its own device or application pre-installed on the devices provided by the company.

The catalogue is updated with predictive valuable content and most demanded content. Accessible media might include videos, music, game patch, and newspapers. Live TV can also be proposed to broadcast for example live TV Show, TV News, or Live Sport program like a champion's league game.

This scenario is tightly related to caching/multicast edge delivery, exploiting the inherent broadcast capabilities of satellite networks and adding the particularity of being addressed to moving platforms. It is considered of high added-value for the moving platform companies being potentially combined with a full broadband access (as described in Scenario 4b).

A standalone satellite backhauling can be envisaged for airplanes and vessels (cruise ships and other passenger' vessels) and in hybrid mode, complementing existent terrestrial connectivity in trains and other vehicles (buses, trucks, or future driverless cars). As such, applications such as live train or bus network schedule and active map updates are part of this scenario.

Particularly, in the case of future driverless cars, satellite role would be to provide live broadcast and multicast streams for the passengers when in remote roads, through a phased-array antenna mounted on the rooftop of the car. In that case, a low-capacity moving platform is envisaged since the computing and storage capabilities for MEC, caching, etc. functions are expected to be smaller or more costly for a car.

2.3.4.2 Scenario 4b: broadband access for passengers and individual media requests

This scenario proposes a bidirectional broadband access for each passenger for private use which is transparent to the moving platform. The network requests are therefore individual and proper to each passenger activities.

As in the precedent scenario (Scenario 4a), a standalone satellite backhauling can be envisaged for airplanes and vessels (cruise ships and other passenger' vessels) and in hybrid mode, complementing existent terrestrial connectivity in trains and other vehicles (buses, trucks, or future driverless cars).

Particularly, in the case of future driverless cars, as in the precedent scenario (Scenario 4a), a low-capacity moving platform is envisaged here as well. The difference here corresponds to the satellite role which would be to provide 5G broadband access for the passengers when in remote roads.

The passenger would use their own device(s) and the whole applications installed as they do on the ground.

2.3.4.3 Scenario 4c: business and technical data transfer for the moving platform company

A moving platform company might use the connectivity to upload the log status of various moving platforms' equipment in order to spend less time on the ground, at the stations or on the dock.

This log data can feed the company data mining and machine learning servers, thus allowing enhancement of predictive maintenance with real-time problem identification.

For instance, the moving platforms are currently using a connectivity provided at their hubs for airline companies [53], at the stations for trains/bus or at the docks for vessels. Using the satellite backhaul would allow a secured real-time upload of data and would mitigate the risk of port destination not offering the required connectivity access to the company business servers.

2.4 Conclusions

This chapter presented initial results available from the SaT5G project [1,26]. It specifically defined how satellite can be seamlessly integrated into the 5G usage scenario of eMBB, by elaborating on selected use cases and scenarios for satellite positioning in eMBB. Note that according to past analyses conducted in [10,11], eMBB appears the most commercially attractive 5G usage scenario for SatCom with respect to mMTC and URLLC and, as such, this chapter focuses only on eMBB-related use cases and scenarios.

Specifically, the selected satellite use cases for eMBB addressed in this chapter are:

- **Use case 1: Edge delivery and offload for multimedia content and MEC VNF software:** Providing efficient multicast/broadcast delivery to network edges for content such as live broadcasts, ad-hoc broadcast/multicast streams, group communications, MEC VNF update distribution;
- **Use case 2: 5G fixed backhaul:** Broadband connectivity where it is difficult or not (yet) possible to deploy terrestrial connections to towers, for example, maritime services, coverage on lakes, islands, mountains, rural areas, isolated areas, or other areas that are best or only covered by satellites, across a wide geographic region;
- **Use case 3: 5G to premises:** Connectivity complementing terrestrial networks, such as broadband connectivity to home/office small cell in underserved areas in combination with terrestrial wireless or wireline;
- **Use case 4: 5G moving platform backhaul:** Broadband connectivity to platforms on the move, such as airplanes or vessels.

For each of the selected satellite use cases for eMBB above, a set of scenarios has been defined which drive the integrated network topology and the architecture design. Specifically, the set of scenarios elaborated in this chapter is the following:

- **Scenarios for use case 1:** (1a) Offline multicasting and caching of video content and VNF software through satellite links, and (1b) online prefetching of video segments through satellite links.
- **Scenarios for use case 2:** (2a) Satellite backhaul to groups of cell towers, (2b) satellite backhaul to individual cell towers, and (2c) satellite backhaul to individual small cells.
- **Scenarios for use case 3:** (3a) Hybrid multiplay (satellite/xDSL) at home/office premises in underserved areas, and (3b) hybrid multiplay (satellite/cellular) at home/office premises in underserved areas.
- **Scenarios for use case 4:** (4a) Updating content for on-board systems and grouped media request by the moving platform company, (4b) broadband access for passengers and individual media requests, and (4c) business and technical data transfer for the moving platform company.

Further work on the requirements definition, business modelling, system architecture definition, research to prototype implementation, validation, and demonstration of the selected satellite use cases and scenarios for eMBB corresponds to SaT5G project currently ongoing and future work, whose results will be reported in future publications.

Acknowledgements

The work presented in this chapter has been conducted as part of the SaT5G (Satellite and Terrestrial Network for 5G) project, which has received funding from the European Union's Horizon 2020 research and innovation programme under Grant Agreement No. 761413. The authors would like to thank their SaT5G consortium partners.

References

[1] SaT5G Project Consortium, "European Commission H2020 5G PPP Project 'SaT5G' (Satellite and Terrestrial Network for 5G)," 2017. [Online]. Available: http://sat5g-project.eu/.

[2] European Parliament, January 2016. [Online]. Available: http://www.europarl.europa.eu/RegData/etudes/BRIE/2016/573892/EPRS_BRI(2016)573892_EN.pdf.

[3] ITU-R, "IMT Vision – Framework and Overall Objectives of the Future Deployment of IMT for 2020 and Beyond" [Online]. Available: https://www.itu.int/dms_pubrec/itu-r/rec/m/R-REC-M.2083-0-201509-I!!PDF-E.pdf. [Accessed September 2015].

[4] GSMA Intelligence, "Understanding 5G," December 2014. [Online]. Available: https://www.gsmaintelligence.com/research/?file=141208-5g.pdf&download.
[5] ITU-T Y.3101, "Requirements of the IMT-2020 Network," January 2018.
[6] 5G PPP, [Online]. Available: https://5g-ppp.eu/. [Accessed July 2017].
[7] European Commission, "H2020 ICT-07-2017 Call," [Online]. Available: http://ec.europa.eu/research/participants/portal/desktop/en/opportunities/h2020/topics/ict-07-2017.html. [Accessed July 2017].
[8] European Commission, "5G for Europe: An Action Plan," European Commission, 14 September 2016. [Online]. Available: http://ec.europa.eu/newsroom/dae/document.cfm?doc_id=17131.
[9] NetWorld2020 SatCom WG, [Online]. Available: http://www.networld2020.eu/satcom-wg/. [Accessed October 2016].
[10] ESA, Avanti (Prime Contractor), "SPECSI (Strategic Positioning of the European and Canadian Satcom Industry)," [Online]. Available: https://artes.esa.int/projects/specsi. [Accessed July 2017].
[11] ESA, Thales Alenia Space France (Prime Contractor), "MENDHOSA (Media & ENtertainment Delivery over Hetnet with Optimized Satellite Architecture)," [Online]. Available: https://artes.esa.int/projects/mendhosa. [Accessed July 2017].
[12] ESA, Eurescom (Prime Contractor), "INSTINCT (Scenarios for Integration of Satellite Components in Future Networks)," [Online]. Available: https://artes.esa.int/projects/instinct. [Accessed July 2017].
[13] INSTINCT Whitepaper, September 2016. [Online]. Available: https://artes.esa.int/news/new-artes-funded-whitepaper-looks-role-satcoms-5g-context.
[14] ESA, Space Hellas (Prime Contractor), "CloudSat (Scenarios for Integration of Satellite Components in Future Networks)," [Online]. Available: https://artes.esa.int/projects/cloudsat. [Accessed July 2017].
[15] SANSA Consortium, "SANSA (Shared Access Terrestrial-Satellite Backhaul Network Enabled by Smart Antennas)," [Online]. Available: http://www.sansa-h2020.eu/. [Accessed July 2017].
[16] VITAL Consortium, "VITAL (VIrtualized hybrid satellite-TerrestriAl systems for resilient and fLexible future networks)," [Online]. Available: http://www.ict-vital.eu/. [Accessed July 2017].
[17] RIFE Consortium, "RIFE (Architecture for an Internet for Everybody)," [Online]. Available: https://rife-project.eu/. [Accessed July 2017].
[18] ESA, Nomor Research (Prime Contractor), "SCORSESE (The Role of Satellite in Collaborative Adaptive Bitrate Streaming Services)," [Online]. Available: https://artes.esa.int/projects/scorsese. [Accessed July 2017].
[19] ESA, SES (Prime Contractor), "HTS-DBS (High Throughput Digital Broadcasting Satellite Systems)," [Online]. Available: https://artes.esa.int/funding/high-throughput-digital-broadcasting-satellite-systems-artes-51-3a073-0. [Accessed July 2017].

[20] ESOA 5G White Paper, "Satellite Communications Services: An Integral part of the 5G Ecosystem," ESOA, 2017. [Online]. Available: https://www.esoa.net/cms-data/positions/ESOA%205G%20Ecosystem%20white%20paper.pdf.

[21] 3GPP SA1, "TR 22.891 V14.2.0 (2016-09), Feasibility Study on New Services and Markets Technology Enablers, Stage 1," 2016.

[22] 3GPP SA1, "SMARTER Technical Reports," 23 July 2016. [Online]. Available: http://www.3gpp.org/news-events/3gpp-news/1786-5g_reqs_sa1.

[23] 3GPP SA1, "TR 22.863 V14.1.0 (2016-09), Feasibility Study on New Services and Markets Technology Enablers – Enhanced Mobile Broadband; Stage 1," 2016.

[24] 3GPP SA1, "3GPP TR 22.864 V14.1.0 (2016-09), Feasibility Study on New Services and Markets Technology Enablers – Network Operation," 2016.

[25] 3GPP RAN1, "3GPP TR 38.811 V0.1.0 (2017-06), 'Study on New Radio (NR) to support Non Terrestrial Networks (Release 15)'," 2017.

[26] SaT5G Consortium, "Deliverable D2.1 'Satellite Reference Use Cases and Scenarios for eMBB'," August 2017.

[27] NSR, "Wireless Backhaul via Satellite, 11th Edition," March 2017. [Online]. Available: http://www.nsr.com/research-reports/satellite-communications-1/wireless-backhaul-via-satellite-11th-edition/.

[28] Research and Markets, "Wireless Network Optimization Research Report 2017: Mobile Edge Computing (MEC) and Content Delivery Network (CDN) Market Outlook, Forecasts, and the Path to 5G Enabled Apps and Services," July 2017. [Online]. Available: http://www.prnewswire.com/news-releases/wireless-network-optimization-research-report-2017-mobile-edge-computing-mec-and-content-delivery-network-cdn-market-outlook-forecasts-and-the-path-to-5g-enabled-apps-and-services-300493587.html.

[29] Cisco, "Cisco Visual Networking Index: Global Mobile Data Traffic Forecast Update, 2016–2021," Cisco Visual Networking Index (VNI), February 2017.

[30] Sandvine, "Sandvine: 70% of Global Internet Traffic Will Be Encrypted In 2016," February 2016. [Online]. Available: https://www.sandvine.com/pr/2016/2/11/sandvine-70-of-global-internet-traffic-will-be-encrypted-in-2016.html.

[31] UK ITRC, "Exploring the Cost, Coverage and Rollout Implications of 5G in Britain," 2016. [Online]. Available: http://www.itrc.org.uk/wp-content/uploads/Exploring-costs-of-5G.pdf.

[32] OECD Directorate for Science, Technology and Industry, "Working Party on Communication Infrastructures and Services Policy – The Development of Fixed Broadband Networks," January 2015. [Online]. Available: https://www.oecd.org/officialdocuments/publicdisplaydocumentpdf/?cote=DSTI/ICCP/CISP(2013)8/FINAL&docLanguage=En.

[33] OECD, "OECD Broadband Statistics – Percentage of Fibre Connections in Total Broadband among Countries Reporting Fibre Subscribers," December 2016. [Online]. Available: https://www.oecd.org/sti/broadband/1.10-PctFibreToTotalBroadband-2016-12.xls.

[34] Euroconsult, "HTS Capacity Lease Revenues to Reach More Than $6 Billion by 2025," June 2017. [Online]. Available: http://www.euroconsult-ec.com/21_June_2017.

[35] YouTube, "YouTube Statistics," [Online]. 2017. Available: https://www.youtube.com/yt/press/statistics.html.

[36] Statista Digital Market Outlook, [Online]. 2017. Available: https://www.statista.com/chart/9539/smartphone-addiction-tightens-its-global-grip/.

[37] C. Brinton, E. Aryafar, S. Corda, S. Russo, R. Reinoso, and M. Chiang, "An Intelligent Satellite Multicast and Caching Overlay for CDNs to Improve Performance in Video Applications," in *AIAA ICSSC*, 2013, Paper No. AIAA 2013-5664.

[38] M. Cha, H. Kwak, P. Rodriguez, Y.-Y. Ahn, and S. Moon, "Analyzing the Video Popularity Characteristics of Large-Scale User Generated Content Systems," *IEEE/ACM Transactions on Networking*, vol. 17, no. 5, pp. 1357–1370, October 2009.

[39] A. Kalantari, M. Fittipaldi, S. Chatzinotas, T.X. Vu, and B. Ottersten, "Off-Line Cache-Assisted 5G Hybrid Satellite-Terrestrial Network," in *IEEE Global Communications Conference (GlobeCom 2017)*, Singapore, December 2017.

[40] ISO/IEC JTC 1/SC 29, "ISO/IEC 23009-1:2014, Information Technology – Dynamic Adaptive Streaming Over HTTP (DASH) – Part 1: Media Presentation Description and Segment Formats," May 2015. [Online]. Available: https://www.iso.org/standard/65274.html.

[41] F. Bronzino, D. Stojadinovic, C. Westphal, and D. Raychaudhuri, "Exploiting Network Awareness to Enhance DASH Over Wireless," in *13th IEEE Annu. Consumer Commun. Netw. Conf.*, pp. 1092–1100, January 2016.

[42] C. Ge, N. Wang, G. Foster and M. Wilson, "Towards QoE-assured 4K Video-on-demand Delivery through Mobile Edge Virtualization with Adaptive Prefetching," *IEEE Transactions on Multimedia*, vol. 19, no. 10, pp. 2222–2237, 2017.

[43] MIT, "Urbanisation in SSA," [Online]. 2001. Available: http://web.mit.edu/urbanupgrading/upgrading/case-examples/overview-africa/regional-overview.html. [Accessed 20 July 2017].

[44] ITU, "Measuring the Information Society Report 2016," [Online]. 2016. Available: https://www.itu.int/en/ITU-D/Statistics/Documents/publications/misr2016/MISR2016-w4.pdf.

[45] IoTONE, "IoTONE "Accelerating the Industrial Internet of Things"," [Online]. 2016. Available: https://www.iotone.com/guide/internet-of-things-designing-smart-villages/g546. [Accessed 31 July 2017].

[46] Huawei, "Small Cell Network White Paper," [Online]. 2016. Available: http://www.huawei.com/minisite/hwmbbf16/insights/small_cell_solution.pdf. [Accessed 5 July 2017].

[47] 3GPP, "HetNet/Small Cells," [Online]. 2014. Available: http://www.3gpp.org/hetnet. [Accessed 5 July 2017].

[48] Small Cell Forum, "Small Cell Forum Release 9.0," [Online]. 2017. Available: http://scf.io/en/themes/Rural__Remote.php. [Accessed 31 July 2017].

[49] EU, “EU action for Smart Villages,” [Online]. 2017. Available: https://enrd.ec.europa.eu/news-events/news/eu-action-smart-villages_en. [Accessed 18 July 2017].

[50] Telecoms, “EE Looks to Satellite Mobile Backhaul with $29 million Avanti Deal,” [Online]. 2016. Available: http://telecoms.com/472384/ee-looks-to-satellite-mobile-backhaul-with-29-million-avanti-deal/. [Accessed 20 July 2017].

[51] ESA, Forsway (Prime Contractor), “Satellite Extension of xDSL Copper Wire Based Networks,” [Online]. 2016. Available: https://artes.esa.int/projects/satellite-extension-xdsl-copper-wire-based-networks. [Accessed 31 July 2017].

[52] ESA, Intecs (Prime Contractor), “SAT4NET: Analysis of Satellite Downstream Boost for xDSL Networks,” [Online]. 2016. Available: https://artes.esa.int/projects/sat4net. [Accessed 31 July 2017].

[53] Airbus, “Airbus Launches New Open Aviation Data Platform, Skywise,” [Online]. 2017. Available: https://youtu.be/D2o–8XzxrI. [Accessed 31 July 2017].

Chapter 3

SDN-enabled SatCom networks for satellite-terrestrial integration

Fabián Mendoza[1], Ramon Ferrús[1], and Oriol Sallent[1]

Key features of satellite communications such as wide-scale coverage, broadcast/multicast support and high availability, together with significant amounts of new satellite capacity coming online, anticipate new opportunities for satellite communications services to become an integral part of upcoming 5G systems. This chapter examines the realization of end-to-end (E2E) traffic engineering (TE) in a combined terrestrial-satellite network embracing software-defined networking (SDN) technologies. The focus is placed on a mobile backhaul network scenario where a satellite component is used to complement the terrestrial infrastructure in a way that E2E paths across both satellite and terrestrial links can be centrally computed and rearranged dynamically at flow-level granularity in front of link congestion and failure events. The chapter describes the architecture of such SDN-enabled satellite ground segment system and presents illustrative TE workflows. Furthermore, sustained in the proposed architectural framework, an SDN-based TE application for hybrid satellite-terrestrial backhaul networks is developed and its performance assessed under diverse scenario conditions.

3.1 Introduction

The role that satellite communications can play in the forthcoming 5G ecosystem is being revisited [1–3]. The satellite communications industry is pushing for better satellite-terrestrial cooperation as part of mobile networks of 2020 [4–6]. Remarkably, a requirement for next-generation 3GPP systems to be able to provide services using satellite access has been included within the normative stage 1 requirements [7] and a study item is on-going to address the support of non-terrestrial networks (i.e. satellite access and other types of access networks based on the use of airborne vehicles for transmission) within the 5G New Radio specifications [8] in order to achieve higher layer operational integration and high degree of radio interface commonality. Indeed,

[1]Department of Signal Theory and Communications, Universitat Politècnica de Catalunya (UPC), Spain

according to [8], non-terrestrial access networks are expected to be an integral part of 5G service deployment by:

- Enabling ubiquitous 5G service to terminals [especially Internet of Things (IoT)/machine type communications, public safety/critical communications] by extending the reach of terrestrial-based 5G networks to areas that cannot be optimally covered by terrestrial 5G network.
- Enabling 5G service reliability and resiliency due to reduced vulnerability of air/space borne vehicles to physical attacks and natural disasters. This is especially of interest to public safety or railway communication systems.
- Enabling connectivity of 5G Radio Access Network (5G-RAN) elements to allow ubiquitous deployment of 5G terrestrial network.
- Enabling connectivity and delivery of 5G services to user equipment (UE) on board airborne vehicles (e.g. air flight passengers, Unmanned Aerial System (UASs)/drones, etc.).
- Enabling connectivity and delivery of 5G services to UE on board other moving platforms such as vessels and trains.
- Enabling efficient multicast/broadcast delivery of services such as A/V content, group communications, IoT broadcast services, software downloads (e.g. to connected cars) and emergency messaging.
- Enabling flexibility in TE of 5G services between terrestrial and non-terrestrial networks.

In addition to achieving high degree of radio interface commonality and higher layer operational integration with 5G terrestrial access, the deployment and operation of networks that combine terrestrial and satellite transmission components are also expected to benefit from the incorporation of network softwarization technologies such as SDN and network function virtualization (NFV) [9–12] into satellite systems. Indeed, during the last decade, the networking community is witnessing a paradigm shift towards the softwarization of communication networks in a quest for improved agility and flexibility, and ultimately cost reduction, in the deployment and operation of networks. In this context, the evolution of satellite ground segment systems (e.g. satellite gateways and terminals) from today's rather closed solutions towards more open architectures based on SDN and NFV technologies arises as a necessary step not only to bring into the satellite domain the benefits associated with the advances in network softwarization technologies being consolidated in the 5G landscape but also to greatly facilitate the seamless integration and operation of combined satellite and terrestrial networks. In particular, the realization of a full E2E networking concept where the whole satellite-terrestrial network behaviour can be programmed in a consistent and interoperable manner is expected to benefit from the introduction within satellite networks of abstraction models, protocols and application programming interfaces (APIs) compatible with the mainstream SDN architectures and technologies being adopted in 5G in the pursue of industry convergence around device-neutral and vendor-neutral SDN solutions.

In this context, this chapter examines the realization of E2E TE in a combined terrestrial-satellite network embracing SDN technologies. To that end, the foundations for SDN-enabled satellite networks are first discussed, outlining the main reference

SDN architectures and delineating a potential SDN-based architecture for satellite ground segment systems. Next, an integration approach for the realization of E2E TE is presented. The focus is placed on a mobile backhaul network scenario where a satellite component is used to complement the terrestrial infrastructure in a way that E2E paths across both satellite and terrestrial links can be centrally computed and rearranged dynamically at flow-level granularity in front of link congestion and failure events. Illustrative workflows for E2E TE are provided. On this basis, the chapter follows with the formulation of an illustrative SDN-based TE application that exploits a combination of control features and criteria, including E2E path computation, satellite capacity resource reservations, allocation criteria depending on the traffic nature, admission control and rate control features and network utility maximization criteria. A performance analysis is finally presented to assess the behaviour of the proposed SDN-based TE application under diverse scenarios, including homogeneous and non-homogeneous load situations, terrestrial link failures in some of the base stations (BSs) and deployment of a number of transportable BSs (TBSs) that exclusively rely on the satellite capacity for backhauling.

3.2 SDN-based functional architectures for satellite networks

This section presents an SDN-based functional architecture for the ground segment of a satellite broadband communications system, delineating the different alternatives for the support of SDN concepts and technologies within both internal and external interfaces of the satellite network. In order to set the ground for the discussion, key foundations on SDN architectures and technologies as well as on satellite broadband system architectures are briefly outlined first.

3.2.1 Foundations on SDN architectures

General principles and reference SDN architectures have been specified by the Open Networking Foundation (ONF) and Internet Engineering Task Force (IETF) in [13,14], respectively. Both SDN architectural models are illustrated in Figure 3.1. Keeping aside some differences in terminology and orientation, both architectures reflect the key principles of SDN: (1) separation of data plane resources (e.g. data forwarding functions) from control and management functions, (2) centralization of the management-control functions and (3) programmability of network functionality through device-neutral and vendor-neutral abstractions and APIs. While the IETF model description is more centred on network devices and control and management abstraction layers, the ONF model is specified around the so-called SDN controller, which is the core functional entity of the SDN architecture. The SDN controller exposes services and resources to clients via application-controller plane interfaces (A-CPIs) and consumes underlying services and resources via data-controller plane interfaces (D-CPIs). A-CPIs and D-CPIs are, respectively, the equivalents of the control plane/management plane southbound interfaces (CP/MP SBIs) and service interfaces within the IETF model. Service interfaces are also commonly referred to as northbound interfaces (NBI).

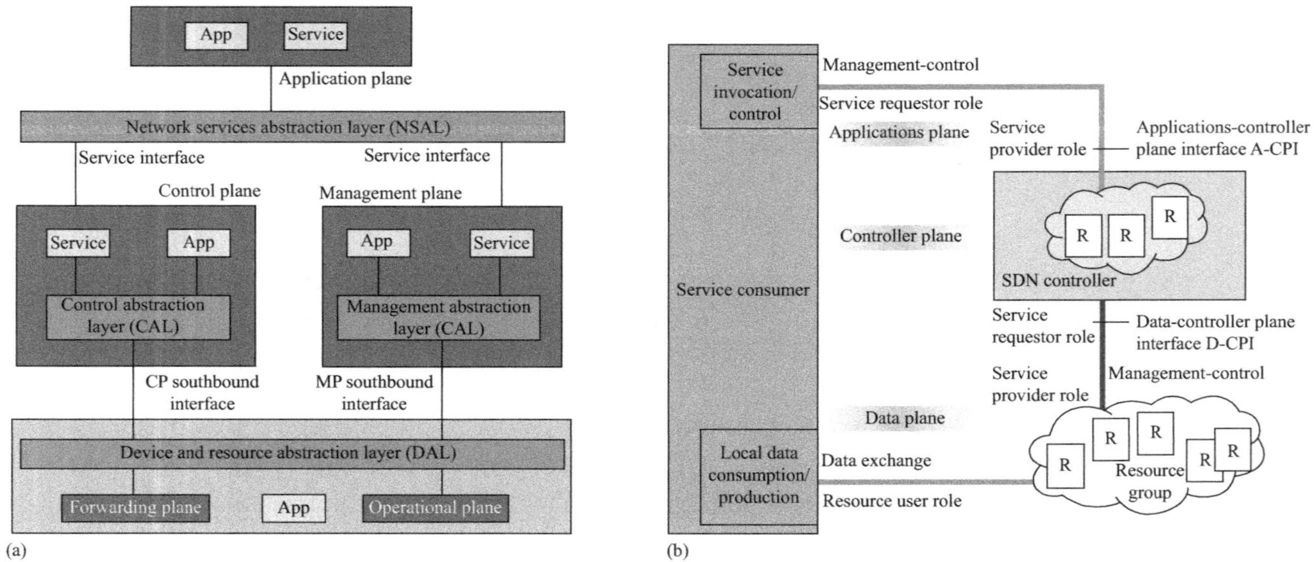

Figure 3.1 (a) IETF RFC 7426 and (b) ONF SDN architectural models

A more purpose specific SDN architecture for transport networks is being developed by the Traffic Engineering Architecture and Signalling Working Group within the IETF, which is responsible for defining multiprotocol label switching (MPLS) and generalized MPLS (GMPLS) TE architectures and protocols. Such SDN architecture, named Abstraction and Control of Transport Networks (ACTN), describes a control framework for operating a TE network (such as an MPLS-TE network or a layer 1 transport network) to provide connectivity and virtual network services for customers of the TE network. The services provided by the ACTN can be tuned to meet the requirements (such as traffic patterns, quality and reliability) of the applications hosted by the customers. An illustration of the ACTN architecture is given in Figure 3.2. The ACTN architecture is well aligned with the previously introduced ONF and IETF SDN architectural principles even though it is represented as a three-tier reference model. Importantly, the ACTN architecture allows for hierarchy and recursion not only of SDN controllers but also of traditionally controlled domains that use a control plane.

With regard to data models, protocols and APIs, the OpenFlow (OF) protocol standardized by ONF is likely the most popular protocol used in the southbound interface (SBI) of SDN architectures. The OF specification [15] currently defines two elements: (1) an abstract model of a switch datapath for packet processing (i.e. the expected behaviour of a switch) and (2) a protocol for the communication between the switch and the SDN controller to program the behaviour of the switch dataplane. While the current scope of OF is basically flow management, the ONF is seeking as future evolutions of the protocol to expand the scope of SDN control, to support a broad spectrum of datapath hardware platforms, including fully programmable packet switches (i.e. switches with no built-in protocol behaviour) [16]. Another important initiative within the ONF is the *Information Modelling Project* (ONF-IMP), which intends to provide a common basis for terminology definition and normalization underpinning SDN API development to facilitate convergence of model-based interface definitions.

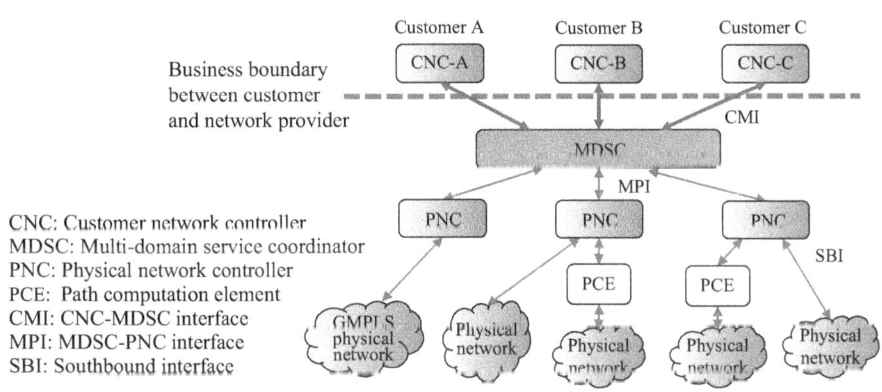

Figure 3.2 IETF Abstraction and Control of Transport Networks (ACTN) architecture

To that end, the ONF-IMP has established the so-called ONF Common Information Model (ONF-CIM) [17], which includes all of the artefacts (objects, attributes and relationships) that are necessary to describe the domain for the applications being developed. The ONF-CIM comprises a core model (ONF Core Information Model [18]), which provides a technology-agnostic representation of network forwarding resources from a management-control perspective, and various specific technology and layer additions (e.g. OTN/OCH/ODU, ETH, MPLS-TP). The ONF-CIM might be continually expanded and refined over time, to add new applications, capabilities or technologies, or to refine it as new insights are gained. Building on the ONF-CIM, the *Open Transport* project within the ONF addresses SDN and OF standard-based control capabilities for transport technologies of different types, including optical and wireless transport. The work includes identifying and addressing different use cases, defining the application of SDN architecture and information modelling to transport networks, and defining standard SDN interfaces for transport networks, including OF protocol extensions and transport controller APIs. Three relevant outputs to consider in our discussion from the Open Transport project are ONF TR-522 [19], which describes the application of the general SDN architecture [13] and the ONF-CIM to transport networks; ONF TR-527 [20], which develops the functional requirements for the definition of a Transport API (T-API); and TR-532 [21], which provides a technology-specific extension to the ONF core information model [18] for the use of the SDN architecture in wireless transport networks. Still within the ONF, it's also worth mentioning the *NBIs* project that develops concrete requirements, architecture and working code for NBIs in order to lower barriers to SDN application development. Thus far, only document ONF TR-523 [22] stating the principles for the definition of intent-based interfaces has been produced. Within the IETF domain, YANG [23] is becoming the data modelling language of choice. YANG can be used to model both configuration and operational states; it is vendor-neutral and supports extensible APIs for control and management of elements. Indeed, YANG data models [23], together with appropriate messaging protocol (e.g. NETCONF [24] or RESTCONF [25]) and encoding mechanisms, have been already adopted and promoted by several industry-wide open management and control (M&C) initiatives (e.g. OpenConfig). YANG data models are also being considered to provide solutions for the ACTN framework [26]. For more information on SDN architectures and technologies along with key developments within ONF, IETF and other standard development organizations and industrial fora, the interested reader is referred to [27,28].

3.2.2 Satellite network architecture

A technology-agnostic reference architecture for Broadband Satellite Multimedia (BSM) communications systems has been established by ETSI [29]. The BSM system architecture is conceived as an overarching architecture consisting of the common components found in an interactive satellite communications network: User Satellite Terminal (ST), Gateway ST, satellite payload, Network Management Centre (NMC) and Network Control Centre (NCC). Importantly, the BSM system architecture is not restricted to any particular satellite air interface (e.g. DVB-S2/RCS2) but intended to support diverse air interface protocols. Indeed, the overall ETSI BSM system architecture is applicable to the different configurations that a satellite network can be implementing in terms of topology (star, mesh) and payload operation (transparent and regenerative) [30].

Figure 3.3 depicts the ETSI BSM architecture in terms of reference interfaces for the user plane (U-plane) and for the control/management planes (C-plane and M-plane). The reference interfaces are divided into physical and logical interfaces, the former referring to physical connections between equipment and the latter referring to logical associations between peer protocol entities. As illustrated in Figure 3.3, one central principle of the BSM system architecture is the logical separation of the satellite independent (SI) layers (e.g. Ethernet/IP layers together with the interworking and adaptation functions needed for the interconnection with external networks) from the satellite dependent (SD) layers, whose interaction is formalized by the definition of an SI-service access point (SI-SAP) interface [31]. Focusing on the U-plane (aka data plane), four physical interfaces are identified at the interconnection points between the premises network and the user ST (T interface), user ST and satellite payload (U/U_{ST} interface), satellite payload and Gateway ST (U/U_{GW} interface) and gateway ST and external network (G interface). The radio interface label U means that the user ST and gateway ST have the same radio interface to communicate among them through the satellite payload while U_{ST} and U_{GW} refer to the case that the radio interface is different in the two sides. On the other hand, three logical interfaces are defined for the U-plane, corresponding to the peer-to-peer interactions of the different layers of the radio interface protocols. One logical interface covers the interaction between SI protocol layers at both sides, i.e. the interworking and adaptation functions. The other two logical interfaces fit within the SD lower layers, one for interfacing with the satellite payload and another for the peer ST. The boundary between these two

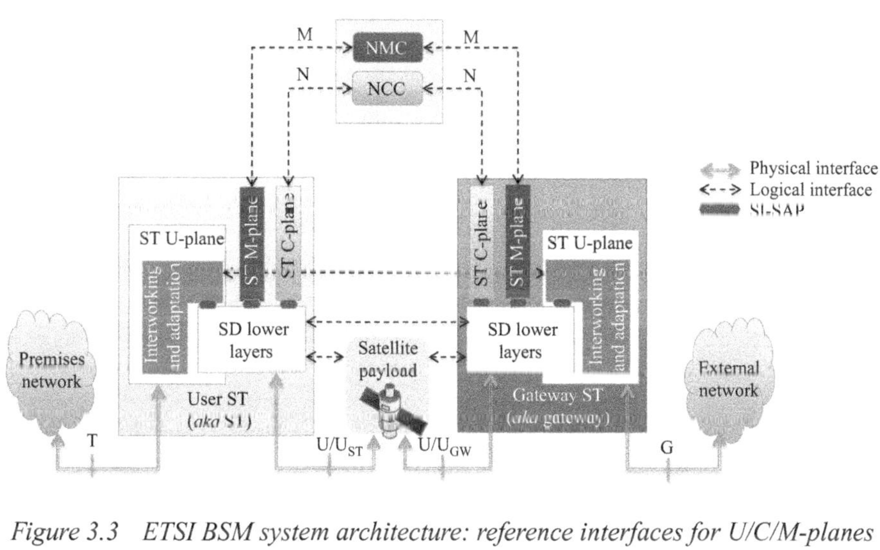

Figure 3.3 ETSI BSM system architecture: reference interfaces for U/C/M-planes

logical interfaces depends on the supported satellite payload capabilities. With regard to the C-plane and M-plane, two logical interfaces named N and M are identified. In particular, interface N is a control interface between the user/gateway STs and the NCC, which is the functional entity that provides the real-time control of the BSM network (e.g. session/connection control, routing, terminals' access control to satellite resources, etc.). And, interface M is a management interface between the STs and the NMC, which is the functional entity in charge of the management of all the system elements in the BSM network (e.g. Fault, Configuration, Performance, Accounting and Security management). Of note is that, currently both N and M interfaces are considered as internal interfaces within the BSM system, not subject to standardization or harmonization between vendors. However, we devise this functional separation established in the BSM reference model as the foundational point to introduce SDN concepts and technologies within the BSM system, as detailed later on in this chapter.

With regard to the BSM service capabilities and QoS support over the satellite links, the BSM system architecture defines *BSM bearer services*. A BSM bearer service includes all aspects to enable the provision of a U-plane data transport service between the user/gateway STs, including the QoS characteristics and other properties such as connectionless or connection-oriented, unidirectional or bidirectional, symmetric or asymmetric and point-to-point/multicast/broadcast nature of the bearer service. The BSM bearer services are defined at SI-SAP interface level and use the services provided by the underlying native bearer services (which depends on the specific implementation of the SD lower layers for link and medium access control). In the same way, the higher layer services (e.g. IP connectivity over the satellite network) are built on the BSM bearer services and can be mapped to different BSM bearer services depending on the particular higher layer service requirements. The abstract representation of the available BSM bearer services at SI-SAP level is done via labels called queue identifiers (QIDs). The QoS properties associated with a given QID are defined by QoS-specific parameters and each QID is mapped onto suitable lower layer transfer capabilities in order to realize that QoS. QIDs are defined in more detail in the SI-SAP specification [31] and SI-SAP guidelines [32]. The QoS model established for BSM systems and the traffic classes used to describe QoS, performance management and resource allocation are defined in detail in [33,34], respectively.

3.2.3 SDN-enabled satellite network architecture

Based on the previously described aspects of the BSM system architecture (i.e. functional components, reference interfaces, bearer services/QIDs and QoS model), Figure 3.4 illustrates the proposed solution for the adoption of an SDN architecture within the satellite network.

This solution relies on the introduction of an SDN controller as part of the satellite network functional architecture to manage the connectivity services between the T and G reference points. In the case of packet switched services (e.g. IP and Ethernet connectivity services), the finest granularity for QoS forwarding treatment is commonly referred to as a *flow*, which can be defined as a sequence of packets between a

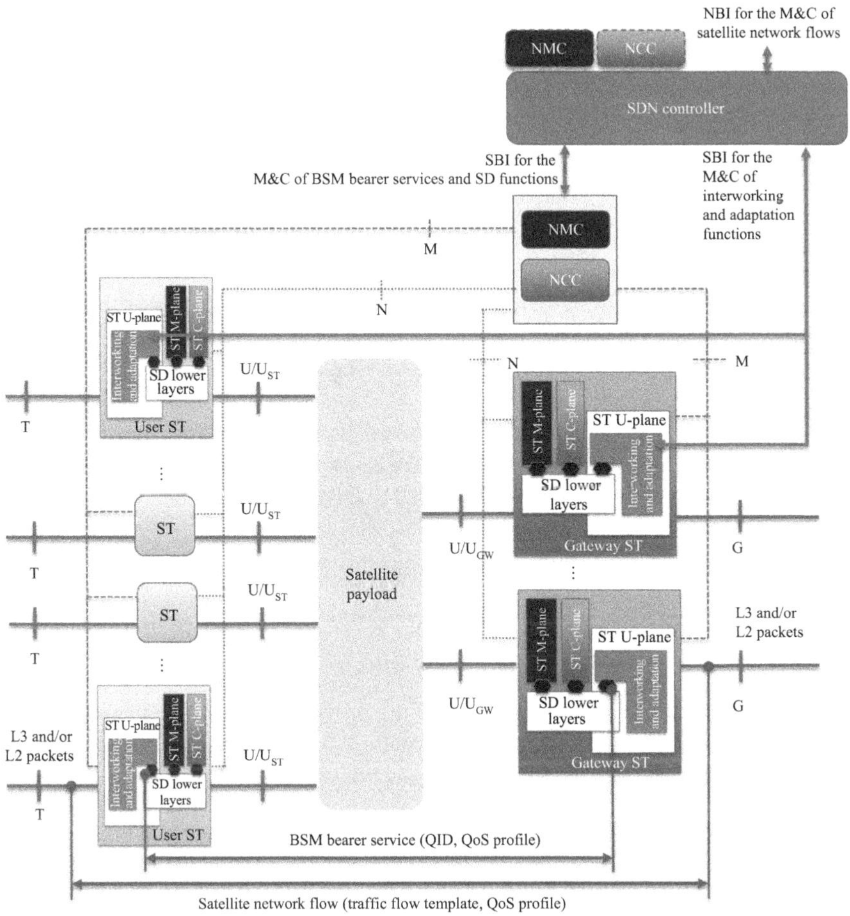

Figure 3.4 SDN-based satellite network architecture

source and a destination intended to receive identical service policies when progressing through the U-plane. A set of packet filters, referred to as traffic flow template in Figure 3.4, shall be used to identify individual data flow belonging to a specific application (e.g. the packet filters for IP flows typically consist of IP five-tuples with source IP address, destination IP address, source port, destination port and protocol type). As depicted in Figure 3.4, the SDN controller directly manages the SI services such as the IP/Ethernet layer QoS and indirectly manages the SD services through the NCC/NMC functions. Accordingly, the following interfaces are then needed:

- SBI for the M&C of the interworking and adaptation functions in the gateway STs and potentially also in User STs. This interface is not satellite dependent so that SDN models and interfaces used in the broad networking domain can be adopted such as *OF* and *YANG models*.

- SBI for the M&C of the BSM bearer services and potentially also of some capabilities within the SD lower layers (satellite resources such as a frequency plan, modulation and coding schemes or other satellite-specific properties) through the interaction with legacy satellite network NCC/NMC functions. This interface may have to consider satellite-specific aspects so that some extension and adaption of existing SDN models and interfaces are necessary. Potential candidate baseline SDN data models and interfaces for the realization of this interface are *OF* and the *Microwave Information Model* [21]. In case NCC/NMC functions could be eventually implemented as network applications on top of the SDN controller, another potential solution for this interface could be based on an extension of the *ETSI SI-SAP interface* for the realization of the N and M interfaces directly serving as SBIs from the SDN controller viewpoint.
- NBI for the M&C of the satellite network flows by network applications running on top of the SDN controller or from external controllers within an upper level control domain. Potential candidate SDN data models and interfaces for the realization of this interface are *OF*, the *ONF Transport API* [20] and the *YANG models* as identified for the ACTN architecture.

3.2.4 Candidate SDN data models and interfaces

The main characteristics and pros/cons of the previously mentioned candidate data models and interfaces for consideration within the SDN-based satellite network architecture are discussed in the following sections.

3.2.4.1 ETSI BSM SI-SAP

The SI-SAP interface provides a functional separation between SD and SI layers. The SI-SAP interface is currently specified in terms of the primitives exchanged between the SD and SI layers, following ISO/OSI protocol stack model. The existing specification [35] defines primitives to support U-plane and C-plane functionalities. More specifically, the C-plane services provided by the SI-SAP interface are (1) *logon/logoff services*; (2) *SI layer configuration service*, to provide the SI layer with the necessary information to configure e.g. the addressing plan and different functions of higher protocol layers such as header compression; (3) *address resolution service*, used to perform the address mapping between SI and SD layers; (4) *resource reservation service*, used for resource allocation and overall QoS management; and (5) *multicast group receive and transmit services*, invoked to build multicast groups and to receive the desired multicast data flows.

The SI-SAP could be deployed as an external interface, as considered in [35]. Under such an approach, the SD lower layers and the control entity can be running at different places and be interconnected by point-to-point protocol technology (e.g. Ethernet). As such, the SI-SAP interface service primitives are defined as specific messages transported by the technology implemented by the point-to-point protocol. Message formats and protocol encapsulation options are discussed in [35]. Additional information on the use of the SI-SAP interface can be found in [32].

Therefore, the BSM SI-SAP is a clear candidate for the implementation of SBIs for the M&C of the BSM bearer services and potentially also some capabilities within the SD lower layers. However, to that end, current specifications should be revisited and extended since they are not currently conceived to manage physical radio aspects of the satellite link (e.g. modulation and coding scheme selection) and have limited monitoring and management plane capabilities.

3.2.4.2 ONF OpenFlow

OF fundamentally provides a solution for flow management, allowing a fine-grained control of the forwarding behaviour at packet level of a switch/router node. The OF specification [15] currently defines two elements: (1) an abstract model of a switch datapath for packet processing (i.e. the expected behaviour of a switch) and (2) a protocol for the communication between the switch and a controller to enable the controller to program the behaviour of the switch dataplane. The OF specification features support for a number of commonly used dataplane protocols ranging from layer 2 to layer 4, with packet classification being performed using stateless match tables and packet processing operations (called actions or instructions) ranging from header modification, metering, QoS, packet replication (e.g. to implement multicast or link aggregation) and packet encapsulation/decapsulation. The specification also counts with several artefacts for statistics collection, which can be retrieved on demand or via notifications. A complementary protocol to the OF protocol is OF-CONFIG, also standardized by ONF. OF-CONFIG adds configuration and management support to OF switches. OF-CONFIG provides configuration of various switch parameters that are not handled by the OF protocol.

However, current OF specification has very limited support to cope with physical layer aspects of the switch ports. Thus far, the OF specification has only introduced a set of port properties to add support for optical ports. As to the support of wireless ports, the only consideration has been to define the process for sending a packet through the same port that it was received, a behaviour that was not clearly defined in in earlier versions of the specification and that is typical in wireless links. Therefore, there is no practical support in current OF specification for the configuration and monitoring of wireless links/ports.

Accordingly, within the SDN-based satellite network, OF is a clear candidate to be used internally to control the switching functions within gateways and STs. Remarkably, OF could also be used as an external interface to expose some control for satellite network flow management and so provide the control features necessary for the realization of E2E TE (this approach is the one further elaborated in this chapter; see Section 3.3). Of note is that, the exposure of an OF interface by a satellite network has been also proposed in the context of the realization of virtual network operator (VNO) solutions [36], in which a VNO is provided with an interface to control and manage the satellite segment resources leased from a satellite network operator as if it was programming an OF switch. All in all, OF is an extensible protocol, providing mechanisms for SDN programmers to define additional protocol elements (e.g. new match fields, actions, port properties) to address new network technologies and behaviours (i.e. the protocol defines the expected behaviour of the switch but also how the behaviour can be customized using the interface).

3.2.4.3 ONF Microwave Information Model

The Microwave Information Model (IM) is an effort led by the ONF to define a standard of a common and generic information model for SDN-enabled wireless transport environments in order to simplify the operations and control of microwave/millimetre wave radiolink network elements (NEs) and facilitate the integration of distinct multi-vendor solutions under a common and single control framework. The Microwave IM is provided in ONF TR-532 [21] as a technology-specific extension to the TR-512 ONF CIM that can be implemented as a YANG data model so that the control-management of the microwave device by the SDN controller can be realized via the NETCONF protocol. The Microwave IM provides the necessary attributes for the device informing the SDN controller about its capabilities, the controller configuring the device and the device providing status, problem and performance information. For example, the Microwave IM allows for the configuration of frequency plans (channel arrangement), channel frequencies and transmission bandwidths, used modulation schemes, etc.

The current specification is limited to point-to-point radiolinks. However, this model could be a valid starting point to base a model for the satellite physical layer and be used as an internal SBI for the M&C of the SD lower layers (satellite resources such as a frequency plan, modulation and coding schemes or other satellite-specific properties).

3.2.4.4 ONF Transport API

The ONF T-APIs seek to provide programmable access to transport SDN controller functions by abstracting a common set of control plane functions such as network topology, connectivity requests and path computation to a set of service interfaces. T-APIs are intended to be applicable on the interface between a transport SDN controller 'Black Box' and its client application. The actors involved in the information exchange over this interface include transport network provider domain controllers in the role of producers and the transport network application systems in the role of the consumers. The transport network application systems could be either a business client system (which itself may include some control functions) or the network operator's upper level control, orchestration and/or operations systems. The T-APIs are also intended to be equally applicable between the controllers within a transport controller recursive hierarchy. The expected services delivered by the T-APIs are:

- Topology service: API to retrieve network topology, node, link and node edge points
- Connectivity service: API to request create, update and delete connectivity including point-to-point and multipoint
- Path computation service: API to request computation and optimization of paths
- Virtual network service: API to create, update and delete virtual network topologies
- Notification service: API to support publish/subscribe models for asynchronous notification of events such as failures or degradations

While ONF T-APIs development is still work in progress (thus far, ONF document TR-527 [20] only provides the functional requirements for the specification of

T-APIs), it could be a clear candidate for the realization of a NBI for the M&C of the E2E flow service delivered by a satellite network.

3.2.4.5 YANG models

YANG models have been produced to allow configuration or modelling of a variety of network devices, protocol instances and network services. A classification of YANG data models is given in [37,38], the latter reference more focused on service models. In particular, four types of service YANG models are distinguished:

- Customer service model: A customer service model is used to describe a service as offer or delivered to a customer by a network operator.
- Service delivery model: A service delivery model is used by a network operator to define and configure how a service is provided by the network.
- Network configuration model: A network configuration model is used by a network orchestrator to provide network-level configuration model to a controller.
- Device configuration model: A device configuration model is used by a controller to configure physical NEs.

As previously pointed out in Section 3.2.1, YANG models coupled with the REST-CONF/NETCONF protocol provides solutions for the ACTN framework, which indeed seeks to provide a control hierarchy and interfaces that would enable deployment of multi-domain transport SDN networks. Hence, according to [26], *customer service models* would be applicable to the ACTN CMI interface, *network configuration models* to the MPI and *device configuration models* to the SBI. In this context, and considering that the integration of the proposed SDN controller of the satellite network within an ACTN architecture would likely be realized through an MPI interface, existing YANG models applicable in the MPI interface that are not OTN/WSON technology specific are summarized in Table 3.1. Note that various YANG models

Table 3.1 YANG models for traffic engineering

Function	YANG model
Configuration scheduling	X. Liu, *et al.*, 'A YANG Data Model for Configuration Scheduling', draft-liu-netmod-yang-schedule, work in progress
Path computation	I. Busi, S. Belotti *et al.* 'Path Computation API', draft-busibel-ccamp-path-computation api 00.txt, work in progress
Path provisioning	T. Saad (Editor), 'A YANG Data Model for Traffic Engineering Tunnels and Interfaces', draft-ietf-teas-yangte, work in progress
Topology abstraction	X. Liu, *et al.*, 'YANG Data Model for TE Topologies', draft-ietf-teas-yang-te-topo, work in progress
Tunnel PM telemetry	Y. Lee, D. Dhody, S. Karunanithi, R. Vilalta, D. King, and D. Ceccarelli, 'YANG models for ACTN TE Performance Monitoring Telemetry and Network Autonomics', draft-lee-teas-actn-pm-telemetry-autonomics, work in progress
Service provisioning	No references available yet

are work in progress. Furthermore, there is also IETF Internet Draft [39] aimed to describe use cases that could be used for analysing the applicability of the existing models defined by the IETF for transport networks with a focus on MPI interface.

3.3 Integration approach for E2E SDN-based TE in satellite-terrestrial backhaul networks

A compelling scenario for the exploitation of SDN-based satellite networks is mobile backhauling [11], where satellite capacity can be used to complement the terrestrial backhauling infrastructure (e.g. fibre and radiolinks connecting the BSs sites) not only in hard to reach areas but also for more efficient traffic delivery to RAN nodes, increased resiliency and better support for fast, temporary cell deployments and moving cells. In this context, the exposition of control and management capabilities of the satellite connectivity services through an SDN-based interface would allow a mobile network operator (MNO) to easily integrate and operate the satellite component within a backhauling infrastructure that is progressively relying on SDN technologies for the terrestrial capacity counterpart. Managing both terrestrial capacity and satellite capacity under a centralized and consistently operated SDN framework enables the deployment of E2E SDN-based TE solutions.

TE mechanisms are used to optimize the performance of a data network by dynamically analysing, predicting and regulating the behaviour of the traffic across the network [40]. In integrated satellite-terrestrial backhaul network, TE solutions shall be able to use the satellite capacity in the way that best complements the terrestrial capacity in front of the changing conditions of both traffic demand (e.g. increase of traffic demand for an especial event, spatial demand fluctuations over time) and network situation (e.g. backhaul backup for terrestrial link failures, network rapid roll-out, fast response capacity and cells on wheels). Facing these multiple and diverse conditions in a consistent manner becomes challenging for TE. Compared to the traditional MPLS/TE mechanisms used in today's transport networks, the big advantage of a centralized SDN framework for the realization of TE solutions is that there is a holistic view of the network together with mechanisms to enforce network polices from a single touch point [41].

A network architecture framework for the realization of E2E SDN-based TE in satellite-terrestrial backhaul networks is presented in the following, together with a couple of illustrative TE workflows to validate the proposed integration approach.

3.3.1 Network architecture framework

Several proposals exist for adopting SDN concepts in mobile network architectures [42,43]. In general terms, an illustrative view of an SDN-based mobile that uses SDN-enabled transport from the RAN nodes (e.g. BSs) all the way through the backhaul to the core network is depicted in Figure 3.5. Though this architecture is contextualized for LTE technology, this vision is claimed to be generic and not constrained by the specifics of the LTE standard.

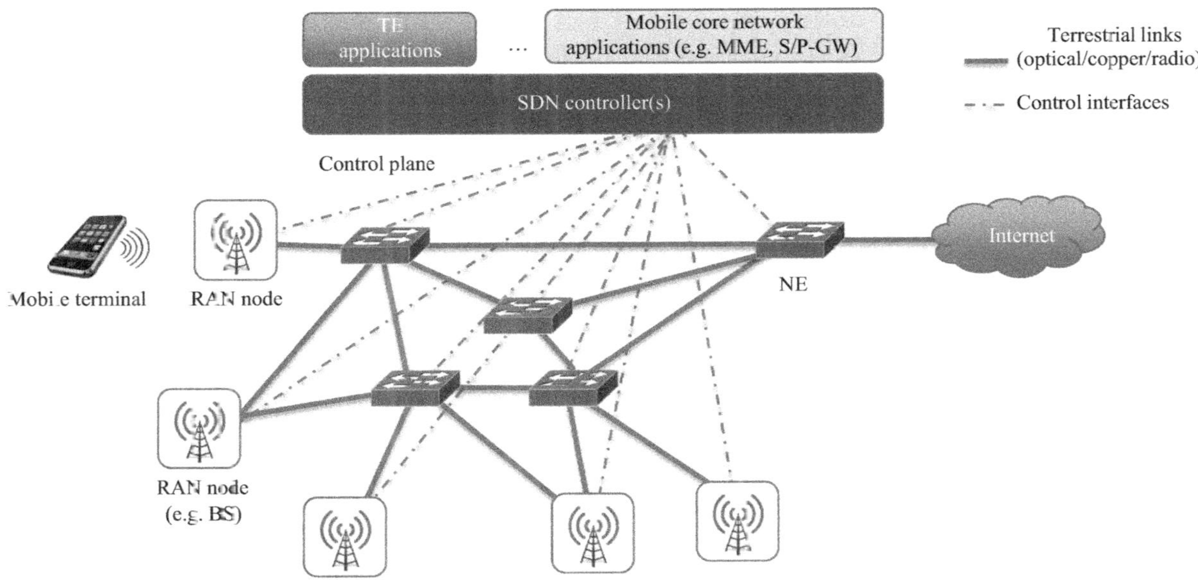

Figure 3.5 Illustrative view of an SDN-based mobile network

As depicted in Figure 3.5, mobile core network (MCN) control functions [e.g. mobility management entity (MME) and serving/packet data network gateways (S/P–GW) functional elements in LTE Evolved Packet Core] together with specific TE functions for the transport network are realized as applications running on top of an SDN controller (represented here as a single functional entity but likely to follow a hierarchical structure of controllers). This SDN controller is responsible for managing the NEs that provide the packet switching and forwarding capabilities within the transport network. In this respect, the underlying transport network infrastructure may involve a number of different physical network equipment, or forwarding devices such as routers, switches and virtual switches, to name a few. Building on the above view of SDN-based mobile networks and on the SDN-based satellite network architecture discussed in Section 3.2.2, Figure 3.6 depicts the functional view of the proposed integration approach, which is founded on two main concepts:

- *Abstraction of the overall satellite network as an SDN-capable 'switch'.* In particular, the OF switch abstraction model [36] is considered to model the operation of the satellite network as seen from the MNO SDN controller entity. This corresponds to one of the candidate solutions discussed in Section 3.2.4 for the realization of the NBI interface for the control and management of the satellite network connectivity services.
- *Use of SDN-based TE applications*, with a central Path Computation Engine (PCE) that supports the operation of the MCN applications for traffic management within the backhaul transport network. It is assumed that the overall transport network is managed as a single logical forwarding domain and that, inside the forwarding domain, a MNO's SDN controller makes the forwarding decisions. As depicted in Figure 3.6, all SDN-capable L2/L3 NEs are connected to the MNO's SDN network controller through OF interfaces, including the 'satellite network switch'. In this way, SDN-based TE mechanisms can seamlessly span the whole network. For the terrestrial connection, no specific technology is assumed rather than considering that traffic flows can also be managed through SDN features.

In order to raise different considerations with regard to the operation of TE procedures, the illustrative network topology depicted in Figure 3.6 considers three RAN nodes with LTE eNB functions, one connected to the transport network only by terrestrial means (RAN node#C), another connected only through the satellite network (RAN node#A) and a third one (RAN node#B) connected to both a terrestrial connection and a satellite connection through an SDN-capable cell switch router (CSR). This third case is used to illustrate the realization of TE mechanisms for multipath optimization. With respect to the terrestrial part of the transport network, three NEs are included in the reference network topology, two of them acting as internal aggregation/core nodes within the transport network (i.e. NE#A and NE#B) and the third one (i.e. NE#C) providing the interconnection with the external networks (e.g. Internet) through a conventional 3GPP Gi interface. Of note is that, in addition to OF interfaces for controlling the forwarding function of the transport network, other control interfaces are likely to be in place in the overall setting for other purposes, such as the

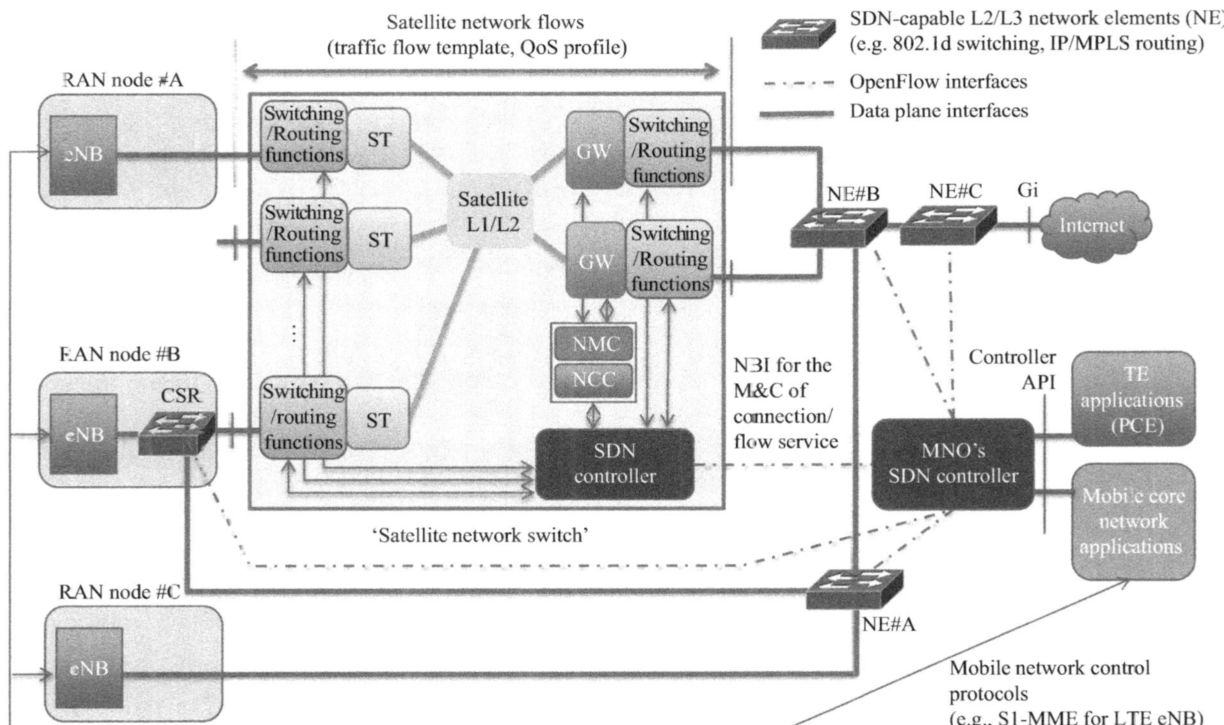

Figure 3.6 Functional view and illustrative network topology used in the TE workflows

3GPP S1-MME interface between the MCN applications and the eNBs within the RAN nodes to manage the activation/deactivation of radio access bearers (RABs) in the eNB for the served mobile terminals.

3.3.2 Illustrative TE workflows

Two illustrative workflows are presented to validate the proposed integration approach. The first one shows the activation of a traffic flow through the satellite-terrestrial network to enforce a mobile network bearer [e.g. so-called evolved packet system (EPS) bearer in the context of LTE] that can benefit from optimal path computation. The second workflow shows the modification of an already established flow as a reaction to a congestion/failure situation in one link within the transport network.

3.3.2.1 Flow activation with optimal path computation

Based on the network topology depicted in Figure 3.6, a message chart with the operation of a path computation mechanism for multipath satellite-terrestrial traffic optimization is provided in Figure 3.7. In particular, the provided workflow covers the case of the establishment of a dedicated EPS bearer that relies on the TE path computation mechanism to activate the traffic path between the RAN node and the external network reachable through NE#C considering the characteristics of the EPS bearer and the load conditions across the whole network. It is assumed that the SDN controller has a global view of network topology which can be represented by means of a graph including all links between OF switches [the links could be found by e.g. leveraging protocols such as LLDP (802.1AB), which is used by network devices to

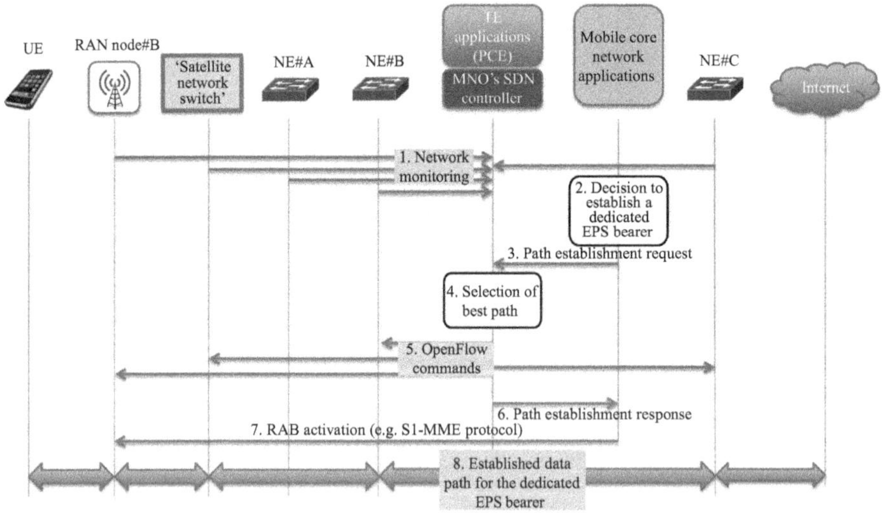

Figure 3.7 Flow activation with optimal path computation

advertise their identity, capabilities and neighbours]. Details of the different steps depicted in Figure 3.7 are given in the following:

- ***Step 1*:** Monitoring of the SDN forwarding elements within the domain, including the CSR, 'satellite network switch' and NEs. Solutions such as the one described in [44] allow for an OF controller to have accurate monitoring of per-flow throughput, packet loss and delay metrics in order to aid TE. In this respect, while a flow is active, the controller and the SDN forwarding element can exchange messages concerning the state of the flow.
- ***Step 2*:** As a result of the activation of a new service (e.g. HD video-streaming service) by a mobile terminal connected in RAN node#B, the MCN decides to establish a new dedicated EPS bearer to support that service. The activation of the dedicated EPS bearer requires the activation of a flow with QoS guarantees across the transport network. The two edge nodes of the EPS bearer are the RAN node#B, where the UE is assumed connected, and the NE#C, which serves as the gateway to the external network.
- ***Step 3*:** The MCN requests to the TE application the computation of the best path between RAN node#B and NE#C. QoS attributes of the EPS bearer are indicated (e.g. Guaranteed Bit Rate).
- ***Step 4*:** Based on the (1) network topology knowledge, (2) the network monitoring information and (3) QoS attributes of flow, the TE application can compute the most appropriate path. Different algorithms could be supported here, including graph searching algorithms for path finding and algorithms for path selection depending on policies with respective of TE or service quality, such as calculating the shortest path forwarding based on a consistent view of network state or provision application-aware routing [45]. Anyway, let us consider that the outcome of this decision is that a path through the satellite network is chosen for this flow.
- ***Step 5*:** Flow entries are installed into the OF switches by the MNO's SDN controller so that traffic associated with the EPS bearer is forwarded through the selected path.
- ***Step 6*:** The MCN gets the path establishment response.
- ***Step 7*:** The EPS bearer activation at the radio layer takes places (i.e. RAB activation), involving the interaction between the MCN functions and the eNB within RAN node#B.
- ***Step 8*:** The data plane for the dedicated EPS bearer gets live and traffic follows the selected path through the satellite network.

The above workflow assumes that the path is established to support a single EPS bearer. However, the same approach would be used in case of deciding the best path for traffic aggregates with common QoS requirements. This is well supported in OF by just establishing the corresponding matching conditions (e.g. IP prefixes to identify a traffic aggregate in front of particular IP addresses of the individual flows).

3.3.2.2 Flow update to overcome congestion/failures

Figure 3.8 shows how the failure of a path, or simply the congestion of a path that could cause QoS degradation, could be handled within the proposed integration approach. In particular, the message chart depicted in Figure 3.8 is a TE mechanism that will update an already established flow in order to overcome a congestion/failure event. Details of the different steps depicted in Figure 3.8 are given in the following:

- ***Step 1*:** The starting point considers that traffic from/to RAN node#B and from/to RAN node#C, called Traffic B and Traffic C, respectively, are both flowing through NE#A, NE#B and NE#C. This could be assumed to be the optimal traffic path for a moderate traffic load scenario.
- ***Step 2*:** Monitoring of the SDN forwarding elements is conducted by the MNO's SDN controller, as described in the previous workflow.
- ***Step 3*:** An event that puts at risk the QoS of the established flows occurs. This could be, for example, a considerable traffic increase in RAN node#C at certain time of the day that overloads the link among NE#A and NE#B, which is shared by Traffic B and Traffic C.
- ***Step 4*:** The TE application detects the congestion situation. For example, the TE application could have set a high utilization threshold of 60% and low utilization threshold of 20% for the traffic load on the shared link. If this high threshold is exceeded, high utilization is observed and e.g. a part of Traffic B could be switched to pass through the satellite network.
- ***Step 5*:** Flow entries are installed to OF switches along the path by the MNO's SDN controller to reroute part of the traffic B through the satellite connection.
- ***Step 6*:** While the path for traffic C remains unchanged, now part of traffic B is served through the satellite network, reducing congestion in the link between NE#A and NE#B.

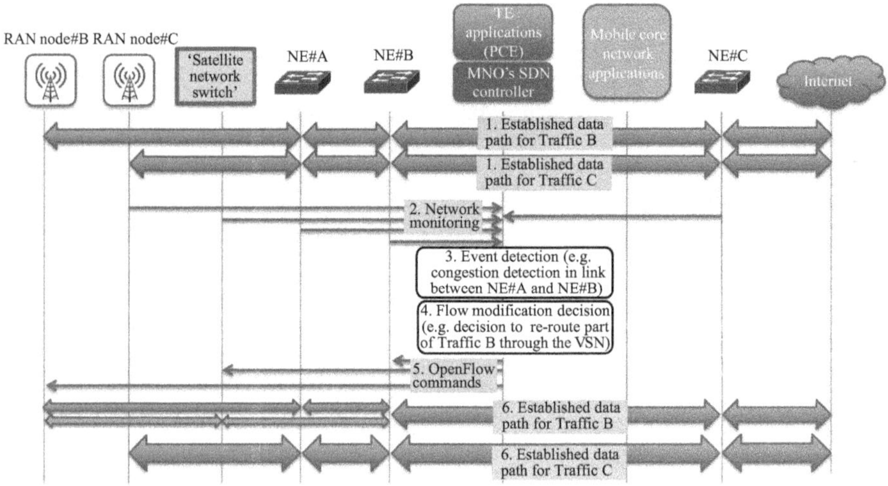

Figure 3.8 Flow update to overcome congestion/failures

Flow updates can also be driven by connection protection in case of failure. Indeed, path protection and network recovery from failure are critical aspects of TE. While these aspects are well understood in conventional MPLS/IP networks, work is still needed to mature these concepts in the context of SDN networks [46].

3.4 Illustrative SDN-based TE application

SDN-based TE applications have been already proposed mostly for data centre or enterprise network scenarios (see [47] for a detailed survey and [48] for a discussion on techniques for flow management, fault tolerance, topology update and traffic characterization). In the context of mobile backhauling scenarios, the proposed SDN-based satellite-terrestrial network integration approach can be used to develop TE applications that exploit the following sort of features and criteria:

- E2E path computation with selection of the terrestrial or satellite link for backhauling, considering multiple optional link utilization rates and flow characteristics more comprehensively.
- Resource reservations mechanisms to protect or give preferential treatment to applications/users/locations such as BSs with no or limited terrestrial link backhaul capacity that are fully reliant on the use of a pool of shared satellite capacity.
- Different allocation criteria depending on the traffic nature (e.g. guaranteed bit rate or not, unicast or multicast).
- Admission control and rate control to face overload and guarantee resources and minimum (committed) transmission rates per flow and group of flows.
- Use of network utility maximization criteria, where the adequacy of handling specific flows over the terrestrial or satellite component, as well the effect of allocating more or less data rate, can be accounted.
- Exploit bandwidth on demand (BoD) features [49].
- Control the activation/deactivation of networking functions for traffic optimization (e.g. trigger the deployment of virtualized network functions for compression, TCP optimization, etc. in NFV-enabled satellite networks [50]).

In this context, the TE solution formulated and assessed in this section combines the following control features as part of its decision-making logic: path selection, admission control, rate control and reservation management, as illustrated in Figure 3.9.

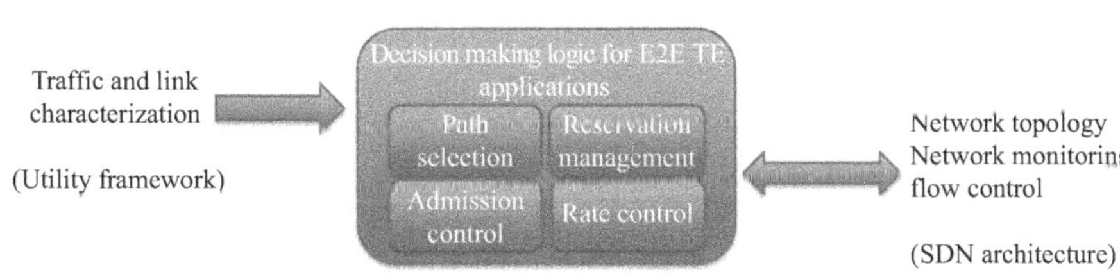

Figure 3.9 Components of the E2E traffic engineering application

On this basis, next subsections detail the specific traffic and link characterization approach established for the specification of the TE decision-making logic, a description of the optimization problems and algorithms used behind the TE decision-making components and, finally, a numerical assessment of the proposed TE application.

3.4.1 Traffic and link characterization for TE

The conception of the TE logic requires that a specific traffic and link characterization is first established [51]. This is necessary to determine, if applicable, the QoS requirements such as maximum tolerable latency and jitter, minimum required bandwidth, etc. per type of service/user that need to be fulfilled in order to achieve a given QoE/satisfaction level. To that end, we resort to the use of utility functions to describe the QoE/satisfaction level that is achieved when a particular flow is served across the hybrid satellite-terrestrial backhaul. In our case, utility functions are formulated to account for two main aspects: (1) the bit rate that the flow can be allocated across the E2E path and (2) whether the E2E path traverses a satellite link or not (i.e. the higher delay incurred when using a satellite link can result in some level of service degradation that is reflected with a lower utility).

Moreover, the formulation of the utility functions is also dependent on the nature of the services. In our analysis, we consider mixes of stream and elastic traffic as well as unicast and multicast traffic. Inelastic/Stream traffic is generated by time-sensitive applications, like e.g. Voice over IP and Videostreaming on Demand (VSOD), and typically has strict bandwidth and/or delay requirements. Elastic traffic on the other hand is generated by applications such as web browsing and file-transfers where the delivered bit rate and/or the download time are more important than inter-packet or E2E delays. Indeed, this traffic classification is captured in the QoS model established for LTE systems by considering two types of bearer services that can be enforced in the network: guaranteed bit rate bearers (GBR) and non-GBR bearers. Thus, the unicast traffic flows served through GBR bearers (called UG flows in the following) are given a minimum guaranteed bit rate to operate satisfactorily; otherwise, the quality might be severely affected. On the other hand, the unicast traffic flows served through non-GBR bearers (called UN flows in the following) do not get such a minimum bit rate reservation but can see a wide variability of the achieved bit rate, being more exposed to congestion-related packet losses and/or delay variability (without necessarily a noticeable impact on QoS). Our traffic model also includes multicast GBR services (called MG services in the following). The consideration of MG services in the analysis allows us to exploit the intrinsic broadcast/multicast transmission capacity of the satellite component, assessing its impact on the network in terms of QoS. Unlike unicast services, one particular MG session can consist of multiple MG flows simultaneously forwarded to multiple BSs simultaneously [52].

Based on the above considerations, the utility functions considered in our analysis for the characterization of UG, MG and UN services are provided in Table 3.2. All the utility functions account for the delivered bit rate (r) and consider whether the flow is served through satellite ($x = 0$) or terrestrial backhaul ($x = 1$). In particular, a two-level step function is used for UG flows [53], reflecting two possible bit rates/quality

Table 3.2 Utility functions

Utility function	Equation	Graph
	UG services	
$U^{UG}(r,x) = U_o^{UG}(x) \cdot U_r^{UG}(r)$ where	(3.1)	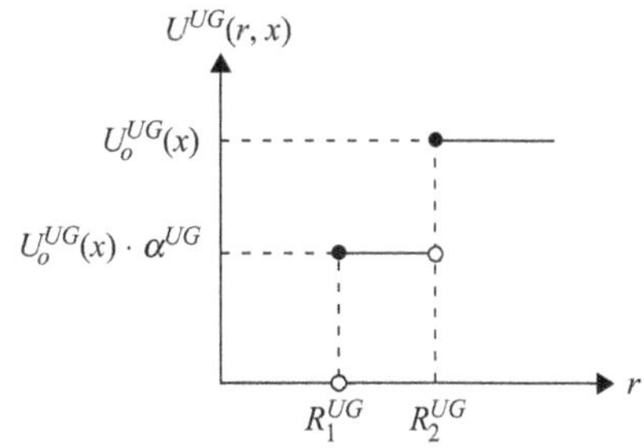
$U_o^{UG}(x) = p^{UG} + x(1 - p^{UG})$ and	(3.2)	
$U_r^{UG}(r) = \begin{cases} 0 & 0 < r < R_1^{UG} \\ \alpha^{UG} & R_1^{UG} < r < R_2^{UG} \\ 1 & r \geq R_2^{UG} \end{cases}$	(3.3)	
	MG services	
$U^{MG}(r,x) = U_o^{MG}(x) \cdot U_r^{MG}(r)$ where	(3.4)	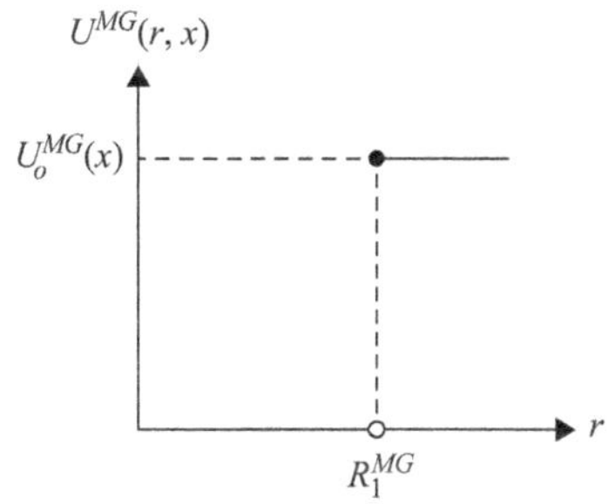
$U_o^{MG}(x) = p^{MG} + x(1 - p^{MG})$ and	(3.5)	
$U_r^{MG}(r) = \begin{cases} 0 & 0 < r < R_1^{MG} \\ 1 & r \geq R_1^{MG} \end{cases}$	(3.6)	
	UN services	
$U_{UN}(r,x) = U_o^{UN}(x) \cdot U_r^{UN}(r)$ where	(3.7)	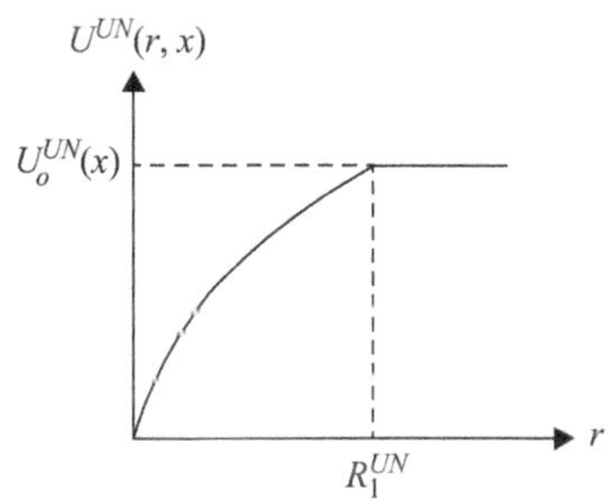
$U_o^{UN}(x) = p^{UN} + x(1 - p^{UN})$ and	(3.8)	
$U_r^{UN}(r) = \begin{cases} \dfrac{\log(r+1)}{\log(R_1^{UN}+1)} & \text{if } 0 < r \leq R_1^{UN} \\ 1 & \text{if } r > R_1^{UN} \end{cases}$	(3.9)	

levels that could be on offer (e.g. standard and high definition VSODs). This UG utility function, defined by (3.1)–(3.3) in Table 3.2, is parameterized by the bit rates R_1^{UG} and R_2^{UG} to be delivered for the standard/high-quality offerings, respectively; a utility reduction factor p^{UG} to account for the potential quality/satisfaction degradation when using satellite links instead of terrestrial one; and a utility reduction factor α^{UG} to account for the impact of rate selection between R_1^{UG} and R_2^{UG}.

For the characterization of MG service flows, a one-level step utility function is used, as defined by (3.4)–(3.6) in Table 3.2. In this case, R_1^{MG} is the minimum bit rate to be delivered for the high quality and the parameter p^{MG} is a utility reduction factor to account for the potential quality/satisfaction degradation when using satellite links instead of terrestrial.

With regard to UN service flows, the utility functions can be more diverse [54], depending on which specific aspects/service characteristics one wants to stress. In our case, we have adopted a logarithmic utility function [55], which is one of the most commonly used and already serves our needs. The normalized utility function of UN service flows is defined by (3.7)–(3.9) in Table 3.2 where R_1^{UN} is used to establish the bit rate for which it is considered that the service is already provided with a good quality (so no utility gain is envisioned by serving UN service flows with higher bit rates) and the parameter p^{UN} is the utility reduction factor for UN service flows.

3.4.2 TE decision-making logic

The TE decision-making logic consists of a combination of processes, some executed when there is some trigger (e.g. new flow request) and others executed periodically (e.g. performance metric computation and flow adjustments). Figure 3.10 shows the TE decision-making logic to handle a UG flow request. Whenever there is a new UG flow request, it is verified whether the new UG flow must be served through a BS that has operational terrestrial and satellite links or a BS that only has available satellite capacity (e.g. terrestrial failure, TBS with no terrestrial backhaul). In the former case, the TE algorithm continues by checking whether there is sufficient capacity across any of the paths to serve the new flow without compromising the quality of the already

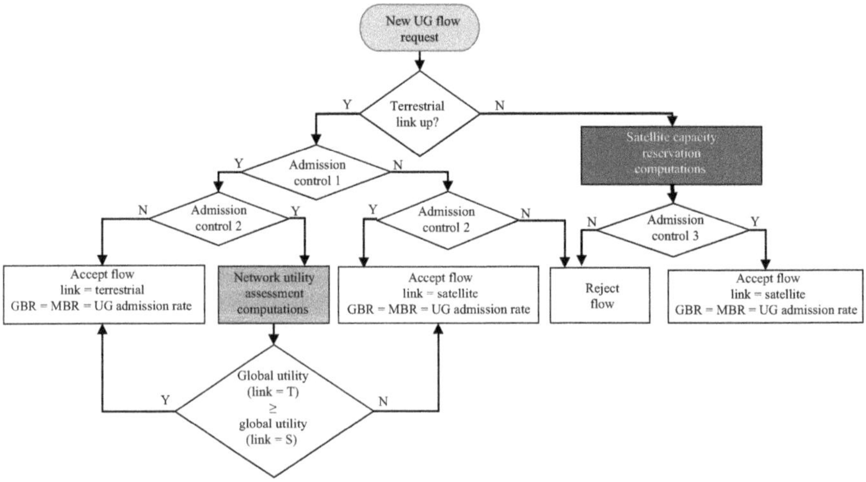

Figure 3.10 TE decision-making logic to handle new UG flow arrivals

established GBR flows (UG and MG active flows). This is achieved by establishing a *GBR admission load threshold* to limit the maximum capacity occupation of a link allowed for use of GBR traffic. This parameter is used by Admission controls 1 and 2 logic (detailed in Table 3.3). If there is sufficient backhaul capacity across both satellite and terrestrial links, the flowchart continues by computing the achievable global network utility (i.e. aggregate of the utility of established flows plus the utility of the new flow) for each of the two candidate paths, selecting the one leading to the higher utility increase. Note that the utility computation is not conducted when there is only a candidate option or when none of them is available, leading the latter case to the rejection of the UG flow request. For the admitted UG flow requests, the GBR and maximum bit rate (MBR) are both set to the rate that gives the maximum utility for UG services.

As previously noted, the flowchart in Figure 3.10 also captures the case where the UG flow is to be served through a BS where the terrestrial link is not available. In this regard, as seen on the right side of the flowchart, a resource reservation management mechanism is introduced in the decision-making process. This mechanism is used to enforce a preferential treatment for the use of the shared satellite capacity to the BSs without an operational terrestrial link. Therefore, at a new UG flow request arrival, now the TE logic first goes through Admission control 3 (detailed in Table 3.3) that takes into account the amount of reserved satellite capacity that is dynamically adjusted over time for the serving BS. The computations needed to manage such satellite capacity reservations are detailed later on in this section.

The TE decision-making logic to handle UN flow requests is depicted in Figure 3.11. Similar to the treatment of UG flows, the TE algorithm first checks if the new UN flow is to be served through a BS with terrestrial and satellite links both

Table 3.3 Admission control computations

Admission control	Description
Admission control 1	(GBR terrestrial load at BS + UG admission rate) < (GBR admission load threshold · terrestrial link capacity at BS)
Admission control 2	(GBR satellite load at BS + UG admission rate) < (GBR admission load threshold · satellite link capacity at BS) *AND* (Global GBR satellite load + UG admission rate) < [GBR admission load threshold · (satellite system capacity − satellite reserved capacity)]
Admission control 3	(GBR satellite load at BS + UG admission rate) < (GBR admission load threshold · satellite reserved capacity at BS)

UG admission rate: Rate that is considered in the admission process. It is selected from the rates specified for the definition of the utility functions in Table 3.2.
Satellite system capacity (C_S): Total amount of satellite capacity shared by a group of BSs
Satellite reserved capacity (*Cr*): Satellite capacity reserved for preferential use of a given BS
GBR admission load threshold: Maximum percentage of the available (satellite, terrestrial and reserved) capacity that can be used to serve GBR traffic

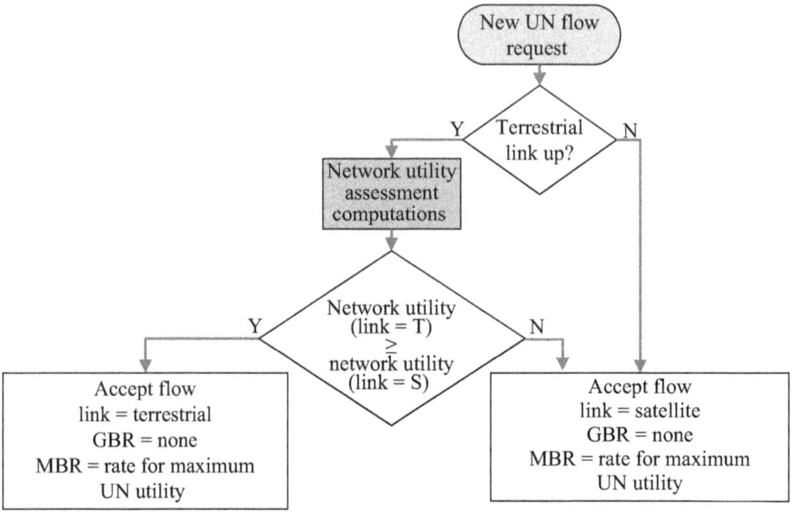

Figure 3.11 TE decision-making logic to handle new UN flow arrivals

operational or only with satellite capacity available. In the former case, the next step is to compute the overall utility increase that would be achieved if the flow is enforced through the terrestrial or satellite links, selecting the option that turns into the higher network utility increase. In the latter case, shown on the right side of Figure 3.11, the flow is always enforced through the satellite connection and the reservation amount is updated accordingly. Note that, unlike UG flows processing, no admission control is enforced for UN flows because of its elastic traffic nature (i.e. the rates achieved per flow are variable and depend on overall number of flows simultaneously served in the network). Therefore, no GBR rate is established for the admitted flows and the MBR parameter, used for rate control purposes, is set to the rate that achieves the maximum utility for UN services.

Even though network utility maximization is sought after each flow arrival, traffic variations (e.g. termination of established flows) and changes in capacity conditions (e.g. changes in reservations, terrestrial link failures) might turn into situations that the achieved network utility is not optimal. To face this situation, a mechanism to reassess the network utility of the established flows and, if necessary, carry out any reallocations is considered. This process is illustrated in Figure 3.12. As seen in the figure, network utility reassessment and reallocation are triggered periodically as well as due to the occurrence of specific events such as a change in the amount of capacity in a network link. Figure 3.12 also shows that after the execution of the network utility reassessment and reallocation process, capacity reservations are also revisited to account for any changes enforced to the ongoing flows.

The reservation management mechanism aims to ensure that some amount of satellite capacity remains available for the BSs that do not count with terrestrial

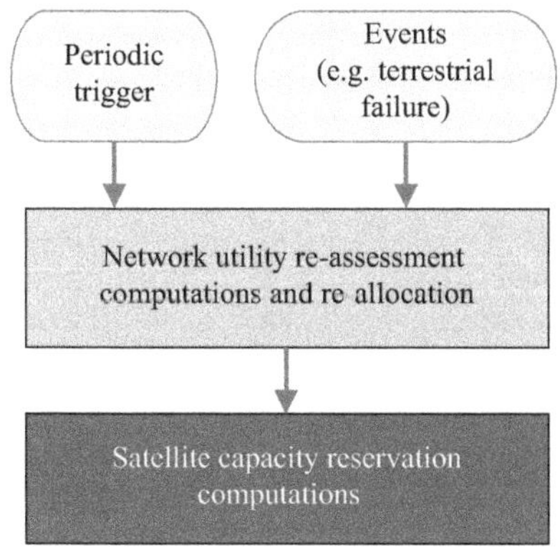

Figure 3.12 Logic for continuous network utility reassessment, reallocation and reservation update

Table 3.4 Satellite reservation computations

Reservation control parameters	Computation
Satellite reserved capacity at BS_m (Cr_m)	Cr_m = UG satellite load at BS_m + UN satellite flows at $BS_m \cdot$ UN global average flow rate
Satellite non-reserved Capacity (Cnr)	Cnr = Satellite system capacity (C_S) $- \sum_{BS_m} Cr_m$
Constraints: Total satellite reserved capacity $\leq$ Maximum capacity reservation Satellite reserved capacity at $BS_m \leq$ Maximum capacity reservation per BS Satellite reserved capacity at $BS_m \leq$ Terrestrial link capacity at BS_m	

capacity. Indeed, considering that one of the conditions that lead to global utility maximization is a fair distribution of the rates delivered to UN flows, this reservation mechanism helps in achieving fairness in terms of the overall capacity distribution among BSs (i.e. BS without terrestrial capacity will get a higher share of the satellite capacity). To that end, the satellite reserved capacity (Cr) variable is introduced. This parameter is initialized with a default reservation value and periodically updated over time based on the evolution of the traffic load served through the corresponding BS (details are given in Table 3.4). In particular, Cr is computed to account for the UG traffic load supported at the BS plus an additional capacity for UN traffic that would allow to deliver an average bit rate as that achieved across the whole network for UN flows. The value of Cr is constrained by the terrestrial link capacity at the BS, the *maximum capacity reservation per BS* and the *maximum capacity reservation* applicable to the total satellite reserved capacity. The remaining satellite system resource available for BSs with terrestrial capacity is defined as satellite non-reserved capacity (Cnr).

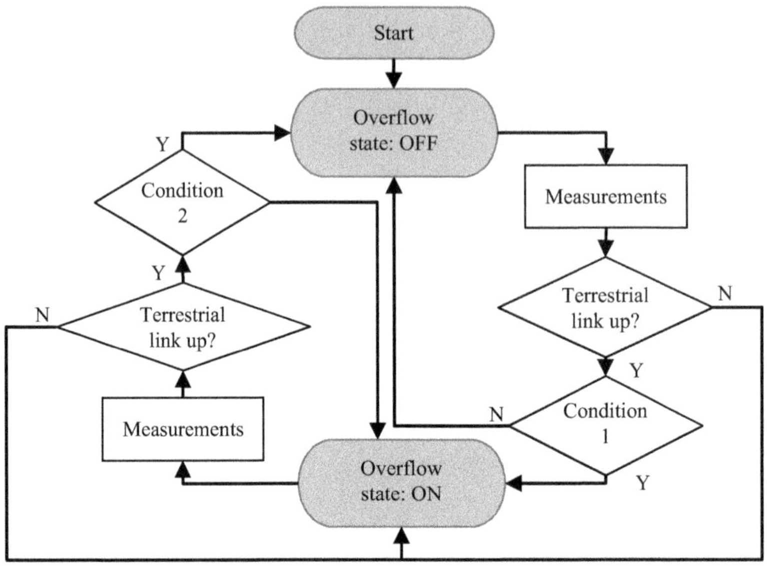

Figure 3.13 State diagram of the overflow strategy

For comparison purposes, the assessment presented in the following section considers also a more conventional overflow strategy that is executed locally at each BS and lacks of any centralized control. An state diagram that describes the operation of the overflow strategy is depicted in Figure 3.13. It is considered that each BS with both terrestrial and satellite capacity can switch between two operational overflow states: OFF and ON. In OFF state, all generated backhaul traffic is handled through the terrestrial link. Otherwise, backhaul traffic generated when the BS is in ON state is always directed through the satellite link. As captured in Figure 3.13, when terrestrial capacity is not available, the operation mode remains in the ON overflow state.

The transition between the OFF and ON state is established based on a twofold condition (Condition 1 in Figure 3.13): the amount of GBR load (UG and MG flows) has started exceeding a given threshold (*overflow GBR load activation threshold*) or the average rate being delivered to UN flows has fallen below a given threshold (*overflow UN rate activation threshold*). The change is executed if this condition holds for an overflow decision interval (ΔT). Similarly, the transition from the ON to OFF states (Condition 2 in Figure 3.13) is determined by the counterpart twofold condition: the GBR load has decreased below a given threshold (*overflow GBR load deactivation threshold*) and the average rate being delivered to UN flows is above a given threshold (*overflow UN rate deactivation threshold*). Both conditions are detailed in Table 3.5.

Figure 3.14 depicts the flowcharts to handle a UG/MG and UN flow requests under the overflow strategy. Admission control applied in the case of GBR traffic

Table 3.5 Overflow states switching conditions and parameters

State transition condition	Computation
From OFF to ON state (Condition 1)	GBR load level > overflow GBR load activation threshold OR Average UN flow rate < overflow UN rate activation threshold within $[t, t - \Delta T]$
From ON to OFF state (Condition 2)	GBR load level < overflow GBR load deactivation threshold AND Average UN flow rate > overflow UN rate deactivation threshold within $[t, t - \Delta T]$

Parameters:
Overflow GBR load activation threshold
Overflow GBR load deactivation threshold
Overflow UN rate activation threshold
Overflow UN rate deactivation threshold
ΔT = *Overflow decision interval* (s)

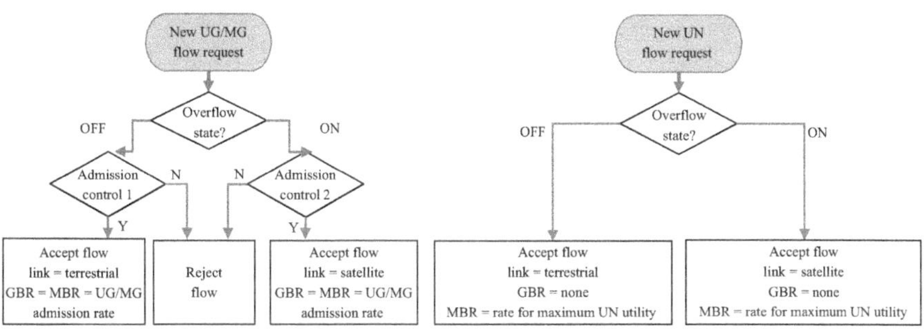

Figure 3.14 Logic to handle flow requests under the overflow strategy

follows the same principles used for the SDN-based TE application. The corresponding admission control computations for the overflow strategy are detailed in Table 3.6.

3.4.3 Performance assessment

In this section, the behaviour of the proposed SDN-based TE application is assessed by means of numerical simulations under diverse scenarios, including homogeneous and non-homogeneous load situations, terrestrial link failures in some of the BSs and deployment of a number of TBSs that exclusively rely on the satellite capacity for backhauling.

The simulation scenario considers a set of BSs with terrestrial and/or satellite backhaul capacity that serve a mix of UG, MG and UN flows. As illustrated in

Table 3.6 Admission control computations for the overflow strategy

Admission control	Description
Admission control 1	(GBR terrestrial load at BS + UG admission rate) < (GBR admission load threshold · terrestrial link capacity at BS)
Admission control 2	(GBR satellite load at BS + UG admission rate) < (GBR admission load threshold · satellite link capacity at BS) *AND* (Global GBR satellite load + UG admission rate) < (GBR admission load threshold · satellite system capacity)

UG admission rate: Rate that is considered in the admission process. It is selected from the rates specified for the definition of the utility functions in Table 3.2.
Satellite system capacity (C_S): Total amount of satellite capacity shared by a group of BSs.
Satellite reserved capacity (Cr): Satellite capacity reserved for preferential use of a given BS.
GBR admission load threshold: Maximum percentage of the available (satellite, terrestrial, reserved) capacity that can be used to serve GBR traffic.

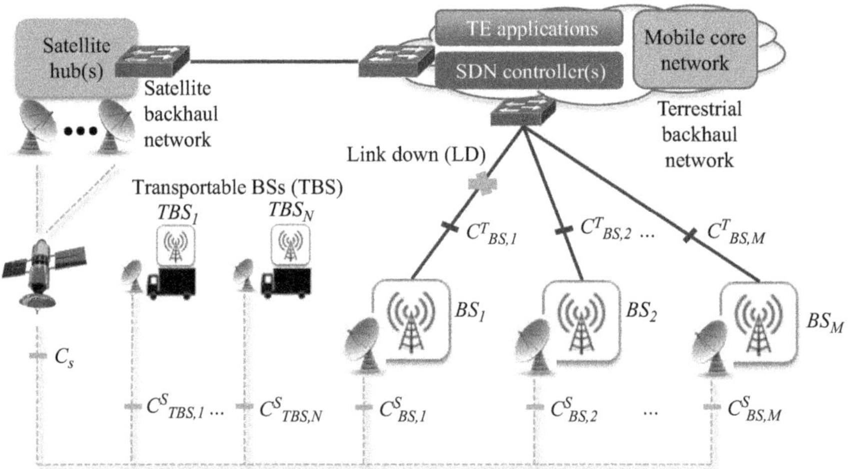

Figure 3.15 Deployment scenario

Figure 3.15, there is a number of M BSs deployed at fixed locations with both satellite and terrestrial backhaul links and a number of N BSs, referred to as TBS, used for temporary deployments/fast network roll-out that only rely on the use of the satellite backhaul links. Table 3.7 provides the range of values considered for the general network deployment settings and the configuration of the overflow and SDN-based TE application in the numerical assessment. Without any loss of generality and for the sake of consistency, the values considered for service characterization as well as backhaul capacities are inspired in current state-of-the-art 4G and satellite broadband technologies. Specifically, the considered setting for the capacity of the terrestrial

Table 3.7 Simulation settings

Parameter	Values
Number of BSs sharing the satellite capacity (N+M)	16
Backhaul capacity	
Terrestrial link capacity per BS_j (C_j^T)	131 Mbps
Maximum satellite capacity per BS_j (C_j^S)	210 Mbps
Satellite system capacity (C_S) (% of terrestrial capacity)	10%–20% (209.6–419.2 Mbps)
Service flows characterization	
Standard quality UG bit rate R_1^{UG}	3 Mbps[a]
High-quality UG bit rate R_2^{UG}	6 Mbps[a]
Utility reduction factor due to rate selection α^{UG}	0.8
MG bit rate R_1^{MG}	6 Mbps[a]
UN bit rate for maximum utility R_1^{UN}	13 Mbps[b]
Satellite utility reduction factors (p^{UG}, p^{MG}, p^{UN})	1.0–0.6
UG rate selection	Only high quality
Overflow strategy parameters	
GBR admission load threshold	90%
Overflow GBR load activation threshold	80%
Overflow GBR load deactivation threshold	70%
Overflow UN rate activation threshold	40% of R_1^{UN}
Overflow UN rate deactivation threshold	60% of R_1^{UN}
Overflow decision interval (ΔT)	5 s
SDN-based TE application parameters	
GBR admission load threshold	90%
Maximum capacity reservation per BS (% of C_j^T)	100%
Initial capacity reservation (% of C_S)	20%
Maximum capacity reservation (% of C_S)	95%
Reassessment update interval	1 s

[a]Typical mobile Video Resolution and Bit Rates [57].
[b]The global average for LTE download speeds (*Source:* 'The State of LTE', OpenSignal, February 2016).

links (131 Mbps) is based on the dimensioning analysis presented in [56] to cope with the 90th percentile of the traffic demand when considering a realistic traffic model that exhibits a log-normal distribution with an average load of 100 Mbps per BS. This value is then considered to establish the range of values of C_S. On the other hand, the maximum satellite link capacity per BS is also set to 210 Mbps in line with the terrestrial capacity and considering that today's top-of-the-line satellite modems based on DVB-S2X can afford this capacity. All traffic flows are modelled by a Poisson arrivals and exponential session duration distribution. Numerical results have been obtained by running 50 times an event-driven simulation, each representing an execution interval of 1,000 s. Table 3.7 values are used as default values unless stated otherwise.

3.4.3.1 Homogeneous spatial traffic distribution

This first assessment is intended to show the performance of the SDN-based TE application under homogeneous spatial traffic distributions and considering that all

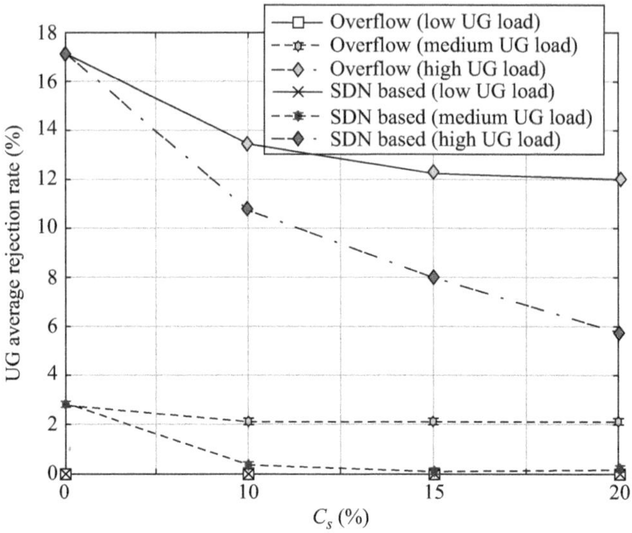

Figure 3.16 Admission rejection rate for UG services

BSs have both terrestrial and satellite backhaul capacity. Traffic load for UG services is set to 30% (low), 60% (medium) and 90% (high) of the terrestrial link capacity in each BS. Considering that UG flows are served with the high-quality UG bit rate R_2^{UG} and average session duration is 30 s, the corresponding flow arrival rates λ_{UG} for the low, medium and high UG load conditions are, respectively, 0.2183, 0.4366 and 0.655 flow/s.

With respect to UN traffic load, the UN service flow arrival rate λ_{UN} is varied between 0.25 and 1.0 at each BS. This results in an average number of active UN flows per BS between 7.5 and 30 per BS considering an average session duration of 20 s. Note that if UN flows could all be served at R_1^{UN}, this would represent an average UG load per BS between 65 and 260 Mb/s. No multicast traffic is considered in this first result.

Figure 3.16 shows the admission rejection rate experienced by the UG traffic under the SDN-based and overflow strategies for different amounts of C_S and when considering a satellite utility reduction factor given by $p^{UG} = p^{UN} = 1$. It can be seen how the availability of the satellite capacity leads to a considerable reduction of the rejection rate for UG traffic and how the SDN-based solution clearly outperforms the overflow strategy. For medium UG load, the SDN-based TE application keeps the blocking ratio well below 0.5% with only $C_S = 10\%$ while the overflow strategy is not able to reduce it from 2.0%.

Focusing now on UN service performance indicators, Figure 3.17 shows the mean and standard deviation of the data rate delivered per UN flow for different UN loads and considering values of $C_S = 10\%$ and $C_S = 20\%$. Results are obtained for a UG medium traffic load, $p^{UG} = p^{UN} = 1$ and the case with $C_S = 0\%$ is added

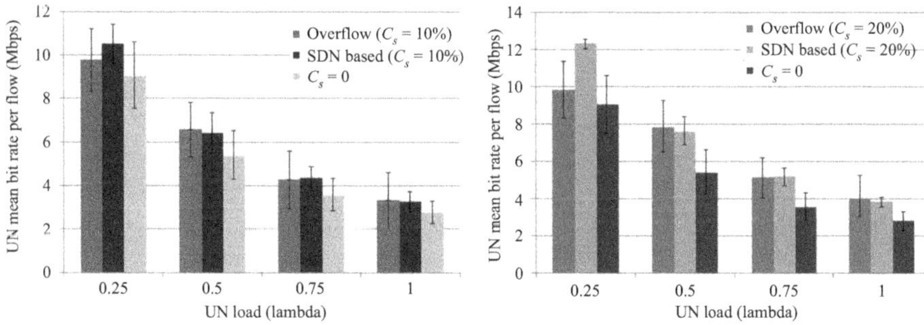

Figure 3.17 UN mean bit rate per flow for $C_S = 10\%$ (left) and $C_S = 20\%$ (right)

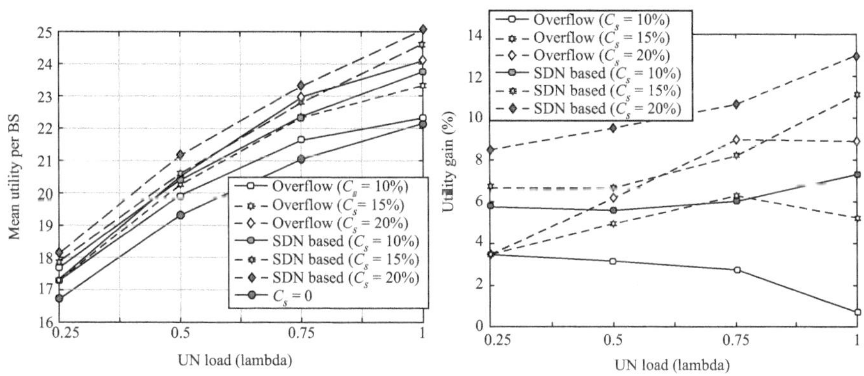

Figure 3.18 Utility (left) and utility gain (right) per BS

for comparative purposes. As seen in the figure, the achieved mean bit rates do not change significantly when comparing the SDN-based and the overflow strategy, though the SDN-based approach clearly outperforms in the less loaded situations. This is due to the fact that under high traffic loads, almost all backhaul capacity (satellite and terrestrial) is being used since the UN traffic flows end up using all the available capacity and, on average, the capacity share per flow is practically the same. However, the most notorious difference comes when observing the standard deviation values, which are considerably reduced by the SDN-based strategy. This is due to the fact that this strategy distributes the traffic based on the global occupation of both satellite and terrestrial links, seeking fairness among all the established UN flows that, in the end, turns in higher network utility.

The network performance in terms of network utility is presented in Figure 3.18. On the left side, average utilities per BS are given in absolute terms for the SDN and overflow strategies under different values of C_S. Results are obtained for a UG medium traffic load, $p^{UG} = p^{UN} = 1$ and the case with $C_S = 0\%$ is added for comparative purposes. It can be seen that the SDN strategy provides the highest utility in all the situations. Deeping in details, the utility gain, computed as the per cent increase of the global utility achieved by the SDN and overflow strategies with respect to that achieved for the case with $C_S = 0$, is represented on the right side of the figure. Here, it could be observed that for instance, the SDN strategy can deliver the same or even higher utility gain when operating under $C_S = 10\%$ (or 15%) than the overflow strategy for $C_S = 15\%$ (or 20%). Additional results not depicted in the figure show that SDN strategy is still able to bring some utility gain when considering utility reduction factors far below 1.0 (e.g. gain of 4% for $C_S = 20\%$ when $p^{UG} = p^{UN} = 0.6$).

The higher utility achieved by the SDN strategy is partly due to the reallocation mechanism considered as part of the TE application (see Section 3.4.2). In this regard, it's been assessed that the number of reallocations that, on average, a UN flow could experience, is kept in the range of 0.26–0.65, depending on UG and UN traffic loads and showing a tendency to decrease as UN traffic increases.

Finally, performance results are provided considering multicast traffic. In this regard, it is assumed that a MG session is forwarded, on average, to six BSs. Unicast traffic load is set to medium load for UG services ($\lambda_{UG} = 0.43$) and UN flows are generated with $\lambda_{UN} = 0.75$ flows/s. Multicast load is fixed as a percentage of the UG load. The satellite utility reduction factor is set to $p^{UG} = p^{UN} = p^{MG} = 0.8$ for all the services and two multicast traffic allocation strategies are considered within the SDN-based TE applications: one strategy seeks to maximize MG utility while the other strategy is intended to minimize the resource consumption of MG flows. Figure 3.19 shows the average utility achieved per BS (left side) and the average mean data rate delivered to UN flows (right side). As it can be observed from the figures, the strategy seeking resource consumption minimization for traffic performs much better in the two performance indicators. The reason is that resource consumption minimization enforces most MG traffic to be delivered over satellite, letting more resources available for UG and UN services that can ultimately get higher utilities and bit rates. While not reported in figures, the obtained UG average rejection rate is in the range of 0.2%–0.5% for the resource consumption minimization strategy in contrast with 0.4%–1.6% for the strategy that maximizes the MG utility.

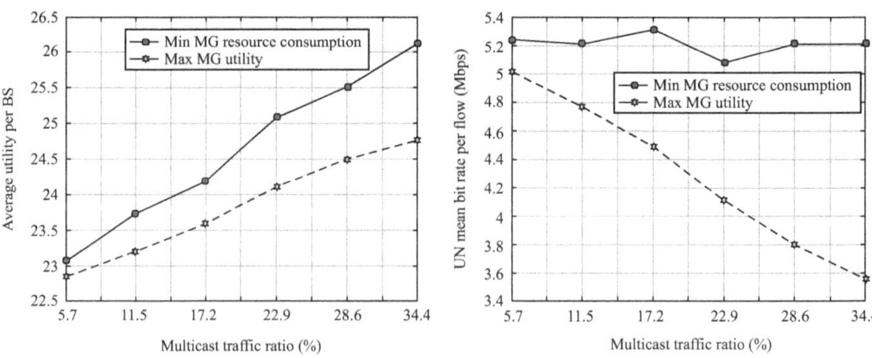

Figure 3.19 Multicast traffic handling strategies – network utility (left) and UN mean bit rate (right)

3.4.3.2 Heterogeneous spatial traffic distribution

Let us now consider the case that traffic is not homogeneously distributed among BSs. In particular, we assume that half of the BSs, denoted as group 1, are exposed to a UN load characterized with $\lambda_{UN} = (1/2)\cdot 0.75$ flows/s and, the other half, denoted as group 2, to a UN load with $\lambda_{UN} = (3/2)\cdot 0.75$ flows/s. In addition, all the BSs support a medium UG load and the C_S is set to 20%. Under this load configuration, Figure 3.20 provides the mean (left) and standard deviation (right) of the data rate achieved per UN flow. Results are given separately for the two groups of BSs, for all the BSs in the scenario and, for comparison purposes, for all the BSs under homogeneous load with $\lambda_{UN} = 0.75$ flows/s. It is observed that in group 1, the mean bit rate provided by the overflow strategy is slightly higher than the one achieved by the SDN strategy. On the other hand, this situation is reversed for BSs in group 2 and when considering the overall set of BSs. This mainly reflects the more fair distribution of satellite capacity enforced by the SDN-based TE application, which is even more evident when comparing the data rate standard deviation values.

Figure 3.21 presents the network performance for the different sets of BSs in terms of network utility per BS. As observed, SDN-based TE application can achieve

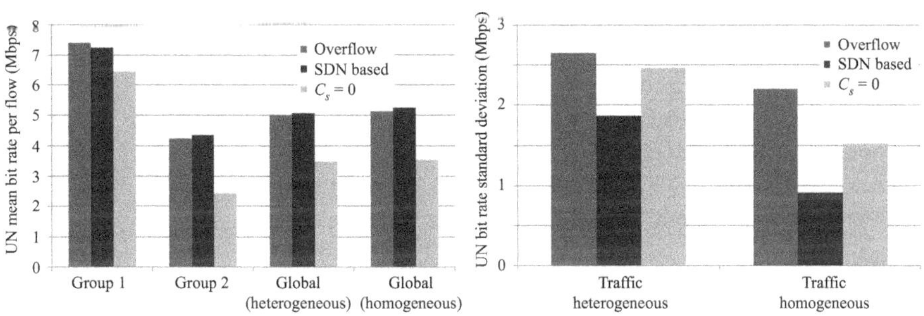

Figure 3.20 Mean (left) and standard deviation (right) of the data rate achieved per UN flow

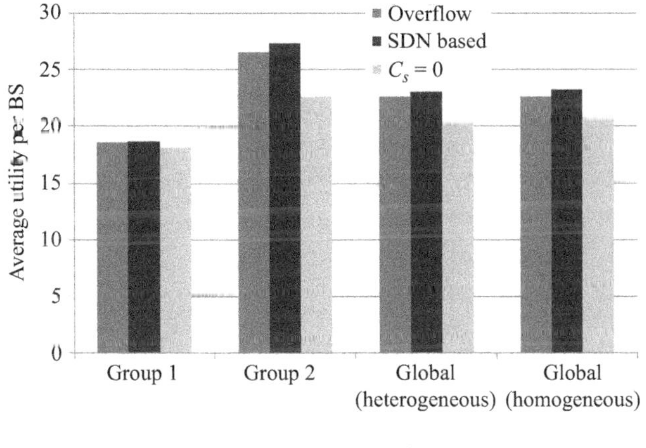

Figure 3.21 Average utility per BS

a higher utility in the most loaded BSs (group 2) and, as a result, a higher performance in the global scenario.

3.4.3.3 Satellite backup for terrestrial link failures and transportable BSs

This assessment shows the performance of the SDN-based TE application when there is one BS with no terrestrial link availability in the set of N+M=16 BSs that share the same pool of satellite capacity. This could be the case of a BS that temporarily faces a lack of terrestrial link availability and the satellite capacity is used as a backup, or the case of a TBS that exclusively relies on the satellite capacity for backhauling. Simulation conditions consider medium UG load and the UN load characterized by $\lambda_{UN} = 1$ flows/s. C_S is set to 20% and satellite utility reduction factor is 1.0 for all services.

Obtained results in terms of admission rejection ratio for UG services are given in Figure 3.22 (left). Results are provided separately for the BS without terrestrial capacity and for the rest of BSs in the scenario. Moreover, for comparison purposes, Figure 3.22 (left) also accounts for the case with $C_S = 0\%$ and the case where there all terrestrial links are fully operational. As seen from the figure, thanks to the reservation management scheme embedded in the SDN-based TE application, the rejection rate can be fully mitigated in the BS without terrestrial capacity, while under the overflow strategy, the rejection rate only decreases slightly.

One of the most noticeable differences is observed in the UN mean bit rate showed in Figure 3.22 (right). The figure presents the mean of data rates delivered to the UN flows served through the BS with no terrestrial capacity, through the rest of BSs with terrestrial capacity and in the scenario without any link failure. As seen in Figure 3.22 (right), the UN mean bit rate achieved by the SDN-based TE application doubles that obtained under an overflow strategy in the impaired BS. This improvement is due to the applicability of the reservation management scheme in the SDN-based TE application that assures that the BS without terrestrial capacity can secure enough satellite capacity to offer an average UN bit rate comparable to the delivered through the rest of BSs.

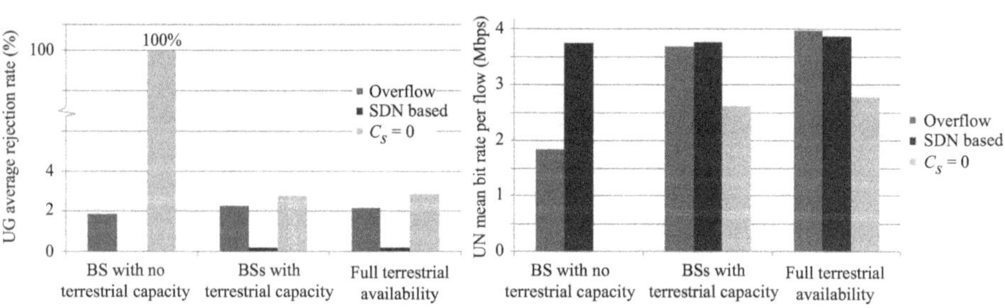

Figure 3.22 Admission rejection ratio for UG services under a terrestrial link failure (left), mean data rate per UN flow under a terrestrial link failure (right)

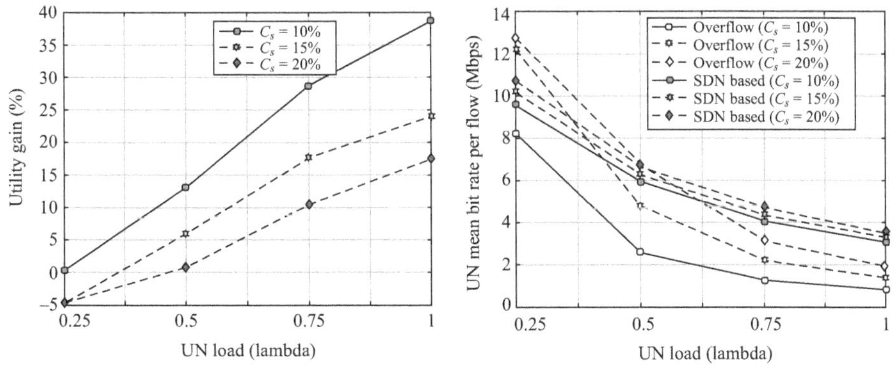

Figure 3.23 Utility gain by SDN-based strategy over overflow strategy at the BS with no terrestrial link availability (left), UN mean bit rate per flow at the BS with no terrestrial availability (right)

Focusing now on the performance of the BS without terrestrial capacity, Figure 3.23 provides further results for different UN loads ($\lambda_{UN} = 0.25$, 0.5, 0.75 and 1.0 flows/s) and different amounts of C_S (10%, 15%, 20%). UG load is set to medium load and satellite utility reduction factor is set to 1.0 for all services. Under these conditions, Figure 3.23 (left) shows the utility gain, computed here as the per cent increase of the global utility obtained by the SDN-based strategy over the overflow strategy. Note that utility gains as high as 40% are realized for high-load conditions. This improvement could even reach up to 85% when UG load conditions are set to high instead of medium. Likewise, Figure 3.23 (right) shows that the SDN-based strategy can keep the UN data rates well above the overflow strategy when the UN load increases.

3.5 Concluding remarks and future recommendations

The evolution of satellite networks towards open architectures based on SDN and NFV technologies arises as a necessary step not only to bring into the satellite domain the benefits associated with the advances in network softwarization technologies being consolidated in the 5G landscape but also to greatly facilitate the seamless integration and operation of combined satellite and terrestrial networks. This chapter has elaborated on the support of SDN concepts and technologies within satellite networks and developed a case study for the applicability of SDN-based TE solutions for the management of a satellite component integrated in next-generation mobile backhaul networks. The study has covered architectural aspects for the realization of such solutions together with the specification and performance assessment of an illustrative SDN-based TE application.

With regard to architectural aspects, building upon the ETSI functional architecture for BSM systems, a solution has been proposed for the adoption of an SDN architecture within the satellite network. The solution relies on the introduction of an SDN controller that manages the connectivity services across the SDN-based satellite network and makes use of the following interfaces: (1) SBIs for the M&C of the interworking and adaptation functions in gateway STs and potentially also user STs; (2) SBIs for the M&C of the BSM bearer services and potentially also some capabilities within the SD lower layers (satellite resources such as a frequency plan, modulation and coding schemes or other satellite specific properties) and (3) NBIs for the M&C of the satellite network flow services from network applications on top of the SDN controller or from external controllers. Candidate SDN data models and protocols for the realization of the SDN-based satellite network architecture have been discussed, namely, ETSI BSM SI-SAP, ONF OF, ONF Microwave Information Model, ONF T-API and IETF YANG models for traffic engineered networks. On this basis, an integration approach for the realization of E2E SDN-based TE in satellite-terrestrial backhaul networks has been presented in which the satellite component has been abstracted as an OF switch. Two central TE workflows have been developed to validate the proposed integration approach.

Next, an SDN-based TE application has been formulated that building on a global view of the hybrid terrestrial-satellite network resources exploits a combination of control features and criteria such as (1) E2E path computation with terrestrial or satellite link selection; (2) satellite capacity resource reservations to deal with BSs with no or limited terrestrial link backhaul capacity; (3) different allocation criteria depending on the traffic nature (GBR and non-GBR services, unicast/multicast); (4) admission and rate control to face overload and guarantee resources and minimum (committed) transmission rates per flow and group of flows and (5) utility maximization criteria, where the adequacy of handling specific flows over the terrestrial or satellite component, as well the effect of allocating more or less data rate, are accounted. A detailed performance analysis has been conducted to assess the behaviour of the proposed SDN-based TE application in multiple and diverse scenarios, including homogeneous and non-homogeneous load situations with BSs that exploit both satellite and terrestrial backhaul capacity, terrestrial link failures in some of the BSs and deployment of a number of mobile TBSs that exclusively rely on the satellite capacity for backhauling. A more traditional overflow strategy has been considered for comparison purposes. As general trends, it's been demonstrated how the proposed SDN-based TE application is able to provide a higher network utility in most of the analysed cases, greatly improving the admission rejection ratio for GBR services and achieving higher fairness in the distribution of delivered data rates for non-GBR flows.

Summing up, concepts and results presented in this chapter clearly advocate for the need to outfit next-generation satellite networks with a set of control and management functions and interfaces (API and/or network protocols) compatible with the mainstream SDN architectures and technologies being adopted in 5G in order to realize a full E2E networking concept where a combined satellite-terrestrial network service can be deployed and operated in a flexible and consistent manner. This sets the stage for the deployment of innovative SDN applications, targeting, for example, enhanced resource efficiency, efficient and fast protection and restoration, and automation of network planning and operation in network infrastructures spanning terrestrial and satellite resources. Moreover, while the SDN-based solution explored in this chapter has exclusively focused on the ground segment of satellite communications systems, dynamic interactions with flexible satellite payloads for more efficient resource management across the whole satellite communications chain should also be explored.

References

[1] NetWorld2020 – SatCom Working Group, "The Role of Satellites in 5G", Version 5, 31st July 2014. Available online at https://www.networld2020.eu/wp-content/uploads/2014/02/SatCom-in-5G_v5.pdf.

[2] B. Evans, O. Onireti, T. Spathopoulos and M. A. Imran, "The role of satellites in 5G", 23rd European Signal Processing Conference (EUSIPCO), September 2015.

[3] C. Sacchi, K. Bhasin, N. Kadowaki and F. Vong, "Toward the 'space 2.0' Era [Guest Editorial]", IEEE Communications Magazine, vol. 53, no. 3, pp. 16–17, March 2015.

[4] M. Corici, A. Kapovits, S. Covaci, *et al.*, "Assessing Satellite-terrestrial Integration Opportunities in the 5G Environment", September 2016. Available online at https://artes.esa.int/sites/default/files/Whitepaper%20-%20Satellite_5G%20final.pdf.

[5] 5G-PPP, "5G Vision – The 5G Infrastructure Public Private Partnership: The Next Generation of Communication Networks and Services", February 2015. Available online at https://5g-ppp.eu/wp-content/uploads/2015/02/5G-Vision-Brochure-v1.pdf.

[6] EMEA Satellite Operators Association, "Satellite Communication Services: An Integral Part of the 5G Ecosystem", June 2017. Available online at https://www.esoa.net/cms-data/positions/ESOA%205G%20Ecosystem%20white%20paper.pdf.

[7] 3GPP TS 22.261, "Service Requirements for Next Generation New Services and Markets; Stage 1 (Release 15)", September 2017.

[8] 3GPP RP-171450, "Study on NR to support Non-Terrestrial Networks", 3GPP TSG RAN WG1 Meeting 88bis, West Palm Beach, USA, 5th–9th June 2017.

[9] H2020 VITAL Research Project, 2015. Available online at http://www.ict-vital.eu/. Last Accessed 1 September 2017.

[10] L. Bertaux, S. Medjiah, P. Berthou, *et al.*, "Software Defined Networking and Virtualization for Broadband Satellite Networks", IEEE Communications Magazine, March 2015.

[11] R. Ferrús, H. Koumaras, O. Sallent, *et al.*, "SDN/NFV-enabled Satellite Communications Networks: Opportunities, Scenarios and Challenges", Physical Communication, pp. 95–112, November 2015.

[12] T. Rossi, M. De Sanctis, E. Cianca, C. Fragale, M. Ruggieri and H. Fenech, "Future Space-based Communications Infrastructures based on

High Throughput Satellites and Software Defined Networking", IEEE International Symposium on Systems Engineering (ISSE), 2015.
[13] ONF TR-521, "SDN Architecture", Issue 1.1, February 2016.
[14] E. Haleplidis and K. Pentikousis (Editors), "Software-Defined Networking (SDN): Layers and Architecture Terminology", IRTF RFC 7426, January 2015.
[15] ONF TS-025, "OpenFlow Switch Specification", Version 1.5.1, March 2015.
[16] ONF TR-535, "ONF SDN Evolution", Version 1.0, September 2016.
[17] ONF TR-513, "Common Information Model (CIM) Overview 1.2", September 2016.
[18] ONF TR-512, "Core Information Model (CoreModel) 1.2", September 2016.
[19] ONF TR-522, "SDN Architecture for Transport Networks", March 2016.
[20] ONF TR-527, "Functional Requirements for Transport API", June 2016.
[21] ONF TR-532, "Microwave Information Model", Version 1.0, December 2016.
[22] ONF TR-523, "Intent Definition Principles", October 2016.
[23] M. Bjorklund (Editor), "The YANG 1.1 Data Modeling Language", IETF RFC 7950, August 2016.
[24] M. Bjorklund, "YANG – A Data Modeling Language for the Network Configuration Protocol (NETCONF)", RFC 6020, October 2010.
[25] A. Bierman, Bjorklund M. and Watsen K., "RESTCONF Protocol", IETF RFC 8040, January 2017.
[26] Y. Lee, X. Zhang, D. Ceccarrelli, B. Y. Yoon, O. G. de Dios and J. Y. Shin, "Applicability of YANG models for Abstraction and Control of Traffic Engineered Networks", June 2017, draft-zhang-teas-actn-yang-05.
[27] B. A. A. Nunes, M. Mendonca, X.-N. Nguyen, K. Obraczka and T. Turletti, "A Survey of Software-Defined Networking: Past, Present, and Future of Programmable Networks", IEEE Communications Surveys & Tutorials, Third Quarter 2014.
[28] C. Janz, L. Ong, K. Sethuraman and V. Shukla, "Emerging Transport SDN Architecture and Use Cases", IEEE Communications Magazine, October 2016.
[29] ETSI TR 101 984, "Satellite Earth Stations and Systems (SES); Broadband Satellite Multimedia (BSM; Services and Architectures", December 2007.
[30] ETSI TR 102 187, "Overview of BSM Families", May 2003.
[31] ETSI TS 102 357, "Common Air Interface Specification; Satellite Independent Service Access Point SI-SAP", May 2005.
[32] ETSI TR 103 444, "Guide to Satellite Independent Service Access Point (SI-SAP) Use", December 2016.
[33] ETSI TS 102 462, "QoS Functional Architecture", June 2015.
[34] ETSI TS 102 295, "BSM Traffic Classes", February 2004.
[35] ETSI TS 103 275, "Satellite Independent Service Access Point (SI-SAP) interface: Services", May 2015.
[36] S. Abdellatif, P. Berthou, P. Gelard, T. Plesse and S. El-Yousfi, "Exposing an Openflow Switch Abstraction of the Satellite Segment to Virtual Network Operators", 2016 IEEE 83rd Vehicular Technology Conference (VTC Spring), Nanjing, 2016.

[37] D. Bogdanovic, B. Claise and C. Moberg, "YANG Module Classification", IETF Internet Draft, October 2016, draft-ietf-netmod-yang-modelclassification.

[38] W. Liu and A. Farrel, "Service Models Explained", IETF Internet Draft, June 2017, draft-ietf-opsawg-service-model-explained-01.

[39] I. Busi (Editor), "Transport Northbound Interface Use Cases", July 2017, draft-tnbidt-ccamp-transport-nbi-use-cases-02.

[40] D. Awduche, J. Malcolm, J. Agogbua, M. O'Dell and J. McManus, "Requirements for Traffic Engineering over MPLS", IETF RFC 2702, September 1999.

[41] M. Conran, "OpenFlow Traffic Engineering", September 2015. Available online at http://network-insight.net/2015/09/openflow-traffic-engineering/. Last Accessed 1 July 2017.

[42] D. Bojic, E. Sasaki, N. Cvijetic, *et al.*, "Advanced Wireless and Optical Technologies for Small-cell Mobile Backhaul with Dynamic Software-defined Management", IEEE Communications Magazine, pp. 86–93, September 2013.

[43] M. R. Sama, L. M. Contreras, J. Kaippallimalil, I. Akiyoshi, H. Qian and H. Ni, "Software-defined Control of the Virtualized Mobile Packet Core", IEEE Communications, pp. 107–115, February 2015.

[44] N. L. M. van Adrichem, C. Doerr and F. A. Kuipers, "OpenNetMon: Network Monitoring in OpenFlow Software-Defined Networks", IEEE Network Operations and Management Symposium (NOMS), Krakow, 2014.

[45] Aricent White Paper, "Demystifying Routing Services in Software-defined Networking", November 2014. Available online at http://www.aricent.com/sites/default/files/pdfs/Aricent-Demystifying-Routing-Services-SDN-Whitepaper.pdf.

[46] R. Pujar and I. Camelo, "Path Protection and Failover Strategies in SDN Networks", Inocybe Technologies, Open Networking Summit, March 2016.

[47] D. Kreutz, F. M. V. Ramos, P. E. Veríssimo, C. E. Rothenberg, S. Azodolmolky and S. Uhlig, "Software-Defined Networking: A Comprehensive Survey", Proceedings of the IEEE, pp. 14–76, January 2015.

[48] I. F. Akyildiz, A. Lee, P. Wang, M. Luo and W. Chou, "A Roadmap for Traffic Engineering in SDN-OpenFlow Networks", Computer Networks, pp. 1–30, October 2014.

[49] T. Ahmed, R. Ferrus, R. Fedrizzi and O. Sallent, "Towards SDN/NFV-enabled Satellite Ground Segment Systems: Bandwidth on Demand Use Case", 1st International Workshop on Satellite Communications – Challenges and Integration in the 5G ecosystem (IEEE ICC 2017), Paris, France, 25 May, 2017.

[50] R. Ferrús, H. Koumaras, O. Sallent, *et al.* "On the Virtualization and Dynamic Orchestration of Satellite Communication Services", Proc. IEEE 84th Vehicular Technology Conference: (VTC'16-Fall), Montréal (Canada), 18–21 September 2016.

[51] C. Niephaus, M. Kretschmer and G. Ghinea, "QoS Provisioning in Converged Satellite and Terrestrial Networks: A Survey of the State-of-the-Art", IEEE Communications Surveys & Tutorials, vol. 18, no. 4, pp. 2415–2441, Fourth Quarter 2016.

[52] N. Cassiau and D. Kténas, “Satellite Multicast for Relieving Terrestrial eMBMS: System-level Study”, IEEE 82nd Vehicular Technology Conference (VTC Fall), Boston, 2015.

[53] S. Shenker, “Fundamental Design Issues for the Future Internet”, IEEE Journal on Selected Areas in Communications, vol. 13, no. 7, pp. 1176–1188, September 1995.

[54] Z. Jiang, Y. Ge and Y. Li, “Max-utility Wireless Resource Management for Best-effort Traffic”, IEEE Transactions on Wireless Communications, vol. 4, no. 4, pp. 100–111, January 2005.

[55] E. Kelly, “Charging and Rate Control for Elastic Traffic”, European Transactions on Telecommunications, vol. 8, pp. 33–37, 1997.

[56] F. Mendoza, R. Ferrús and O. Sallent, “Flexible Capacity and Traffic Management for Hybrid Satellite-Terrestrial Mobile Backhauling Networks”, International Symposium on Wireless Communication Systems (ISWCS), Poznan, September 2016.

[57] Huawei, “Video as a Basic Service of LTE Networks: Mobile vMOS Defining Network Requirements”, July 2015. Available online at http://www.huawei.com/minisite/4-5g/en/industryjsdc-j.html.

Chapter 4
NFV-based scenarios for satellite–terrestrial integration

H. Koumaras[1], G. Gardikis[2], Ch. Sakkas[1], G. Xilouris[1], V. Koumaras[1], and M.A. Kourtis[1]

During the last years, the telecom/network community is pursuing a paradigm shift toward the virtualization/"softwarization" of infrastructure components, enabling a novel "cloud networking" model, which allows the flexible management of network resources and functionalities in a cloud-like manner. Future networks are envisaged to consist of heterogeneous wireless and wired physical infrastructures, whose resources are abstracted via virtualization mechanisms, unified, dynamically pooled and offered as a Service to multiple tenants.

The foundation of current networking infrastructures (wired/wireless and also satellite) on fixed, hardware components with vendor-specific management interfaces, although achieving satisfactory performance and reliability, significantly constrains management flexibility and resource federation, while also hampering the rapid introduction of new network services. This "ossification" is even more visible in the case of satellite networks, where the resource-demanding procedure of hardware prototyping of network technologies and protocols into on-board processors, as well as the delay and costs associated with satellite manufacturing and launch, introduce considerable delays in the adoption of new technologies.

In order to be able to benefit from such a progress and also seamlessly integrate with future networks, satellite communication platforms need to follow this transformation which is currently occurring in the terrestrial segment. This chapter focuses on this issue, reviewing the applicability of cloud networking technologies to satcom platforms and determining the benefits and the challenges associated with the integration of satellite infrastructures into future cloud networks.

4.1 Brief introduction to cloud computing

Network functions virtualization (NFV) appears as an emerging aspect in the networking domain, which has the potential to radically redefine the substance of what

[1]NCSR Demokritos, Institute of Informatics and Telecommunications, Greece
[2]Space Hellas, Greece

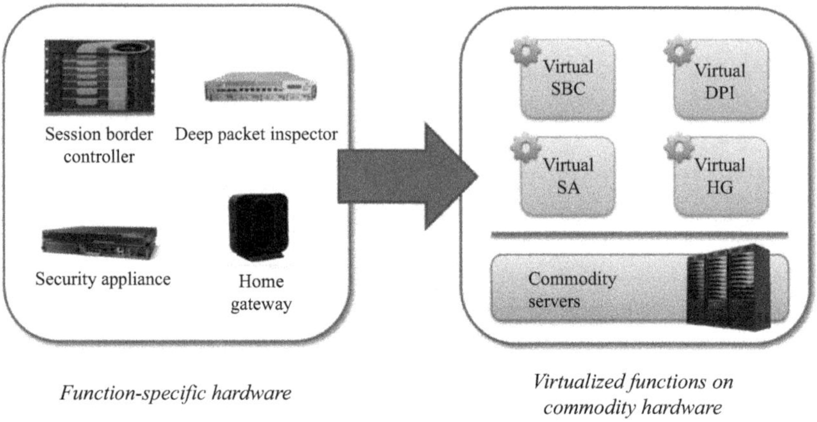

Figure 4.1 The network functions virtualization (NFV) concept

is referred to as "network infrastructure." NFV refers to the virtualization of network functions (NFs), as Figure 4.1 depicts, carried out by specialized hardware devices and their migration as software-based appliances, which are deployed on top of commodity IT (including cloud) infrastructures.

The NFV approach introduces key benefits for network operators/service providers (SPs), such as:

- Consolidation of hardware resources, leading to reduced equipment investment and maintenance costs [reduction of both Capital expenditures (CAPEX) and operating expenses (OPEX)] and power consumption,
- Sharing of resources among different NFs and users,
- Up- and downscaling of resources assigned to each function,
- Rapid introduction of novel NFs (including upgrading of existing ones) at much lower cost and lower risk, leading to significant decrease of time to market for new solutions; new experimental services can coexist in the same infrastructure with "production" ones and
- Promotion of innovation, by opening a part of the networking market and transforming it to a novel virtual appliance market, facilitating the involvement of software entrants, including Small Medium Enterprise (SMEs) and even academia.

While, leveraging the aforementioned benefits, several vendors are already offering virtualized appliances and middleboxes as commercial products, NFV presents several critical challenges when it comes to the automated and large-scale deployment of virtualized appliances within an operational infrastructure. While state-of-the-art IaaS cloud management platforms have proved very effective in deploying virtual machines (VMs) for hosting user applications, the automated deployment of virtualized network appliances instead is a much more challenging task, since it implies joint management of IT and networking resources within the same infrastructure, in order to couple the existing network connectivity services with the deployed NFs.

A scalable and at the same time efficient management solution should achieve NF deployment and resource management, while also taking into account the established network topology. Fault resilience and availability are also critical issues, since the malfunction of a virtual NF may affect the entire network service (NS). What is more, an NFV solution should be compatible with existing network management infrastructures [including operational support systems (OSS)/business support systems (BSS) platforms] enabling a smooth migration path toward a fully virtualized infrastructure. It should also be as generic and universal as possible, supporting both virtual appliances and underlying hardware assets from different vendors. Last but not least, scalability and performance of NFs are also crucial, since software appliances should achieve performance comparable to their hardware counterparts.

Toward addressing all the aforementioned challenges and accelerating the adoption of NFV, a dedicated Network Functions Virtualization Industry Specifications Group was launched by ETSI in 2012, triggered by a joint initiative by telecom operators. Europe is now driving the first standardization effort worldwide in the NFV area, presenting a unique opportunity for European industrial leadership.

Network virtualization is a key enabler technology to escape from the current well-known limitations of the Internet. Moreover, it is also seen as a viable tool for experimenting novel network protocols on production networks without affecting other critical services, running of the same substrate network. It is widely proposed to be an integral part of the Future Internet.

In the past years, network virtualization has received significant attention, as surveyed in [1]. Future Internet initiatives, such as 4WARD [2], Cabernet [3] and GEYSERS [4], presented network virtualization architectures with emphasis on the business roles and the interfaces required for the provisioning and management of virtual networks (VNs) across multiple domains. References [5,6] presented early prototype implementations which realize several components of the 4WARD network virtualization architecture, while their work continued in [7] shows that this architecture is technically feasible and robust. Several platforms have been deployed, assisting network operators to deploy VNs on their own infrastructure [8]. Also, [9] project proposes a network-infrastructure-as-a-Service architecture but without accommodating in-network services. Other initiatives are also addressing network virtualization, via the so-called Network Information and Control (NetIC) Generic Enabler [10]. NetIC is intended to provide access to network operation to higher layer entities. It is more focused on VN provisioning, while programmability is supported by applying the software-defined networking (SDN) paradigm to allow users to develop applications for network management. Most recent projects, such as H2020 VITAL [11] and 5G-SAT [12], are researching the integration of the satellite network with advances of SDN/NFV within the 5G ecosystem.

In terms of VN embedding, most existing algorithms (e.g., [13–16]) consider a single substrate provider and require full knowledge of the available resources and the underlying network topology. Recent work [17] presents a multi-domain VN-embedding framework. This approach consists in relaying VN requests across infrastructure providers till the embedding has been completed. However, this VN-embedding approach lacks algorithms for resource assignment and allocation and it has not been evaluated. Hence, it is unclear how fast it converges to a complete VN embedding. Reference [18] provides a set of algorithms for multi-domain VN embedding. Resource planning becomes more complicated if computing constraints on the network elements are also to be taken into account [19,20].

With regard to node virtualization, advances on server (e.g., [21]) and link (e.g., [22]) virtualization provide the technological ingredients needed to deploy VNs at global scale. In addition, [23] showed that virtual routers on commodity hardware have the capability to forward minimum-sized packets at several Gbps, while offering a high level of programmability [24]. Platforms, such as VINI [25] and Trellis [26], synthesize server and link virtualization technologies to build simple VNs, mainly used for experimentation. In most cases, a virtual router provides an illusion of isolation, rather than a real isolation, as it lacks dedicated memory, processing and I/O resources [27].

Some relative activities also expand the network virtualization concept to cover also the operations of the lower layers (PHY/MAC). In this context, the EU-funded projects iJoin [28] and TROPIC [29] projects focus on the cloud-based virtualization of the cellular radio access network (RAN) aiming at efficient resource management for small cells. Similarly and with a more extended scope, Mobile Cloud Networking (MCN) proposes a framework to fully virtualize a mobile network on end-to-end basis, from the RAN to the application server domains. For this purpose, MCN embraces the concept of NFV, however, specifically targeting to mobile network components.

NFV adds new capabilities to communications networks and requires a new set of management and orchestration functions to be added to the current model of operations, administration, maintenance and provisioning. In legacy networks, NF implementations are often tightly coupled with the infrastructure they run on. NFV decouples software implementations of NFs from the computation, storage and networking resources they use. The virtualization insulates the NFs from those resources through a virtualization layer.

The decoupling exposes a new set of entities, the virtualized network functions (VNFs), and a new set of relationships between them and the NFV Infrastructure (NFVI). VNFs can be chained with other VNFs and/or physical network functions (PNFs) to realize a NS. Since NSs (including the associated VNF forwarding graphs, virtual links, PNFs, VNFs, NFVI and the relationships between them) did not exist before the emergence of NFV, their handling requires a new and different set of management and orchestration functions.

The following sections aim to survey a number of integrated NFV/SDN orchestration solutions, as proposed by R&D projects currently running, industry frameworks and solutions as well as efforts from standardization bodies related to NFV.

4.2 NFV orchestration overview

NFV decouples NFs from underlying hardware so they run as software images on commercial off-the-shelf and purpose-built hardware. It does so by using standard virtualization technologies (compute, network and storage) to virtualize the NFs. The objective is to reduce the dependence on dedicated, specialized physical devices by allocating and using the physical and virtual resources only when and where needed. With this approach, SPs can reduce overall costs by shifting more components to a common physical infrastructure while optimizing its use, allowing them to respond more dynamically to changing market demands by deploying new applications and services as needed.

Simple examples demonstrating the benefit of a NFV service are a virtualized firewall or a load balancer. Instead of installing and operating a dedicated appliance to perform the NF, NFV allows operators to simply load the software image on a VM on demand. In a mobile network, examples include virtualizing the mobile packet core functions such as packet data network gateway, serving gateway, mobile management entity and other elements.

NFV decouples the NF from the hardware. However, extracting maximum value from NFV-based services requires new orchestration capabilities.

Traditional orchestration, in the broader context of service fulfillment, is the process of coordinating and aligning business and operational processes in designing, creating and delivering a defined service. This orchestration process involves the use and management of complex systems and tools such as order, inventory and resource management systems; configuration and provisioning tools; and OSSs combined with the processes associated with these tools and systems. Orchestration solutions play a critical role for SPs by automating tasks across technologies and organizations by integrating with BSSs and customer-relationship-management systems orchestration and by ultimately reducing order-to-revenue time.

NFV orchestration has unique requirements based on the need to automate the highly dynamic delivery of virtualized NSs based on service intent, compared to traditional orchestration for services on physical appliances. These requirements include the following:

- The rapid configuration, provisioning and chaining of virtual NFs in addition to other resources required for the service. The ability to chain several VNFs together is an important and differentiating feature to create innovative and customized services.
- Intelligent service placement: Automating the determination and selection of an optimal physical location and platform on which to place the VNFs, depending on various business and network parameters such as cost, performance and user experience, is a key benefit. A VNF can be placed on various devices in the network—in a data center, in a network node or even on the customer premises.
- Dynamic and elastic scaling of services: The orchestration process maps the instantiation of virtual NFs against real-time demand. This capability frees up physical capacity to be used for other services. In doing so, SPs use their

infrastructure more efficiently. They can also achieve a more predictable and optimized return on investment (ROI) by deploying additional NSs without unnecessary equipment costs. This ROI is especially beneficial for SPs with limited subscriber populations faced with having to add hardware that may significantly exceed the demand for services in the foreseeable future.

- Full lifecycle management of the VNFs: This management includes the creation, instantiation and monitoring of the VNF until it is decommissioned.

The goal of NFV is to enable SPs to better meet their business objectives of agility, to reduce costs and to enable faster service delivery. To do so, it will have to interwork closely with existing OSSs.

NFV requires the implementation of a completely new level of management—not only of cloud infrastructure and the virtual resources that make up that infrastructure but also of the consumption of those resources by individual VNFs. At the very least, NFV will require existing OSS to interact with cloud resource management systems such as OpenStack. In the future, a cloud management and orchestration function and associated data center management systems may supersede "legacy" OSS functionality and systems.

4.3 Integration scenarios

The scenarios elaborated in this chapter correspond to promising cases for the integration of cloud networking techniques into satellite networks [30]. These integration scenarios have been derived using concepts from the terrestrial NFV use cases and adapting them to the satcom context [31]. We have also taken into account the services with currently the highest market share for satcom, i.e., content delivery, broadband access and M2M, and oriented the integration scenarios to correspond to these services [32].

For each integration scenario, we identify:

- the actors/roles involved;
- the high-level description;
- the technical added value for satcom with regard to existing services and technologies;
- the aspects and challenges associated with the implementation of the scenario, also including an assessment of the readiness of the required technological framework.

Regarding the value chain and the business roles involved, Figure 4.2 depicts a generic model including most of the roles which are associated with satellite/terrestrial cloud NS offerings.

Satellite operators offer the satellite platform as well as the raw capacity to be used for the establishment of the satellite network. In most cases, the application of cloud networking techniques is transparent to them.

Satcom and terrestrial network operators/SPs possess a virtualization-capable networking infrastructure, able to offer cloud NSs. SPs fulfill the customers' service

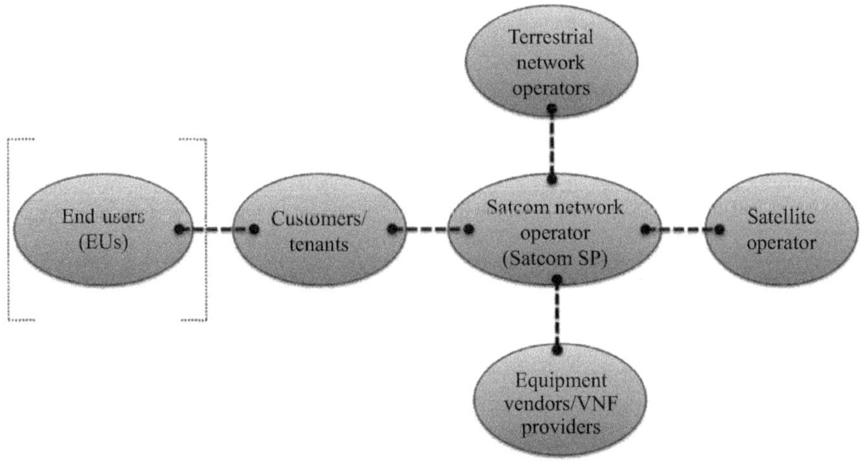

Figure 4.2 Generic value chain for satellite/terrestrial cloud network services

requests by allocating and orchestrating infrastructure resources in order to compose the virtualized service.

Customers or tenants are the "operators" of the virtual tenant service. Commonly, customers establish Service Level Agreements (SLAs) with the SPs for the desired service level and have specific management, control and monitoring rights on the provisioned slices. In case of federated satellite/terrestrial services, customers maintain a unified view of the provisioned slice, regardless of the multiple infrastructure domains on which it may be built.

The customers may exploit the network slice for own internal use (e.g., in the case of an enterprise user establishing a corporate VPN). Moreover, customers may also in turn act as SPs themselves and exploit the slice for offering a service to their customers (e.g., in the case of a content provider leasing the slice to distribute an Internet Protocol Tele Vision (IPTV) service). In this case, the model also includes *end users (EUs)*, who receive the application/content over the slice. The existence of the slice is totally transparent to the EUs, who interact only with the offered application/content.

Finally, in a cloud network model, the role of the *equipment vendor* is expanded in order to encompass also the *VNF providers*, i.e., the developers of virtual NFs, which constitute crucial components of the NS, along with their hardware counterparts.

4.3.1 Scenario 1: virtual CDN as a Service

The Virtual Content Delivery Network (CDN)-as-a-Service (vCDNaaS) scenario involves the virtualization, abstraction and offering of slices of the satellite network—enhanced with in-network functionalities such as content caching and transcoding—as a virtual CDN (vCDN) infrastructure, to be used for efficient content distribution over satellite.

4.3.1.1 Actors and roles

The scenario involves a **satcom network operator** employing virtualization mechanisms to facilitate the deployment of a vCDN service over its infrastructure. The latter is offered as a service to one or more **vCDN providers**. The **customers** are the eventual consumers of the content; they commonly have contract with the satcom SP—in this case, the CDN service is transparent to them.

Additionally, vCDN provider(s) may in turn offer the content handling service to one or more **content providers**. However, the latter are not expected to actually interact with the satcom infrastructure, so their participation in the scenario is rather limited.

4.3.1.2 Description and added value

CDNs are widely used to improve the distribution of content (mostly web and media) over the Internet, allowing content providers to provide high-quality live and on-demand content to EUs with quality similar than—and often superior to—EUs. Integrating CDN nodes into networks has been an effective and cost-efficient way to boost customers' quality of experience (QoE), mostly by caching content close to the consumers, thus relieving core and backhaul links from unnecessary retransmissions of highly popular content. CDN providers either exploit the CDN infrastructure to deliver their own content or offer these capabilities as a wholesale service to third parties (e.g., content providers).

Currently, a CDN provider who seeks to extend their coverage using satellite access would have to physically install CDN nodes i.e., dedicated physical appliances into the satellite infrastructure. This installation would require an agreement with the satcom network operator, who would also (optionally) offer some dedicated capacity for the delivery of the content, if network QoS is desired. This traditional approach, besides requiring significant CAPEX from the CDN provider to acquire and install equipment, would be quite inflexible, mainly because

- physical devices would need to be over-provisioned to match peak demand requirements,
- upgrades and modifications on the CDN node operations (e.g., updates on video formats, installation of new protocols, etc.) would be costly and resource-demanding.

Another very important limitation specifically associated with satellite CDN is that, in the traditional approach, CDN nodes could only be installed in the satellite gateway side (i.e., before the satellite access segment). This limitation would significantly hamper the efficiency of caching, since there would be no saving on the valuable satellite link capacity, cached content would still be served over satellite every time it is consumed. Instead, it would be desirable that caching be also possible after the satellite access, by the satellite terminal. This deployment could also exploit the broadcast satellite capabilities for content distribution in a "push" manner. However, with the traditional hardware-based approach, this is particularly complex, inflexible and costly, especially when many CDN providers share the same satellite infrastructure.

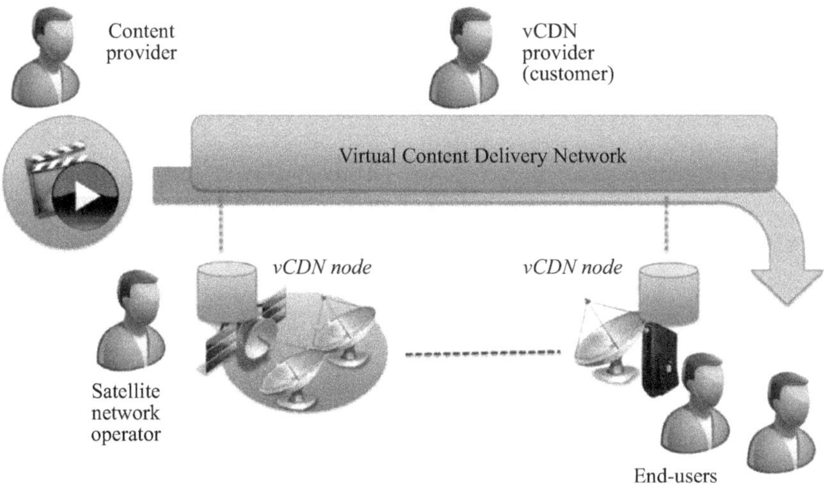

Figure 4.3 *vCDN as a Service over satellite scenario*

Virtualization technologies promise to alleviate most of the aforementioned limitations by completely virtualizing the CDN infrastructure. The application of the vCDNaaS paradigm to satcom would mean that:

- the vCDN nodes are instantiated as software entities within the satcom infrastructure, while still fully managed by the vCDN provider like physical devices.
- the vCDN nodes would be able to scale up/down on-demand, rather than rely on statically allocated resources.
- the vCDN nodes would be able to be instantiated also at the terminal equipment, thus allowing content caching mechanisms to partially relieve the satellite network from multiple transmissions of the same content—as well as radically reducing access latency for popular content. This approach would make sense when multiple customers are served by a single terminal and would greatly benefit from the inherently broadcast nature of satellite, since popular content could be simultaneously pushed to hundreds or thousands of remote caches and served locally.
- the vCDN provider would very easily deploy (and offer to content providers) additional added-value services, such as media transcoding, content pushing or digital rights management, in addition to passive caching.
- the vCDN provider would be able to acquire network resources on-demand for content delivery (e.g., bandwidth and QoS on-demand), rather than operating on a best effort basis. This capability would be particularly useful for maintaining an acceptable customer QoE level during peak hours.

Apart from the vCDN nodes, the centralized CDN controller could also be a target for virtualization. In this approach, the entire vCDN service would be completely virtual and could be deployed with minimal upfront investment (Figure 4.3).

Last but not least, the vCDN scenario, which was described to apply to a single satellite infrastructure, could be expanded to address multi-domain deployments. In a federated concept, the vCDN service could span across multiple satellite and terrestrial domains, in order to reach a wide range of customers.

4.3.1.3 Implementation aspects and challenges

Since the virtualization of CDN functions is the core concept of this scenario, NFV appears as the most prominent enabling technology. In order that vCDN functionalities (not only caches but also transcoders, security appliances, etc.) be deployed as VNFs, the satellite network infrastructure needs to be NFV enabled. That is, the satellite gateway must also feature private cloud infrastructures for VNF hosting and management. Moreover, an NFV management mechanism must be in place, supporting among others multi-tenancy, i.e., allowing each vCDN provider to manage his/her own vCDN nodes.

Additionally, if network resource management is also desired, i.e., elastic bandwidth-on-demand (BoD) and QoS for content delivery, then SDN-based network control would greatly assist.

If the virtual resources (both computing and network) allocated to each vCDN providers are not fixed but dynamically resized, then appropriate metering/accounting/billing mechanisms should be established in order to properly bill the resources used so that the vCDN provider can be charged in a pay-as-you-go model.

Last but not least, while the instantiation of vCDN nodes at the satellite gateway seems quite straightforward, the deployment of vCDN functions at the satellite terminal poses some technical as well as business challenges. Technical challenges are associated with the potentially limited computing resources at the terminal, which need to be carefully managed, especially when shared among various vCDN providers. Business challenges arise when the satellite terminals are not owned by the vCDN provider or the satellite network operator but by the customer. In these cases, the business model must elaborate specific benefits for the customer as a compensation for borrowing local resources in order to support the vCDN service.

4.3.2 Scenario 2: satellite virtual network operator (SVNO)

This scenario is inherited from the concepts of VNOs in terrestrial wired infrastructures and Mobile Virtual Network Operators (MVNOs) in cellular networks. The Satellite Virtual Network Operator (SVNO) scenario involves the partitioning of the satcom infrastructure into logically isolated end-to-end slices with dedicated network, IT and radio resources. These slices, in the form of "virtual hubs," are leased as a Service to several SVNOs, who are offered full control of the virtual infrastructure, as if it were a physical network.

4.3.2.1 Actors and roles

The main interactions of this scenario take place between the satcom network operator, who will be called **Satcom Infrastructure Provider (InP)** in this scenario to be

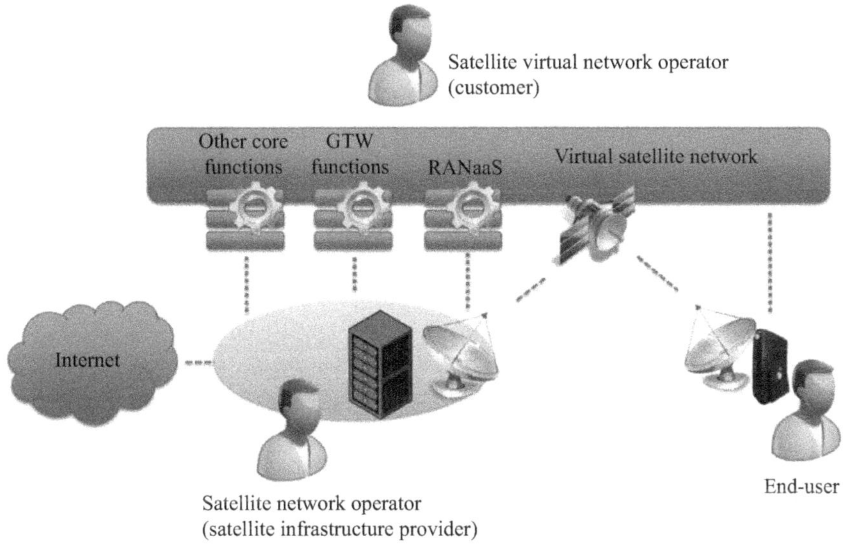

Figure 4.4 SVNO service scenario

clearly distinct from the virtual operator, and the **SVNO**, who corresponds to the customer in this case, leases the slice and consumes the SVNO service. In this scenario, the **EUs** are assumed to maintain relationships only with the SVNO.

Terrestrial network virtualization value chains often also include the role of the Virtual Network Provider (VNP). The VNP uses the resources of the InP to provide the virtualized service to the VNO. However, in a single-domain satcom context, it would make sense to assume that this role is also undertaken by the InP.

4.3.2.2 Description and added value

With the advent of virtualization technologies and enablers, the concept of Virtual Network Operators (VNOs) and especially MVNOs is gaining ground, and the VNO business case is becoming more and more attractive.

During the last years, the VNO concept has extended to encompass the satellite segment and SVNO offerings have emerged. The DVB-RCS2 technology supports SVNO by dividing the capacity into several logical and independent networks—Operator Virtual Networks (OVNs). Each OVN is assigned a set of customer terminals and dedicated capacity, staying logically isolated from the rest OVNs.

By exploiting the virtualization paradigm, the scenario described herein extends the SVNO concept from the plain slicing of capacity, to the full virtualization of the entire hub—i.e., the core gateway and front-end functions, including traffic control [caching, firewalling, Performance Enhancing Proxy (PEP), etc.], multiplexing, multiple access and also radio (coding and modulation). Each of these functions is implemented in logically isolated virtualized appliances (VNFs) and is chained together to become components of a "virtual hub"—and eventually of an end-to-end SVNO service (Figure 4.4).

Key added-values stemming from this approach, compared to current SVNO offerings, are the full administrative privileges which are offered to the SVNO, who is able to manage all the virtual appliances involved in the service independently, as if he/she was managing physical devices. For example, he/she could configure the PEP, change scheduler priorities, manage the multiplexing process and even fine-tune the modulation/coding parameters—respecting of course the satellite power and link budget constraints. That is, he/she can enjoy (almost) the same administrative freedom as a physical satcom network operator. However, depending on the operating model and also on the technical competence of the SVNO, the latter might decide to outsource some management functionalities to the InP.

Another benefit, which can be potentially offered to the SVNO under this scenario, is the capability to choose among multiple virtual appliances and combine (chain) them as desired. For instance, the SVNO service could combine the virtual firewall of vendor A with the virtual multiplexer of vendor B and the virtual modulator of vendor C. In this mix-and-match case, it would make sense to extend the value chain to also include the role of the virtual appliance vendors (VNF developers), since they play a more active role in the scenario.

The fast set-up time as well as the resource elasticity are also advantages to be considered. According to the traffic served and the customer density and demand, the SVNO might request to scale up or down the resources assigned to the virtual network; however, this scaling would not be considered highly dynamic.

Last but not least, it would be also possible (although with several technical and business considerations) that a SVNO combines resources from several satcom infrastructures to form a federated virtual infrastructure. In this case, the virtual NS would span across several administrative domains. This approach would achieve, e.g., increased capacity (via bandwidth aggregation from multiple satellites) and/or extended footprints (via exploiting multiple satellites covering diverse areas).

For all these purposes, under several business and operational models, the SVNO paradigm could be suitable for a wide variety of actors, including but not restricted to

- small data SPs who wish to enter the market with low CAPEX investment,
- terrestrial Internet Service Provider (ISPs) who wish to add a satellite "branch" to reach certain customers—or to offer hybrid access,
- M2M SPs who also own M2M application platforms and wish to offer turn-key and to end M2M solutions via satellite,
- large enterprise users who want the virtual network for internal use and seek a service more "owned" and self-managed.

4.3.2.3 Implementation aspects and challenges

SDN and NFV appear as key enabling technologies for the SVNO scenario. In order to fully support the SVNO offering, with the capabilities described, the satcom infrastructure needs to be fully SDN and NFV enabled.

As described in the previous scenarios, SDN can be used to (i) reserve SVNO capacity within the infrastructure, (ii) establish network tunnels where necessary and

(iii) implement the service chaining, interconnecting the various virtual appliances of the "virtual hub."

In addition, while current SVNO offerings provide specific—often limited—management capabilities based on protocols such as SNMP or even on proprietary protocols, an SDN-driven SVNO may (optionally) expose an SDN northbound interface for network control; in this sense, the virtual operator can control the service by any standard SDN controller, even developing his/her own control applications. This capability paves the way toward fully programmable satellite VNs.

SDN-based control also means that SVNOs can make the provisioning process of the services delivered to their customers fully automated. Indeed, a provisioning engine can be used to orchestrate and perform all the required configurations via SDN. In other words, services such as the elastic BoD can now be offered over the VN, rather than the physical one.

In turn, NFV is needed for the virtualization and unified management of the virtual appliances which are the components of the "virtual hub," assuming that all VNFs will expose a common, standards-compliant interface for management.

Although the technological enablers are in place, the SDN/NFV-driven SVNO remains a highly challenging scenario. As with any infrastructure virtualization approach, two main considerations are security and resilience. Since the virtual service has the same availability requirements as the physical one, any malfunctions (accidental or deliberate) should be rapidly mitigated—by means of, e.g., live migration of virtualized appliances—and should not affect the SVNO services of other tenants using the same infrastructure.

Another challenge concerns the dynamicity of the SVNO resources. Although, thanks to SDN, the resources among the customers within the VN can be rapidly reallocated, the scaling of the SVNO service as a whole would be rather limited and would not be assumed to take place often. Especially—in realistic conditions—the RF bandwidth offered to the virtual radio front-end would not be considered a dynamically scalable resource.

Concluding, although L2/L3 logical network partitioning mechanisms are already well established, the application of the radio access virtualization concept can only be considered for the long term.

4.3.3 Scenario 3: dynamic backhauling with edge processing

The dynamic backhauling with edge processing as-a-Service scenario investigates the dynamic extension of terrestrial networks via satellite links, in cases where terrestrial coverage is inadequate. Beyond allocating capacity on-demand and providing the necessary QoS per service, it becomes possible to also deploy instances of specific services of the terrestrial network, such as Long Term Evolution (LTE) Evolved Packet Core (EPC) components as VNFs on the satellite access segment. This is the concept of satellite edge processing, which is in-line with the emerging paradigm of Multi-Access Edge Computing (MEC).

Apart from backhauling support for mobile networks, this scenario also aims to augment the typical satellite M2M service by dynamically deploying data processing components as VNFs at the satellite access segment i.e., at the gateways providing satellite connectivity to the local M2M network. This capability allows local preprocessing of the M2M traffic at the aggregation point (e.g., data aggregation, statistical processing, video feature extraction, etc.) in a reprogrammable/reconfigurable manner.

4.3.3.1 Actors and roles

Although this scenario has considerable technical implications, the value chain is simple. The **satellite network operator** offers the dynamic backhauling service, also providing satellite terminals with edge computing/processing capabilities. The **customers** are expected to be, e.g., mobile operators (using the satellite segment to extend network coverage), M2M platform operators, institutional users, etc. Generally, this type of service is not targeted to retail/residential customers.

4.3.3.2 Description and added value

Mobile backhauling (e.g., for 2G/3G/4G networks) has been one of the typical use cases for satcom. Integrating satellite in the cellular infrastructure by feeding remote base stations via satellite allows mobile network operators to extend their services to areas and cases not covered by terrestrial backhauls (e.g., fiber or microwave). These cases include remote, isolated locations, where the extension of terrestrial backhauls is not technically feasible or economically advisable. Satellite backhauling is also used where the terrestrial infrastructure has suffered considerable damage (e.g., after a natural disaster).

As also explained in the previous scenarios, the use of network programmability technologies greatly facilitates the allocation, management and optimization of the backhaul capacity. Thus, short service set-up time and resource elasticity are key benefits to be introduced.

However, in a virtualization-enabled world, backhauling can mean much more than capacity. Specifically, one of the envisaged key elements of the 5G technological framework is the capability to deliver intelligence directly to network's edge, in the form of VN appliances, jointly exploiting the emerging paradigms of NFV and MEC. Novel edge infrastructures promise to offer dynamic processing capabilities on-demand, optimally deployed close to the user. Following this direction, novel business cases will produce added value from any kind of infrastructure or application that has the potential to be offered "as a Service."

The satellite edge processing scenario assumes the extension of this paradigm to the satellite domain; specifically, it foresees that the backhauling service is coupled with virtualization capabilities at the satellite terminal, able to host virtual traffic processors close to the EUs (Figure 4.5). Such local traffic processing can achieve significant savings in satellite capacity.

Two examples of this scenario are provided below.

- In 3G/4G mobile backhaul services, some IP Multimedia Subsystem (IMS) or EPC components could be deployed at the edge so that user traffic is processed and rerouted locally, without the need to traverse the satellite segment.

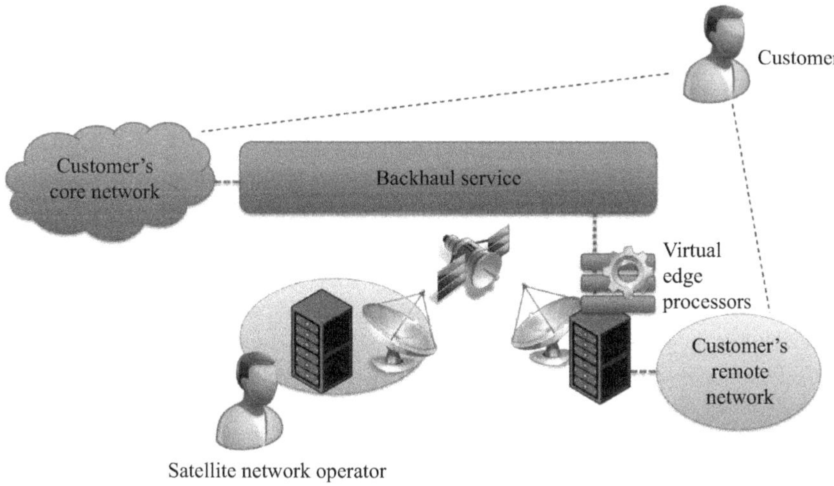

Figure 4.5 Dynamic backhauling with edge processing scenario

- In M2M services, sensor data can be aggregated and processed locally at the virtual processor(s) of the terminal, for example
  - measurements from multiple sensors can be aggregated, and only aggregates and possibly detected events are transmitted back over satellite
  - video streams can be dynamically transcoded, features can be extracted and only the features/processing results are transmitted back over satellite.

The NFV agility allows customers to deploy such traffic processing functionalities on-demand in professional satellite terminals, upgrade them and configure/manage them in a unified manner. Resources of virtual appliances can be scaled up and down on-demand, matching the traffic characteristics and customer requirements.

This concept eventually results in a totally new service mix, in which traditional backhauling is coupled with edge processing resources, offered on-demand, as a Service. The terminal is essentially transformed to a virtualization-capable remote head-end, able to serve a wide range of use cases.

Last but not least, although the scenario, as described, assumes the use of the satellite terminal by a single customer, virtualization technology allows also multi tenancy at the edge segment; this means that the professional terminal itself may be partitioned into multiple "virtual terminals," offered to different customers. This capability can be exploited in scenarios where the satcom operator has already deployed a network of terminals and leases portions of the terminals to different customers. For example, a set of terminals covering a remote village can be leased and shared among two or more mobile operators. This interesting and novel approach demonstrates the power of virtualization technology to introduce new market opportunities and to transform the typical Telco value chains.

4.3.3.3 Implementation aspects and challenges

With regard to reserving bandwidth capacity for the backhaul service, the use of SDN greatly simplifies network control and facilitates QoS assurance with per-flow or per-application granularity. It is advisable that SDN capabilities are integrated in both the satellite gateway and the remote terminal, which are centrally controlled by the satellite operator who uses SDN management to allocate bandwidth on-demand. Although, as explained, BoD is already feasible with legacy technologies, SDN allows elasticity, per-application differentiation and flexible SLAs and pricing—specifically suited to more dynamic use cases. However, it is essential to couple SDN with radio resource management in order to efficiently control and share the satellite capacity, especially for the return link.

When it comes to edge processing, then NFV, coupled with emerging MEC concepts for deployment of cloud resources at the network edge, are the key enabling technologies. The satellite terminal needs to encompass virtualized IT resources in order to host the traffic processors, as VNFs. When it comes to management, since it is not advisable to deploy an entire cloud system (e.g., OpenStack) on the terminal, it could be assumed that the cloud controller is located centrally at the satellite gateway, controlling remote compute nodes at the terminals. In a more lightweight approach, the terminals can encompass plain IT virtualization (e.g., via a KVM hypervisor or even via Docker containers), without any cloud framework. This approach has the cost of reduced elasticity and management features. However, it saves IT resources and also relieves the satellite segment from excessive signaling, thus it would be more appropriate for edge VNFs (rather than for VNFs hosted at the gateway, where OpenStack-based management is still advisable).

The technology readiness level is considered **medium**, mostly due to low maturity of edge computing mechanisms.

4.3.4 Scenario 4: customer functions virtualization

This scenario is based on the VNF-as-a-Service (VNFaaS) paradigm and assumes the dynamic offering of VN appliances to satcom customers in the form of VNFs (e.g., firewalls, traffic filters, home gateway functionalities, media storage and processing, etc.). According to their nature, these VNFs can be instantiated either at the satellite gateway or at VNF-enabled satellite terminals. It must be noted that this scenario focuses on consumer use.

4.3.4.1 Actors and roles

This scenario assumes that the **satellite network operator** also undertakes the role of the NFV SP and offers VNFaaS as added-value services along with satellite connectivity to **customers**. In a more pluralistic scenario, the **VNF providers** (developers) play a more active role, advertising and dynamically pricing their services which are published in a catalogue. The customers may select the services that best suit their needs. In some business models, the VNF providers may receive direct profit from the customers, either indirectly as a share of the satcom service fee or directly, as a license fee for using the VNF.

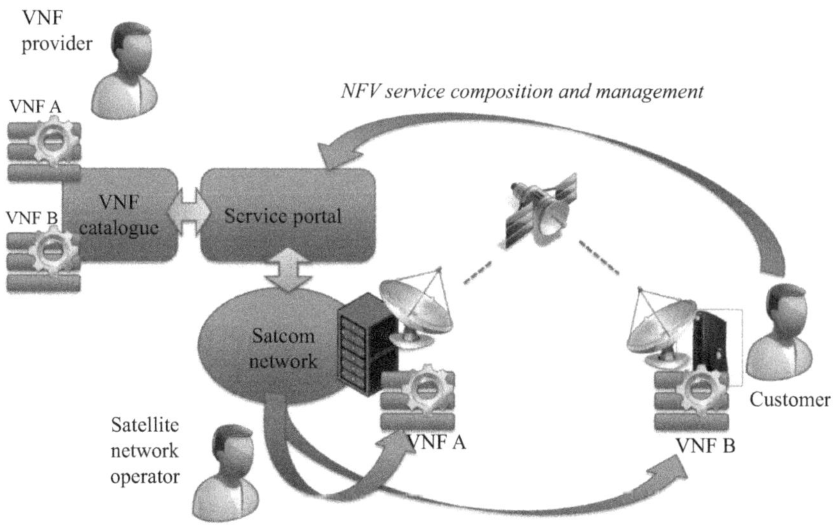

Figure 4.6 Customer functions virtualization scenario

4.3.4.2 Description and added value

In the most common scenario of satellite broadband access, the satellite terminal itself exposes some basic network functionalities to the customer, such as firewalling, NAT, port forwarding, etc. If more capabilities are needed, the customer has to acquire and install additional physical appliances.

Furthermore, there are some capabilities that would be advisable to be present before the satellite segment, for the sake of saving satellite capacity. For example, firewalling should be conducted at the satellite gateway to avoid transmitting over satellite traffic which will be eventually blocked at the terminal. Same with media transcoding; it would be advisable to transcode streams before they are transmitted, so they occupy less satellite capacity. However, such capabilities cannot be currently provided per customer; the network functionalities at the gateway apply to the entire traffic and of course cannot be managed by the customer.

The VNFaaS scenario promises to alleviate these limitations by allowing NFs in the form of VN appliances to be acquired on demand by the customer and instantiated either at the satellite terminal or in a shared resource space (mostly private cloud infrastructure) at the gateway (Figure 4.6). Some functions, such as PEP and application classification, could be installed at both ends.

In a more static scenario, the satellite network operator manually deploys the VNFs and interconnects them, following a customer request. In a more interactive and dynamic approach, the customer composes the NFV service in a completely automated manner by accessing a service portal, browsing the VNF catalogue, selecting the VNFs which best match his/her needs and integrating them into a satcom service package.

The same service portal could then be used for the monitoring and the management of the service. VNFs may be managed either via the portal or via individual management interfaces.

Examples of VNFs which would bring added value when offered as a Service in a satcom context would be as follows:

- Firewalling and content filtering (GW side)
- Application classification (GW and terminal sides)
- Caching (terminal side to cache traffic from external networks; GW side to cache traffic stemming from the terminal)
- Media transcoding (GW side for media streams consumed by the customer)
- Performance enhancement proxy (GW and terminal sides).

4.3.4.3 Implementation aspects and challenges

For the implementation of the use case, a VNFaaS platform needs to be integrated into the satellite infrastructure. Commonly, the NFV management entities are deployed at the gateway side, controlling NFV resources both local (at the gateway) and remote (at the terminals). In order not to pose excessive capacity overhead in the satellite segment, the remote management of NFV resources at the terminals should involve as little signaling as possible.

The NFV management is expected to carry out procedures for controlling the entire NFV lifecycle, including

- NFV service mapping, i.e., allocating the resources which match the service requirements and characteristics
- VNF instantiation, i.e., launching of the VNF images in the host machines
- service chaining, i.e., controlling the network to interconnect the various VNFs of the service and directing the customers' traffic through the VNFs
- service monitoring, i.e., collecting and aggregating metrics from VNFs and VNs
- service rescaling, including rescaling of VNF resources and network resources
- service starting/stopping and teardown.

Apart from the aforementioned management procedures, the NFV platform also needs to accommodate interactions with the customers, allowing them to select, deploy, manage and monitor VNFs. An NFV service catalogue is essential in order to allow customers to customize the services according to their needs. Proper SLA and billing mechanisms must also be in place.

In order to allow deployment of VNFs in the satellite terminals, the latter need to offer generic computing resources, as well as the proper management interfaces, in order to accommodate VNFs. Given that terminals have generally constrained hardware resources, it is of particular interest to exploit novel virtualization techniques for non-x86 processors (e.g., suitable for ARM processors) as well as lightweight virtualization schemes (e.g., Linux containers or Docker containers), rather than full virtualization based on VMs. This approach would allow the deployment of multiple VNFs chained together in a single terminal with minimal resource overhead.

Furthermore, SDN support within the satellite network (at least in the gateway local network) is considered essential, since other emerging NFV architectures are based on SDN for network management.

The technology maturity for this scenario is considered medium, closely associated with the foreseen progress of the NFV architectures in the years to come.

4.4 Conclusions

This chapter presented a review on candidate integration scenarios among satellite communications infrastructures and cloud networking technologies, based on NFV. Current NFV solutions offer rich features as well (and more are planned for the future), and there does not seem to be a fundamental new requirement for NFV from a satcom point of view regarding its applicability to the ground segment. It is concluded that the interplay of satcom with NFV can result in quite attractive use cases, with considerable added value, from both a technical and a business view.

Vendors benefit from improved ease of product evolution, acceleration of assembly, integration and tests (AIT) and better lifecycle support. In turn, SPs benefit from widening their service portfolio, CAPEX and OPEX reduction and better resource management.

It would be considered quite beneficial—and also safe from both a technical and business point of view—to adopt NFV in the short term some virtualization strategies, especially at the edges of the network (terrestrial interface and later also at the terminals). Longer term evolutions should be carefully planned given also the evolution and adoption of the NFV technology in general.

Further evolutions of NFV technology to facilitate integration with satcom would include as follows:

- VNF deployment in compute nodes with very limited resources (e.g., payload or terminal).
- Reconsideration of the SDN/NFV paradigm to allow efficient distribution to multiple gateways at a very long distance (i.e., to relax bandwidth requirements for backhaul links), for multi-GW configurations. In such scenarios, only specific functionalities should be centralized and not the entire baseband processing chain.
- Better integration with satcom OSS/BSS functions, practices and workflows.

In order to facilitate the foreseen technical achievements in a most efficient and effective manner, several possible interactions with the software network and satcom community maybe be pursued.

References

[1] Boutaba, N., and Chowdhury, R. (2009). Network Virtualization: State of the Art and Research Challenges. *IEEE Communications Magazine*, *47*(7), 20–26.

[2] 4WARD. (2014). *4WARD Project*. Retrieved 7 14, 2014, from http://www.4ward-project.eu/.

[3] Zu, Y., Zhang-Shen, R., Rangarajan, S., and Rexford, J. (2008). Cabernet: Connectivity Architecture for Better Network Services. Madrid, Spain: in Proc. ACM ReArch'08.

[4] GEYSERS. (2014). *Generalised Architecture for Dynamic Infrastructure Services*. Retrieved 7 21, 2014, from http://www.geysers.eu/.

[5] Schaffrath, G., Werle, C., Papadimitriou, P., *et al.* (2009). Network Virtualization Architecture: Proposal and Initial Prototype. Barcelona: in Proc. ACM SIGCOMM VISA.

[6] Werle, C., Papadimitriou, P., Houidi, I., *et al.* (2011). Building Virtual Networks Across Multiple Domains. Toronto: in Proc. ACM SIGCOMM 2011, Poster Session.

[7] Papadimitriou, P., Houidi, I., Louati, W., *et al.* (2012). Towards Large-Scale Network Virtualization. Santorini: IFIP WWIC 2012.

[8] Nogueira, J., Melo, M., Carapinha, J., and Sargento, S. (2011). A Platform for Operator-driven Network Virtualization. in Proc. IEEE EUROCON—International Conference on Computer as a Tool.

[9] Peng, B. (2011). A Network Virtualisation Framework for IP Infrastructure Provisioning. in Proc. 3rd IEEE Int. Conf. on Cloud Computing Technology and Science.

[10] FI-WARE. (2014, 7 14). *FI-WARE Interface to Networks and Devices (I2ND)*. Retrieved 7 14, 2014, from http://forge.fi-ware.eu/plugins/mediawiki/wiki/fiware/index.php/FI-WARE_Interface_to_Networks_and_Devices_(I2ND).

[11] VITAL. (2018, 01 13). *VITAL Project*. Retrieved from VITAL Project: http://www.ict-vital.eu/.

[12] SAT-5G. (2018, 01 13). *SAT-5G Project*. Retrieved from SAT-5G Project: http://sat5g-project.eu/.

[13] Zhang, S., Qian, Z., Guo, S., and Lu, S. (2011). FELL: A Flexible Virtual Network Embedding Algorithm with Guaranteed Load Balancing. in Proc. 2011 IEEE International Conference on Communications (ICC).

[14] Masti, S., and Raghavan, S. (2012). VNA: An Enhanced Algorithm for Virtual Network Embedding. in 21st International Conference on Computer Communication Networks (ICCCN).

[15] Yu, J. (2012). Solution for Virtual Network Embedding Problem based on Simulated Annealing Genetic Algorithm. in The 2nd International Conference on Consumer Electronics, Communications and Networks (CECNet).

[16] Sarsembagieva, K., Gardikis, G., Xilouris, G., Kourtis, A., and Demestichas, P. (2013). Efficient Planning of Virtual Network Services. in IEEE Region 8 EuroCon Conference.

[17] Chowdhury, M., Rahman, M., and Boutaba, R. (2012). ViNEYard: Virtual Network Embedding Algorithms With Coordinated Node and Link Mapping. *IEEE/ACM Transactions on Networking*, *20*(1), 203–226.

[18] Houidi, I., Louati, W., Bean-Ameur, W., and Zeghlache, D. (2011). Virtual Network Provisioning Across Multiple Substrate Networks. *Computer Networks*, *55*(4), 1011–1023.

[19] Nogueira, J., Melo, M., Carapinha, J., and Sargento, S. (2011). Network Virtualization System Suite: Experimental Network Virtualization Platform. Proc. International Conf. on Testbeds and Research Infrastructures for the development of Networks and Communities, Shanghai, China, April.

[20] Nogueira, J., Melo, M., Carapinha, J., and Sargento, S. (2011). *Virtual Network Mapping into Heterogeneous Substrate Networks.* Corfu: in Proc. IEEE ISCC-11.

[21] Barham, P., Dragovic, B., Fraser, K., *et al.* (2003). Xen and the Art of Virtualization. New York, NY: in Proc. 19th ACM Symposium on OS Principles.

[22] Rekhter, E., and Rosen, Y. (1999). *BGP/MPLS VPNs, RFC 2547.* IETF.

[23] Egi, E., Greenhalgh, A., Handley, M., Hoerdt, M., Huici, F., and Mathy, L. (2008). Towards High Performance Virtual Routers on Commodity Hardware. Madrid: in Proc. ACM CoNEXT 2008.

[24] Egi, N., Greenhalgh, A., Handley, M., *et al.* (2011). A Platform for High Performance and Flexible Virtual Routers on Commodity Hardware. *ACM SIGCOMM Computer Communication Review Archive*, *40*(1), 127–128.

[25] Bavier, A., Feamster, N., Huang, M., Peterson, L., and Rexford, J. (2006). In VINI Veritas: Realistic and Controlled Network Experimentation. Pisa: in Proc. ACM SIGCOMM.

[26] Bhatia, S., Motiwala, M., Muhlbauer, W., *et al.* (2008). Trellis: A Platform for Building Flexible, Fast Virtual Networks on Commodity Hardware. Madrid: in Proc. 3rd ACM Workshop on Real Overlays and Distributed Systems.

[27] Carapinha, J., and Jimenez, J. (2009). Network Virtualization—A View from the Bottom. Barcelona: in Proc. ACM SIGCOMM VISA'2009 Workshop.

[28] IJOIN. (n.d.). Retrieved from http://www.ict-ijoin.eu/.

[29] TROPIC. (n.d.). Retrieved from http://www.ict-tropic.eu/.

[30] Ferrús, R., Koumaras, H., Sallent, O., *et al.* (2016). SDN/NFV-enabled Satellite Communications Networks: Opportunities, Scenarios and Challenges. *Physical Communication*, *18*(2), 95–112.

[31] Bertaux, L., Medjiah, S., Berthou, P., *et al.* (2015). Software Defined Networking and Virtualization for Broadband Satellite Networks. *IEEE Communications Magazine*, *53*(3), 54–60.

[32] Gardikis, G., Costicoglou, S., Koumaras, H., *et al.* (2016). NFV Applicability and Use Cases in Satellite Networks. European Conference on Networks and Communications (EuCNC), Athens, pp. 47–51.

Chapter 5

Propagation and system dimensions in extremely high frequency broadband aeronautical SatCom systems

Nicolas Jeannin[1], Barry Evans[2], and Argyrios Kyrgiazos[2]

The growing interest in the exploration of frequency bands above Ka band (20–30 GHz) for fixed satellite systems is mainly driven by the bandwidth needs of feeder links. Indeed at Q/V band (40–50 Hz), there is theoretically a bandwidth available of 5 GHz for uplink and 5 GHz for downlink. For W band (70–80 GHz), the situation is somewhat similar with up to 5 GHz of bandwidth for up and downlink, respectively.

At those frequency bands, the propagation impairments become so high in presence of adverse weather conditions (cloud and rain) that commercially interesting availabilities can be offered only if large antennas or spatial diversity are used in most climatic areas. It prevents thus a usage of those bands for user links despite the significant bandwidth amount that can be used for the fixed satellite service. Indeed ground-based user terminals can usually not draw benefits from site diversity nor be associated to large antennas. Thus, the usage of those frequency bands for satellite applications is almost exclusively considered for feeder links with operator gateways.

The usage of EHF (extremely high frequency) bands can however constitute a promising solution for the provision of services to aeronautical terminals, to cope with the significant demand increase for in flight connectivity. In fact, satellite-aircraft links, for aircrafts flying in the upper troposphere, are less subject to atmospheric propagation impairments, as they are mainly occurring in the lowest part of the troposphere. The residual propagation margins are sufficiently low to ensure a good availability level even at high frequency. In addition, considering the same satellite and terminal antenna sizes as at lower frequency bands the system can benefit from a more favourable link budgets. In this chapter, the main challenges linked to the establishment of an EHF SatCom system dedicated to the provision of communications to aircrafts are discussed.

In the first section of this chapter, an overview of existing or planned systems dedicated to broadband communication with aircrafts is presented. Projected commercial

[1]ONERA/DEMR, Université de Toulouse, France
[2]Institute for Communication Systems, University of Surrey, United Kingdom

aviation generated traffic demand is then analysed, considering current commercial aviation traffic and the forecasted data usage.

In the next stage, the characteristics of the aeronautical to satellite channel at EHF bands are presented. In particular, the impact of the altitude on the tropospheric impairments is analysed. The latest ITU-R standards to assess the impact of the troposphere on an aircraft-space link are detailed. The characteristics of the propagation channel are also discussed.

Then, a possible system sizing at EHF bands to match the demand is presented. The challenges linked to the different elements of the transmission chain are described with an analysis of the satellite and airborne antennas and some discussion of the payload aspects. In the last stage, a preliminary assessment of the performances of plausible systems dedicated to the service of aircrafts is presented for the different frequency bands.

5.1 Traffic demand and characterization

One of the currently largest growing sources of growth for SatCom industry is the provision of services to passenger aircrafts. In fact, the demand for ubiquitous connectivity combined with the development of data rate demanding applications yields to a tremendous increase of data demand per plane [1]. A traffic demand of 125–200 Mb/s per single-aisle aircraft is forecasted around 2020. A fraction of the capacity will be provided by ground based LTE (long-term evolution) infrastructure at S band like the one of Gogo [2] in the United States or Inmarsat and Deutsch Telekom in Europe [3]. It is possibly complemented by a satellite segment (Europasat payload for European systems). Other solutions based on a meshed network in which the planes are nodes of an ad-hoc network with line of sight connections between the aircrafts are currently under investigations ([4] or [5]). It could be a promising solution in areas in which plane density is sufficiently high to offer a connectivity over the oceans.

However, satellite is likely to remain the main connectivity provider at least for long-haul flights as it could provide the connectivity almost worldwide for GEO (Geostationary Earth Orbit) constellations. Future Non-Geostationary Orbit satellite constellations targeting high data rate could also be involved in the provision of capacity to commercial aircrafts. Currently, services are provided to aircraft using either L band mobile satellite services, Ku band high throughput satellite systems or Ka band systems like global Xpress. It is however unlikely that the bandwidth available to those systems will be sufficient to match the capacity demand. Indeed, the capacity per beam ranges from less than 1 Mbps for L band systems to some tens of Mbps for Ka band ones. To match the majority of the demand, a significant increase of the bandwidth offer will be required. This will be especially sensitive in areas in which plane density is high.

An example of aircraft position worldwide is illustrated in Figure 5.1. The largest concentration of aircrafts can be found around the main hubs in Northern America, Europe and Eastern Asia. Significant aircraft concentrations can also be found along main transcontinental routes whose position can vary depending on jet-stream regime.

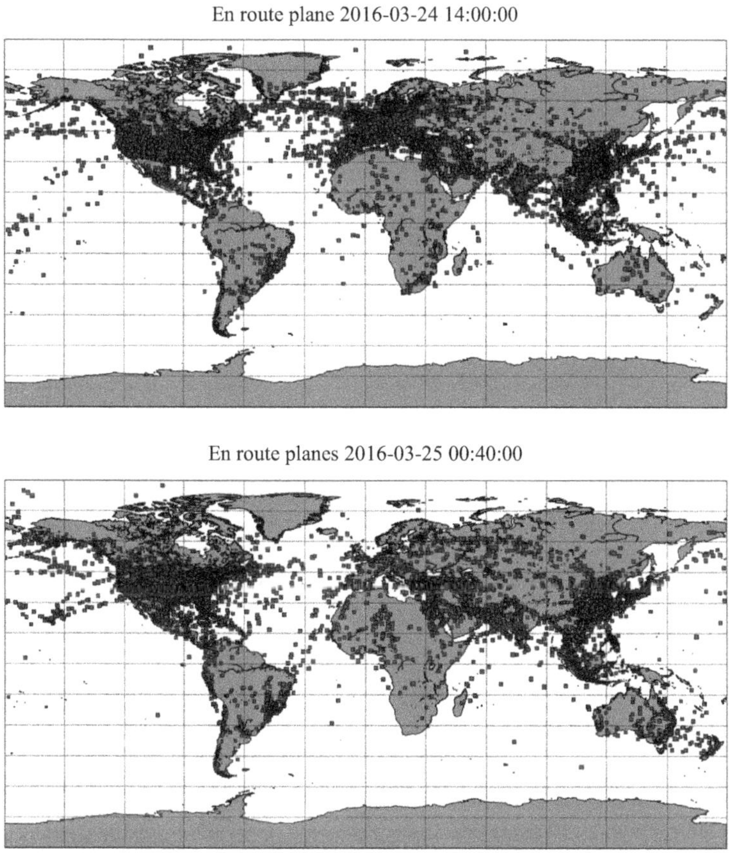

Figure 5.1 Worldwide distribution of aircrafts tracked by ADS-B at 14:00 UTC and 0:40 UTC. The position of the flights not tracked continuously has been interpolated taken from [6]

There are marked diurnal patterns in the traffic demand, with the largest fractions of flights landing or taking-off in the morning or evening local-time.

From the map of aircraft position, a passenger density map can be established by performing the average of the number of passengers present in every considered geographical area (removing flights shorter than 3,000 km). An example of such a map is illustrated in Figure 5.2.

Assuming a capacity demand per passenger of 0.5 Mbps and an aircraft load rate of 0.7 [7], it gives the possibility to establish a demand map as illustrated in Figure 5.3. This capacity map shows capacity demands higher than 1 Gbps per 100,000 km^2 in some areas and could thus almost saturate the beam of a high throughput satellite. Considering the increase of the capacity demand per passenger as well as the increase of air traffic and of the associated number of passengers, this figure could increase by one order of magnitude within some years [1]. Thus, a significant part of

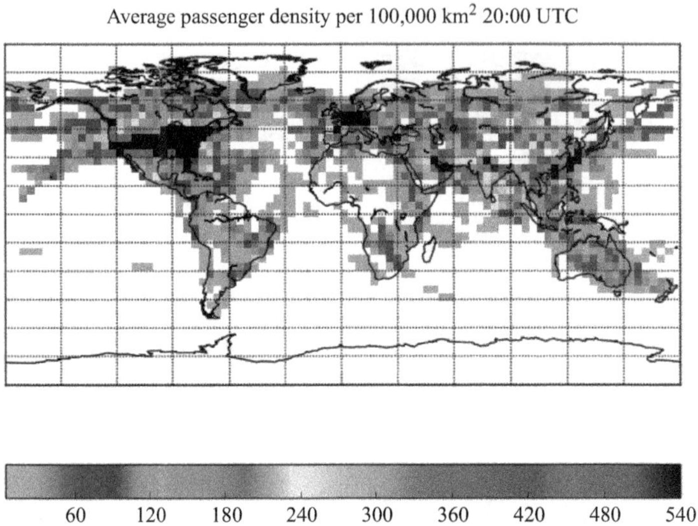

Figure 5.2 Passenger concentration at 20:00 UTC time

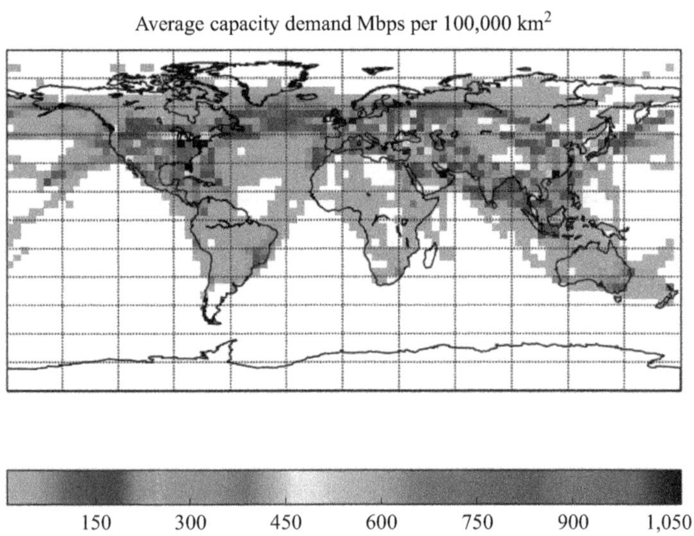

Figure 5.3 Traffic demand map computed from aircraft position at 20:00 UTC time

the bandwidth available at Ka band could be dedicated exclusively to the provision of data to aeronautical terminals.

The histogram of altitude considering all the aircrafts in one day once in flight is illustrated in Figure 5.4.

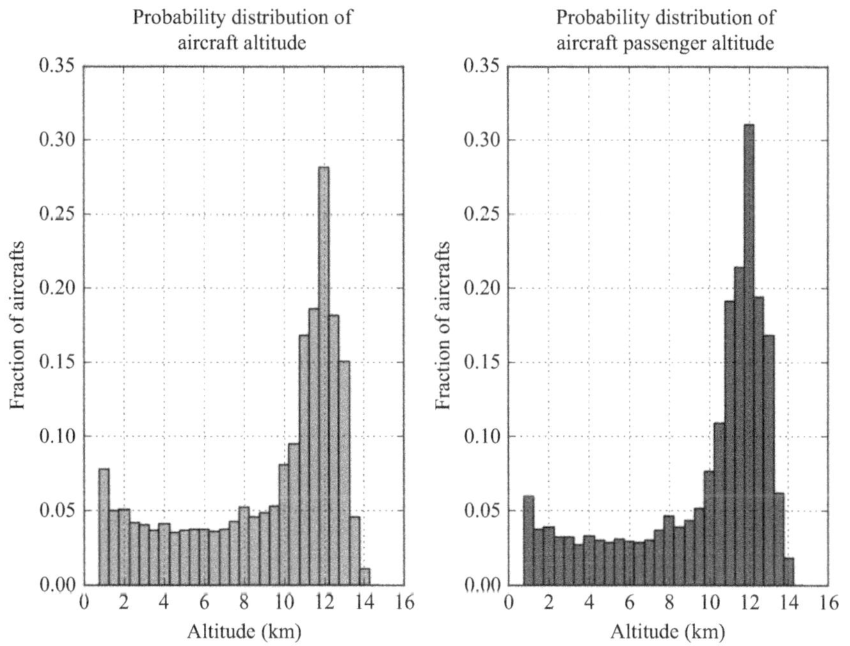

Figure 5.4 Aircraft altitude probability distribution function and passenger altitude

It can be noticed that most of the time aircrafts are above 5 km of altitude. This is even more true for the repartition of passengers' altitude as the lowest flight levels are usually reserved to small aircrafts with a low number of passengers. As what will be discussed more in detail in Section 5.3, the consequence will be that, considering this high altitude, most of the links for the majority of the passengers will undergo a very low impact of the atmosphere.

5.2 Regulatory environment

Unlike in Ka band, the ITU frequency allocations in Q/V and W bands do not provide for satellite exclusive bands. Thus, satellite systems operating in EHF will be in shared bands with fixed service (FS), broadcast services (BS) and mobile services and hence operate on an unprotected basis. Within Europe, the CEPT (European Conference for Postal and Telecommunications) has made some provision for the operation of satellites in Q/V bands as shown in Figures 5.5 and 5.6.

The most promising bands appear to be

- Downlink – 39.5–40.5 GHz.
- Uplink – 48.2–50.2 GHz.

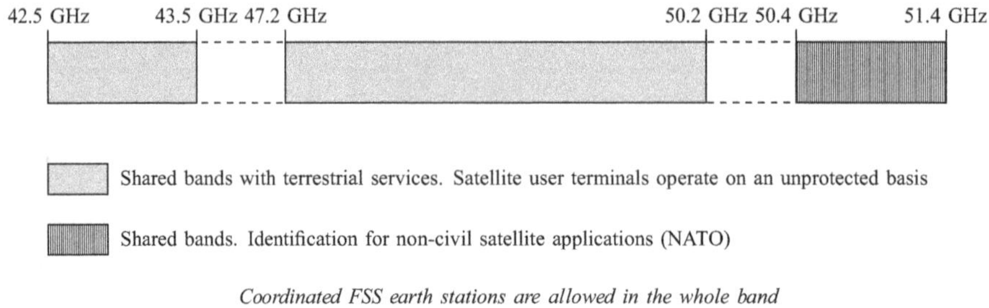

Figure 5.5 CEPT sharing of the Q/V band (uplink) from 42.5 to 51.4 GHz

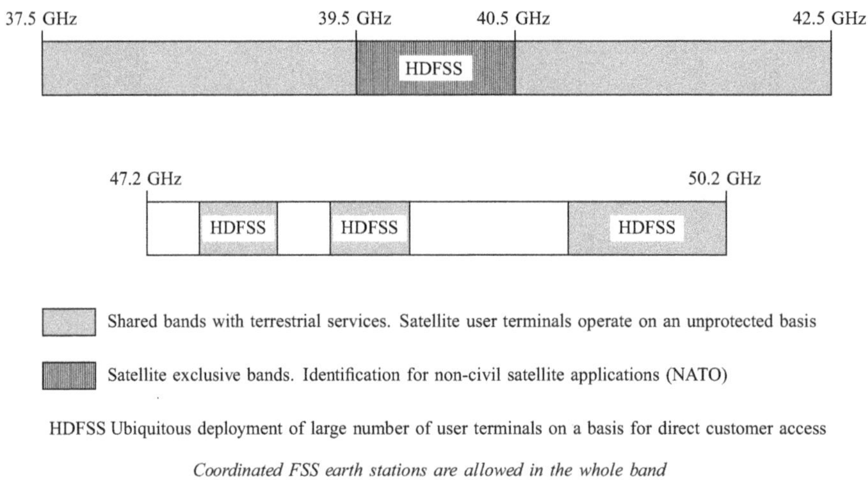

Figure 5.6 CEPT sharing of the Q/V band (downlink) from 37.5 to 42.5 GHz

However, co-existence with other services will still need further evaluation. In W band, the situation is less clear as in some countries (e.g. United Kingdom and France), there are large numbers of low power FS links but due to the improved directivity, they are not considered to be so significant. The most promising bands for satellites are:

- Downlink – 74–76 GHz
- Uplink – 84–86 GHz.

The situation within the millimetre bands is further complicated by current studies within World Administrative Radio Conference for spectrum allocation for 5G systems which will report at WRC 19 (World Radio-communication Conference). For Aeronautical satellite systems, implications would mainly be around airports for the take-off and landing phases.

Earth stations in motion (ESIM), erstwhile known as Earth Stations On Moving Platforms (ESOMPS), have been operated at C and Ku bands for sometime, and there are CEPT regulations associated with them. More recently, in 2013, there have been considerations for Ka band ESOMPS [8] operating to GEO satellites as these are just coming into existence. As yet, there are no regulations applying to higher frequencies; ESIM have been treated to date in a similar fashion to uncoordinated FSS earth stations. Thus, ESIM are merely treated as an application of FSS and thus cannot claim protection from FS or BSS. In the downlink Ka band, they operate on a non-protected basis as with the FSS (fixed satellite service). In the uplink Ka band, they operate in the protected parts of HDFSS (high density FSS) in Europe, but in other regions, they have to co-exist with FS. Thus, aeronautical ESIM when over land have to adhere to FSS regulations in the country that they overfly. Aeronautical ESIM are different because of the particular geometry concerned and the fuselage shielding. ECC has studied interference from aeronautical to FS and has adopted a power flux density (PFD) mask (see details in [9] but essentially 124.7 dBW/m^2 in a 14-MHz bandwidth with angular adjustments). Within the CoRaSat project [10], studies have also been conducted on the interference from FS to the aeronautical terminals in 17.7–19.7 GHz band and shown that dependence on altitude, interference above the recommended criterion for FSS can be exceeded and mitigation is needed.

As stated, no evaluations of ESIM in Q and W bands have been made to date, but it is assumed that the regulatory situation will continue to treat them as applications of FSS.

5.3 Propagation channel

5.3.1 Distribution of tropospheric margins

As mentioned in the introductory section, the tropospheric propagation losses for an aircraft to satellite link will be limited if the aircraft is flying at its cruise level, even considering EHF frequency bands. In fact, at cruise level around 11 km above mean sea level, most of the meteorological impairments are below the aircraft and do not alter the aircraft satellite path. The largest part of atmospheric gases is below the aircraft cruise altitude as are the precipitations and the majority of clouds. Even if the impact of the propagation channel is less significant than at ground level, the residual losses have to be assessed. In addition, the probability of outage during ascent and descent phases needs to be evaluated.

Towards achieving this aim, a specific model addressing the issue of obtaining the propagation losses CCDF (complementary cumulative distribution function) for an aircraft at a given altitude and geographical position has been standardized in ITU-R (International Telecommunication Union Radio-communication sector) Recommendation P.2041 [11]. It relies largely on the models used for the evaluation of the propagation margin for fixed earth-space links of ITU-R Recommendation P.618-12 [12]. The main difference is that there is guidance to account for the altitude of the aircraft in the computation of the margins. The adaptation of the models to compute rain attenuation, cloud attenuation, gaseous attenuation and scintillation is detailed in the following of this section.

5.3.1.1 Rain attenuation

The main change in order to obtain the rain attenuation CCDF is that the portion of the link that is affected by rain, is the portion of the slant path comprised between the aircraft and the rain height altitude and not between the ground and the rain height. Thus, if the aircraft is higher than the rain height, there is no attenuation. This approach should give moderately accurate results especially on the altitude above which there is no attenuation due to rain. In fact, the vertical structure of the precipitation is more complex. For stratiform precipitations, the rain height is effectively driven by the height of the 0 °C isotherm altitude, but the height of this isotherm may exhibit significant fluctuations throughout the year. For convective precipitation, liquid rain may exist well above the 0 °C isotherm under the shape of supercooled water uplifted by ascending air fluxes. Additional studies using radar data or numerical weather forecast model at high resolution would be needed to establish a more realistic dependence between the aircraft altitude and the rain attenuation distribution [13]. However, the trend given by the recommendation P.2041 [11] should be sufficient to get an order of magnitude of the propagation margins. The rain attenuation exceeded p% of the time for an Earth space link given by ITU-R Recommendation P.618-11 for a ground station located on the ground at a latitude ϕ_g, at an height h_g above mean sea level is denoted by $A_r^g = f_R^{-1}(p, R_{001}, h_r, h_g, f, \theta, \phi_g, \pi)$ for a link at frequency f, an elevation θ, a polarization π knowing the rain rate exceeded 0.01% of the time R_{001}, and a rain height h_r. Those two parameters can be derived from the position of the ground station from ITU-R Recommendation P.837-6 [14] and ITU-R Recommendation P.839 [15], respectively.

To obtain the attenuation for an aircraft link at an altitude h_a and a latitude ϕ_a, the Recommendation ITU-R Recommendation P.2041 simply advocates to get the rain attenuation exceeded p% of the time through

$$A_r^a = f_R^{-1}(p, R_{001}, h_r, h_a, f, \theta, \phi_a, \pi) \tag{5.1}$$

The use of this kind of statistical model for the description of the propagation effects on an aircraft satellite link is dubious as it implicitly assumes that the aircraft will encounter poor weather conditions with the same probability as a fixed receiver. However, aircrafts will tend to avoid hazardous meteorological events for obvious safety issues (especially thunderstorms). The latter are the highest source of impairments. Thus, the use of this kind of statistical approach will tend to be pessimistic for margin design as it is implicitly assumed that the position of the aircraft is independent of the weather.

5.3.1.2 Cloud attenuation

It is difficult to predict cloud attenuation from an airborne platform to space since different cloud types occur at different altitudes with different vertical extents and liquid water content. However, a conservative approach employed in ITU-R Recommendation P.2041 [11] is to assume that the cloud base is at the rain height specified in ITU-R Recommendation P.839 [15] and the cloud top is at 6 km. The cloud attenuation

A_c exceeded for a fraction of time p for an aircraft at an altitude h_a is computed as advocated in ITU-R Recommendation P.840 [16] by

$$A_c^a = f_c^{-1}(L(p, h_a), f, \theta) \tag{5.2}$$

where $L(p, h_a)$ is the columnar liquid water content above an altitude h_a. f and θ are the frequency and elevation of the link. $L(p, 0)$ can be computed from ITU-R Recommendation P.840 [16] and is the integrated liquid water content. The model of ITU-R Recommendation P.2041 [11] proposes to take $L(p, h_a)$ equal to $L(p, 0)$ between the ground level and the rain height. It is taken as null above 6 km and has a linear evolution in between. Using those assumptions, the columnar liquid water content above the altitude h_a can be expressed as

$$L(p, h_a) = \begin{cases} L(p, 0), \text{ if } h_a < h_R \\ L(p, 0)\frac{6-h_a}{6-h_r}, \text{ if } h_r \leq h_a < 6 \text{ km} \\ L(p, 0) = 0, \text{ if } h_a > 6 \text{ km} \end{cases} \tag{5.3}$$

Here also, the approach can be considered as rather conservative as liquid clouds may exist well below the rain height and that the decrease in the columnar liquid water content is taken only from the rain height. Traces of liquid water may still exist above the freezing height under the shape of supercooled water droplets that requires a freezing nucleus to reach the solid state. The attenuation caused by ice clouds is usually neglected due to the much lower value of the imaginary part of the dielectric constant. The use of cloud radar data [17] or of numerical weather forecast model outputs [13] could also allow the inclusion of a more realistic dependence between the cloud attenuation CCDF and the altitude of the aircraft, but here also, the accuracy should be sufficient to get a reasonable order of magnitude of the losses.

5.3.1.3 Gases

The gaseous attenuation for an Earth-space path can be predicted by ITU-R Recommendation P.676-12 [18], the contributions from oxygen and water vapour have to be considered. The computation of those atmospheric losses is based on the integration of the gaseous specific attenuation along a simplified (but area dependent) atmospheric profile. Considering the low variability of oxygen concentration, the oxygen attenuation can be approximated by a constant value depending only on the altitude. This dependence on the altitude is introduced by

$$A_o^a = A_o \exp\left(-\frac{h_a}{h_O}\right) \tag{5.4}$$

where A_O is the oxygen attenuation given by ITU-R Recommendation P.676-12 that depends on the mean average ground temperature, frequency of the link and on the elevation. h_O is characteristic scaling height of the oxygen that reflects the exponential decay of the atmospheric pressure with the altitude.

Water vapour is concentrated in the lowest layers of the troposphere, and it decays relatively rapidly with altitude. Recommendation ITU-R P.676-12 gives a mean to estimate the gaseous attenuation for a link at a frequency f and an elevation θ for integrated water vapour content V. ITU-R Recommendation P.836-5 [19] gives a model to obtain the integrated water vapour content V at an altitude h exceeded for a percentage of the time p, knowing the position of the receiver. Thus, the water vapour attenuation CCDF for an aircraft can be obtained through the combination of those two stages. Recommendation ITU-R P.836-5 [19] proposes a model of water vapour CCDF through

$$V(p,h) = f_V^{-1}(p, h_a, \phi_a, \psi_a) \tag{5.5}$$

where h_a, ϕ_a, ψ_a denote, respectively, the altitude, latitude and longitude of the aircraft, and f_V is the integrated water vapour content CCDF. The water vapour attenuation exceeded p% of the time can then be expressed as

$$A_{wv}(p) = f_{wv}^{-1}(V(p,h), f, \theta) \tag{5.6}$$

where the methodology to obtain the integrated water vapour content function of the altitude has been developed from the analysis of vertical profiles of water vapour content data and should hence be relatively accurate.

5.3.1.4 Scintillation

Scintillation is caused by refractive index fluctuations induced by atmospheric turbulence. The fluctuations of the refractive index are triggered by the fluctuations of water vapour and thus mostly located in the lowest layers of the troposphere. For links with terminals at ground level, the fading A_S^g due to tropospheric scintillation exceeded p% of the time is assessed using the guidance of Recommendation ITU-R Recommendation P.618-12:

$$A_S^g(p) = f_s^{-1}(p, f, \theta, N_{wet}) \tag{5.7}$$

where N_{wet} represents the median value of the wet term of surface refractive index that can be obtained using ITU-R Recommendation P.453 [20].

ITU-R Recommendation P.2041 proposes a methodology to account for the platform altitude when computing the scintillation losses $A_s^a(p,h)$:

- If the airborne platform is at an altitude below the rain height specified in Recommendation ITU-R P.839, tropospheric scintillation is calculated assuming that the airborne platform is located at the surface of the Earth.
- If the airborne platform is at an altitude above the rain height specified in Recommendation ITU-R P.839, tropospheric scintillation is ignored.

In this respect, $A_s^a(p,h)$ can be expressed as

$$A_s^a(p,h) = \begin{cases} A_S^g(p), \text{ if } h < h_R \\ 0, \text{ if } h \geq h_R \end{cases} \tag{5.8}$$

This approach is a bit asymptotic as the decay should in reality be smoother and that there may still be some scintillation above the rain height. However, its impact should be negligible on the overall margin design. To obtain the total attenuation exceeded p% of the time $A^a_{tot}(p,h)$, the advocated methodology in ITU-R Recommendation P.2041 is to combine the various components by using the following equation.

$$A^a_{tot}(p,h) = A^a_O(h) + A^a_{wv}(h,p) + \sqrt{(A^a_R(h,p) + A^a_C(h,p))^2 + A^a_S(h,p)^2} \tag{5.9}$$

The use of this method should be sufficient to provide an estimation of the attenuation undergone by an aircraft satellite link and of the underlying availability. There are, as discussed previously, several sources of inaccuracies, especially due to the rough description of the vertical structure of rain and clouds. When addressing specifically the question of the EHF bands, there are additional problems due to the lack of experimental data and of validation at those frequency bands [21]. Some physical assumptions (especially on scattering regimes for rain) are at their limit of validity and will induce an additional inaccuracy. However, this impact should be limited considering aircraft at high altitude for which there is almost no impact of rain. A further refinement can be to include the marked diurnal variability in the attenuation [22,23], considering for instance that short- and medium-range flights will occur mainly during daytime.

5.3.1.5 Examples of results using ITU-R Rec P.2041

The scaling in altitude of the results of ITU-R Recommendation P.2041 is illustrated in Figure 5.7 for a temperate and an equatorial location. The main difference between those two kinds of climate is linked to the average altitude of the 0° isotherm and consequently of the rain height. The losses remain significant in equatorial areas up to higher altitude than in temperate areas considering the higher rain height in those locations.

The distributions of tropospheric impairments at 50 and 80 GHz given by ITU-R Recommendation P.2041 for a link between an aircraft and a satellite, with an elevation of 35° are illustrated in Figure 5.8 for the same location as in Figure 5.7.

It can be seen from Figures 5.7 and 5.8 that the decrease of propagation margins with altitude is fast. In temperate areas, the propagation margin at 3 km for a W band link becomes less important than margins for a fixed receiver at Ka band. Above 6 km, there is only a slight residual gaseous attenuation. Thus, as what could be anticipated, the propagation effects are not a significant obstacle to establish links in EHF bands between aircraft at cruise altitude and satellites.

In order to associate an availability to an attenuation margin for a given flight path, the availability can be computed by integrating along the flight path, in the time interval $[t_1, t_2]$, the probability of outage at every position $P_{\psi(t),\phi(t),h(t)}(A_{tot} > A^*)$ weighted by the time spent at this position according to the following equation

$$P_{flight}(A_{tot} > A^*) = \frac{1}{t_2 - t_1} \int_{t_1}^{t_2} P_{\psi_a(t),\phi_a(t),h(t)}(A_{tot} > A^*)dt \tag{5.10}$$

In (5.10), $\phi_a(t)$, $\psi_a(t)$ and $h_a(t)$ are denoting, respectively, the latitude, the longitude and the altitude of the aircraft at time t. The probability to exceed a given total

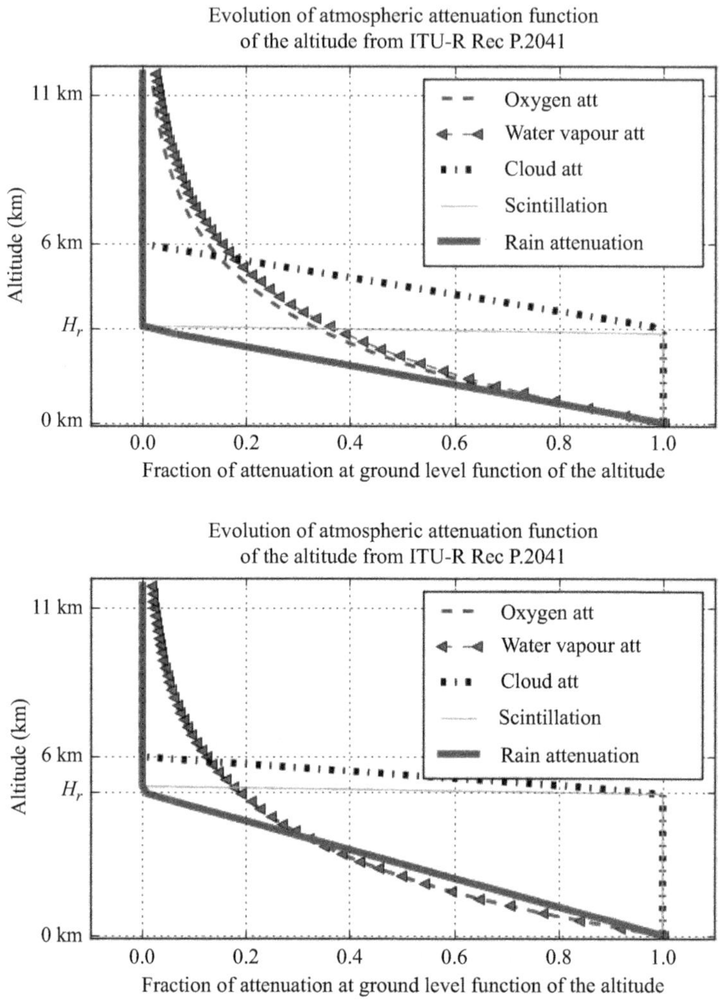

Figure 5.7 Scaling of atmospheric impairments with altitude for a temperate location (Toulouse) and an equatorial location (Kourou)

attenuation value $P_{\phi_a(t),\psi_a(t),h(t)}(A_{tot} > A^*)$ at a known position $(\phi_a(t), \psi_a(t), h(t))$ can be computed through the inversion of (5.9).

This margin determination holds only if the other parameters of the link budget as satellite G/T or EIRP (equivalent isotropically radiated power) are not experiencing significant fluctuations during the flight. Otherwise, those fluctuations need to be included in the margin calculation.

An evaluation of this flight outage margin for various flight scenarios is presented in Figure 5.9 for link frequencies of 40 and 70 GHz. For each flight, the satellite is assumed to be positioned at a longitude corresponding to the middle of the flight path.

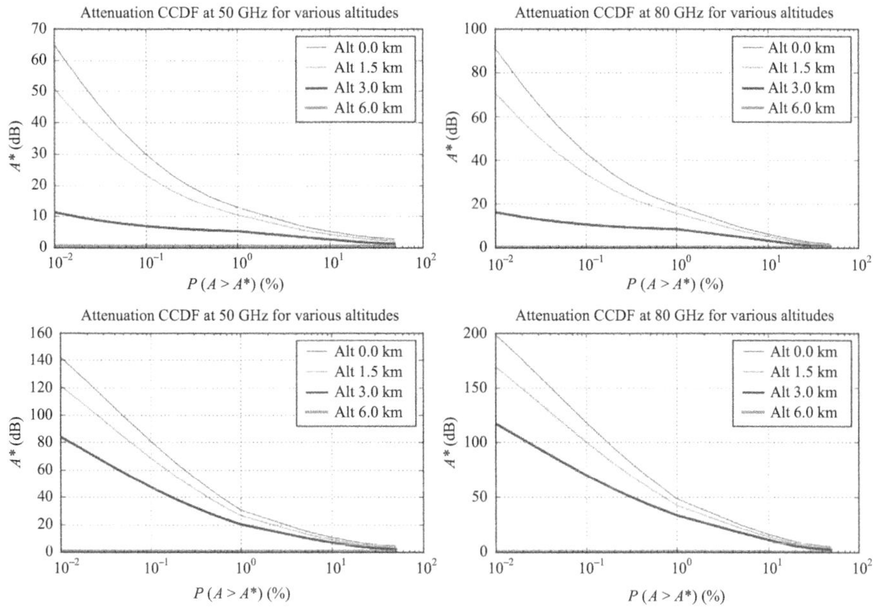

Figure 5.8 Attenuation CCDF for different altitudes at V and W band from ITU-R Recommendation P. 2041 for Toulouse, France (pictures on top) and Kourou, French Guiana (pictures on the bottom line). The elevation of the link is 35°

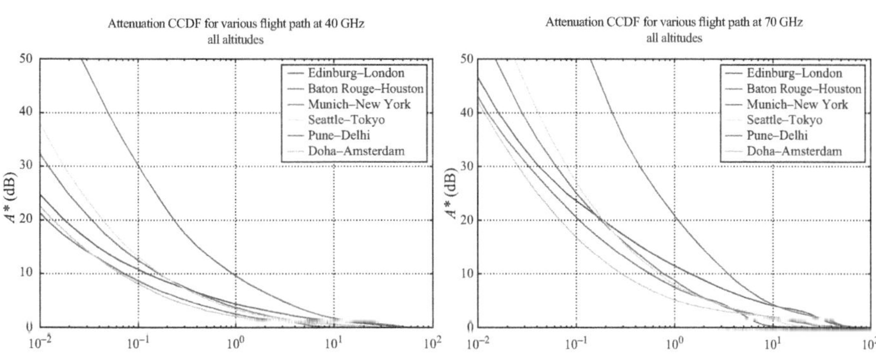

Figure 5.9 Attenuation CCDF for various flight paths at 40 and 70 GHz considering all the flight phases

As what can be noticed in Figure 5.9, an availability over 99% of the time can be offered at Q and W band for various flight trajectories even between tropical locations. Different trends can however be noticed:

- the shorter the flights, the lower the overall availability or the larger the required margin. For short flights, the fraction of the time spent by the aircraft at low

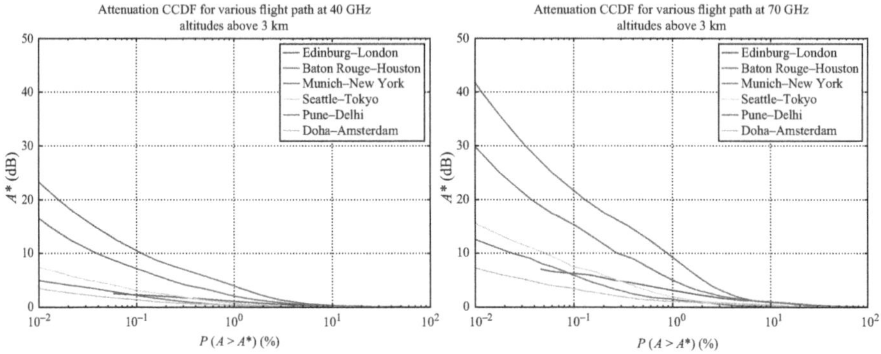

Figure 5.10 Attenuation CCDF for various flight paths at 40 and 70 GHz considering flight phases over 3 km of altitude

altitude in landing and taking off phases, during which propagation impairments are potentially significant, is larger than for long-haul flights.

- links for flights between unfavourable regions from propagation point of view (as tropical locations) would require a larger margin or would have a lower availability than links for flight between temperate locations.

It has also to be kept in mind that the results presented in Figure 5.9 are comprising all the flight phases (but not taxiing phases). The large attenuation margins to obtain availabilities larger than 99.9% are mainly needed for low altitude flight phases. However, it is likely that the communication system will not be operational during taking off and landing phase. To account for this absence of operation near the ground, (5.10) can be applied only to part of the trajectory during which the link is operational. Assuming a limit height of 3 km below which the system must be switched off yields to the result of Figure 5.10.

As shown in Figure 5.10, the trends are the same as when computing the availability for the whole flight path but the requested margins are much lower, with margins of 10 dB enabling an availability of almost 99.9 % whatever the flight path.

5.3.1.6 Flight path channel model

For further system analyses, time series representation of the temporal evolution of the channel is required. This is needed for instance to size optimally ACM control loops. In order to be able to generate time series of propagation impairments for an aircraft satellite link, the model to generate propagation time series for Earth Space links with terminal on the ground described in Rec ITU-R P.1853-1 [24] has been adapted to the aeronautical case with an approach similar to the one reported in [25].

- The first modification has been the change of the models to convert meteorological-related parameters into attenuation according to the methodology proposed by ITU-R Rec P.2041 [11].

- The second modification has been the change of the correlation parameters in order to account for the motion of the vehicle (in particular the rate of change of the attenuation can be larger in the case of an aircraft than in the case of a fixed terminal).

The inputs are flight trajectories defined by the longitude $\psi_a(t)$, the latitude $\phi_a(t)$ and the altitude $h_a(t)$ as well as link parameters such as frequency, satellite position and polarization. Time series from meteorological parameters can be constructed from such trajectories. The outputs are the attenuation time series indexed by time for the various propagation effects.

The adjustment of the correlation parameters for time series generation of ITU-R Recommendation P. 1853-1 [24] has been discussed in various previous works with different parametrization (see [26] for instance). The general idea is to assume that the temporal fluctuations of the channel are due to the advection (translation under the influence of the wind) of a spatially heterogeneous attenuation field. Thus, in the presence of a mobile receiver, the temporal fluctuations will be due to the combinations of the advection of the field and of the displacement of the mobile receiver in a spatially heterogeneous meteorological field. Considering the specific case of an aircraft, its speed is much larger than the advection speed of the meteorological fields (usually much less than 100 against 800 km/h). In this respect, the correlation of the time series can be assumed to be a time contracted replica of the one used for a fixed receiver. The contraction ratio is taken equal to V_0/V_a where V_0 is the average advection velocity that is around 50 km/h and V_a is the aircraft velocity.

The time series synthesis methodology in ITU-R Recommendation P.1853-1 relies on the conversion of a correlated Gaussian random process into a process distributed according to the distribution of the considered impairments. The various impairments are generated separately, but the random noise is correlated to introduce a dependence on the various effects. (Thus, there is cloud when it rains and the water vapour content tends to be higher in this case.)

To generate rain attenuation time series for the aircraft satellite configuration, a correlated Gaussian process $G_R^a(t)$ is generated. The correlation function used to generate the process is defined by:

$$c_{G_r}^a(t) = c_{G_r}^g\left(\frac{t\,V_0}{V_a}\right) = \exp\left(-\beta\frac{|t|V_0}{V_a}\right) \tag{5.11}$$

where β is a constant characterizing the autocorrelation function of the rain attenuation time series. The advocated value in Rec ITU-R P. 1853-1 [24] is $\beta = 2 \times 10^{-4}\ \mathrm{s}^{-1}$. The process can be generated through a first order linear filtering with a time varying coefficient. In the generation process, the filter coefficient β has to be replaced by a coefficient $\beta(t) = \beta(V_0/V_a(t))$. This time varying property requires the simulation of the process in the temporal domain unlike what is described in [24].

The rain attenuation process $A_R(t)$ is obtained by applying the following transformation:

$$A_R^a(t) = f_R^{-1}\left(\frac{1}{2}\,erfc\left(G_R^a\frac{t}{\sqrt{2}}\right), R_{001}, h_r, h_a, f, \theta, \phi_a, \pi\right) \tag{5.12}$$

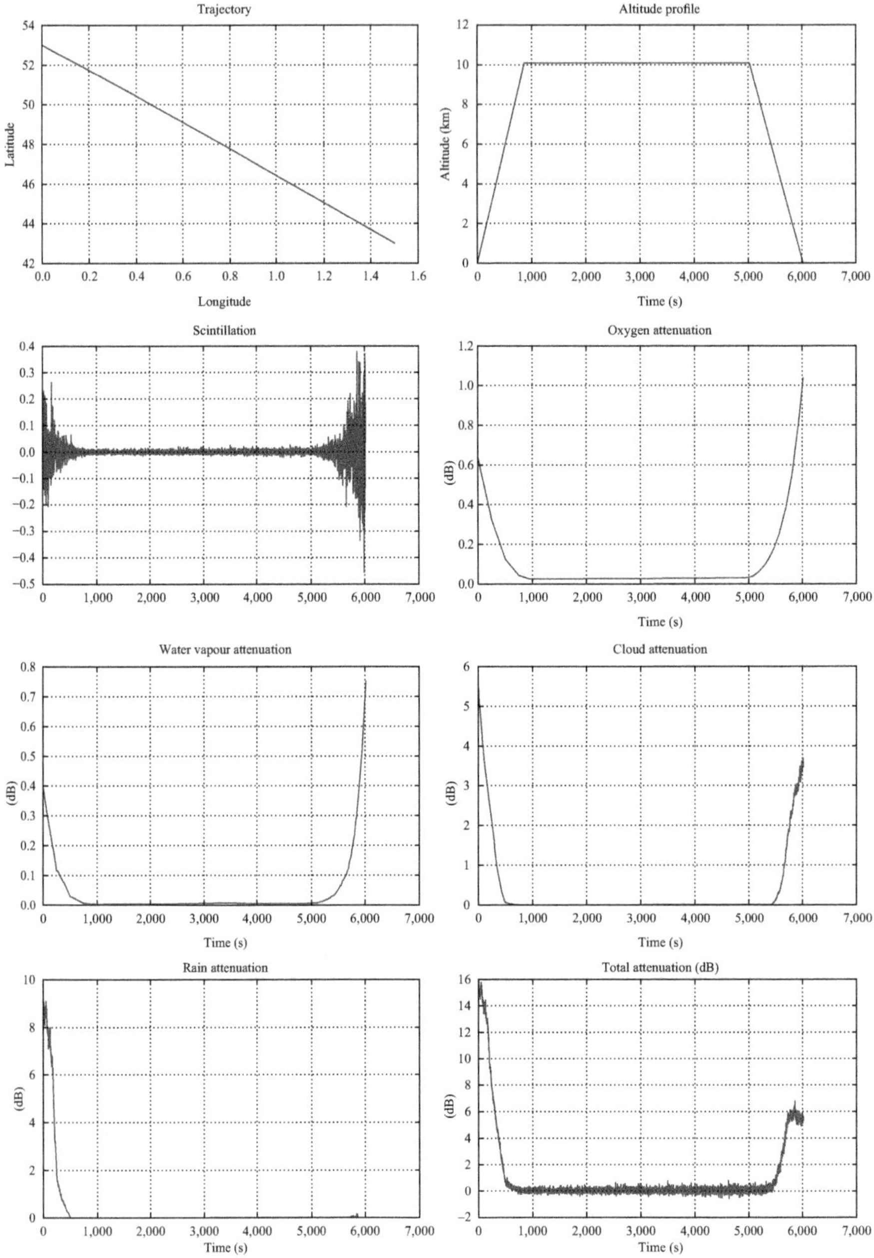

Figure 5.11 Example of time series generated for moderate rain conditions. The random draw is made accounting for the distribution of the impairments depicted previously

This converts a Gaussian distributed process into a process distributed according to the rain attenuation distribution. The elevation θ, rain rate exceeded 0.01% of the time R_{001}, h_r, h_a are taken as dependent on the aircraft trajectory. Similar methodologies are used to generate cloud and water vapour attenuation time series denoted as $A_C^a(t)$ and $A_{wv}^a(t)$, using the same contraction factor for the autocorrelation functions of the underlying Gaussian processes $c_{G_C}^a(t) = c_{G_C}^g(tV_0/V_a)$ and $c_{G_{WV}}^a(t) = c_{G_{WV}}^g(tV_0/V_a)$.

The transformation of the Gaussian processes into attenuation can be made through

$$A_c^a(t) = f_c^{-1}\left(L\left(\frac{1}{2}\,erfc\left(\frac{G_C^a(t)}{\sqrt{2}}\right), h_a\right), f, \theta\right) \tag{5.13}$$

$$A_{wv}^a(t) = f_{wv}^{-1}\left(V\left(\frac{1}{2}\,erfc\left(\frac{G_{wv}^a(t)}{\sqrt{2}}\right), h\right), f, \theta\right) \tag{5.14}$$

To correlate the processes, the random noises used to generate the random processes are correlated using the mechanism described in ITU-R Recommendation P.1853-1. For scintillation fading, the autocorrelation of the process is also contracted temporally by the velocity ratio. Considering the filtering methods advocated in ITU-R Recommendation P.1853-1, it amounts to the multiplication of the corner frequency of the filter by the ratio of the velocities. The scintillation time series is then computed from the underlying Gaussian process $G_S^a(t)$ as

$$A_S^g(t) = f_s^{-1}\left(\frac{1}{2}\,erfc\left(\frac{G_S^a(t)}{\sqrt{2}}\right), f, \theta, Nwet\right) \tag{5.15}$$

The oxygen attenuation margin is computed as a fixed value A_O, dependent only on the position of the aircraft. The total impairments time series is the sum of all these contributions.

For moderate-to-heavy rain, an example of generated time series at 70 GHz for a flight between Toulouse and London is illustrated in Figure 5.11.

On this example, the effect of the altitude on the propagation impairments can easily be noticed. The gaseous losses are quickly decreasing in the ascent phase. The rain attenuation disappears once the aircraft is above the rain height and the cloud attenuation once the plane is above 6 km. The fluctuations of the channel are fast due to the fast displacement of the aircraft with regards to the spatial correlation of the impairments.

5.4 System sizing

To get an estimation of data rate that could be possibly achieved with the usage of Q/V and W band on the user link, a comparative analysis of the data rate that could be achieved with regards to a Ka band system is performed. The focus is set on the forward link only as usually much more demanding for the provision of multimedia content in terms of data rate. In a first stage, plausible performances of terminals adapted for the aeronautical case are presented. Then, the service area and payload model are established. In the last stage, a comparative evaluation of projected performances is presented.

5.4.1 Aero terminals

5.4.1.1 Technological aspects

When considering satellite antennas on-board of aircrafts, the crucial aspect is to minimize the drag induced by the radome. In fact, aerodynamic disturbances brought by the radome are causing an increase of the drag and therefore an increase of fuel consumption. Far from being negligible, the overall fuel consumption can be increased by more than 0.20% [27,28]. Integrating this excess consumption in the overall cost of the SatCom service makes it extremely cost-intensive. To reduce the fuel consumption induced by the radome, its dimensions have to be limited. It has two different possible outcomes for the antenna design. Either the antenna has a small physical aperture and therefore a small gain, or the antenna is conformal and electronically steerable. Flat panels electronically steerable along one axis and mechanically steered along the other constitute an intermediate solution. This solution is currently available and is widely spread among currently used antennas for aeronautical terminals [29].

The use of a small gain antenna is detrimental to the overall spectral efficiency and thus impacts negatively the cost of the service and the achievable data rate.

Electronically steerable antennas are still extremely expensive, relatively heavy and power inefficient. It could however change quickly with the advent of metamaterial antennas or antennas with integrated numerical processing. Various companies are currently promising flat or conformal antennas at Ku and Ka band [30,31]. The thickness of the antennas can be around 1 cm, strongly reducing the fuel overconsumption. Frequency bands currently targeted by those technologies are Ku and Ka band, but a further extension of those devices towards Q/V or W band could be foreseen (using for instance technologies close to the ones used for Wi-Fi HD electronically steering antennas operating around 60 GHz using CMOS technologies). Integrated power amplifiers at W band are currently under development [32] and should be available within 2020.

5.4.1.2 Projected performances

In order to obtain realistic characteristics for the aeronautical terminals, current characteristics of Ka band terminals of global Xpress have been considered and extrapolated to Q/V and W band. In doing so, it has been assumed that the mechanical tracking antenna will be replaced by conformal arrays using meta-materials, digital beam forming or advanced MMIC technologies. Current manufacturer data (Kymeta, Phasor, Thinkom, etc.) have been used and extrapolated accounting for an additional degradation to take into account the lower performance of the RF components at higher frequency bands. For the LNA (low noise amplifier), noise figure data from currently existing LNA have been used across the bands. This resulted in the Q/V and W band terminal projections as given in Table 5.1.

Table 5.1 Aeronautical terminals, extrapolated performances for various frequency bands function of equivalent aperture diameter

	Gain dBi for downlinks and uplinks					
Diameter (m)	**Ka**		**Q/V**		**W**	
0.5	38.5	42.1	44.6	46.5	49.4	50.6
0.7	41.5	45.0	47.5	49.4	52.3	53.5
1	44.6	48.1	50.6	52.5	55.4	56.6

	LNA performances		
	Ka	**Q/V**	**W**
Noise factor (dB)	2.0	2.2	3.0
Noise temperature (dB K)	22.3	22.8	24.6

	G/T dB/K		
Diameter (m)	**Ka**	**Q/V**	**W**
0.5	16.2	21.7	24.8
0.7	19.2	24.8	27.7
1	22.3	27.8	30.8

	EIRP dBW		
Diameter (m)	**Ka**	**Q/V**	**W**
0.5	49.1	53.5	57.6
0.7	52.0	56.4	60.5
1	55.1	59.5	63.6

5.4.2 Satellite model

The satellite is modelled as an HTS multi-beam configuration with fourfold frequency reuse (considering also the use of orthogonarization) with TWTA (travelling wave tube amplifier) powers taken from manufacturer's data up to W band and a DVB-S2X air interface is assumed. It is proposed to carrry out a performance comparison on a per TWTA basis considering the available TWTA at each frequency band. For comparison purposes, it is also assumed that the coverage area for the different systems is identical.

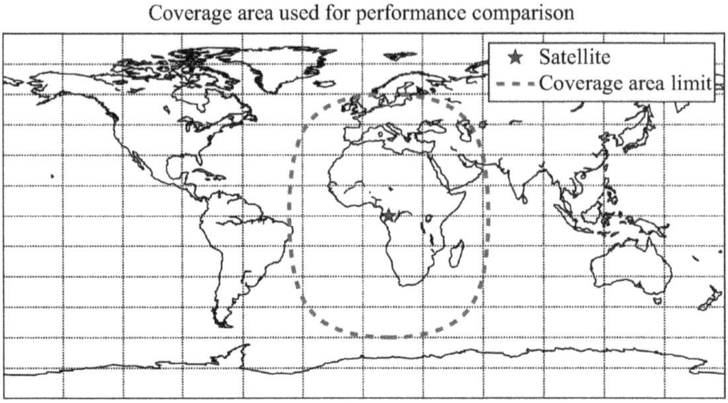

Figure 5.12 Considered coverage for the aeronautical service

Table 5.2 Beamwidth, beam surface and number of beams to fill the coverage area

Parameter/system	Ka	Q	W
Beamwidth (°)	1	0.57	0.304
Area per beam (km^2)	566,796	183,062	51,962
Number of beams	78	242	851

Figure 5.12 shows a potential coverage of a satellite for aeronautical services. The satellite is positioned at 13° of longitude east and its service area is down to 22° elevation angle. This was used as a common base for all the payloads.

For the antenna radiation pattern, the model described in [33,34] has been considered, which is further developed in [35]. This model with parameters taper roll off 1.6, edge of taper −10 dB and crossover points between three adjacent beams −4 dB has been adopted. The antenna gain to off-axis angle is denoted as θ, D is the antenna diameter, f is the frequency, n the taper roll off and ET represents the edge taper in dB. D, f, n, and ET are all given. Thus, the objective is to find θ that results to $g(\theta, D, f, n, ET) = crossover$. This is done iteratively.

The area per beam is calculated according to the methodology described in [36]. The beam's centre is considered at 48 °N 8 °E.

From [37], the number of beams N within a circle with diameter θ_c allowing 21% overlap between beams can be approximated by

$$N = 1.21 \left(\frac{1 - \cos \theta_c}{1 - \cos \theta} \right) \tag{5.16}$$

As what can be noticed in Table 5.2, the number of beams required to cover an area of equivalent size at Ka or W band change by one order of magnitude if the size of the antenna is left unchanged on the satellite. Therefore assuming that the number

Table 5.3 Main parameters used for system comparison

	Ka	Q	W
Available spectrum	19.7–20.2 GHz = 500 MHz	2.5 GHz around 40 GHz	5 GHz = 71–76 GHz
Number of colours		4 (2 Freq. 2 Pol.)	
Bandwidth per beam	250 MHz	1,250 MHz	2,500 MHz
TWTA sat power	50 W	50 W	40 W
TWTAs/beam	1 per beam	1 every 3 beams	1 every 10 beam
No. of spot beams	80	240	800
Total # of TWTAs	80	80	80
Waveform		DVB-S2x	
Cosine filter roll-off		5%	
Satellite antenna diameter	1 m	1 m	1 m
Terminal Aperture size	0.5 m	0.5 m	0.5 m
Terminal G/T	16.2 dB/K	21.7 dB/K	24.8 dB/K
Co-channel C/I	16 dB	18 dB	21 dB
Adjacent satellite C/I	12% increase of thermal noise	No interference	No interference

of amplifiers on the Ka, Q/V and W band payload can be kept identical, a Q/V band TWTA has to serve three times more beams than a Ka band TWTA and a W band TWTA has to serve 10 times more beams than a Ka band TWTA. This can be done either using FDM either using beam hopping [38] or a combination of both options.

The main parameters used for the benchmark are shown in Table 5.3.

A terminal with an aperture corresponding to a 0.5-m dish is considered on board of the aircraft for the different frequency bands. The interferences are assumed to be higher at Ka band than at Q/V band considering current spectrum occupation. Q/V and W band systems are assumed to make use of beam hopping with a switch throw count of 3 and 10, respectively (the output of the transponder is connected iteratively to a group of beams by a switch).

Using the terminal and satellite data from above, the performances at the three frequency bands are evaluated in Table 5.4 for one TWTA at each frequency band. The forward uplink is assumed to have a fixed identical C/N+I for the different frequency bands. It is assumed that it has not a significant impact on the overall link budget (considering that the gateways can have large antenna and that diversity can be used to counteract tropospheric fading). The link budget is established for an aircraft altitude of 10 km; therefore, the tropospheric attenuation is extremely limited as discussed in Section 5.3.

To account for the lower performances of the RF components at Q/V and W band, lower HPA power and larger IMUX/OMUX losses have been considered.

Table 5.4 Clear sky link budget – aircraft 10 km, different frequency bands

Parameters/system		Ka	Q/V	W
Satellite longitude	(°)	13	13	13
RX station				
Latitude	(°)	45	45	45
Longitude	(°)	8	8	8
Satellite elevation	(°)	36.54	36.54	36.54
Satellite link				
Occupied bandwidth	(MHz)	250	1,250 (5 × 250)	2,500 (10 × 250)
Roll off	0.2	0.2	0.2	
System implementation margin	(dB)	1	1	1
Downlink frequency	(GHz)	20	40	74
Uplink (SAS/LES to satellite)				
(C/N+I) uplink	(dB)	24	24	24
Downlink (satellite to station)				
Tx satellite power				
Sat antenna gain	(dBi)	44.6	50.6	55.9
Saturation power/HPA	(W)	50	50	40
Saturation power/HPA	(dBW)	17.0	17.0	16.0
Output back-off	(dB)	1	1	1
Total TX EIRP	(dBW)	60.5	66.6	70.9
Propagation losses				
Free space losses	(dB)	210.1	216.1	221.
Sat depoint loss	(dB)	1	1	1
IMUX/OMUX losses	(dB)	1	3	3
AERO terminal depoint	(dB)	0.1	0.1	0.1
Gaseous attenuation	(dB)	0.01	0.04	0.1
Total losses	(dB)	212.2	220.2	225.6
Rx parameters				
Rx station: G/T	(dB/K)	16.2	21.7	24.8
Downlink link budget				
(C/N_0) downlink	(dBHz)	93.0	96.4	98.5
(C/N) downlink + adjacent satellite C/I	(dB)	8.5	5.5	4.5
(C/I) downlink	(dB)	16	18	21
Total (C/N+I)	(dB)	7.7	5.2	4.4
Modulation		**8APSK 2/3**	**QPSK 3/4**	**QPSK 2/3**
Margin	(dB)	**1**	**1**	**1**
Data rate/HPA		**412 Mbps**	**1540 Mbps**	**2750 Mbps**

Note: The bold values are the final results resulting from the analysis of previous data.

The comparison made on a per HPA (high power amplifier) basis in Table 5.4 shows a clear improvement of the data rate achievable in Q/V and W band with, respectively, 4–10 times the data rate achievable at Ka band. This can be related to the assumptions made. In fact, the use of the same satellite antenna size for Ka, Q/V and W band with approximately the same TWTA power compensates the increase of the free space losses. The use of the same aperture size for the terminal lead to a more favourable situation at Q/V and W band in terms of C/N_0 than at Ka band, even taking into account the degradation of the performances of Q/V and W band RF components. Considering the large available modulation bandwidth at Q/V and W band, there is a possibility to exploit this favourable link budget to increase significantly the capacity per amplifier. It has nevertheless to be noticed that the comparison assumes the use of beam hopping at Q/V and W band and that it will significantly increase the complexity of the payload but will add a large level of flexibility to match the heterogeneity of the traffic illustrated previously in Figure 5.2. In addition the bandwidth per amplifier will likely be split into several carriers to accommodate for terminal modem speed. This will lead to additional losses in terms of output back-off to avoid a significant increase of the intermodulation (that have been neglected in Table 5.4). The optimal point of operation can however be determined knowing the number of carriers per HPA using for instance the methodology described in [39].

5.5 Conclusion

The potentialities of using EHF frequencies on a satellite for aeronautical broadband communication provision have been discussed in this chapter. Currently used Ka band frequencies will soon not be able to cope with the increased Internet demands from aircraft passengers. There do not appear to be any major regulatory barriers to adopting Q/V and W bands, except perhaps around airports. It has been shown that the propagation impairments in the troposphere that are preventing for now the use of those bands for satellite user links are not a major issue for aeronautical applications as the magnitude of those impairments is significantly decreasing with altitude. They are almost negligible at cruise level. The various tools available to size the propagation margins have been detailed. An outcome of the analysis is that the margins required to ensure more than 99.9% of availability could be lower than 10 dB for most of the flight configurations at Q/V and W band.

In order to get an idea of the improvement of the performances brought by the use of those higher frequency bands, current aeronautical terminals and satellites characteristics' have been extrapolated to EHF. It has been shown that the capacities provided can be enhanced by use of conformal antennas and provide from 4 to 10 times increases over current Ka band systems. These would appear to accommodate the predicted requirements of around 200 Mbps per aircraft made for 2020 and beyond. This demonstrates the feasibility of EHF satellite systems to meet future Aero passenger requirements, letting bandwidth for ground-based applications at lower frequency bands.

Acknowledgement

The authors strongly acknowledge European Space Agency for funding most of the research work presented in this chapter under the Satnex IV framework.

References

[1] Luecke O, Buechter KD, Moll F. Future broadband aeronautical communication–opportunities and challenges for SatCom. In: 21st Ka band and Broadband Communications Systems; 2015.

[2] Gogo ATG4 | Gogo Commercial Aviation; 2018. Available from: https://www.gogoair.com/commercial/atg4.

[3] European Aviation Network. 2018. Available from: https://www.europeanaviationnetwork.com/.

[4] Medina D, Hoffmann F, Rossetto F, *et al.* North Atlantic inflight internet connectivity via airborne mesh networking. In: Vehicular Technology Conference (VTC Fall), 2011 IEEE. IEEE; 2011. p. 1–5.

[5] Newton B, Aikat J, Jeffay K. Analysis of topology algorithms for commercial airborne networks. In: Network Protocols (ICNP), 2014 IEEE 22nd International Conference on. IEEE; 2014. p. 368–373.

[6] ADS-B Exchange World's Largest Co-op of Unfiltered Flight Data; 2018. Available from: https://www.adsbexchange.com/.

[7] IATA Annual Review. 2017. Available from: http://www.iata.org/publications/Documents/iata-annual-review-2017.pdf.

[8] ECC. Report 184: The Use of Earth Stations on Mobile Platforms Operating with GSO Satellite Networks in the Frequency Range 17.3–20.2 GHz and 27.5–30.0 GHz; 2013. Available from: http://www.erodocdb.dk/docs/doc98/official/pdf/ECCRep184.pdfasat1/11/2015.

[9] ECC. Decision(13)01: The Harmonised Use, Free Circulation and Exemption from Individual Licensing of Earth Station On Mobile Platforms (ESOMPs) within the Frequency Bands 17.3–20.2 GHz and 27.5–30.0 GHz; 2013. Available from: http://www.erodocdb.dk/does/doc98/official/pdf/ECCDec1301.pdf as at 1/11/2015.

[10] CoRaSat. The CoRaSat EU FP-7 Project; 2012. www.ict-corasat.eu.

[11] ITU-R P 2041-0. Prediction of Path Attenuation on Links Between an AirBorne Platform and Space and Between an Airborne Platform and the Surface of the Earth; 2013.

[12] ITU-R P 618-12. Propagation Data and Prediction Methods Required for the Design of Earth-Space Telecommunication Systems; 2015.

[13] Jeannin N, Outeiral M, Castanet L, *et al.* Atmospheric channel simulator for the simulation of propagation impairments for Ka band data downlink. In: The 8th European Conference on Antennas and Propagation (EuCAP 2014); 2014. p. 3357–3361.

[14] ITU-R P 837. Characteristics of Precipitation for Propagation Modelling; 2012.

[15] ITU-R P 839. Rain Height Model for Prediction Methods; 2013.

[16] ITU-R P 840. Attenuation Due to Clouds and Fog; 2013.

[17] Luini L, Capsoni C. Scaling cloud attenuation statistics with link elevation in earth-space applications. IEEE Transactions on Antennas and Propagation. 2016 March;64(3):1089–1095.

[18] ITU-R P 676. Attenuation by Atmospheric Gases; 2016.
[19] ITU-R P 836. Water Vapour: Surface Density and Total Columnar Content; 2013.
[20] ITU-R P 453. The Radio Refractive Index: Its Formula and Refractivity Data; 2013.
[21] Riva C, Capsoni C, Luini L, *et al.* The challenge of using the W band in satellite communication. International Journal of Satellite Communications and Networking. 2014;32(3):187–200. Available from: http://dx.doi.org/10.1002/sat.1050.
[22] Riva C. Seasonal and diurnal variations of total attenuation measured with the ITALSAT satellite at Spino d'Adda at 18.7, 39.6 and 49.5 GHz. International Journal of Satellite Communications and Networking. 2004 July;22(4):449–476. Available from: http://onlinelibrary.wiley.com/doi/10.1002/sat.784/abstract.
[23] Fiebig UC, Riva C. Impact of seasonal and diurnal variations on satellite system design in V band. IEEE Transactions on Antennas and Propagation. 2004 April;52(4):923–932.
[24] ITU-R P 1853. Tropospheric Attenuation Time Series Synthesis; 2012.
[25] Arapoglou PD, Liolis KP, Panagopoulos AD. Railway satellite channel at Ku band and above: Composite dynamic modeling for the design of fade mitigation techniques. International Journal of Satellite Communications and Networking. 2012;30(1):1–17. Available from: http://dx.doi.org/10.1002/sat.991.
[26] Graziani A, Vanhoenacker-Janvier D, Pereira C, *et al.* Synthetized tropospheric total attenuation time series for satellite-to-aeronautical link from L to Q band. In: 2016 10th European Conference on Antennas and Propagation (EuCAP); 2016. p. 1–4.
[27] Boeing Radome Solutions: Tri-band; 2017. Available from: http://www.boeing.com/resources/boeingdotcom/commercial/services/assets/brochure/boeing-radome-solutions.pdf.
[28] Gogo 2Ku Brochure. 2018. Available from: http://static1.squarespace.com/static/57b203af5016e15b4c5dfac1/t/57b20ec215d5db405ffc1f6d/1471286985901/Gogo+2KU.pdf.
[29] Vaccaro S, Diamond L, Runyon D, *et al.* Ka-band mobility terminals enabling new services. In: The 8th European Conference on Antennas and Propagation (EuCAP 2014); 2014. p. 2617–2618.
[30] Silvestri F, Benini A, Gandini E, *et al.* DragOnFly—Electronically steerable low drag aeronautical antenna. In: Antennas and Propagation (EUCAP), 2017 11th European Conference on. IEEE; 2017. p. 3423–3427.
[31] Sato H, Miyashita H. Heritage of Mitsubishi's phased array antennas development for mobile satellite communications. In: Antennas and Propagation (EUCAP), 2017 11th European Conference on. IEEE; 2017. p. 1521–1524.
[32] Zhao D, Reynaert P. An E-band power amplifier with broadband parallel-series power combiner in 40-nm CMOS. IEEE Transactions on Microwave Theory and Techniques. 2015;63(2):683–690.

[33] Sciambi AF. The effect of the aperture illumination on the circular aperture antenna pattern characteristics. RC Microwave Scanning Antennas. 1964;I(Academic Press):71.

[34] Strutzman W, Terada M. Design of offset-parabolic-reflector antennas for low cross-pol and low sidelobes. IEEE Antennas and Propagation Magazine. 1993;35(6):46–49.

[35] Kyrgiazos A, Evans B, Thompson P, *et al.* A terabit/second satellite system for European broadband access: a feasibility study. International Journal of Satellite Communications and Networking. 2014;32(2):63–92. Available from: http://dx.doi.org/10.1002/sat.1067.

[36] Salmasi AB, Rahmat-Samii Y. Beam area determination for multiple-beam satellite communication applications. IEEE Transactions on Aerospace and Electronic Systems. 1983 May;AES-19(3):405–412.

[37] Lutz E, Werner M, Jahn A. Satellite systems for personal and broadband communications. Springer; 2000. Available from: http://books.google.co.uk/books?id=LgBTAAAAMAAJ.

[38] Anzalchi J, Couchman A, Gabellini P, *et al.* Beam hopping in multi-beam broadband satellite systems: System simulation and performance comparison with non-hopped systems. In: Advanced Satellite Multimedia Systems Conference (ASMS) and the 11th Signal Processing for Space Communications Workshop (SPSC), 2010 5th. IEEE; 2010. p. 248–255.

[39] Aloisio M, Angeletti P, Casini E, *et al.* Accurate characterization of TWTA distortion in multicarrier operation by means of a correlation-based method. IEEE Transactions on Electron Devices. 2009 May;56(5):951–958.

Chapter 6

Next-generation non-geostationary satellite communication systems: link characterization and system perspective

Charilaos Kourogiorgas[1], Apostolos Z. Papafragkakis[2], Athanasios D. Panagopoulos[2], and Spiros Ventouras[1]

Non-geostationary (NGSO) satellites on a geocentric orbit include the low Earth orbit (LEO), medium Earth orbit (MEO) and highly elliptical orbit (HEO) satellites. These orbits are classified according to the altitude of the satellites above Earth. Apart from the HEO satellites, LEO and MEO satellites are orbiting constantly at a much lower altitude than that of Geostationary Earth Orbit (GEO) satellites. Therefore, the link losses are less, and the latency due to signal propagation is lower, thus making these orbits attractive for services which are tolerant to delays up to certain milliseconds, such as real-time data services. NGSO satellites have been already used in numerous applications, such as telecom applications (Globalstar and Iridium), positioning systems (Global Positioning System) and Earth Observation (EO) systems (Sentinel mission). The last few years, new satellite communication (SatCom) systems based on the NGSO satellites have started to operate, and more constellations are planned for the future. MEO satellites are used by O3b at Ka-band to deliver data services at equatorial areas [1]. The O3b constellation consists of, at the moment, 12 satellites, and more satellites are planned to be launched. Moreover, Laser Light Communications plan the use of MEO constellation with optical frequencies [2]. Considering the LEO satellites, new concepts have been emerged employing a great number of cross-linked LEO satellites, creating a mega-constellation such as the IRIDIUM NEXT, LEOSat, OneWeb and ORBCOMM systems. Depending on the provided services, for example trunking and last mile services, direct-to-home solutions or machine-to-machine communications, the frequency of operation varies from lower to very high frequency bands.

[1]RAL Space, Science and Technology Facilities Council, United Kingdom
[2]School of Electrical and Computer Engineering, National Technical University of Athens, Greece

6.1 Next-generation NGSO satellite systems

GEO satellites are orbiting at the equatorial plane at an altitude of 35,678 km with an almost zero-inclination angle. Although they provide large coverage, these satellites cannot cover the high-latitude regions. Moreover, the communication links between GEO satellites and ground stations are susceptible to high propagation losses and therefore, large antennas and higher emitted power are required. Moreover, due to the long propagation path, the propagation delay is high, thus making the GEO systems less attractive for delay intolerant services. Although lower orbit satellites may serve a smaller coverage area, they can provide communications to high-latitude regions. However, a constellation of satellites is needed for providing a global coverage.

Two of the first developed systems of lower than GEO satellite orbits for communication purposes were the first generation IRIDIUM and first generation GLOBALSTAR satellites for telephone services. These systems were using L-/S-bands for the communications between the satellite handhelds on ground and the satellites. The fleet of Globalstar was composed of 48 LEO satellites at an altitude of 1,400 km and this of IRIDIUM of 66 satellites at an altitude of about 770 km. For the case of IRIDIUM due to its high-inclination angle, the fleet could also provide coverage to polar areas. The coverage of an IRIDIUM satellite with 5° elevation contour plots is shown in Figure 6.1 [3].

These were among the first generation LEO constellations used for communication purposes. However, since demands for data volumes and data rates are increasing, the next-generation NGSO satellite communication systems have emerged using

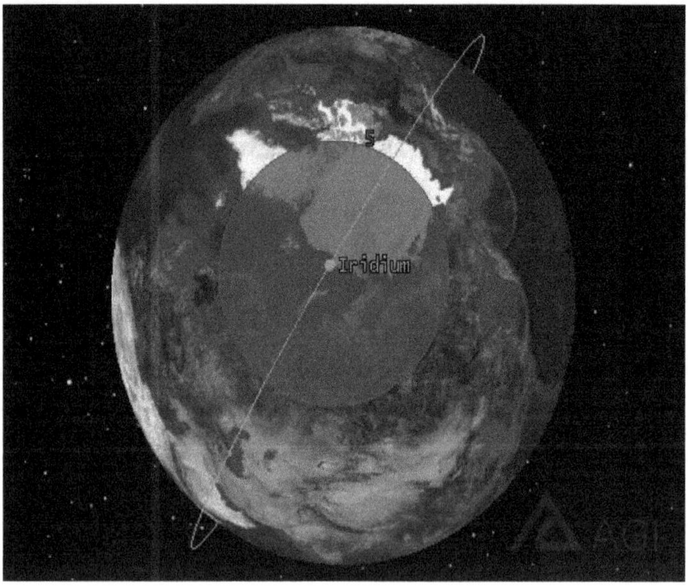

Figure 6.1 5° Contour plot of a single-IRIDIUM satellite above North Pole designed with [3]

L-/S- and Ka-bands in order to support the various multimedia and data applications and Internet services. The first systems have been already developed and or planned such as the O3b system [1], IRIDIUM NEXT [4], second generation Globalstar system, LEOSat [5] and OneWeb [6] among others. These systems were or are developed for providing data, voice and/or satellite trunking services (backhauling through satellite). Moreover, Boeing has already asked for licenses to operate a MEO constellation of more than 1,000 satellites operating at C- and V-bands.

In all the aforementioned satellite networks, multiple satellites are deployed in order to provide a global or quasi-global coverage. The constellations of a very large number (hundreds or thousands) of NGSO satellites are also called mega-constellations. LEOSat is planning to deploy 108 LEO satellites, while IRIDIUM in IRIDIUM NEXT will deploy 77 LEO satellites, OneWeb plans to use more than 600 LEO satellites and O3b already uses 12-MEO satellites and designs the next generation MEO fleet (O3b mPower) [1]. Depending on the inclination angle and satellite altitude, different regions can be covered. For example, O3b has launched 12-MEO satellites equally spaced with an inclination angle less than 0.1° (equatorial plane) at an altitude of 8,062 km. To have an example of the coverage regions, in Figure 6.2, the 12-MEO constellation and the 5° contour plots are shown.

Considering the applications targeted by the NGSO communications system, different frequencies can be used. For mobile applications, usually the L-/S-bands are used. However, the Ku- and/or Ka-band can be also employed especially for maritime and aeronautical applications, e.g. the Aeronautical Ku-band Mobile Satellite Systems of INTELSAT [7]. One of the already operating systems at Ka-band with NGSO satellites is using 1.3 GHz of spectrum for the downlink. Most of the future planned NGSO systems will use Ku- or Ka-band. However, in order to further increase the available bandwidth mostly for feeder links, Q/V- or higher bands can be used. From [8], as this is reported in [9], there is an available bandwidth at the moment of 5 GHz at

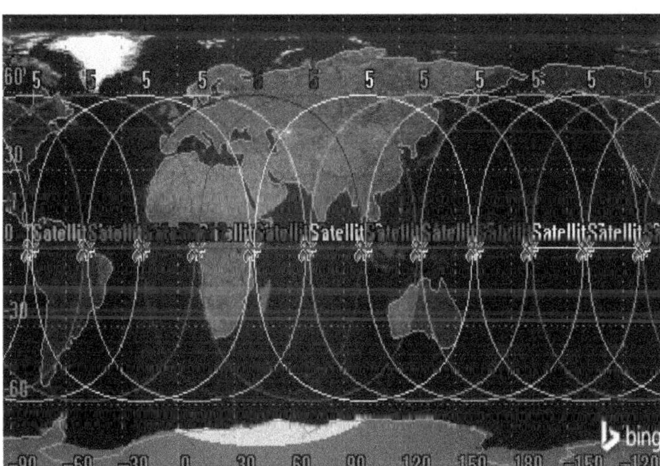

Figure 6.2 The 12-MEO satellite constellation along with the 5° contour plots

Q-band between 37.5 and 42.5 GHz. However, the spectrum is still under investigation due to its use from 5G mobile communication systems. In ITU preparatory studies for World Radio Communications Conference of 2019, item 1.6 refers to studies for the potential use of a part of Q-band spectrum for NGSO systems [10]. Moreover, the use of W-band has been investigated for the moment in the frame of GEO high-throughput satellite (HTS) systems [11]. Finally, the use of optical range can be beneficial for the feeder links considering that the frequency range is unlicensed, the bandwidth is large and the security is higher due to narrow beams [12].

Another technical advancement that may be employed in the next-generation NGSO systems is the use of more than one antennas at the user terminal (operating at Ka-band) in order to provide seamless connection, since NGSO satellites pass over the visibility area of the user and have a limited contact time. The two-antenna terminals have also been incorporated by O3b networks. From [13], two handover techniques have been identified for equatorial MEO constellations. In the make-before-break technique, seamless connection is guaranteed. It refers to the case in which a single receiver has two antennas. In every instance, there is the primary and secondary antenna, and it is considered that there are two satellites in the visibility area – one is setting and the other is rising. For the handover, the primary antenna is communicating with the setting satellite. Then the secondary antenna starts to communicate with the rising satellite. When the connection between the secondary antenna and the rising satellite is established, then the former primary antenna may stop communicating with the setting satellite and the former secondary antenna becomes the primary one for the next handover. However, whatever handover mechanism is used, the antennas must be equipped with a tracking mechanism in order to be able to follow the satellite as it passes through the station's visibility area.

Furthermore, apart from the use of multiple antennas on the ground segment, the next-generation NGSO satellites will be equipped on-board with multiple antennas, thus enabling multi-beam satellites and capacity improvement. For example, in O3b and LEOSat 10 Ka-band antennas are used for the communications to user terminals, while two for the communications to gateways (GWs).

Whenever large- or mega-constellations are employed, the communications between the NGSO satellites or the NGSO and GEO satellites could be realized through inter-satellite links (ISLs). ISLs may use RF or optical links for the realization of the communications. More particularly, IRIDIUM NEXT uses Ka-band for the transmission between satellites. In [14], the use of optical frequencies for ISL with beaconless tracking is studied. It is reported that a duplex data rate of 5.6 Gbps with a bit error rate of 10^{-9} is achieved with full communication entered at less than 20 s using a master–slave approach.

Moreover, in order to further decrease the communications latency, according to the LEOSat plans, on-board processing of the traffic is used. As this is explained in [5], the traffic may not pass through the GWs and travel directly through one or multiple satellites from one user to another, thus reducing further the communications latency.

Considering the spectrum availability and the need for high data rates for satellite trunking services, data applications and the 5G traffic offload, higher than Ka-band frequencies can be used, especially for the feeder links, i.e. links between the GWs and the satellite, thus leading to NGSO HTS systems. Moreover, due to the architecture of the NGSO systems, additional techniques can be applied for further increase of the throughput of the system. Such techniques which are analysed on the third section of this chapter include the use of adaptive coding and modulation (ACM) or variable coding and modulation (VCM). Since multiple antennas are used on ground and space segment, spatial diversity techniques can be used for further increase of availability, such as site diversity and/or orbital diversity. In order to increase the bandwidth of the systems, spectrum could be shared with terrestrial and GEO satellite networks. In any case, through ISL links, the NGSO and GEO satellite systems can cooperate to provide global services on ground.

6.2 Propagation characteristics and models

In this section, the propagation characteristics for links between ground stations and NGSO satellites will be presented. Depending on the frequency band used, the propagation phenomena that are considered in the design of the system are different, and therefore the design and modelling of channel conditions differ. In L-/S-bands, the phenomena that mainly affect the signal are the local environment effects which are close to the ground station. The local environment, i.e. buildings, road signs, cars, cause the reflection and diffraction of the signal. In Ka- and Q/V-bands, line-of-sight conditions prevail due to the use of directional antennas, and therefore the local environment effects are not the most severe, but the atmospheric phenomena are the dominant ones. The same holds for the optical links. However, the main effects on optical links and RF links are different.

One great difference between the communication of NGSO or GEO links and ground terminals is that in the former case, the satellite's position as seen by a ground observer changes with time, and therefore the elevation angle and the azimuth of the link vary with time. An example of elevation angle time series is shown in Figure 6.3 for an 8-MEO constellation and a ground station at Hawaii, considering that the ground station is always communicating with the satellite at the maximum elevation angle. Moreover, in Figure 6.4, the probability density function at each elevation angle is shown for three stations of this system located at Lima, Peru, Nemea, Greece and Hawaii, United States.

6.2.1 Local environment effects

At the L- and S-bands, the local environment at the ground station causes the attenuation of the signal. The local environment effects include the impacts from the natural and man-made objects such as buildings, trees, road signs and cars. The main effects are the reflections from the buildings, the diffraction at the environment's objects and the scattering causing slow and fast variations of the signal's amplitude and its spread on temporal domain. One of the main considerations in the land mobile satellite (LMS) channels is whether the channel is considered flat fading or frequency selective.

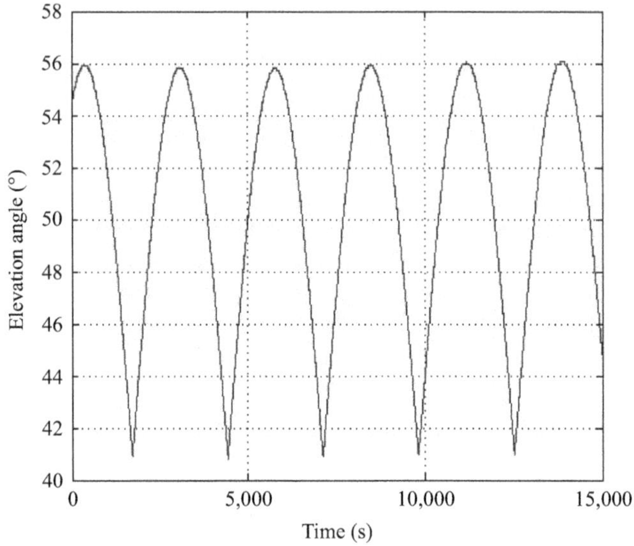

Figure 6.3 Elevation angle time series for a station located at Hawaii and communicating with an 8-MEO constellation

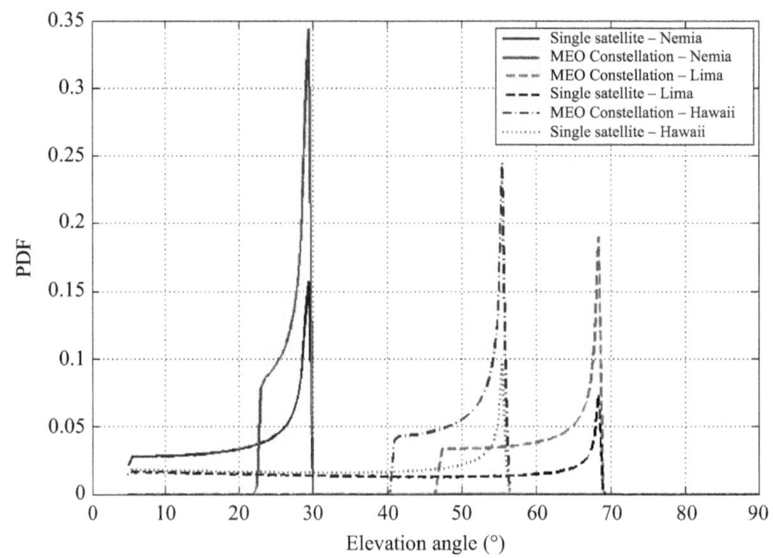

Figure 6.4 PDF of elevation angles for an 8-MEO constellation for three stations: (1) Lima, Peru, (2) Nemea, Greece and (3) Hawaii, United States

A variety of LMS channels have been identified and categorized as deterministic, physical statistical and statistical models [15]. The statistical models are based on the purely statistical characteristics of the channel derived from channel measurements and are based on the description of the received signal's envelope with a statistical distribution. Various single distributions have been identified, such as the lognormal, Rayleigh and Rice distributions and composite channel models have been developed, Loo distribution [16], Corazza-Vatalaro [17] among others. In [17], also the parameters of the distribution are given as a function of the elevation angle. A similar analysis was made in [18], where the Loo distribution, given in (6.1), was fitted to the 2×2 dual-polarization multiple-input–multiple-output (MIMO) channel components for various intervals of elevation angles.

$$f(r) = \frac{8.686r}{\sigma_L^2 \Sigma \sqrt{2\pi}} \int_0^{\infty} \frac{1}{a} \exp\left(-\frac{r^2 + a^2}{2\sigma_L^2}\right) \exp\left[-\frac{(20\log a - M)^2}{2\Sigma^2}\right] I_0\left(\frac{ra}{\sigma_L^2}\right) da \tag{6.1}$$

where Σ is the standard deviation in dB of the direct signal and reflects shadowing effects, and M is the mean value of the direct signal in dB and $MP = 10\log(2\sigma_L^2)$ in dB also. The function $I_0(.)$ is the zero-order-modified Bessel function of the first kind. The fitted values were derived through fitting to measurements and the experimental setup included an airship which was constantly moving, while the receiver was stationary. An example of the channel gain from the measured received signal at two orthogonal circular polarizations is shown in Figure 6.5. It was found that Loo distribution gave the best fit. Using the same measurements, the Inverse Gaussian distribution was tested for the modelling of shadowing effects in various intervals of elevation angles in [19].

For the system evaluation, time series of the received signal are required. In [20], a method is proposed for the first time for the generation of time series of received signal for LMS channels with NGSO satellites. Due to the high values of Doppler shift (several tens of kHz [20,21]) due to movement of both ground terminal and NGSO satellites, it is assumed that fading bandwidth is equal to maximum Doppler shift and filters are used for the incorporation of Doppler effects. In [21], a three-state Markov chain is proposed for the generation of time series of LMS channels. The three states represent the line-of-sight conditions, moderate shadowing and deep shadowing events. In every state, the distribution of the received envelope is described through Loo distribution.

As discussed in [21], one difference between the GEO and NGSO LMS channel is that the parameters of Loo distribution for a given state may change due to the change of elevation angle of the link. Therefore, the triggering of the Markov chain can be enabled either due to the movement of the mobile ground terminal or in case that elevation angle changes. For the separation of elevation angle intervals, a 10° step is used. As for the total Doppler spectrum, since a geometrical–statistical model is used through the positioning of scatterers, the total Doppler shift is divided into the one due to the movement of mobile terminal and the one due to the movement of the satellite.

The deterministic models are used in order to have an accurate description of the received power. Using detailed inputs such as the city maps, the electromagnetic

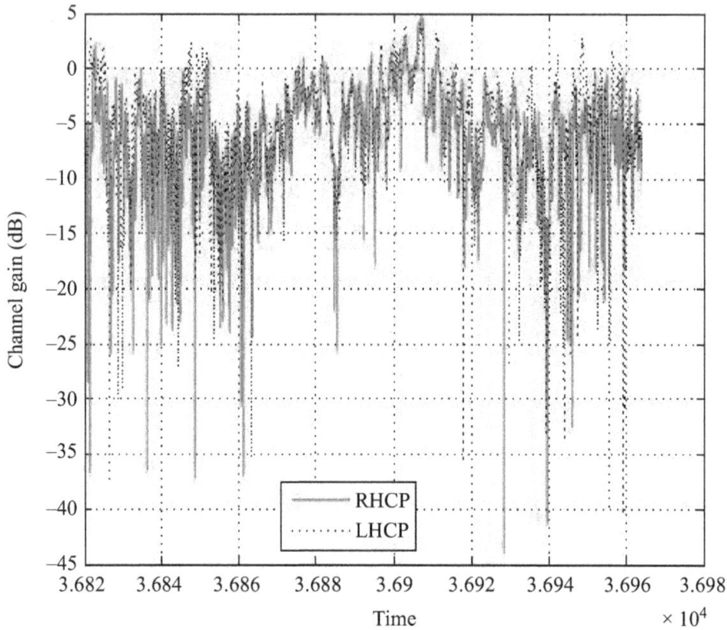

Figure 6.5 Time series of channel gain (with respect to free space losses) for an airship emulating the movement of a LEO satellite at L-band

parameters of the objects and buildings and theoretical electromagnetic equations (e.g. Maxwell's equations), the electromagnetic waves arriving from all possible directions to a single receiver in a given area, e.g. urban city, a park, are simulated. The most widely used technique is the ray tracing combined with physical optics [22,23]. In this latter technique, the electromagnetic fields of all the possible rays transmitted from the satellite and received on ground are calculated and then the received signal envelope is calculated. The received power can be calculated at any point of the area for a spatio-temporal resolution of interest as high as it is required. However, deterministic modelling of the channel requires a great computational power, in order to provide accurate simulations, especially for the NGSO satellites where geometry changes continuously.

In the physical–statistical models, such as [24], the objects in the local environment of the user are modelled as canonical shapes, such as boxes and cylinders, in order to analyse the propagation channel. However, various distributions are used as input in order to describe the height of the buildings or the distances between the buildings as random variables. The physical statistical models can then be used with electromagnetic theory in order to derive the received power of the signal.

6.2.2 Propagation characteristics through atmosphere

In next-generation NGSO links for maritime applications, ES on mobile platforms in general, and for fixed users, the use of Ka-band has been already adopted for delivering high data rate services. However, for the feeder links of NGSO HTS systems, apart from Ka-band the Q/V-, W-bands and optical frequencies are proposed as solutions.

In case that high RF, i.e. Ka- and Q-bands are used, or optical range, line-of-sight is always guaranteed. Moreover, due to the use of higher frequencies, the beam at the Ground terminal antenna is more narrow and, therefore, the local environment effects do not contribute to the losses. At these bands, the atmospheric phenomena greatly affect the signal, although different mechanisms are the main causes of signal attenuation and links quality degradation at high-RF and optical systems.

The great difference on atmospheric propagation between NGSO and GSO systems is that the position of NGSO satellites constantly changes in relation to an Earth station. The NGSO satellite movement above an observer on Earth affects the length of the link to the satellite, the elevation angle and the azimuth of the link. Therefore, for a given atmospheric phenomenon or weather front (clouds, rain, increased water vapour, turbulence) which is shaped and may also move towards a certain direction, the time that an NGSO link is affected by it is different from the time that a GSO link may be affected. Furthermore, the path length through atmosphere constantly changes due to variation of elevation angle and so there are no constant statistical characteristics of attenuation. Due to the change of azimuth, the path through the atmosphere would always be different to this for a GSO satellite even for the same elevation angle. Moreover, as the NGSO satellite systems may employ a great number of satellites handover of a ground station from one satellite to another is obligatory, and therefore for the simulation of the channel of an NGSO satellite system, spatial correlation must be always considered.

6.2.2.1 Propagation characteristics for RF systems at Ka-band and above

At Ka-, Q/V- and W-bands clouds, precipitation, atmospheric gases and turbulence affect severely the signal [25] with rain being the dominant fading mechanism.

Electromagnetic waves are scattered when the raindrops size is comparable to the size of the wavelength and therefore the extinction of the power is large [26]. Clouds consist of liquid water and ice particles. Although ice particles mostly depolarize the signal, liquid water particles attenuate the signal power. On the contrary of the raindrops, and especially for Ka- and Q/V-bands, the liquid water particles in the clouds are much less than the signal's wavelength [27–29]. Therefore, the extinction of signal power is much less than the attenuation due to rain. Moreover, atmospheric gases which highly affect the signal level are the water vapour and the oxygen. Two methodologies for calculating the attenuation due to gases are given in ITU-R. P. 676 [30]. As for turbulence, the variations of the refractive index due to wind shear cause the scintillation of the signal's amplitude.

The total attenuation due to the propagation through atmosphere for RF systems operating at Ka-band has been modelled for GEO links. ITU-R has recommended a methodology for calculating the exceedance probability for total attenuation for frequencies up to 55 GHz [31].

For the modelling of total attenuation first-order statistics, i.e. exceedance probability, a methodology is recommended by ITU-R. P. 618 [31] based on defining

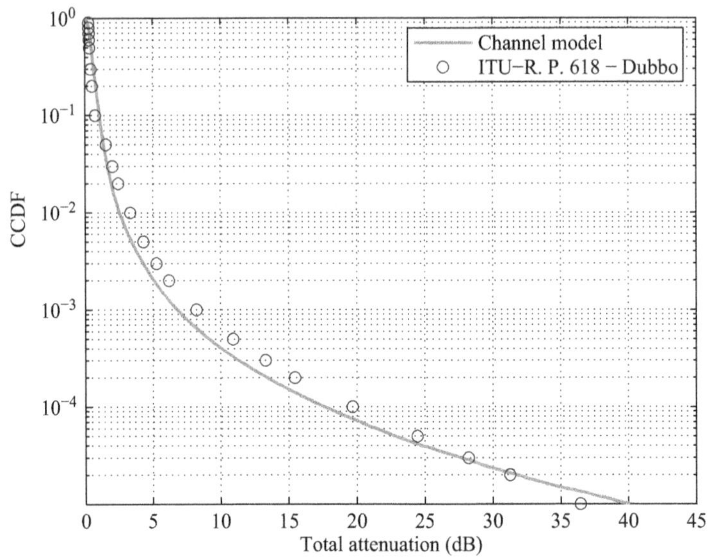

Figure 6.6 Exceedance probability of total atmospheric attenuation for an 8-MEO constellation and a ground station at Dubbo, Australia at 19 GHz

intervals of elevation angle of the link. In [32], a mathematical expression is given for the calculation of the atmospheric attenuation for NGSO links with varying elevation angle:

$$P(A_{tot} \geq A_{th}) = \int_{\theta_{min}}^{\theta_{max}} P(A_{tot} \geq A_{th}|\theta)P(\theta)d\theta \tag{6.2}$$

where $P(A_{tot} \geq A_{th})$ is the exceedance probability of total atmospheric attenuation for a NGSO link, $P(A_{tot} \geq A_{th}|\theta)$ the exceedance probability of total atmospheric attenuation at a given elevation angle and $P(\theta)$, the probability density function of elevation angles for a NGSO link. In Figure 6.6, the exceedance probability of total attenuation is shown for a ground station at Dubbo, Australia for an 8-MEO constellation at Ka-band using the ITU-R. P. 618-12 model and the model proposed in [9].

Apart from the use of (6.2) or ITU-R recommendation, a number of models have been developed for the generation of maps of the main meteorological metrics and quantities (such as rain rate or liquid water content) which can then be used for the calculation of atmospheric attenuation induced in GSO and NGSO links. One of these models is presented in [33] another model is proposed using numerical weather products from ERA-40 database of European Centre of Medium Range Weather Forecasting for the generation of spatial maps of rain rate, liquid water content in clouds and water vapour. At first, the rain effects are modelled through the use of MultiExcell model for having rain rate maps. Then the 3D cloud fields synthesizer (stochastic modelling of clouds – SMOC) is used for obtaining the 3D fields of cloud liquid water content [28] and the stochastic modelling of water vapour for the water vapour attenuation [34]. Since the physical quantities of rain rate, liquid water content and water vapour content are generated, the total attenuation is calculated through numerical integration and numerical expressions. The inter-correlation between rain and cloud fields is considered through the use of ERA-40 database with an additional pattern matching. The pattern matching algorithm is used in order to identify the highest correlation between the underlying 2D Gaussian field of the cloud fields with the rain rate generated by MultiExcell [33]. Another model which is based on the generation of space–time fields of the meteorological metrics on which the total attenuation is calculated is presented in [35]. In this model, the Weather Research Forecasting (WRF) algorithm is used for the downscaling of ERA-Interim data in order to obtain high resolution meteorological products. Then the propagation effects can be calculated.

For the evaluation of total attenuation in NGSO satellite systems using fade mitigation techniques, space–time synthesizers are needed in order to obtain time series of attenuation. A total attenuation synthesizer used for EO datalinks has been presented in [35]. Considering rain attenuation, in [36], the rain fade slope is investigated for LEO-to-Ground links, while the rain cells are modelled through EXCELL model [37]. Then, in [38], the fade slope is also investigated for modelling the rain cells using HYCELL model [39]. Rain cell models strive to capture the rainfall inhomogeneity through the modelling of rainfall rate in a single rain cell (using probability distribution of rain cell diameter) and the aggregation of rain cells in an area. Attenuation due to rain can then be calculated through the numerical integration of rain rate on the slant path.

Radar data have been analysed for NGSO satellite links for evaluating the rain attenuation induced in links between mobile terminals and MEO satellites [40]. Moreover, the synthetic storm technique (SST) [41] has also been used for generating rain attenuation time series for links with time varying elevation angles [42]. The SST makes use of Taylor hypothesis and a storm speed in order to convert rain rate time series measured by rain gauges to rain attenuation time series. In [42], the two-layer model is used for the calculation of rain attenuation from rain rate. According to [42], the elevation angle is sampled and kept constant or a given period of time. The sampling time for Lagrangian L1 orbit as found in [42] could be chosen equal to 6 min since in that interval of time the elevation angle changes slightly. Then for this interval, the rain rate time series are transformed to rain attenuation time series considering that the elevation angle of the link is constant and equal to the elevation angle observed at the first sample of the interval.

In [43], a model is presented using stochastic differential equations (SDEs) with time-variant parameters for generating rain attenuation time series. The same model has also been used in a single-MEO satellite scenario [44]. In Figure 6.7, a snapshot of rain attenuation time series for a Ka-band link between an IRIDIUM satellite and a ground station located in Athens is shown. Regarding the evaluation of scintillation, a model based on Kalman filters has been proposed in [45] for generating time series of amplitude scintillation, while in [46], a model based on WRF and a linear time varying filter for shaping the spectrum of scintillation.

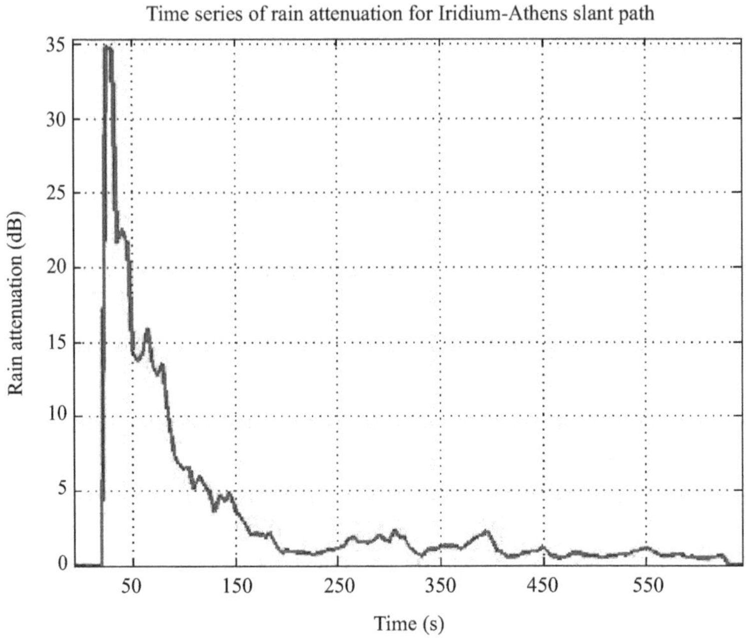

Figure 6.7 Snapshot of time series of rain attenuation at 20 GHz between Athens and IRIDIUM satellite

In [9], a time series synthesizer based on multi-dimensional SDEs is proposed for the generation of time series of total attenuation for communication to NGSO constellations. The synthesizer is developed in order to capture the temporal and spatial characteristics of the attenuation. In [47], an expression for the calculation of total atmospheric attenuation at a given time instance is given as a function of the attenuation components for the same time instance:

$$A_{atm}(t) = A_{rain}(t) + A_{cl}(t) + A_{wv}(t) + A_{oxygen}(t) + S(t) \tag{6.3}$$

where A_{rain}, A_{cl}, A_{wv}, A_{oxygen} and S are the rain attenuation, cloud attenuation, attenuation due to water vapour, oxygen attenuation and scintillation, respectively, all in decibels.

Since the above expression refers to a certain time instance, the same expression can be also used for the calculation of total atmospheric attenuation as a function of the attenuation factors for NGSO links. For rain attenuation, the methodology presented in [43] had to be extended to multi-dimensional SDEs ([48–50]) in order to consider the spatial correlation of rain attenuation between the links of ground station to two or three different MEO satellites. Since rain attenuation can be assumed that follows a lognormal distribution at a given elevation angle, rain attenuation is linked to an underlying Gaussian process. The general expression of the multi-dimensional SDE for the underlying Gaussian process is

$$\mathbf{U}_t = e^{\int_0^t \mathbf{B}_y dy}\mathbf{U}_0 + e^{\int_0^t \mathbf{B}_y dy}\int_0^t \exp\left(-\int_0^y \mathbf{B}_{y'}dy'\right)\mathbf{S}_y d\mathbf{W}_y \tag{6.4}$$

where $\mathbf{B}_t$ is an $n \times n$ matrix with elements $b_{ij,t} = -\beta_{i,t}\delta_{ij}$ with δ_{ij}, the Kronecker delta function and $-\beta_{i,t}$, the dynamic parameter of rain attenuation as defined in [43]. The matrix $\mathbf{B}$ is time dependent since, as shown in [43], its elements depend on the elevation angle and therefore for MEO slant paths, the dynamic parameter is time dependent. The main assumptions of the rain attenuation model is that rain attenuation follows lognormal distribution and has an exponential decaying autocorrelation function as proposed in [51] for a given elevation angle. The spatial correlation for converging links as a function of the separation angle is calculated through [52]. In Figure 6.8, the block diagram for generating rain attenuation time series induced in multiple NGSO satellite links is shown.

For cloud attenuation, the methodology recommended in ITU-R. P. 1853-1 [47] is extended for multiple and spatially separated links. The synthesizer for attenuation due to clouds is based on generating time series of integrated liquid water content (ILWC) on a point. Then, using the recommendation of ITU-R. P. 840 [27], the time series of attenuation due to clouds are generated from the time series of ILWC depending on the elevation angle. In order to extend the above methodology, to multiple links, multi-dimensional correlated Gaussian noise is used. To include the spatial correlation, the formula proposed in SMOC model [28] as a function of separation distance (d) is used:

$$\rho_C(d) = 0.35\,e^{-(d/7.8)} + 0.65\,e^{-(d/225.3)} \tag{6.5}$$

Therefore, in the case of converging links, the separation distance is set equal to the distance between the two converging links at the low cloud base, which is set equal to 1 km a.m.s.l. (above mean sea level) [28]. So, the correlation matrix that is created is time dependent.

For the attenuation due to atmospheric gases, the oxygen attenuation is considered constant as also proposed in the ITU-R. P. 1853 for a given elevation angle and the same oxygen attenuation value is considered for the spatially separated links, due to the very high spatial correlation that oxygen distribution exhibits [53]. For generating attenuation due to water vapour on multiple links, the methodology presented in [47] is extended for multiple links. First, the time series of integrated water vapour content (IWVC) are generated for a single point, and then using ITU-R. P. 676 [30], the attenuation due to water vapour time series induced in a single link are calculated. The correlation coefficient is derived from [53] and depends on the separation distance between the links. For the case of site diversity, this distance is equal to the distance between the ground stations, while for orbital diversity the separation distance is the distance between the links at a height of 1-km a.m.s.l.

For the generation of amplitude scintillation time series, the methodology presented in [54] is modified for time-dependent parameters. In the latter model, SDEs

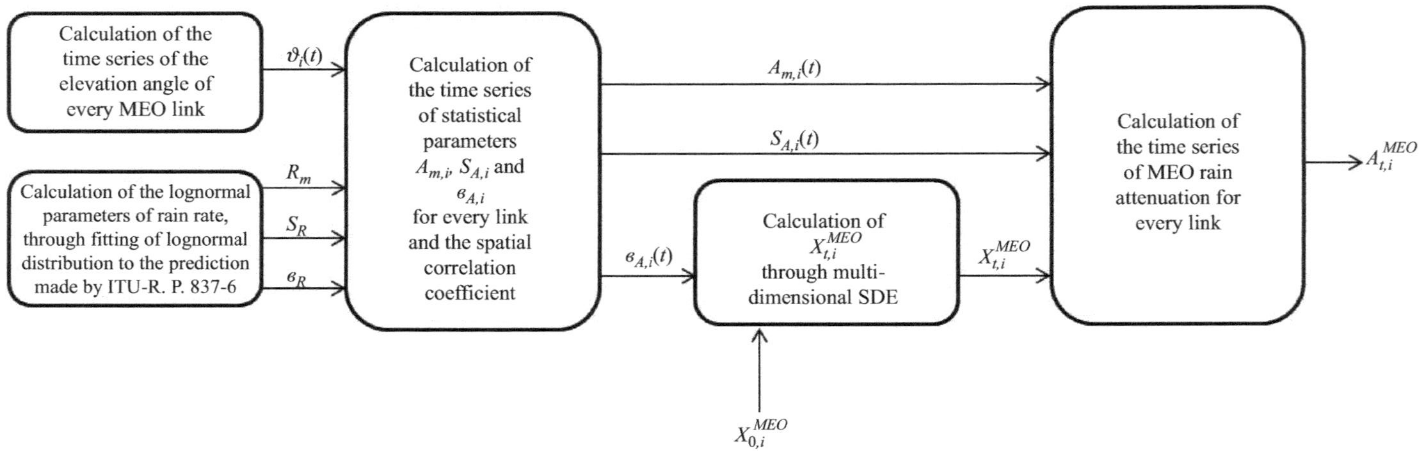

Figure 6.8 Block diagram of rain attenuation time series generation for multiple NGSO links

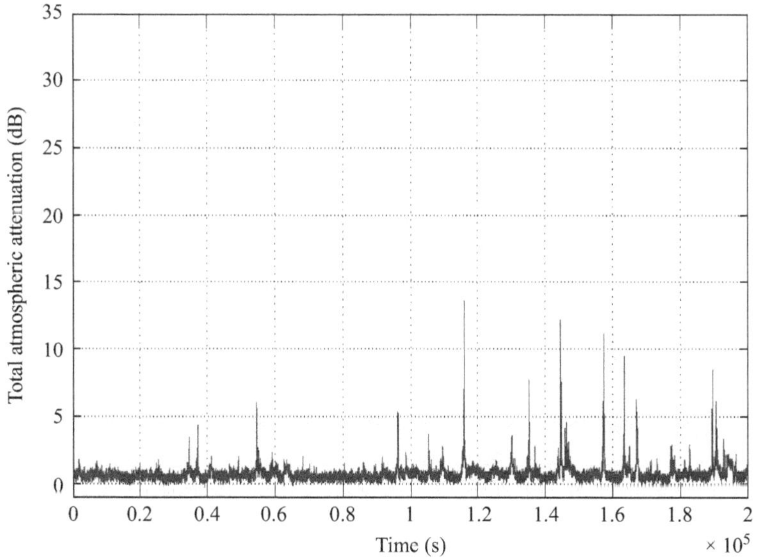

Figure 6.9 Snapshot of time series of total atmospheric attenuation at 20 GHz between Sintra and an 8-MEO constellation

driven by fractional Brownian motion are used in order to generate time series following a Gaussian distribution with a lowpass power spectrum with a slope of −80/3 dB/decade.

In ITU-R. P. 1853-1 [47], a correlation between the Gaussian processes which are used for the generation of attenuation is enforced in order to obtain the interdependency of the various factors. The same white Gaussian noise is used to synthesize both rain attenuation and cloud attenuation, while the correlation between the white Gaussian noise of rain and IWVC is 0.8. For the correlation of scintillation with rain, the variance is calculated according to

$$\sigma_{scint} = \begin{cases} \sigma_{sc}, & A_{rain} < 1\ \text{dB} \\ CA_{rain}^{5/12}, & A_{rain} \geq 1\ \text{dB} \end{cases} \tag{6.6}$$

where σ_{sc} is derived from ITU-R. P. 618 [31] and C is set equal to 0.039 and 0.056 for Ka- and Q-band links [55], respectively.

Using the above methodology for an NGSO constellation, the time series of total atmospheric attenuation can be generated. An example is given in Figure 6.9 for an 8-MEO constellation and a ground station in Sintra, Portugal.

6.2.2.2 Propagation characteristics for optical NGSO systems

For optical systems, the main effects are the clouds which include either liquid water particles or ice water particles, the aerosols in the atmosphere which cause the extinction of signal, the molecules whose resonant frequency is close to the optical frequency used and turbulence [56].

Table 6.1 Number of ground stations considered for each region and the derived CFLOS probability

Region	Number of stations	CFLOS probability (%)
Nemea	8	99.906
Karachi	5	99.94
Vernon	8	99.94
Lima	4	99.93

More particularly, clouds can cause hundreds of dB of attenuation with their presence along the slant path. This occurs due to the comparable or even higher water particles size compared to the signal's wavelength [29]. Therefore, for the description of clouds effects, the probability of cloud-free line of sight (CFLOS) is used. CFLOS is the probability that no clouds are present along the slant path of the link. The only way to combat the clouds presence is through the use of multiple spatially separated sites, i.e. site diversity. Models for the prediction of CFLOS have been proposed through the use of database [57], through analytical expressions [58] or through the spatial modelling of cloud fields [29]. In [9], a study on the use of multiple ground stations for optical feeder links is presented using the model in [29] and few results are reproduced from [9] in Table 6.1. In the table, the number of stations required to reach the CFLOS probability shown in the third column is given for the various regions. The geometry chosen is either a circle or a rectangular with a radius or side, respectively, of length higher than 300 km. For Nemea, Greece due to the country's topology, the cities chosen are Nemea, Mytillini, City of Rhodes, Kalamata, Heraklion, City of Corfu, Volos, Thessaloniki.

Apart from the clouds with liquid water particles, there is another type of cloud named cirrus clouds which are transparent and so it does not cause the total blockage of the link but it causes signal losses [59].

Another significant effect for optical links is the turbulence which has different effects on downlink and uplink. The difference between downlink and uplink effects is that in the former case, the EM waves enter the turbulent layer after traveling thousands of kilometres, while in the second case the optical waves enter the turbulent layer at the same instance the beam is transmitted. In downlink, depending also on the elevation angle the scintillation index can be approximated by this of infinite plane wave and the off axis scintillations are constant. Moreover, the correlation width of turbulence is small and therefore the aperture on ground averages the amplitude scintillation. In the uplink, the fluctuations are stronger and beam wander also exists. Moreover, the satellite is seen as a point receiver, since the spatial correlation of turbulence is much higher [60].

In [61], a study on the use of MEO satellites for very HTS systems is presented. The turbulence effects are modelled for the uplink studied scenario. Turbulence causes the scintillation of signal amplitude, the wander of the maximum of the Gaussian beam and the beam spreading [60]. For the compensation of turbulence effects on optical links, pre-compensation techniques can be used. In pre-compensation techniques, the downlink signal is used for the compensation of turbulence on the uplink [62]. However, in order that such techniques be used, the isoplanatic angle, i.e. the angle in which turbulence remains constant (or highly correlated), must be higher or equal to the point ahead angle. Therefore, it is more difficult pre-compensation techniques be used for the full compensation of all the atmospheric turbulent effects due to the larger movement of the satellite in comparison to GSO systems.

6.3 NGSO satellite communication systems capacity enhancement through transmission techniques

The use of high frequencies along with the demand of high throughput necessitates the use of techniques for capacity improvement. Such techniques are ACM, spatial diversity and multiple antenna techniques and in general MIMO techniques [63].

Before moving to the other techniques, a special attention is required for the smart GW techniques. Recently, in the context of GEO systems and the HTS systems, the smart GW concept has gained a lot of interest due to its scalability, flexibility and optimized resource allocation [64–66]. According to the context of smart GW diversity for multi-beam satellites, two architectures are defined: (a) N+P scheme – a number of redundant GWs are used in case that a GW goes in outage and (b) N-active scheme – if a GW goes in outage, its traffic is rerouted to the other operating GWs. Such techniques can be also applicable in the NGSO HTS systems. However, in order to be sufficient either, more GW beams exist or redundant GWs are established in every service region.

6.3.1 Variable and adaptive coding and modulation

Independently of the frequency band used, one technique which helps increase the capacity of the system is the use of multiple modulation and coding schemes (ModCods) pre-programmed based on geometry of the system or adaptively depending on the signal-to-noise ratio (SNR). The former case is called VCM and the ModCod table is pre-programmed based on certain criterion. VCM has been proposed for EO downlinks in [67]. In this case, the ModCods are chosen based on the elevation angle of the link. However, the ModCods can be selected adaptively based on the measured or estimated SNR. This technique is known as ACM. ACM has been already adopted in the ETSI DVB standards, DVB-S2 and DVB-S2x [68,69]. The use of ACM at NGSO MEO constellations has been already used by O3b for the Ka-band (also investigated in [70] for Ka-band with single antennas) and research results have been obtained in [71] for Q-band MEO system and in [9] for spatial diversity systems operating at Q-band. For Q band and according to [9], the ModCod tables used are shown in Table 6.2. Finally, based on the time that a link is affected by tropospheric attenuation, a method based on varying the symbol rate in order to deliver the same data volume as in clear sky conditions is proposed in [72].

Table 6.2 Details of the MODCODs used for Q-band MEO satellite communications system

#	Mod.	Rate	E_s/N_0 (dB)	SE (bit/s/Hz)
15	8PSK	100/180	6.36	1.7105
16	8PSK	104/180	6.77	1.7789
17	8PSK	3/5	7.13	1.8474
18	8PSK	2/3	7.97	2.0526
19	8PSK	13/18	8.97	2.2237
20	16APSK	100/180	8.22	2.2807
21	16APSK	104/180	8.63	2.3719
22	16APSK	28/45	9.37	2.5544
23	16APSK	116/180	9.77	2.6456
24	16APSK	2/3	10.12	2.7368
25	16APSK	25/36	10.59	2.8509
26	16APSK	3/4	11.66	3.0789

6.3.2 Diversity techniques

Spatially separated links have the advantage that the channel gain or the attenuation induced is not highly correlated with increasing the distance between the receivers and/or transmitters. Therefore, the probability that attenuation is high for both links is smaller than for a single link.

In LMS channels, this spatial separation of transmitted and/or received antennas has been used as orbital diversity scheme and its modelling is presented in [73] in which apart from the channel model an image lens for defining the path towards the satellite in the sky is also considered. Moreover, spatial separation of the links can be used in LMS channels through MIMO techniques. MIMO techniques have gained a lot of attention to the SatCom industry due to their success in terrestrial communications [63]. In general, MIMO techniques are actually taking advantage of the digital-processing techniques of the baseband signal and the different (best case independent) conditions of the channel to provide a gain in capacity or availability of the system.

Moving to higher frequencies (Ka-band and above), line-of-sight must be always guaranteed with narrow beamwidths and therefore propagation through atmosphere cause the most severe losses. For the compensation of atmospheric effects, spatial diversity can be used considering either multiple-ground stations communicating with a single-NGSO satellite or a single-ground station with two or more equipped antennas communicating with multiple satellites. The first scenario is also called site diversity, the geometry of which is shown in Figure 6.10, while the latter orbital diversity, the geometry of which is shown in Figure 6.11.

Although more equipment on ground or more payload resources are needed for the realization of the diversity schemes, it is worth noting that for NGSO satellite systems the ground terminals should be equipped with at least two antennas, as also

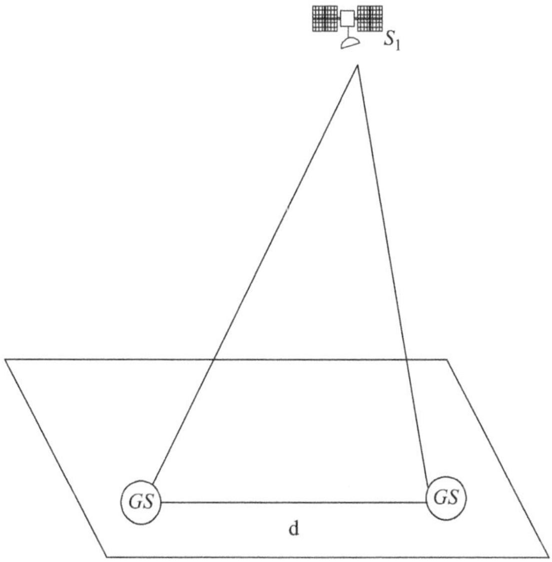

Figure 6.10 System geometry for the site diversity system

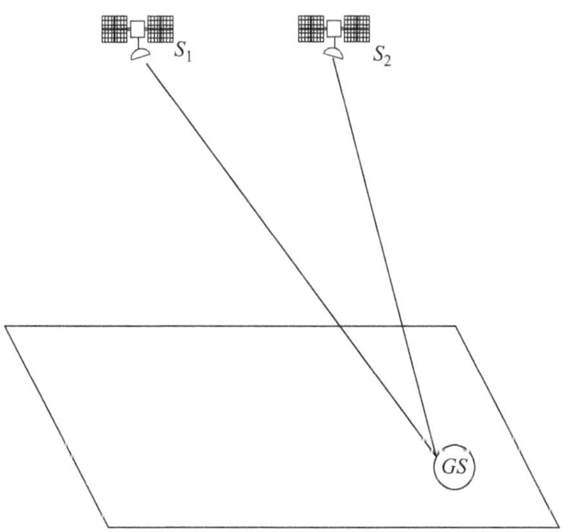

Figure 6.11 System geometry for the orbital diversity system

indicated in the first section, thus making orbital diversity feasible. However, both scenarios require the communication between the multiple-ground stations or the communication between the different satellites. Therefore, this process adds a delay on the signal propagation. Considering an 8-MEO satellite constellation system with

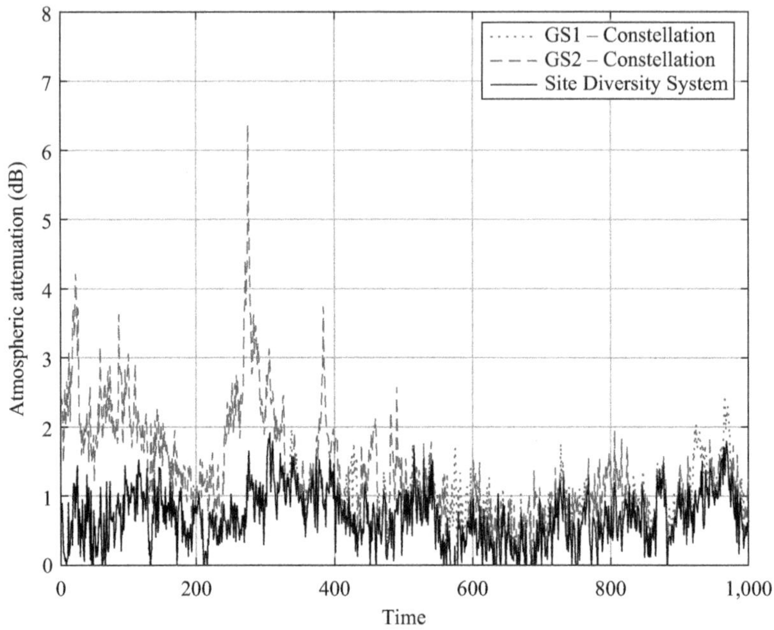

Figure 6.12 Snapshot of time series of total atmospheric attenuation at 20 GHz between Hawaii and an 8-MEO constellation for a site diversity system

equally spaced satellites an example of time series for site diversity system operating at Q-band using [9] and the Complementary Cumulative Distribution Function (CCDF) for orbital diversity are shown in Figures 6.12 and 6.13, respectively. The gain can be observed for both cases. More particularly, for a target availability of 99.7%, the exceeded attenuation for an orbital system is 15 dB while for a single link is 17.4 dB. Therefore, a gain of close to 3 dB can be observed.

The site-diversity technique is also obligatory for optical feeder links for NGSO satellite systems. However, due to the higher correlation of clouds in comparison to rain, the stations must be placed in much greater distances between them (macro-diversity schemes) [9]. In order to combat turbulence a solution could be to use more optical apertures on ground optical terminal in small distances between them (micro-diversity schemes).

Moreover, for X-, Ku- and Ka-band communication links with GEO satellites, it has been shown that the coherent uplink and downlink arraying techniques may improve the SNR [74]. The improvement comes from the gain on the use of multiple antennas as an antenna array. However, the reception must be coherent and the main errors on phase and phase fluctuations come from the RF unit location, the hardware and the atmospheric phase fluctuations [75–77]. Since the ground terminals of NGSO satellites can be equipped with two or more antennas very close to each other, such distributed techniques can be applied.

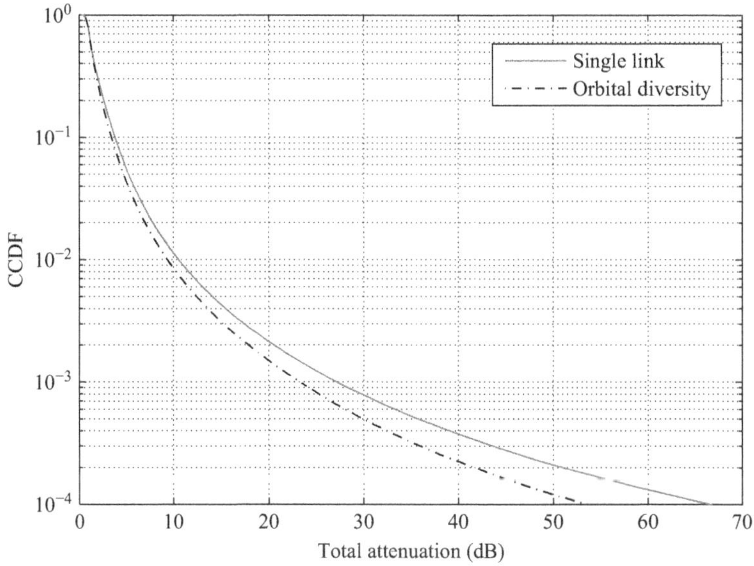

Figure 6.13 Exceedance probability of total atmospheric attenuation for an 8-MEO constellation system at Q-band, ground station at Vernon with and without orbital diversity

6.3.3 Interference issues and NGSO–GSO cooperation

An issue which may arise in the future is the interference between the NGSO HTS and GEO HTS systems. A study presented in [78] has shown the impact of differential atmospheric attenuation for RF systems in radio interference and the co-existence between GEO and NGSO satellite systems. Inter-system interference for ground-to-satellite or satellite-to-ground links is highly affected by propagation conditions. As also presented in [78], the free space loss is highly different and much smaller for NGSO systems comparing to GEO systems. Moreover, rain and in general atmospheric phenomena may affect highly the interference, since the elevation angle changes and the elevation angle between ground station and a LEO satellite can take values from 10° up to higher than 80°. In Figure 6.14, the time series of total differential attenuation are depicted for a single-MEO and GEO satellite. It can be observed that although the losses, which include free space losses, are higher in GEO links, due to the lower elevation angles that an MEO link may experience, the atmospheric attenuation can imbalance the differential attenuation. In Figure 6.15, the cumulative distribution (or outage probability) for carrier over interference ratio for a single MEO at equatorial plane and GEO satellite is shown as a function of carrier-over-interference at clear-sky conditions.

A solution which has been proposed by OneWeb is the progressive pitch. When the beam coming from the LEO satellite is aligned with the beam of the GEO satellite, then the beam of the LEO is switched off. However, due to the potential employment

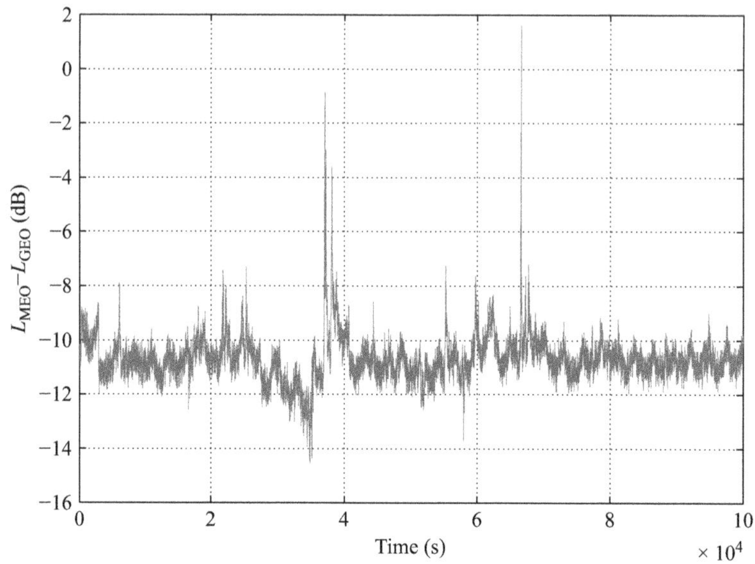

Figure 6.14 Differential total attenuation (atmospheric and free space losses) between an MEO and GEO link

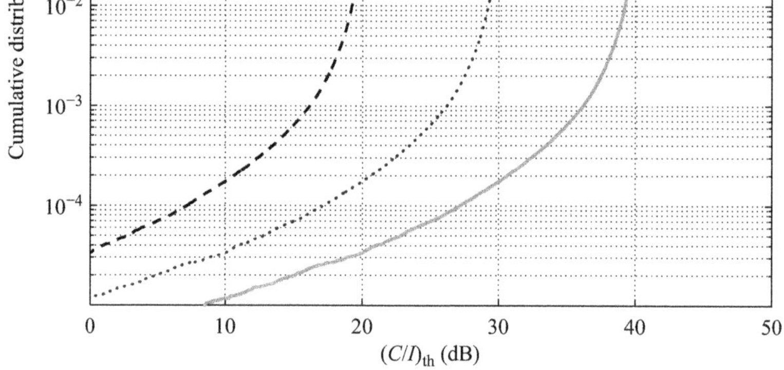

Figure 6.15 Cumulative distribution function of carrier over interference ratio for a single MEO and a GEO satellite

of different mega-constellations, interference issues may arise between the different NGSO systems. IRIDIUM NEXT has proposed and developed a system which use the NGSO satellites and GEO satellites in cooperation, thus enabling the truly global coverage with satellite communications. Finally, cognitive techniques and analytical model for the evaluation of the co-existence of terrestrial and satellite networks have been proposed [79–81].

6.4 Conclusions

In this book chapter, the next-generation NGSO satellite communication systems and the propagation link characteristics induced are presented and discussed. NGSO satellite systems are not a very recent idea, since NGSO systems have been operating since the end of the 1990s of the previous century. In Table 6.3, the advantages and disadvantages of NGSO systems in comparison to GSO networks are briefly given. The advantages of NGSO systems are the lower latency, smaller size and lower losses in comparison to GEO satellite systems and that when a constellation is shaped a global coverage can be achieved. Now, new systems have been put in operation and are planned which are using NGSO satellites. The next-generation systems will make use of the high-frequency bands and higher data rates could be delivered. Depending on the application, service provided and kind of link (feeder or user links), different frequency bands will be used. Different bands experience different propagation characteristics. Lower bands (L-/S-bands) are mostly affected by the local environment while in high-RF bands and optical range, atmospheric effects must be considered for the system design. Moreover, the use of fade mitigation techniques, such as ACM or diversity techniques, increase system's throughput and availability as has been shown in recent studies. However, an issue which must be tackled is the inter-system interference not only for the NGSO and GEO systems but also for the different mega-constellations, if all these planning systems will be set during launch and operation.

Table 6.3 Advantages and disadvantages of NGSO systems compared to GSO systems

Advantages	Disadvantages
Can offer truly global coverage (even poles)	Require a higher number of satellites to provide even quasi-global coverage
Lower losses, emitted power and latency	Higher number of gateways to serve all beams globally
Smaller antenna terminals	Tracking antennas are required and larger number of antennas for seamless handover
Lower atmospheric attenuation	Propagation modelling is more complex

References

[1] O3b mPower; Accessed Sep. 2017. Available from: https://www.ses.com/networks/o3b-mpower.

[2] Laser Light Communications; Accessed Sep. 2017. Available from: http://www.laserlightcomms.com/.

[3] AGI STK – Academic Version; Accessed Sep. 2017. Available from: https://www.agi.com/products/by-product-type/applications/stk/.

[4] IridiumNEXT; Accessed Sep. 2017. [cited 2017 Sep]. Available from: https://www.iridiumnext.com/.

[5] LEOSat; Accessed Sep. 2017. [cited 2017 Sep]. Available from: leosat.com.

[6] OneWeb; Accessed Sep. 2017. [cited 2017 Sep]. Available from: oneweb.world.

[7] INTELSAT AMSS; Accessed Dec. 2017. Available from: http://www.intelsat.com/media-resources/high-throughput-ku-band-for-aero-applications/.

[8] ITU-R. Radio Regulations – Articles. International Telecommunications Union; 2012.

[9] Kourogiorgas CI, Tarchi D, Ugolini A, *et al.* Capacity statistics evaluation for next generation broadband MEO satellite systems. IEEE Transactions on Aerospace and Electronic Systems. 2017;53(5):2344–2358.

[10] ITU-R Preparatory Studies for WRC-19; Accessed Dec. 2017. Available from: http://www.itu.int/en/ITU-R/study-groups/rcpm/Pages/wrc-19-studies.aspx.

[11] Riva C, Capsoni C, Luini L, *et al.* The challenge of using the W band in satellite communication. International Journal of Satellite Communications and Networking. 2014;32(3):187–200.

[12] Giggenbach D, Lutz E, Poliak J, *et al.* A High-Throughput Satellite System for Serving whole Europe with Fast Internet Service, Employing Optical Feeder Links. In: Broadband Coverage in Germany. 9th ITG Symposium. Proceedings; 2015. p. 1–7.

[13] Blumenthal SH. Medium Earth Orbit Ka-Band Satellite Communication Systems. In: MILCOM 2013 – 2013 IEEE Military Communications Conference; 2013. p. 273–277.

[14] Smutny B, Kaempfner H, Muehlnikel G, *et al.* 5.6 Gbps Optical Intersatellite Communication Link; 2009.

[15] Arapoglou PD, Michailidis ET, Panagopoulos AD, *et al.* The land mobile earth-space channel. IEEE Vehicular Technology Magazine. 2011 Jun;6(2):44–53.

[16] Loo C. A statistical model for a land mobile satellite link. IEEE Transactions on Vehicular Technology. 1985 Aug;34(3):122–127.

[17] Corazza GE, Vatalaro F. A statistical model for land mobile satellite channels and its application to nongeostationary orbit systems. IEEE Transactions on Vehicular Technology. 1994 Aug;43(3):738–742.

[18] Kourogiorgas C, Kvicera M, Skraparlis D, *et al.* Modeling of first-order statistics of the MIMO dual polarized channel at 2 GHz for land mobile satellite systems under tree shadowing. IEEE Transactions on Antennas and Propagation. 2014 Oct;62(10):5410–5415.

[19] Kourogiorgas C, Kvicera M, Panagopoulos AD, *et al.* Inverse Gaussian-based composite channel model and time series generator for land mobile satellite systems under tree shadowing. IET Microwaves, Antennas Propagation. 2016;10(6):612–616.

[20] Ko SC, Kim J, Yang CY. Practical Channel Simulation Model for the Non-GEO Land Mobile Satellite (LMS) Communications. In: 1997 IEEE 47th Vehicular Technology Conference. Technology in Motion. vol. 1; 1997. p. 411–415.

[21] Fontan FP, Vazquez-Castro M, Cabado CE, *et al.* Statistical modeling of the LMS channel. IEEE Transactions on Vehicular Technology. 2001 Nov;50(6):1549–1567.

[22] Catedra MF, Perez J. Cell Planning for Wireless Communications. 1st ed. Norwood, MA, USA: Artech House, Inc.; 1999.

[23] Dottling M, Jahn A, Didascalou D, *et al.* Two- and three-dimensional ray tracing applied to the land mobile satellite (LMS) propagation channel. IEEE Antennas and Propagation Magazine. 2001 Dec;43(6):27–37.

[24] Saunders SR, Evans BG. A Physical–Statistical Model for Land Mobile Satellite Propagation in Built-Up Areas. In: Tenth International Conference on Antennas and Propagation (Conf. Publ. No. 436). vol. 2; 1997. p. 44–47.

[25] Panagopoulos AD, Arapoglou PDM, Cottis PG. Satellite communications at KU, KA, and V bands: Propagation impairments and mitigation techniques. IEEE Communications Surveys Tutorials. 2004 Third;6(3):2–14.

[26] Olsen R, Rogers D, Hodge D. The aRb relation in the calculation of rain attenuation. IEEE Transactions on Antennas and Propagation. 1978 Mar;26(2):318–329.

[27] ITU-R P 840-7. Attenuation Due to Clouds and Fog. International Telecommunications Union; 2017.

[28] Luini L, Capsoni C. Modeling high-resolution 3-D cloud fields for earth–space communication systems. IEEE Transactions on Antennas and Propagation. 2014 Oct;62(10):5190–5199.

[29] Lyras NK, Kourogiorgas CI, Panagopoulos AD. Cloud attenuation statistics prediction from Ka-band to optical frequencies: integrated liquid water content field synthesizer. IEEE Transactions on Antennas and Propagation. 2017 Jan;65(1):319–328.

[30] ITU-R P 676. Attenuation by Atmospheric Gases; 2012.

[31] ITU-R P 618-13. Propagation Data and Prediction Methods Required for the Design of Earth -Space Telecommunication Systems. International Telecommunications Union; 2017.

[32] Arapoglou PD, Panagopoulos AD. A Tool for Synthesizing Rain Attenuation Time Series in LEO Earth Observation Satellite Downlinks at Ka Band. In: Proceedings of the 5th European Conference on Antennas and Propagation (EUCAP); 2011. p. 1467–1470.

[33] Luini L. A comprehensive methodology to assess tropospheric fade affecting earth space communication systems. IEEE Transactions on Antennas and Propagation. 2017 Jul;65(7):3654–3663.

[34] Luini L. Modeling and synthesis of 3-D water vapor fields for EM wave propagation applications. IEEE Transactions on Antennas and Propagation. 2016 Sep;64(9):3972–3980.
[35] Jeannin N, Outeiral M, Castanet L, *et al.* Atmospheric Channel Simulator for the Simulation of Propagation Impairments for Ka Band Data Downlink. In: The 8th European Conference on Antennas and Propagation (EuCAP 2014); 2014. p. 3357–3361.
[36] Liu W, Michelson DG. Fade slope analysis of Ka-band earth-LEO satellite links using a synthetic rain field model. IEEE Transactions on Vehicular Technology. 2009 Oct;58(8):4013–4022.
[37] Capsoni C, Fedi F, Magistroni C, *et al.* Data and theory for a new model of the horizontal structure of rain cells for propagation applications. Radio Science. 1987;22(3):395–404.
[38] Liu W, Michelson DG, Malm J, *et al.* Effect of rain cell shape on fade slope statistics over simulated Earth-LEO Ka-band links. In: 2010 14th International Symposium on Antenna Technology and Applied Electromagnetics the American Electromagnetics Conference; 2010. p. 1–4.
[39] Feral L, Sauvageot H, Castanet L, *et al.* HYCELL—A new hybrid model of the rain horizontal distribution for propagation studies: 1. Modeling of the rain cell. Radio Science. 2003;38(3):1056.
[40] Matricciani E, Selva SP. Attenuation statistics estimated from radar measurements in MEO satellite communication systems for mobile terminals. International Journal of Satellite Communications. 2002;20(3):167–185. Available from: http://dx.doi.org/10.1002/sat.719.
[41] Matricciani E. Physical-mathematical model of the dynamics of rain attenuation based on rain rate time series and a two-layer vertical structure of precipitation. Radio Science. 1996;31(2):281–295. Available from: http://dx.doi.org/10.1029/95RS03129.
[42] Matricciani E. Space communications with variable elevation angle faded by rain: Radio links to the Sun Earth first Lagrangian point L1. International Journal of Satellite Communications and Networking. 2016;34(6): 809–831. SAT-15-0005.R1. Available from: http://dx.doi.org/10.1002/sat.1134.
[43] Kourogiorgas C, Panagopoulos A. A rain attenuation stochastic dynamic model for LEO satellite systems above 10 GHz. IEEE Transactions on Vehicular Technology. 2015 Feb;64(2):829–834.
[44] Kourogiorgas C, Panagopoulos AD, Arapoglou PD. Rain Attenuation Time Series Generator for Medium Earth Orbit Links Operating at Ka Band and above. In: Antennas and Propagation (EUCAP), 2014 8th European Conference on; 2014. p. 3506–3510.
[45] Liu W, Michelson DG. Effect of turbulence layer height and satellite altitude on tropospheric scintillation on Ka-band earth LEO satellite links. IEEE Transactions on Vehicular Technology. 2010 Sep;59(7):3181–3192.
[46] Pereira C, Vanhoenacker-Janvier D, Jeannin N, *et al.* Simulation of Tropospheric Scintillation on LEO Satellite Link Based on Space–Time Channel

Modeling. In: The 8th European Conference on Antennas and Propagation (EuCAP 2014); 2014. p. 3516–3519.
[47] ITU-R P 1853-1. Tropospheric Attenuation Time Series Synthesis. International Telecommunications Union; 2012.
[48] Karagiannis GA, Panagopoulos AD, Kanellopoulos JD. Multidimensional rain attenuation stochastic dynamic modeling application to earth space diversity systems. IEEE Transactions on Antennas and Propagation. 2012 Nov;60(11):5400–5411.
[49] Karatzas I, Shreve SE. Brownian Motion and Stochastic Calculus. Berlin: Springer-Verlag; 2005.
[50] Kourogiorgas CI, Panagopoulos AD. Space–time stochastic rain fading channel for multiple LEO or MEO satellite slant paths. IEEE Wireless Communications Letters. 2017;PP(99):1.
[51] Maseng T, Bakken P. A stochastic dynamic model of rain attenuation. IEEE Transactions on Communications. 1981 May;29(5):660–669.
[52] Panagopoulos AD, Kanellopoulos JD. Prediction of triple-orbital diversity performance in Earth–space communication. International Journal of Satellite Communications. 2002;20(3):187–200. Available from: http://dx.doi.org/10.1002/sat.720.
[53] Jeannin N, Carrie G, Castanet L, Lacoste F. A Space Time Channel Model for the Simulation of Total Attenuation Fields. In: ESA Workshop on Radiowave Propagation; 2011.
[54] Kourogiorgas C, Panagopoulos AD. A Tropospheric Scintillation Time Series Synthesizer based on Stochastic Differential Equations. In: 2013 Joint Conference: 19th Ka and Broadband Communications, Navigation and Earth Observation Conference and 31st AIAA ICSSC; 2013.
[55] Matricciani E, Mauri M, Riva C. Scintillation and Simultaneous Rain Attenuation at 49.5 GHz. In: Antennas and Propagation, 1995. Ninth International Conference on (Conf. Publ. No. 407). vol. 2; 1995. p. 165–168.
[56] Hemmati H, editor. Near-Earth laser communications. No. 143 in Optical Science and Engineering. Boca Raton: CRC Press; 2009.
[57] Fuchs C, Moll F. Ground station network optimization for space-to-ground optical communication links. IEEE/OSA Journal of Optical Communications and Networking. 2015 Dec;7(12):1148–1159.
[58] Lyras NK, Kourogiorgas CI, Panagopoulos AD. Cloud free line of sight prediction modeling for optical satellite communication networks. IEEE Communications Letters. 2017 Jul;21(7):1537–1540.
[59] Degnan JJ. Millimeter accuracy satellite laser ranging: A review. In: Smith DE, Turcotte DL, editors. Geodynamics Series. vol. 25. Washington, D. C.: American Geophysical Union; 1993. p. 133–162.
[60] Andrews LC, Phillips RL. Laser Beam Propagation through Random Media. vol. PM152. SPIE; 2005.
[61] Mengali A, Lyras N, Kourogiorgas C, *et al.* Optical Feeder Links Study towards Future Generation MEO VHTS Systems. In: 35th AIAA International Communications Satellite Systems Conference (ICSSC). Trieste, Italy; 2017.

[62] Dimitrov S, Barrios R, Matuz B, *et al.* Digital modulation and coding for satellite optical feeder links with pre-distortion adaptive optics. International Journal of Satellite Communications and Networking. 2015 Jan;34:625–644.
[63] Arapoglou PD, Liolis K, Bertinelli M, *et al.* MIMO over satellite: a review. IEEE Communications Surveys Tutorials. 2011 First;13(1):27–51.
[64] Kyrgiazos A, Evans BG, Thompson P. On the gateway diversity for high throughput broadband satellite systems. IEEE Transactions on Wireless Communications. 2014 Oct;13(10):5411–5426.
[65] Kyrgiazos A, Evans B, Thompson P, *et al.* A terabit/second satellite system for European broadband access: a feasibility study. International Journal of Satellite Communications and Networking. 2014 Mar;32(2):63–92. Available from: http://onlinelibrary.wiley.com/doi/10.1002/sat.1067/abstract.
[66] Jeannin N, Castanet L, Radzik J, *et al.* Smart gateways for terabit/s satellite. International Journal of Satellite Communications and Networking. 2014 Mar;32(2):93–106. Available from: http://onlinelibrary.wiley.com/doi/10.1002/sat.1065/abstract.
[67] Toptsidis N, Arapoglou PD, Bertinelli M. Link adaptation for Ka band LEO Earth observation systems: A realistic performance assessment. International Journal of Satellite Communications and Networking. 2012 May/Jun;30(3):131–146.
[68] Second Generation Framing Structure, Channel Coding and Modulation Systems for Broadcasting, Interactive Services, News Gathering and Other Broadband Satellite Applications, 2014. Part I: DVB-S2.
[69] Second Generation Framing Structure, Channel Coding and Modulation Systems for Broadcasting, Interactive Services, News Gathering and Other Broadband Satellite Applications, 2014. Part II: S2-Extensions (DVB-S2X).
[70] Kourogiorgas C, Tarchi D, Ugolini A, *et al.* System Capacity Evaluation of DVB-S2X Based Medium Earth Orbit Satellite Network Operating at Ka-Band. In: in ASMS/SPSC Conference; 2016.
[71] Kourogiorgas C, Tarchi D, Ugolini A, *et al.* Performance Evaluation of DVB-S2X Based MEO Satellite Networks Operating At Q-Band. In: IEEE GlobeCom 2016; 2016.
[72] Matricciani E. A method to achieve clear-sky data-volume download in satellite links affected by tropospheric attenuation. International Journal of Satellite Communications and Networking. 2016;34(6):713–723. SAT-14-0078.R1. Available from: http://dx.doi.org/10.1002/sat.1126.
[73] Akturan R, Vogel WJ. Path diversity for LEO satellite-PCS in the urban environment. IEEE Transactions on Antennas and Propagation. 1997 Jul;45(7): 1107–1116.
[74] Suzaki K, Suzuki Y, Kobayashi K. A novel earth station antenna concept for Ku-band mobile satellite communication systems – Distributed array antenna and key technologies. In: 29th AIAA International Communications Satellite Systems Conference (ICSSC); 2011.
[75] Ishimaru A. Wave Propagation and Scattering in Random Media. London: Academic Press; 1978.

[76] Nessel JA, Acosta RJ, Morabito DD. Phase Fluctuations at Goldstone Derived from One Year Site Testing Interferometer Data. In: 14th Ka and Broadband Communications Conference. Matera, Italy; 2008.
[77] Matricciani E. A relationship between phase delay and attenuation due to rain and its applications to satellite and deep-space tracking. IEEE Transactions on Antennas and Propagation. 2009 Nov;57(11):3602–3611.
[78] Kourogiorgas C, Panagopoulos AD. Spectral Coexistence of GEO and MEO Satellite Communication Networks: Differential Total Atmospheric Attenuation Statistics. In: 2017 11th European Conference on Antennas and Propagation (EUCAP); 2017. p. 10–14.
[79] Vassaki S, Poulakis MI, Panagopoulos AD, *et al.* Power allocation in cognitive satellite terrestrial networks with QoS constraints. IEEE Communications Letters. 2013 Jul;17(7):1344–1347.
[80] Vassaki S, Poulakis MI, Panagopoulos AD. Optimal iSINR-based power control for cognitive satellite terrestrial networks. Transactions on Emerging Telecommunications Technologies. 2017;28:e2945.
[81] Kourogiorgas C, Panagopoulos AD, Liolis K. Cognitive Uplink FSS and FS Links Coexistence in Ka-Band: Propagation Based Interference Analysis. In: 2015 IEEE International Conference on Communication Workshop (ICCW); 2015. p. 1675–1680.

Chapter 7
Diversity combining and handover techniques: enabling 5G using MEO satellites

Nicolò Mazzali[1], *Bhavani Shankar M. R.*[1], *Ashok Rao*[2], *Marc Verheecke*[3], *Peter De Cleyn*[3], *and Ivan De Baere*[3]

In this chapter, we provide a thorough review of medium Earth orbit (MEO) satellites, highlighting their applications, peculiarities, and the role that they may play in the implementation of 5G for satellite networks. In particular, we will explain why MEO satellites are a new paradigm, how to tackle the challenges related to their usage, and how they fit into the 5G context. In this perspective, we will show how diversity combining and handover are key functionalities for their successful integration.

Towards describing the 5G paradigm with MEO satellites, the chapter first provides a high-level description of the satellite characteristics, the services that have been deployed, and also possible future applications. Further, a high-level description of all the atmospheric effects affecting typical MEO communications is included. Finally, a critical review of handover techniques (state-of-the-art, trade-offs, and future challenges), as well as a review of combining techniques (theoretical performance in a MEO scenario, advantages, drawbacks, and trade-offs), will be presented.

7.1 Introduction

5G, the next generation of wireless networking, is expected to bring about a new era of ubiquitous, high-bandwidth, low-latency communications. However, the cost of deploying terrestrial radio access networks (RANs) and backhaul circuits to enable access to 5G for a large portion of the populace will be prohibitive even in developed countries. Satellites are expected to play a key role in bridging the digital divide either by providing high-bandwidth backhaul for 5G terrestrial RAN or by direct end user access to high-bandwidth links.

[1]Interdisciplinary Centre for Security, Reliability and Trust, University of Luxembourg, Luxembourg
[2]SES, Luxembourg
[3]Newtec, Belgium

Geostationary Earth Orbit (GEO) satellites have been traditionally considered for providing requisite services due to a wider coverage, a simpler network management, and user terminal processing. However, the use of non-geostationary satellites for very high-speed data services has been taking momentum in the past years. In particular, the launch of the first MEO constellation satellites is a recent example of a new mission concept, taking advantage of a lower latency and a higher radiated power to cover underserved geographical areas compared to the existing broadband coverage in GEO. By virtue of their lower latency, higher throughputs, and wide coverage, MEO satellites can play an important role in enabling ubiquitous access to 5G. The atmospheric conditions, particularly at Ka-band, create new challenges in maintaining the link availability targets. The use of multiple satellites and multiple spot beams (with possible ability to steer the beams) implies new access scenarios with multiple reception and transmission paths at each end node. Towards this, support from the air interface in addition to advanced receiver signal processing is needed. Because elevation and azimuth of MEO satellites vary, seamless handover between MEO satellites is an important requirement, which necessitates a terminal modem with multiple demodulators and at least two antennas, one tracking the satellite in view and the other positioned to acquire the rising satellite. Indeed, using only one very fast tracking antenna would generate a repointing outage disrupting the communication. The two antennas can be pointed at the same satellite outside the handover period, and diversity combining can be employed to increase the signal-to-noise ratio (SNR) and further improve throughputs for both backhaul and direct access scenarios. Diversity combining is a very scalable and well-known technique, and the use of software-defined radio architectures can allow for high order diversity combining at a relatively modest increase in hardware complexity. However, care has to be taken that the adaptive coding and modulation loop is robust to sudden changes in the SNR at the beginning and at the end of the handover phase, when combining is switched off and on, respectively.

7.2 Medium Earth orbit satellites: architectures, services and applications, challenges

The MEO orbit, also referred to as an intermediate circular orbit (ICO), refers to equatorial and inclined circular orbits that are above 2,000 km from the Earth's surface and below the geostationary orbit that is 35,786 km above the Earth's surface. To reduce the damage to the electronic systems that can be caused by radiation, MEO satellites are usually located in between the two Van Allen belts. The inner Van Allen belt consists of high-energy protons and extends between 1,000 and 6,000 km from the Earth's surface while the outer Van Allen belt has high-energy electrons and extends from 13,000 to 60,000 km above the Earth's surface with the maximum intensity being in the areas closest to the Earth. Consequently, MEO satellites are either located between the inner and outer belts (between 6,000 and 13,000 km from the Earth's surface) or in the lower intensity areas of the outer Van Allen belt (beyond 20,000 km from the Earth's surface).

MEO satellites that have been launched include communications satellites like the ICO S-band system and O3b Ka-band system, and navigational satellites such as the global positioning system (GPS), Glonass, and Galileo. The MEO orbit is attractive for telecommunications because of the lower path delay and because the lower path loss (compared to GEO) allows satellites with smaller antennas with lower gain and power to be deployed. The lower path delay reduces the latency for voice, video, and data applications which greatly improves the quality of experience [1]. Since MEO satellites are located closer to the Earth, their field of view is limited (due to the curvature of the Earth) compared to GEO satellites, and their coverage for terminals is limited to latitudes of about ±45° for MEO constellations like O3b which are located at an altitude of 8,062 km. Beyond these latitudes, the elevation angle for ground terminals is very low and it is difficult to obtain an unobstructed view of the MEO arc. The limitation in coverage is not an issue in practice because most of the world's population and an even larger fraction of the population that does not have robust terrestrial access are located within this ±45° band. Aeronautical terminals and ships in open sea, which do not face obstruction from natural and manmade structures, can continue to operate at much higher latitudes.

Of late, broadband low Earth orbit (LEO) constellations have been proposed that would orbit at around 1,000 km altitude, offering even lower latency than MEO satellites. Unlike most MEO communications satellites which are in equatorial orbit, LEO satellites need to be in highly inclined or polar orbits in order to provide coverage to a sufficiently large portion of the Earth. Also, due to the rotation of the Earth, inclined or polar orbit LEOs need to be in multiple planes to provide that uninterrupted coverage. Thus, because of the lower altitude, the smaller field of view, and the inclined orbit, many more satellites are needed to provide uninterrupted coverage. The larger number of satellites combined with higher launch costs make such systems more expensive than MEO or GEO systems.

7.2.1 The O3b satellite network

The O3b satellite network consists of a constellation of 12 communications satellites in a circular equatorial MEO orbit at 8,062 km above the surface of the Earth with an orbital period of 288 min (or 24/5 h). Thus, each satellite orbits the Earth five times in the 24 h that it takes for the Earth to orbit once and so, each satellite passes over the same location on Earth four times a day. The 12-satellite constellation went into commercial service in February 2015 with some satellites in the constellation acting as in-orbit spares. Eight more satellites of the current generation are currently under construction and will be launched in 2018, bringing the constellation size to a total of 20 satellites. A significant advantage of the O3b constellation is that satellites can be added to the constellation without trying to acquire orbital slots as is the case for GEO systems. As satellites are added, the capacity is increased, the elevation angle improves, and the path loss decreases, since the terminal can be served by a satellite that is closer in latitude. This increases the achievable terminal throughput in both downlink and uplink directions and makes it easier to get a clear line of sight to the entire arc at some sites that may have previously faced obstructions in a portion of the arc.

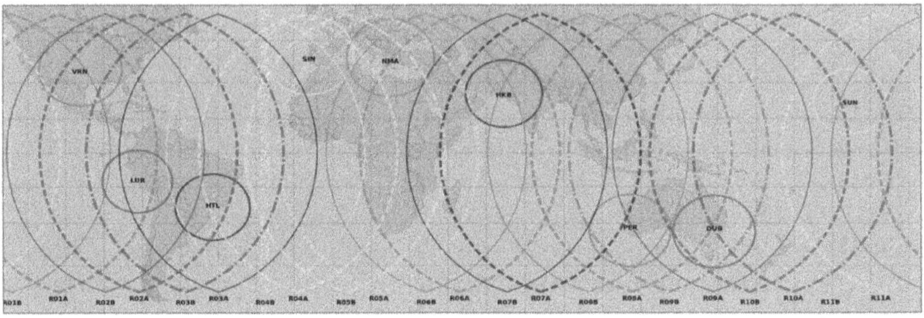

Figure 7.1 The O3b network (Used with the permission from O3b Networks)

For a 12-active-satellite constellation, the Earth can be divided up into 12 regions (actually 11 as explained below), with 1 region per satellite as shown in Figure 7.1. In the figure, larger circles are used to show gateway locations and the area of overlap in frequency with customer beams. Smaller circles show some customer beams. The oval regions indicate areas with minimum 15° elevation and are anchored to gateways of the same colour.

A satellite will be overhead in each region for 30 min, after which it moves on to the next region in the East. Meanwhile, the next satellite in the constellation rises in the West to take its place in the region. Maintaining a seamless communications link requires a handover from the setting satellite to the next rising satellite. The handover from one satellite to the next in each region cannot take place instantaneously and the handover in the next region cannot take place until the handover in the previous region has completed and the satellite antennas have moved to point at the gateway and customer locations in the new region. The system allows for a small amount of overlap time, the 'handover interval', to accomplish this. To account for this extra time, the number of active regions is reduced to 11 service regions for this 12-satellite constellation.

Each of these service regions is anchored by a gateway that provides connections to the terrestrial fibre infrastructure. The two gateway beams on each satellite can be independently pointed to two different gateway locations within the service area for greater flexibility. Each of the service regions is split into two subregions which are slightly offset from each other in space and in handover time so as to allow for higher elevation angles through the pass. The gateway sites are equipped with large 7.3-m antennas and provide the anchor point in the region for initiating handover of all of the customer terminals in the subregion from one satellite (the descending satellite) to the next (the ascending satellite).

Each O3b satellite has ten user spot beams with five beams in each subregion. A spot beam has a diameter of about 700 km and corresponds to a transponder (channel) which is 216 MHz in bandwidth. The transmissions are circularly polarized with the beams in one subregion being on the opposite polarity as the beams in the other subregion. In contrast to most GEO Ka-band satellites where the uplink and downlink are opposite in polarity, the O3b satellites use the same polarity for uplink and downlink. Each beam is steerable since it has to be focused on the same area of the Earth while the satellite is moving with respect to the Earth.

The O3b customer terminals enable a wide range of communications services for customers. In most cases, customer terminals communicate over the satellite to a common gateway site in their service region. Customer terminals can also communicate directly with each other over satellite within the same customer beam (a so-called loopback beam). Three tiers of customer terminals have been defined for the O3b system. Tier-1 terminals have 4.5-m diameter antennas with very high-power amplifiers (HPAs) and are capable of transmitting and receiving at gigabit speeds. These terminals typically support hundreds of thousands of end users since the statistical multiplexing gains are large with such high-bandwidth links. Tier-2 terminals have 2.4- or 1.8-m diameter antennas, support tens of thousands of end users, and can sustain rates of 500 Mbps or higher depending on their location within a beam, the elevation angle, and weather conditions. Tier-3 terminals have diameters of 1.2 m and smaller and are capable of hundreds of megabits of throughput and are intended for sites with fewer users (thousands of users or less).

The capabilities of a customer terminal depend on the size of the antenna, wattage of the power amplifier, and the type of modem deployed. Many of the large sites are configured for point-to-point links and the modems are configured for the classical single channel per carrier (SCPC) mode of operation. On the other hand, if there are multiple sites for a customer in a beam, then a common configuration is to use point-to-multipoint (PMP) connectivity in the gateway to terminal direction (forward direction) and point-to-point SCPC mode for the reverse direction. The PMP configuration allows the customer sites in a beam to share the available bandwidth, burst instantaneously to the maximum bandwidth, and provides statistical multiplexing gains. In the return direction, SCPC mode is preferred for sites with a lot of users. For sites with fewer users, time division multiple access (TDMA) can offer benefits; however, the movement of the satellites introduces a time variation in the path length and time-varying Doppler shift, both of which need to be accounted for in the TDMA burst demodulator at the hub.

The original and indeed primary application for O3b satellites is to provide very high-speed data connectivity to telecom operators in the developing world. These so-called trunking services are delivered to 2.4- or 4.5-m customer terminals. The other application that emerged fairly early is high-speed remote connectivity for Enterprise and Government customers in remote areas without access to terrestrial infrastructure. The services can be delivered to fixed terminals, portable terminals, and terminals on boats and ships. Some of the largest cruise ships in the world have 2.2-m maritime terminals that are served by steerable O3b beams which actually follow the ships through their cruise itineraries. Mobile backhaul in which cellular 3G and 4G base stations in rural areas are connected over satellite to the core network is also getting increasingly deployed over O3b.

The biggest challenges for the O3b service are the need for steerable antennas and the limited coverage due to the small number of beams on each satellite. Considerable progress is being made on the antenna front. New lower cost mechanically steered and electronically steered antennas are being developed that will be available in the near future. To address the coverage issue, more satellites are being launched that will increase the area served by O3b. Furthermore, the development of a next-generation MEO constellation called mPOWER has just been announced. This new constellation will have satellites with high-gain phased-array antennas that can form thousands of beams and will enable ubiquitous coverage within ±45° of latitude.

7.3 Channel characterization for MEO satellites

In this section, the key elements of the end-to-end channel (including uplink and downlink) are described for a typical forward link.

7.3.1 Uplink radio propagation effects

Except for the uplink noise, an ideal feeder link is commonly assumed [2]. In MEO scenarios, where the user terminals are provided with multiple antennas and combining capability, the uplink noise cannot be discarded as it is usually done for GEO applications. Indeed, even though the downlink noise is normally dominant, it can be mitigated by the combining. On the other hand, the uplink noise impairing the feeder link cannot be reduced because it is common to all the incoming signals at the receiver. Further details will be provided in Section 7.5.

7.3.2 Downlink radio propagation effects

The downlink channel typically includes propagation impairments, Doppler effects, and variations in the channel gain, as well as additive thermal noise.

Propagation-induced impairments may be labelled as clear-sky effects or rain-and-cloud effects. Clear-sky effects include attenuation by atmospheric gases, change in the elevation angle due to the refraction by the atmosphere, polarization effects, and tropospheric/ionospheric scintillation. On the other hand, rain-and-cloud effects also include attenuation by rain, fog, clouds, and snow. These effects are the only ones that contribute significantly to the overall attenuation below 70 GHz. Details on modelling and generation of propagation effects can be found in [3–5].

When the user terminal is provided with multiple antennas, each received signal is impaired differently by the radio propagation effects. However, since the antennas are typically co-located, such impairments result to be correlated in both the spatial and the temporal domains.

Variations in the channel gains may occur as a consequence of the mobility of both the satellite and the user terminal. Shadowing and blockage (e.g. caused by buildings nearby) may prevent the link to be in line-of-sight; signal reflections caused by water and ice may also introduce multipath. As in terrestrial wireless communications, the mobility of the user terminal (and/or the satellite, in this case) generates Doppler effects such as Doppler shift and rate.

7.3.3 Payload effects

The satellite transponder model consists of an input multiplexer (IMUX) filter, a HPA unit, and an output multiplexer (OMUX) filter. A travelling wave tube amplifier (TWTA) is typically used as HPA. In the transponder, the desired single carrier (or the bunch of multiple carriers) is band-pass filtered by the IMUX. Thereafter, the filtered signal is amplified by the TWTA, whose operational working point has to be selected by introducing a proper input back-off. The closer the operative point to the saturation point, the higher the nonlinear distortion introduced by the amplifier. Typical distortions include a spectral regrowth in the tail regions of the signal power spectral density, and memory effects. The OMUX filter is a channel filter which is designed to shape the nonlinearly distorted signal for reducing the interference to adjacent transponders.

The HPA characteristics are defined using the AM/AM and AM/PM characteristics. For relatively narrowband applications, these are assumed to be frequency independent and memoryless. On the other hand, memory effects may arise in wideband amplifiers. Typical amplitude response and the group delay of an IMUX filter are illustrated in Figure 7.2. These response curves are not expected to model accurately the transfer functions of any practical OMUX and IMUX since a relatively large variation of characteristics may occur. Typical AM/AM and AM/PM characteristics for linearized TWTAs are also reported in Figure 7.2.

7.3.4 User terminal effects

The effects caused by the imperfections of the user terminal are usually less detrimental than the payload and propagation effects. However, the gain slopes introduced by the LNB and cabling, phase noise, and frequency offsets are still relevant. On top of this, there is of course the thermal noise, which is usually much higher at the user terminal than at the transponder or the gateway. The thermal noise is typically modelled as additive white Gaussian noise (AWGN), whose power is determined by the noise temperature of the receiving element.

7.4 Handover: satellite switching for MEO

While service provisioning by MEO satellites benefits from higher throughputs (due to the increased Effective Isotropic Radiated Power (EIRP)) and lower latencies, the nature of the mobile infrastructure brings in novel aspects in architecture, network management, and ground segment. The number of satellites and their orbital planes determine the coverage area and the minimum elevation angle during the visibility, amongst others. The payload architecture determines the number of user and gateway (GW) beams, the ability to steer them on ground, on-board processing, etc. The network management (management and control server) deals with effective intra-satellite and inter-satellite handover while ensuring minimal link loss. The ground

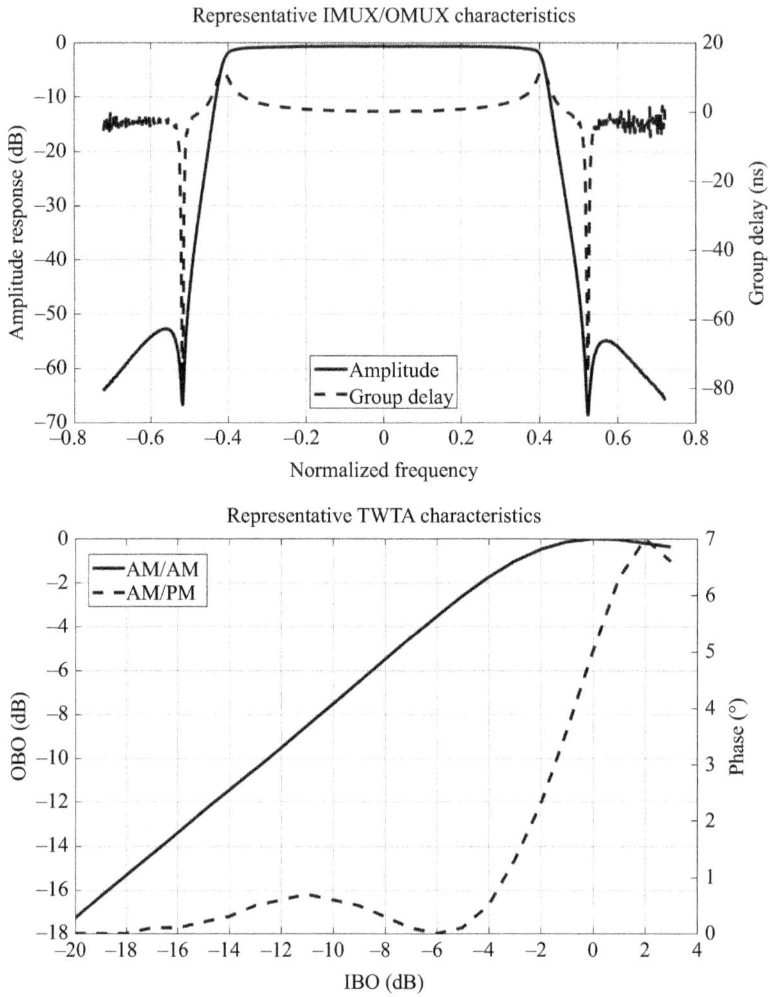

Figure 7.2 Typical payload characteristics: amplitude response and group delay of a transponder filter (IMUX and OMUX), AM/AM, and AM/PM characteristics of a linearized TWTA

system components now need to track the satellites and need to cater to the need for satellite switching in a link from a hub to a terminal through their support for seamless handover.

In order to offer seamless handover functionality for MEO applications, packet loss and throughput reduction have to be avoided as much as possible, while off-the-shelf modems should be used to mitigate the cost increase. This section presents an overview of existing handover techniques for MEO applications and elaborates in detail on the seamless handover concept, which is part of the current technological improvements in ground segment equipment (e.g. Newtec Dialog platform).

7.4.1 Literature

7.4.1.1 Concepts of handover

Besides synchronization and diversity combining, handover is key for the QoS of MEO systems. In general, handover can be achieved by using the so-called make-before-break (make a link over a second satellite before breaking the link over the first satellite) and break-before-make (break the link over the first satellite before making the link over the second satellite) strategies. During the latter, packets are often lost or duplicated. There is a strategy to deal with duplicate or eliminated packets on the higher layers in the network protocol stack. However, this increases delay, which is not desirable. Therefore, the general requirement is to have no packet loss and to have no duplicate packets. The break-before-make strategy is therefore clearly not desirable, as retransmission or loss of data is the outcome. Make-before-break is thus absolutely preferred.

7.4.1.2 Physical layer handover mechanisms

The advantage of performing the handover at the physical layer is that it is transparent to higher layers in the network protocol stack. Inputs to such a handover are satellite ephemeris/GPS data, which are known by the Monitor and Control (M&C) server and used to instruct the antenna control unit to track the satellite.

One implementation is to perform a handover in a separate device before the modem, at RF level, e.g. an L-band device. The advantage is that off-the-shelf modems can be used. That device then typically also applies diversity combining (outside of the handover phase), an SNR-improving technique illustrated in Section 7.5. One way to avoid time delay jitter is to perform the handover at RF level when, for example, the propagation delays of the ascending (having a negative Doppler shift) and descending (having a positive Doppler shift) satellite are exactly equal (up to 10 ns). Such a device typically applies a positive and negative Doppler shift to ascending and descending satellite, respectively. The disadvantage of this implementation is that a dedicated handover device at RF level is required, which can be very expensive. Second, the diversity combining is very complex from a resource point of view and suboptimal as it works with noisy signals and it has to combine on the signal level. Finally, concerning the diversity combining, this implementation is less robust to sudden drops of one of the links.

Another implementation is to perform the handover within the modem, which is commercially available with some equipment manufacturers. In order to solve the time-delay difference between the two satellite paths, a correlation can be taken, but this is not sufficient as two identical packets can be sent (e.g., an ACK). Typically, the air interface is changed and BBF numbers are included, which gives some overhead. Note that in both examples above, the same packets must be sent over both satellites during the handover.

7.4.1.3 Higher layer handover mechanisms

Instead of performing the actual handover at the physical layer, one could also perform the handover on the two output streams of the modem(s). The output streams could come out of one or two modems. In this section, we describe the two-modem solution, but a single modem solution is certainly a candidate for a product as well. At least two demodulators, including the forward error correction (FEC) decoder, are used during handover, one to lock on the first satellite and one to lock on the second satellite.

In the next sections, we will provide a high-level overview of existing protocols and how they could map to the MEO case. We consider here existing mobility management solutions to solve the MEO satellite switching problem.

L3 IP protocols

Mobility or handover at L3 (IP layer) can be achieved in several ways. It is possible to use Mobile IP (MIP) or any of the associated protocols (Hierarchical Mobile IP, Fast Mobile IP, Seamless Handoff architecture for Mobile IP). They typically use tunnelling mechanisms to isolate the mobile's node IP address and corresponding routing decisions from the actual networks participating in the handover. Both the signalling overhead and transport tunnelling overhead make them less suitable for actual handover over satellite. In addition, the networking topology can be designed specifically for our handover needs, allowing the design of a much more efficient handover concept.

Another approach could be to use existing dynamic routing protocols to connect a modem with its associated IP address and associated network(s) to another attachment point in the Internet. Examples are Border Gateway Protocol, Open Shortest Path First (OSPF), and Routing Information Protocol. If the remote location is equipped with two modems, carrying dynamic routing functionality, local VRRP (Virtual Router Redundancy Protocol) signalling between both will move the remote connectivity of networking equipment to the proper modem or gateway able to provide satellite network connectivity.

L2 Ethernet protocols

Looking at terrestrial technology capable of supporting link handover, also L2 Ethernet offers potential candidate protocols with a potential fit with our needs. We mainly distinguish between link aggregation protocols (IEEE 802.3ad Link Aggregation Protocol) and link protection protocols (IEEE 802.1ag and ITU-T SG15/Q9 G.8032), each of which has certain advantages and disadvantages in relation to our purpose.

The advantages of Ethernet link protection and ring protection are that they are widely supported in existing Ethernet switches (carrier Ethernet) and that they operate fully independently and automatic, not requiring any operator intervention or reconfiguration during satellite handover. The disadvantages are that the actual switching has quite some latency and is not seamless, involving packet loss. If the satellite link is unstable, the link or ring protection will also become unstable with potential link flapping as a result.

SDN-based protocols

Software-defined networking (SDN) mechanisms provide a flexible way of switching each direction of a bidirectional link over satellite independently, moving traffic across in sync with satellite connectivity and link availability. It serves as a perfect starting point for seamless handover.

The paper [6] describes a solution for LEO satellites based on SDN technology. The involved control plane designed focuses on the redirection of traffic between multiple LEO satellites and between LEO satellites and ground equipment. The presented approach in this chapter abstracts LEO satellite capabilities and assumes intelligent ground equipment and satellite link management. The resulting handover can be done with existing satellites and is independent from whether these are LEO, MEO, or GEO.

7.4.1.4 Summary on handover

Since the goal is a seamless handover, only make-before-break concepts come into consideration. Indeed, they allow zero packet loss because of the capability to drain packets in transit over the satellite. Handover at physical layer with a purpose-built L-band device or within a purpose-built modem would both result in a less cost-efficient solution. On the other hand, L2 and L3 mechanisms that involve instantaneous bidirectional switching will suffer from packet loss: in particular, the L3 IP dynamic routing protocols (e.g. OSPF) and L2 Ethernet protocols (e.g. IEEE 802.1ag) would not guarantee a handover without packet loss. Also, the family of MIP protocols adds considerable overhead and complexity.

This motivates the presented solution where, making use of the SDN technology, the construction of an infrastructure combining the concepts of 'make-before-break' and 'unidirectional switching' is accomplished. This allows for the achievement of the goals regarding optimal performance and zero packet loss with off-the-shelf modems.

7.4.2 Handover architecture

The overall network topology and architecture are presented in Figure 7.3. It consists of two bidirectional satellite connections, serving as alternative paths between a gateway

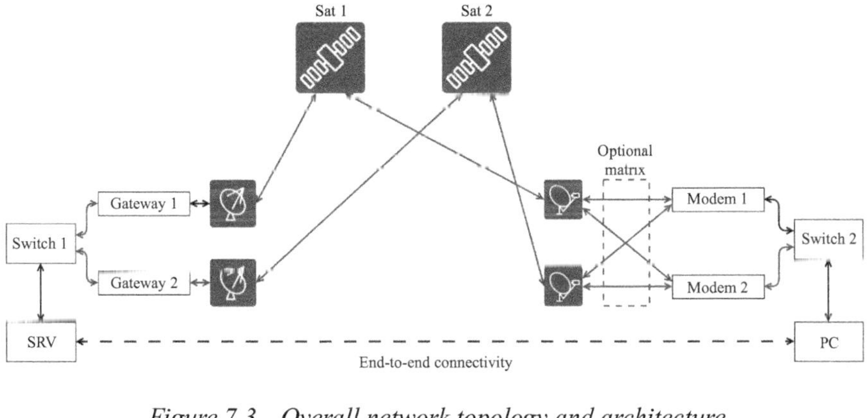

Figure 7.3 Overall network topology and architecture

location and a modem location. Each path consists of a gateway or hub section, a satellite section, and a modem section. At each location, an Ethernet switch connects a device to the local satellite infrastructure. It should be noted that those switches may be connected to any local network requiring 'network extension' over satellite. In the proposed architecture, both satellite paths are implemented via dedicated equipment and brought together via the Ethernet switch. This is not strictly necessary. Equipment can be shared at the gateway location by using an integrated hub capable of supporting multiple satellite connections. At the modem location, one modem could contain multiple receivers and a single transmitter, facilitating an integrated deployment.

7.4.3 Dynamic interactions

The satellite handover is established by Layer 2 unidirectional switching on the carrier gigabit Ethernet switches Switch1/Switch2 in a make-before-break manner.

Assume that a link is established with the first modem pair (over Sat 1). During the handover, the second modem pair will make a connection over the second satellite (Sat 2). When the connection has been established, the traffic should be rerouted over the second modem pair. This will be done with L2 switches using unidirectional switching. Layer 2 unidirectional switching can be achieved in SDN. Unidirectional switching implies that the switch can distinguish between the transmit and receive direction of the traffic flow. By doing this, the switch at transmit side can initiate the redirection of the traffic. The switch at the receive side, however, can listen to a configurable number of paths (in this case two paths). When two Ethernet packets arrive at the same time, one Ethernet packet is buffered until the other Ethernet packet is processed. The switches process packets at a very high rate, so there is no risk of overflow in these cases. So, no packet is lost and we guarantee that no duplicates will be sent over the link (the switching at transmit side guarantees that traffic is only sent once and only over one satellite path). Packet reordering mainly due to transmission delay differences between the two satellite paths in this scenario is likely during handover.

7.4.3.1 Flows

For clarity, only one direction for the data packets is shown. The other direction is similar. The handover of both directions is independent of each other.

Handover phase 1

In the first step of the handover (shown in Figure 7.4), traffic flows via Switch 1, Gateway 1, Sat 1. The second satellite dish is repointed to Sat 2. A path is also established between Gateway 2, Sat 2 (i.e. the rising satellite), and Modem 2. Switch 2 accepts traffic flows both from Modem 1 and Modem 2. The control plane of the Gateway 2/Modem 2 pair is operational. However, no real traffic is sent via this pair (only dummy DVB-S2 Baseband Frames).

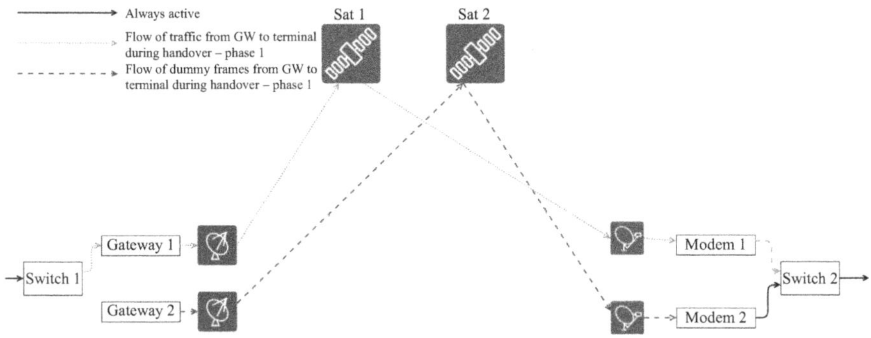

Figure 7.4 Handover phase 1

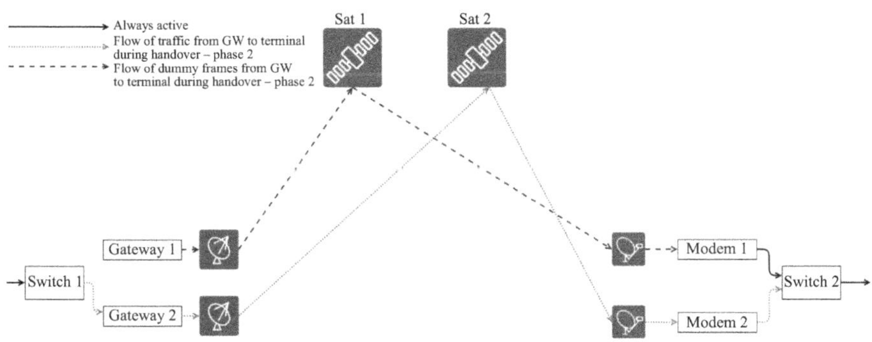

Figure 7.5 Handover phase 2

Handover phase 2

In the second step of the handover (shown in Figure 7.5), traffic is switched in Switch 1 to flow via GW 2 instead. The traffic goes now via Sat 2, Modem 2, and Switch 2.

7.4.4 Proof of concept and results

In the devised series of handover tests, an abstraction is made for Es/N0 variations and diversity combining (see Section 7.5 for a detailed analysis of combining techniques). Also, no delay compensation is considered in the following tests, in order to quantify the maximum packet reordering and its effects.

In a nutshell, end-to-end tests are performed between a source traffic generator and a traffic sink for different satellite link delay variations between the setting and the rising satellite link. The case where the traffic is handed over from a link with a high delay towards a lower delay link is of paramount importance. In this case, the physical medium acts as a buffer delaying the advanced packets and introducing reordering. However, by design, no packet is lost over the physical medium. Even if the delay difference is large, the over-the-air buffered packets will be received at some

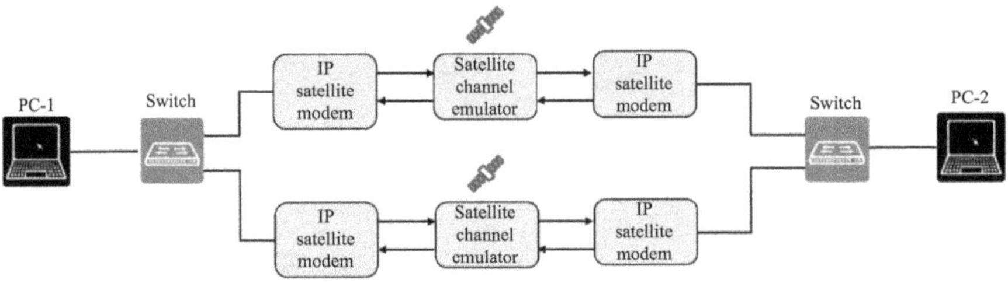

Figure 7.6 Basic handover test setup

Test	ping	iperf	scp
Description	A software utility used to test the reachability of a host on an Internet Protocol (IP) network	A tool for network performance measurement and tuning	Secure copy – a means of securely transferring computer files between a local host and a remote host
Purpose	Detection of packet loss	Detection of packet re-ordering	Transfer duration comparison
Setup/Load	Packet size 300 bytes, packet interval 0.01 s/10 handovers	Packet size 1,500 bytes, 2 Mbps, UDP/10 handovers	File size 736 MB/10 handovers

Figure 7.7 Basic handover tests

Test result	ping	iperf	scp
No delay difference	No packet loss	10 packets re-ordered	Duration 04:09 mm:ss
100 ms delay difference	No packet loss	10 packets re-ordered	Duration 04:15 mm:ss

Figure 7.8 Results of handover tests

point. The effect of this maximum delay jitter on the higher layers will be investigated in the following.

7.4.4.1 Basic handover tests

The test setup is depicted in Figure 7.6 consisting of a source traffic generator (PC-1) and a traffic sink (PC-2) connected over two separate satellite links, each with a full satellite channel emulation.

The switches receive traffic from both links over satellite. Traffic is sent by a switch over one of the satellite links only. In a satellite handover, the switches are reconfigured independently of each other to direct the traffic over the other satellite link. The basic tests with their description, purpose, and setup/load are listed in Figure 7.7, while the corresponding results are listed in Figure 7.8. Concerning the scp test, the file transfer duration without handovers is 02:20 mm:ss. This test is the most impacted by the handovers. However, in reality only one handover would occur during the file transfer instead of 10. The impact for a real file transfer would thus be far less.

Test	Avalanche		Echotest
Setup			
Delay	Sat 1 has a 75-ms delay / Sat 2 has a 90-ms delay		Sat 1 has a 75-ms delay/Sat 2 has a 90-ms delay
Round trip time	RTT is 150 ms for Sat 1. The difference in RTT between the path via Sat 1 and Sat 2 is 30 ms		RTT is 150 ms for Sat 1. The difference in RTT between the path via Sat 1 and Sat 2 is 30 ms
Throughput	Max throughput per Modem pair set to 40 Mbps		Max throughput per Modem pair set to 200 Mbps
System	Spirent avalanche		Linux Ubuntu 14.04
Load			
	Web clients	250 simultaneous users	100 tcp connections
		1,800 transactions/s	2 Mbps per connection
		Browser HTTP/1.1 compatible	Test duration: 192 s
	Web servers	Microsoft-IIS/6.0	10 handovers
		100 kB page/1 Mb page	

Figure 7.9 Setups for avalanche and echotest tools

100 kB page	No HO	With 2× HO
Test duration	5 min 55 s	6 min
tcp connections	13,936	14,000

1 Mb page	No HO	With 10× HO
Test duration	0.25	0.24
tcp connections	8,557	8,511

Figure 7.10 Results for spirent avalanche test

7.4.4.2 Handover tests with many TCP sessions

In this series of tests, the following tools are used to create many transmission control protocol (TCP) sessions (see Figure 7.9):

- Spirent avalanche: a commercial tool that can simulate many web connections.
- Echo test tool: a tool developed by Newtec whereby a specific number of TCP sessions and a target bitrate can be set. The test consists in verifying that the target bit rate is achieved.

Spirent avalanche test

All TCP connections closed without error. Insignificant difference observed between a situation without a satellite handover or a situation where two satellite handovers took place (see Figure 7.10).

Echo test tool

In the following figures the x-axis denotes the id of the TCP session, whereby the first session has an id 1 and the last session an id 100. The number of received bytes (y-axis) for each TCP session is shown in Figure 7.11, while the throughput for each TCP session is reported in Figure 7.12. The variations between individual TCP sessions are marginal.

7.4.4.3 Summary on the handover tests

The handover tests showed the absence of packet loss with a low, acceptable amount of packet reordering. However, secure copy was impacted. This demonstrates that some applications are quite sensitive to packet reorder. Different delay differences

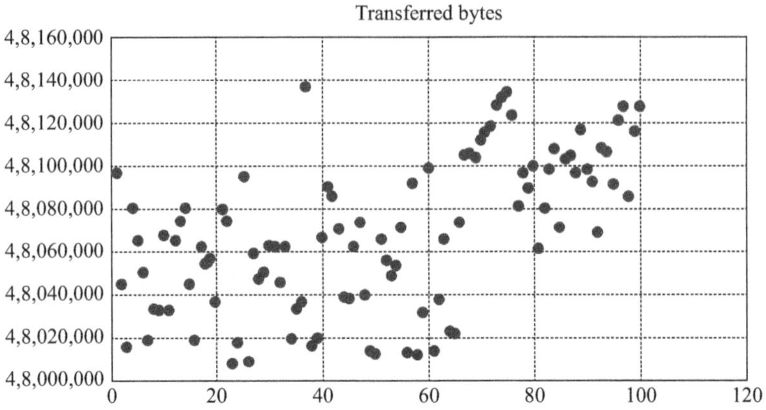

Figure 7.11 Transferred bytes vs. TCP session

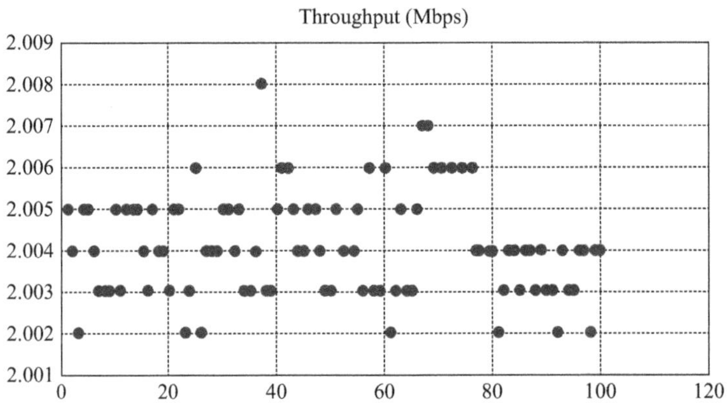

Figure 7.12 Throughput vs. TCP session

between the rising and the setting satellite link yield similar test results. The tests with many TCP sessions were not impacted noticeably by the handovers.

7.5 Diversity combining for MEO satellite applications

Diversity combining has been used in terrestrial radio communications for decades as an effective way to combat fading [7]. The key idea consists in linearly combining the received multiple replicas of the transmitted signal in order to increase the SNR at the receiver, assuming that each replica experiences a different fading realization. In GEO satellite applications, diversity combining typically exploits site diversity to counteract the fading stemming from the atmospheric conditions (e.g. rain events) [8].

MEO applications, on the other hand, require a completely different approach. The need to preserve the data stream continuity during the handover entails the receiver terminal to have two receive antennas. Outside the handover phase (i.e. when only one satellite is in view), the terminal still receives the same data stream twice (one per antenna) from the same gateway and through the same satellite and may perform diversity combining. However, since the two receive antennas are typically located close to one another, there is no significant difference between the realizations of the propagation channel seen by the receiver. Hence, the benefit of the combining performed by the user terminal can be typically construed as a combining gain (in terms of power) rather than a diversity gain. However, in certain cases, the diversity combining enables continuity of services, outside of handover phase, in case of blockage (in ships due to mast) or in case of cable snip. In view of this, we continue to use the term diversity combining.

In this section, the state-of-the-art of the combining techniques is reviewed. In particular, a concise overview of the possible combining techniques is provided, and their possible application to realistic scenarios is considered.

7.5.1 Combining mechanisms: state-of-art-review

The following combining schemes are known in the literature [7,9]:

- maximal ratio combining (MRC)
- equal gain combining (EGC)
- selection combining (SC)
- switching combining (SwC).

Since MRC is derived so as to maximize the SNR at the receiver by linearly processing the received streams, it results to be the optimal scheme (in terms of achieved SNR) when interference is not present. To this purpose, the MRC weights used for the linear combination of the signal replicas are complex coefficients, allowing a rescaling of the amplitudes of the replicas and their co-phasing. EGC, on the other hand, operates only on the phases of the replicas: its weights are still complex but with unit amplitude.

Both MRC and EGC require the acquisition of channel state information and some processing for the computation of their weights. Instead, SC and SwC do not need weights: they simply select the branch with the highest SNR. While SC constantly monitors all the replicas and may instantly switch from one to another, SwC remains on the selected replica until its SNR drops below a predetermined threshold.

In MEO systems, usually, only two branches are available. Moreover, since the two receive antennas are typically located very close to one another, the channel realizations experienced by the two streams will be highly correlated during the combining phase. Therefore, SC and SwC would not provide any gain in the normal mode of operation. However, MRC and EGC would improve the performance by averaging the thermal noise as well as the non-common impairments generated by the HW of the two receive chains. On the other hand, in scenarios where a blockage occurs in one of the antennas, MRC and SC are both optimal in the sense that they prevent the noise injection from the blocked branch by completely discarding the signal coming from the blocked branch (SC) or by setting the corresponding weight to zero (MRC).

In the following, we provide a short summary of MRC and EGC.

MRC requires coherent gain and phase combining, and it is considered optimal in the sense that no other detector leads to a higher SNR after combining the incoming streams. Indeed, with MRC, the useful signal is coherently combined while the noise is not. Having two receive antennas would thus lead to a 3-dB gain in the link budget. However, there is typically an implementation loss leading to a reduced gain. In situations where the magnitude of the channel fluctuates rapidly, the combining weights need to vary accordingly, which means that fast and robust estimation algorithms are required. A more simplified scheme from the implementation point of view is therefore EGC.

Unlike MRC, EGC only co-phases the signal before combining and hence its performance results to be suboptimal with respect to MRC. In practice, assuming quasi-equal noise levels in each branch (e.g. in a practical MEO scenario), the achievable diversity gain with EGC can be only marginally inferior as compared to MRC. In particularly challenging conditions, for example, when one stream has a high SNR while the other is highly dominated by noise, the SNR after combining may be lower than the SNR of the best stream. This means that EGC is injecting more noise in the signal after combining, which is of course an undesirable effect.

7.5.2 Combining position

The receiver has always two inputs coming from the two different antennas. Each input is processed by a separate receive chain until the combining takes place (except during the handover, when combining is not performed). Therefore, different architectures may be implemented to reflect the key differences arising from performing the combining in different positions, typically before or after the matched filtering.

7.5.2.1 Combining before the matched filtering

In this architecture, the only operations to be performed before the combining are the stream alignment and the channel estimation. The delay between the channels arises from the physical path differences between the two streams (e.g. cable lengths), and the alignment can be performed by correlating one stream with the other. This approach also allows compensating for the differential phase offsets between the streams. The alignment is particularly challenging because no pilots are available at this stage. However, a robust blind algorithm based on the correlation function (e.g. the non-coherent post-detection integration algorithm) can be used to correlate the two streams and estimate the relative delay. A relevant sample of the vast literature on this topic can be found in [10] and in the references therein.

7.5.2.2 Combining after the matched filtering

In this configuration, synchronization is performed before the combining, facilitating the required alignment between the two streams. Indeed, the compensation of the relative timing offset between the two streams can be performed inside the synchronization chain by detecting the beginning of the frame on each stream. This diversity technique is known as post-detection combining.

7.5.3 Performance of combining techniques

In this section, an analytical study is performed for MRC and EGC by considering a signal model normalized with respect to the channel gain. The rationale behind this normalization is that practical demodulators usually normalize the received signal by an estimate of the channel gain based on the correlation computed over the pilot fields.

7.5.3.1 Maximum ratio combining

Let us consider a system with a single transmit antenna (at the gateway) and N receive antennas having the following form:

$$\boldsymbol{y} = \mathbf{1}s + \mathbf{1}g + \hat{\boldsymbol{\eta}} \tag{7.1}$$

where

- $\boldsymbol{y} = [y_1, y_2, \ldots, y_N]^T$ is the $N \times 1$ received vector (y_k denotes the signal at the kth receive antenna);
- $\mathbf{1}$ is the $N \times 1$ vector of all ones;
- s is the transmitted data symbol, assumed zero mean and unit variance;
- g is the common noise component between the different received signals, modelled as a Gaussian noise with mean zero and variance α^2;
- $\hat{\boldsymbol{\eta}} = [\eta_1/h_1, \eta_2/h_2, \ldots, \eta_N/h_N]^T$ is the $N \times 1$ normalized receiver noise vector;
- h_k denotes the channel between the transmitter and the kth receive antenna, and $\boldsymbol{h} = [h_1, h_2, \ldots, h_N]^T$. is the $N \times 1$ channel vector;
- η_k denotes the noise at the kth receive antenna. We model $\boldsymbol{\eta} = [\eta_1, \eta_2, \ldots, \eta_N]^T$ as a Gaussian vector with
  - independent components
  - $E[\eta_l] = 0$ for all l
  - $E[|\eta_l|^2] = \sigma_l^2$: note that the noise components can have different variances.

The aforementioned model can be used in the case of the MEO combining scenario with $\boldsymbol{h}$ denoting the downlink channel, g denoting the common uplink noise, and η denoting the downlink receiver specific AWGN front-end noise.

The received signals from the different receive antennas are combined after suitable weighting. Let the weight vector be $\boldsymbol{u} = [u_1, u_2, \ldots, u_N]^T$ so that the combined signal takes the form

$$z = \boldsymbol{u}^H\boldsymbol{y} = \boldsymbol{u}^H\mathbf{1}s + g\boldsymbol{u}^H\mathbf{1} + \boldsymbol{u}^H\hat{\boldsymbol{\eta}}$$

Using the signal model mentioned above, the resulting SNR can be computed as

$$SNR = \frac{|\boldsymbol{u}^H\mathbf{1}|^2}{\alpha^2|\boldsymbol{u}^H\mathbf{1}|^2 + \sum_{l=1}^{N}\sigma_l^2(|u_l|^2/|h_l|^2)} \tag{7.2}$$

The MRC weights are the component of the vector $\boldsymbol{u}$ that maximizes the SNR. Towards obtaining the expression for $\boldsymbol{u}$, we define $w_k = u_k/(h_k)^*$ so that (7.2) becomes

$$SNR = \frac{|\boldsymbol{w}^H\boldsymbol{h}|^2}{\alpha^2|\boldsymbol{w}^H\boldsymbol{h}|^2 + \sum_{l=1}^N \sigma_l^2 |w_l|^2} \tag{7.3}$$

Then, we introduce the following modifications:

- $|\boldsymbol{w}^H\boldsymbol{h}|^2 = \boldsymbol{w}^H\boldsymbol{h}\boldsymbol{h}^H\boldsymbol{w} = \boldsymbol{w}^H\boldsymbol{R}_h\boldsymbol{w}$, where $\boldsymbol{R}_h = \boldsymbol{h}\boldsymbol{h}^H$
- $\sum_{l=1}^N \sigma_l^2|w_l|^2 = \boldsymbol{w}^H\boldsymbol{R}_n\boldsymbol{w}$, where $\boldsymbol{R}_n$ is a diagonal matrix with entries $\{\sigma_l^2\}$
- $\sum_{l=1}^N \sigma_l^2|w_l|^2 + \alpha^2|\boldsymbol{w}^H\boldsymbol{h}|^2 = \boldsymbol{w}^H\boldsymbol{R}_{tot}\boldsymbol{w}$, where $\boldsymbol{R}_{tot} = \boldsymbol{R}_n + \alpha^2\boldsymbol{h}\boldsymbol{h}^H$

The SNR in (7.3) then takes the compact form

$$SNR = \frac{\boldsymbol{w}^H\boldsymbol{R}_h\boldsymbol{w}}{\boldsymbol{w}^H\boldsymbol{R}_{tot}\boldsymbol{w}} \tag{7.4}$$

Denote $\boldsymbol{Q}$ to be the Cholesky square root of $\boldsymbol{R}_{tot}$, i.e. $\boldsymbol{R}_{tot} = \boldsymbol{Q}^2$ and we further assume that $\boldsymbol{R}_{tot}$ and hence $\boldsymbol{Q}$ are invertible. Let us define $\boldsymbol{x} = \boldsymbol{Q}\boldsymbol{w}$. With this substitution, the SNR in (7.4) can be further reduced as

$$SNR = \frac{\boldsymbol{x}^H\boldsymbol{Q}^{-1}\boldsymbol{R}_h\boldsymbol{Q}^{-1}\boldsymbol{x}}{\boldsymbol{x}^H\boldsymbol{x}} \tag{7.5}$$

The SNR in (7.5) is maximized (using the Rayleigh Quotient, standard matrix algebra result) when $\boldsymbol{x} = \boldsymbol{v}$, where $\boldsymbol{v}$ is the eigenvector corresponding to the largest eigenvalue of $\boldsymbol{Q}^{-1}\boldsymbol{R}_h\boldsymbol{Q}^{-1}$. Using the definition $\boldsymbol{x} = \boldsymbol{Q}\boldsymbol{w}$, the optimal combining vector becomes $\boldsymbol{w} = \boldsymbol{Q}^{-1}\boldsymbol{v}$.

It now remains to find an expression for $\boldsymbol{v}$. Noting that $\boldsymbol{R}_h = \boldsymbol{h}\boldsymbol{h}^H$, we can see that the matrix $\boldsymbol{Q}^{-1}\boldsymbol{R}_h\boldsymbol{Q}^{-1} = \boldsymbol{Q}^{-1}\boldsymbol{h}\boldsymbol{h}^H\boldsymbol{Q}^{-1}$ is nothing but an outer product of two column vectors. Hence the matrix is rank-one and the eigenvector corresponding to the non-zero eigenvalue is

$$\boldsymbol{v} = \boldsymbol{Q}^{-1}\boldsymbol{h} \tag{7.6}$$

By using (7.6) along with the definition $\boldsymbol{R}_{tot} = \boldsymbol{Q}^2$, the combining vector takes the form

$$\boldsymbol{w} = \boldsymbol{Q}^{-1}\boldsymbol{Q}^{-1}\boldsymbol{h} = [\boldsymbol{R}_{tot}]^{-1}\boldsymbol{h}$$

From a practical perspective, $\boldsymbol{R}_{tot}$ and $\boldsymbol{h}$ can be estimated from (7.1) by using standard estimation algorithms based on pilot fields. Further noting that $\boldsymbol{R}_{tot} = \boldsymbol{R}_n + \alpha^2\boldsymbol{h}\boldsymbol{h}^H$, and by using the Sherman–Morrison formula (or matrix inversion lemma for rank-one updates), we have

$$[\boldsymbol{R}_{tot}]^{-1} = \boldsymbol{R}_n^{-1} - \frac{\alpha^2\boldsymbol{R}_n^{-1}\boldsymbol{h}\boldsymbol{h}^H\boldsymbol{R}_n^{-1}}{1 + \alpha^2\boldsymbol{h}^H\boldsymbol{R}_n^{-1}\boldsymbol{h}} \tag{7.7}$$

We can further simplify (7.7) as

$$[\boldsymbol{R}_{tot}]^{-1}\boldsymbol{h} = \boldsymbol{R}_n^{-1}\boldsymbol{h} - \frac{\alpha^2\boldsymbol{h}^H\boldsymbol{R}_n^{-1}\boldsymbol{h}}{1 + \alpha^2\boldsymbol{h}^H\boldsymbol{R}_n^{-1}\boldsymbol{h}}\boldsymbol{R}_n^{-1}\boldsymbol{h}$$

leading to

$$w = [R_{tot}]^{-1}h = \frac{1}{1 + \alpha^2 h^H R_n^{-1} h} R_n^{-1} h \tag{7.8}$$

Scaling of the weight vectors does not change the SNR and hence, the expression in (7.8) can be simplified since the scaling factor $1/(1 + \alpha^2 h^H R_n^{-1} h)$ is common to both the channels. In view of this, the MRC weight vector takes the form $w = R_n^{-1}h$, and hence, $u = (R_n^{-1}h) \odot h^*$, where $\odot$ denotes the element-wise product.

Since R_n^{-1} refers only to the downlink noise variance, the MRC weight vector does not need any uplink SNR information. In particular, for independent downlink components, R_n is a diagonal matrix with entries $\{\sigma_l^2\}$ and the weight for the lth channel is $u_l = |h_l|^2/\sigma_l^2$.

The resulting SNR will be the largest eigenvalue of $Q^{-1}R_h Q^{-1}$ and it takes the form $\text{SNR} = h^H[R_{tot}]^{-1}h$. Noting that R_n is a diagonal matrix with entries $\{\sigma_l^2\}$, the SNR expression can be expanded as

$$SNR = \frac{1}{(1/h^H R_n^{-1} h) + \alpha^2} = \frac{1}{\left(1/\left(\sum_{l=1}^{N}(|h_l|^2/\sigma_l^2)\right)\right) + \alpha^2}$$

Clearly, the SNR on the lth downlink channel is $\gamma_{DL,l} = |h_l|^2/\sigma_l^2$; further, the uplink SNR takes the form, $\gamma_{UL} = 1/\alpha^2$. Hence, the resulting SNR for MRC takes the form

$$SNR = \frac{1}{\left(1/\sum_{l=1}^{N}\gamma_{DL,l}\right) + (1/\gamma_{UL})} \tag{7.9}$$

7.5.3.2 Equal gain combining

The model for the received signal is the same as in (7.1). The EGC in this case is a simple average and the combined signal takes the form

$$z = \frac{1}{N}(\mathbf{1}^H y) = s + g + \frac{1}{N}(\mathbf{1}^H \hat{\eta})$$

By using the signal model mentioned above, the resulting SNR can be computed as

$$SNR = \frac{1}{\alpha^2 + (1/N^2)\sum_{l=1}^{N}(\sigma_l^2/|h_l|^2)} \tag{7.10}$$

After some algebra (Cauchy Schwartz inequality) it can be shown that the SNR of EGC is lower than that of MRC.

7.5.4 Switching threshold computation using downlink SNR

Despite the theoretical optimality of MRC, its practical implementation always requires the computation of the combining weights. On the other hand, in the described scenario, EGC becomes a simple average of the streams and requires no extra computation for the weights, making its implementation much simpler. Further, SC can be directly implemented as a simple switch between streams. Hence, from the implementation point of view, EGC and SC are much more appealing than MRC.

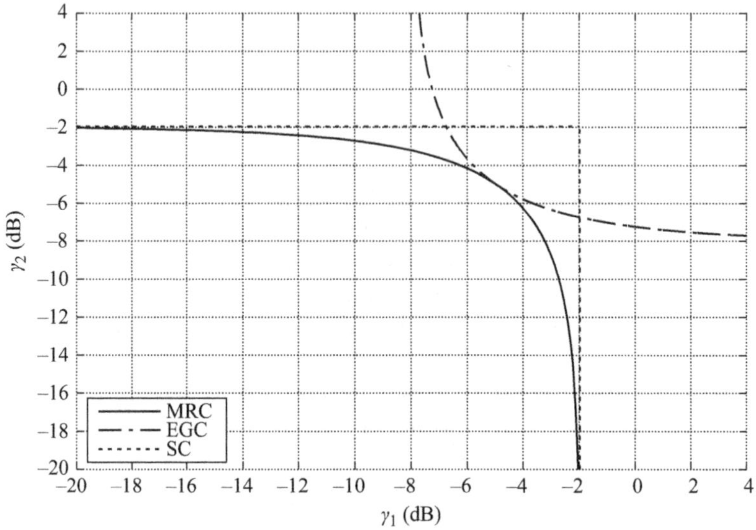

Figure 7.13 SNRs over the two branches for different combining techniques with the same SNR after combining

Since the performance of the different combining algorithms depends on the SNR value on each of the receiver branch, in this section, we address the analysis of the performance-complexity trade-off as a function of the SNR imbalance between the two branches. Such imbalance may be the result of hardware aging, and its magnitude may vary from very small (e.g. statistical fluctuations of fractions of dB) to very big (e.g. when one antenna is blocked or an outage/failure occurs).

In case of SNR imbalance, MRC is always optimal (in terms of maximization of the received SNR), performing better than other techniques. However, under certain operative conditions, the performance of EGC may be very similar to the one of MRC, making it a good candidate for a practical implementation. Nevertheless, in different operative conditions, EGC performs poorly, while SC results to provide gains relatively close to MRC (see e.g. Figure 7.13). Therefore, in a practical implementation, it may be convenient to use both EGC and SC and switch from one to the other depending on the evolution over time of the operative conditions. This choice would improve the SNR after combining (achieving gains close to the ones provided by MRC) while, at the same time, it would minimize the receiver complexity. The optimal switching threshold between EGC and SC can be found at the crossing point of the curves representing the SNR after combining, written as a function of the SNRs over the branches. In the following, we assume that only the downlink SNR is known.

The resulting SNR after combining for EGC reads

$$\gamma_{EGC} = \frac{1}{\alpha^2 + (1/4)((1/\gamma_1) + (1/\gamma_2))} \tag{7.11}$$

For SC, the SNR after combining is simply the maximum SNR over the branches

$$\gamma_{SC} = \max\left(\frac{1}{(1/\gamma_1)+\alpha^2}, \frac{1}{(1/\gamma_2)+\alpha^2}\right) \tag{7.12}$$

The optimal threshold is such that EGC is used only when $\gamma_{EGC} > \gamma_{SC}$, otherwise SC is adopted. By denoting with γ_M the maximum between the two downlink SNRs and with γ the other one, the condition for the switching can be written as

$$\frac{1}{\alpha^2 + (1/4)((1/\gamma_M)+(1/\gamma))} > \frac{1}{(1/\gamma_M)+\alpha^2}$$

Introducing the imbalance $\theta = \gamma/\gamma_M$ and replacing γ with $\theta\gamma_M$ yield to $\theta > 1/3 = -4.77$ dB. This means that when the SNR imbalance (in absolute value) between the two streams is greater than 4.77 dB, SC outperforms EGC in terms of achievable SNR after combining.

7.5.4.1 Performance trade-off

As an example, the couples of SNRs (on a single branch) that provide an SNR after combining equal to −2 dB are shown in Figure 7.13. The single-branch SNRs are reported on the x- and y-axis, respectively, for different combining techniques. The SNR value after combining (chosen as integer for the ease of visualization) is close to the lowest operational point of the DVB-S2X standard (excluding the very low SNR operational mode) for normal length FEC frames [11,12]. Any other operative point would show the same relative behaviour of the three algorithms, resulting in all the curves being shifted of the same quantity. From the figure, the optimality of MRC is evident, since it is the technique that requires the lowest SNRs over the two branches to satisfy the constraint of −2 dB after combining. When there is no SNR imbalance, MRC and EGC are equivalent in terms of performance since they both provide the expected 3 dB gain. When the SNR imbalance grows, EGC starts diverging from MRC, and at a certain point, it is outperformed by SC.

7.5.5 Switching threshold computation using total SNR

The per-stream SNR, $\hat{\gamma}_k, k = 1, 2$ takes the form

$$\frac{1}{\hat{\gamma}_k} = \frac{1}{\gamma_k} + \alpha^2 \tag{7.13}$$

The SNR for SC, denoted by $\hat{\gamma}_{SC}$, takes the familiar form

$$\hat{\gamma}_{SC} = \max(\hat{\gamma}_1, \hat{\gamma}_2) = \hat{\gamma}_M$$

However, the expression for the SNR of EGC, $\hat{\gamma}_{EGC}$, involves $\hat{\gamma}_1, \hat{\gamma}_2$, and the power of the uplink noise α^2. By replacing (7.13) in (7.11), the SNR of EGC becomes

$$\hat{\gamma}_{EGC} = \frac{1}{(\alpha^2/2) + (1/4)((1/\hat{\gamma}_1)+(1/\hat{\gamma}_2))}$$

By imposing $\hat{\gamma}_{EGC} > \hat{\gamma}_{SC}$ and by introducing the imbalance $\hat{\theta} = \hat{\gamma}/\hat{\gamma}_M > 0$, it can be shown that the threshold reads

$$\hat{\theta} > \frac{1}{3 - 2\hat{\gamma}_M\alpha^2} \tag{7.14}$$

The threshold computation in (7.14) depends on the uplink noise power and on the maximum SNR over the branches. When the power of the uplink noise is unknown at the receiver, an exact threshold computation is intractable. However, the optimal threshold results to be greater than the threshold obtained considering only the downlink SNRs, that is

$$\frac{1}{3 - 2\hat{\gamma}_M\alpha^2} > \frac{1}{3} \tag{7.15}$$

when $\hat{\gamma}_M < 3/2\alpha^2$. By using (7.13), (7.15) can be expressed as $\gamma_M > -3/\alpha^2$, which is always satisfied because γ_M is positive.

In general, for any $\theta \leq 1$, by using (7.13), it can be shown that

$$0 < \theta = \frac{\gamma}{\gamma_M} \leq \frac{\hat{\gamma}}{\hat{\gamma}_M} \leq 1 \tag{7.16}$$

The SNR imbalance computed with the total SNR

$$0 < \hat{\theta} = \frac{\hat{\gamma}}{\hat{\gamma}_M} \leq 1$$

is therefore closer to 1 than the SNR imbalance computed with the downlink SNR.

Since the receiver usually can measure only the total SNR, the optimal threshold to be used is (7.14), i.e. the one computed by using the total SNR. However, since estimating the uplink noise power may be challenging, a trade-off between accuracy and complexity would be using the threshold computed by using the downlink SNRs. Since, as shown in (7.16), the value of the optimal threshold is higher than the value of the suboptimal threshold, using the suboptimal threshold means that sometimes EGC would occur when it should not, thus increasing the noise at the demodulator.

7.5.6 Combining gain

In practical scenarios for MEO systems, the two receive antennas are co-located, and thus, there is no SNR imbalance between two streams. With this assumption, the total SNR after combining can be written as

$$\gamma_{post} = \frac{1}{(1/\gamma_{UL}) + (1/2\gamma_{DL})} \tag{7.17}$$

where γ_{UL} is the uplink SNR and γ_{DL} is the downlink SNR. The expression (7.17) holds for both MRC and EGC, as can be seen by inspecting (7.9) and (7.10). Similarly, the total SNR before combining reads

$$\gamma_{pre} = \frac{1}{(1/\gamma_{UL}) + (1/\gamma_{DL})}$$

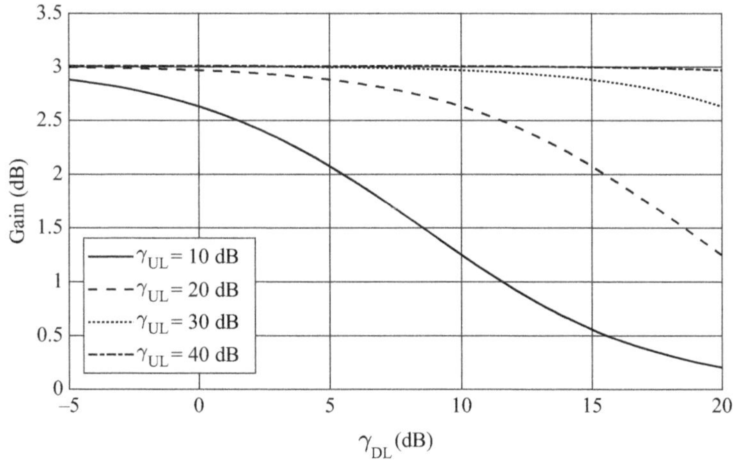

Figure 7.14 Combining gain for different values of downlink and uplink SNRs

The combining gain can be therefore defined as

$$\chi = \frac{\gamma_{post}}{\gamma_{pre}} = 1 + \frac{1}{1 + 2(\gamma_{DL}/\gamma_{UL})} \tag{7.18}$$

and $\chi \in [1, 2]$. From (7.18), one can see that the presence of uplink noise reduces the combining gain. Indeed, if the system is dominated by the uplink noise, then the gain provided by the combining may be very limited since the combining affects exclusively the downlink noise. On the other hand, when no uplink noise is present, then $\chi = 3$ dB. This behaviour is clearly shown in Figure 7.14.

7.6 Roadmap

Since diversity combining is commercially interesting when two antennas are already in place, use cases are driven by dual-antenna use cases for handover purposes. In these cases, diversity combining can offer up to 3 dB of SNR gain when the two signals are corrupted by uncorrelated noise components. MEO applications are the most obvious use case for such dual-antenna system: in order to do a seamless satellite handover, the receiver typically exploits two antennas, which can be linked to two demodulators.

Another use case for dual-antenna systems is the maritime use case, whereby two (or more) antennas are used for backup in case of blockage of one of the antennas. Dual-antenna systems typically use an arbitrator to switch signals between antennas. In case of blockage (which can be preconfigured) or antenna failure, the arbitrator switches automatically between the two antennas. In case of a modem with dual-demodulator, this functionality could be absorbed inside the modem.

In the case of GEO, dual-demodulator but single-antenna systems are used for satellite handover in the context of beam-switching. By having the second demodulator locked to the new beam, several steps of the synchronization can be prepared before the switchover, allowing for a much faster handover. However, since in this case a single antenna is foreseen, diversity combining is not applicable. On the other hand, if seamless GEO satellite handover is considered for mobile terminals, then dual antennas need to come into play to avoid service disruptions during the antenna repointing.

7.7 Conclusions

In this chapter, we presented a cost-effective solution to implement seamless satellite switching or handover in the context of MEO satellites. We have shown that existing mechanisms described in literature have some shortcomings when it comes to achieving optimal performance and zero packet loss. The presented solution, making use of SDN technology, allows to construct an infrastructure combining the concepts of 'make-before-break' and 'unidirectional switching' to achieve the goals put forward.

A prototype has been built to demonstrate the handover performance and the test results clearly indicate that satellite switching with zero packet loss is achievable. Moreover, the impact of potential packet reordering as a result of delay differences between both satellite paths is negligible. TCP communication can be established without errors and User Datagram Protocol (UDP) hardly experiences significant reordering. We can safely state that the mechanisms described in this chapter do not result in 'service degradation' of any kind. SDN can successfully be used for satellite switching and is the technology of choice to realize seamless handover.

Further, in order to take full advantage of the two antennas outside of the handover phase, diversity combining is considered to increase the SNR at the receiver. The performance of three classic algorithms (i.e. MRC, EGC, and SC) has been specialized to MEO applications by resorting to realistic modelling of both channel and signal. Although MRC is the optimal solution in terms of SNR maximization, a simpler implementation that does not require extra computation for the combining weights resorts to a combination of EGC and SC. In particular, while EGC performs very close to MRC when the SNR imbalance between the two received signals is small, SC outperforms EGC and approximates MRC when the imbalance is big. The optimal switching threshold between EGC and SC (in terms of SNR imbalance) has been computed in order to provide a good trade-off between performance and complexity for a realistic implementation.

The proposed concepts are not limited to MEO constellations but can be applied to scenarios where handover is needed and diversity can be exploited. However, MEO constellations, being commercially available, allow for a faster realization of the concepts illustrated herein.

References

[1] ITU-T Recommendation P.10, 'Vocabulary for performance and quality of service', Amendment 5, July 2016.
[2] ETSI TR 102 376-2 V1.1.1, 'Digital Video Broadcasting (DVB); Implementation guidelines for the second generation system for Broadcasting, Interactive Services, News Gathering and other broadband satellite applications; Part 2-S2 Extensions (DVB-S2X)', November 2015.
[3] ITU-R P.618-12, 'Propagation data and prediction methods required for the design of Earth-space telecommunications systems', P Series Radiowave Propagation, July 2015.
[4] ITU-R P.840-6, 'Attenuation due to clouds and fog', P Series Radiowave Propagation, September 2013.
[5] Kourogiorgas C., Panagopoulos A.D., and Arapoglou P.-D., 'Rain Attenuation Time Series Generator for Medium Earth Orbit Links Operating at Ka Band and above,' European Conference on Antennas and Propagation (EUCAP), p. 3506–3510, 2014.
[6] Yang B., Wu Y., and Chu X., 'Seamless Handover in Software-Defined Satellite Networking,' IEEE Communication Letters, vol. 20, no. 9, pp. 1768–1771, 2016.
[7] Brennan D.G., 'Linear Diversity Combining Techniques,' Proceedings of the IEEE, vol. 91, no. 2, pp. 331–356, February 2003.
[8] Panagopoulos A.D., Arapoglou P.-D.M., Kanellopoulos J.D., and Cottis P.G., 'Long-term Rain Attenuation Probability and Site Diversity Gain Prediction Formulas,' IEEE Transactions on Antennas and Propagation, vol. 53, no. 7, pp. 2307–2313, July 2005.
[9] Simon M.K. and Alouini M.-S., *Digital Communication over Fading Channels*, 2nd ed., Hoboken, NJ: Wiley, 2005.
[10] Pedone R., Villanti M., Vanelli-Coralli A., Corazza G.E., and Mathiopoulos P.T., 'Frame Synchronization in Frequency Uncertainty,' IEEE Transactions on Communications, vol. 58, no. 4, pp. 1235–1246, April 2010.
[11] ETSI EN 302 307-1 Ver. 1.4.1, 'Digital Video Broadcasting (DVB); Second generation framing structure, channel coding and modulation systems for Broadcasting, Interactive Services, News Gathering and other Broadband satellite applications – Part 1: DVB-S2', November 2014
[12] ETSI EN 302 307-2 Ver. 1.1.1, 'Digital Video Broadcasting (DVB); Second generation framing structure, channel coding and modulation systems for Broadcasting, Interactive Services, News Gathering and other Broadband satellite applications – Part 2: DVB-S2 Extensions (DVB-S2X)', October 2014.

Chapter 8

Powerful nonlinear countermeasures for multicarrier satellites: progression to 5G

Bassel F. Beidas[1]

The exigent demand for transmitting high data rates over satellites, coupled with the challenge to maximize satellite mass efficiency, has necessitated that multiple carriers with high-order modulation share the same transponder's high-power amplifier (HPA) that is operated close to saturation. Several powerful technological solutions, applied at the transmitter, in the form of predistortion, and at the receiver, in the form of equalization, are explored in this chapter to minimize the resulting nonlinear distortion.

To establish greater commonality with the emerging fifth-generation (5G) ecosystem, the second part of this chapter endeavors to apply orthogonal frequency-division multiplexing (OFDM) signaling for broadband satellite transmission in the forward direction, namely, from the gateway to terminals. 5G terrestrial systems continue to use OFDM air interface. The aforementioned powerful countermeasures are then generalized, utilized, and shown to exhibit excellent performance in allowing OFDM-based satellite systems to be competitive with, and in some cases surpassing, traditional systems that use single-carrier modulation (SCM) when employing high-order constellations and/or having multiple signals share the same transponder.

8.1 Introduction

5G terrestrial wireless networks are advancing at accelerated speed and attracting billions of dollars in research and infrastructure development. Very ambitious goals are to be reached that include significantly higher capacity, more reliable communications, enhanced security, low latency, reduced cost, and broadband service in rural areas. It is envisaged that satellite communications will play a dominant role in the emerging 5G landscape. This is partly due to the satellites' large footprint with massive area coverage, prominent reach to regions with limited infrastructure, and their inherent resilience to natural disasters or wide scale attacks.

[1]Advanced Development Group, Hughes, USA

The key to satisfying the aggressive demand for higher satellite throughput in broadband and broadcasting applications is to maximize the system efficiency at various levels: (1) payload mass efficiency through sharing the satellite HPA by multiple carriers; (2) bandwidth efficiency by employing frequency-compact carriers with high-order modulation with multiple concentric rings, which are packed tightly in the available spectrum; (3) power efficiency by driving the HPA close to saturation; (4) energy efficiency by utilizing adaptive coding and modulation (ACM) using capacity-approaching forward-error correction (FEC) codes from the widely adopted satellite standard DVB-S2 [1] and its extension DVB-S2X [2]. However, due to the inherently nonlinear nature of HPAs, this creates an environment with substantial nonlinear distortion that is severely damaging if left uncompensated.

The first part of this chapter presents a cornucopia of powerful compensation techniques, applied at the transmitter and at the receiver, to effectively minimize the linear and nonlinear distortion for highly efficient satellite operation. Toward this, an analytical framework is described that characterizes distortion among carriers suitable for multicarrier satellite applications. This framework, introduced by Beidas in [3], utilizes Volterra series representation to account for the memory associated with nonlinear distortion, memory that is within the carrier itself and those pertaining to other carriers. When the number of carriers is one, the intermodulation (IMD) analysis simplifies to that of nonlinear intersymbol interference (ISI), as studied by Benedetto *et al.* in the classical work [4]. Another special case of this analytical characterization is when the number of carriers is two, an investigation of which is conducted by Beidas and Seshadri [5].

Further enhancement to the spectral efficiency of a communication system can be achieved through the use of faster-than-Nyquist (FTN) signaling [6,7]. FTN offers several advantages for nonlinear satellite systems. For example, FTN increases the symbol rate without being adversely affected by the transponder's multiplexing filters as FTN does not alter the signal spectral content or shape. Also, FTN provides a degree of freedom to increase the spectral efficiency without the need to introduce additional rings in the signal constellation, which is beneficial in the presence of nonlinear HPA. Advanced receivers are developed in [8,9] to realize the gains of FTN for nonlinear satellite links.

The second part of this chapter covers the intersection of 5G and satellite communications by endeavoring to apply OFDM-like signaling for broadband satellite transmission in the forward direction, namely, from the gateway to user terminals. The OFDM air interface continues to be used in 5G on the downlink. This follows the recent progress in the standardization of 5G new radio access technology as specified by the leading third-generation partnership project (3GPP) [10]. The greater commonality in the air interface can facilitate integration between satellite and terrestrial broadband networks. This integration leads toward future-proofing satellite use cases and provides resilience and ubiquity to 5G services by extending their coverage to regions only satellites can penetrate.

Additionally, OFDM, a special form of multicarrier modulation (MCM) and due to its many well-documented advantages [11], can deliver wide-ranging benefits to broadband satellite systems, among them (1) resistance to narrowband interference from terrestrial microwave signals as service providers are increasingly demanding a share of the radio frequency spectrum traditionally occupied by satellites, promoting coexistence; (2) robustness to frequency-selective distortion from on-board transponder multiplexing filters, i.e., amplitude distortion appears flat over narrowband OFDM subcarriers; (3) flexible and highly efficient spectrum utilization by using adaptive loading of the best power allocation and modulation selection, in a location-dependent manner, when channel state information is exploited.

The major obstacle to adopting OFDM in satellite systems is its inherent sensitivity to nonlinear distortion, due to high peak-to-average power ratio (PAPR) levels, requiring inefficiently operating the on-board HPA at a large output back-off (OBO). The aforementioned powerful countermeasure strategies are then generalized and applied to efface the nonlinear distortion. The common aspect is the successive application of distortion cancellation, distortion that incorporates not only intercarrier interference (ICI) from the narrowband OFDM subcarriers within the signal itself, but also those belonging to other signals that share the same HPA. These techniques are applied at the transmitter or at the receiver, and a combination of both for extracting additional performance gain. With advanced compensation, it is possible to use OFDM-like signaling over satellite, while being competitive with, and in some cases surpassing, traditional systems with SCM employing state-of-the-art enhanced receiver architecture from [12].

The remainder of the chapter is organized as follows. The system model is described in Section 8.2, wherein a single nonlinear transponder relays multiple FEC-encoded high-order modulation carriers. In Section 8.3, a theoretical characterization of IMD is described using multicarrier Volterra representation, along with a modular formulation and a reduced-complexity implementation. Section 8.4 investigates several powerful countermeasures that minimize linear and nonlinear distortion for satellite systems which use SCM. Section 8.5 explores OFDM-like signaling in broadband satellite applications and evaluates methods of successive correction to progressively reduce nonlinear distortion. Section 8.6 contains concluding remarks.

8.2 System description

8.2.1 Signal model

The system architecture considered here models a multicarrier satellite system, depicted in Figure 8.1, that includes M_c independent carriers amplified by a single HPA on-board the satellite transponder. Each carrier applies FEC encoding, followed by interleaving and Gray mapping onto constellation associated with high-order modulation such as amplitude phase-shift keying (APSK) with alphabet size M, yielding complex-valued symbols $\{u_{m,k}; m = 1, 2, \ldots, M_c\}_{k=-\infty}^{\infty}$ at the symbol rate of T_s^{-1}. The composite signal $s_c(t)$ at the transmitter output is described in complex-valued form as

$$s_c(t) = \sum_{m=1}^{M_c} \frac{1}{\sqrt{M_c}} \cdot s_m(t) \cdot e^{j(2\pi f_m t+\theta_m)}. \tag{8.1}$$

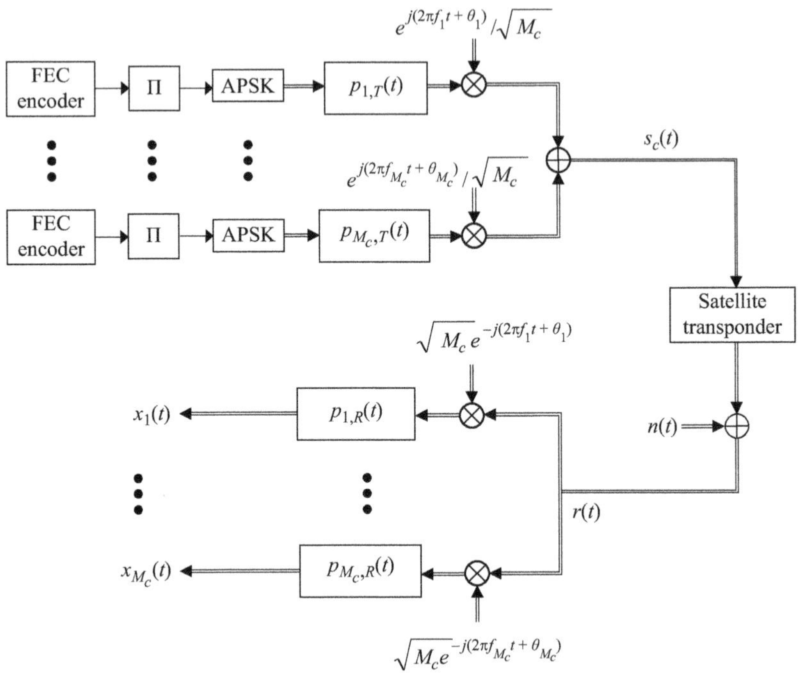

Figure 8.1 Block diagram of multicarrier satellite system

The individual waveforms $s_m(t)$ in (8.1) are digitally modulated and given by

$$s_m(t) = \sum_{k=-\infty}^{\infty} a_{m,k} \cdot p_{m,T}(t - kT_s - \varepsilon_m T_s), \tag{8.2}$$

where $\{\varepsilon_m, \theta_m\}$ represents the normalized difference in signal time and carrier phase, respectively, and f_m is the mth center frequency. For better utilization of bandwidth, the case of uniform spacing in frequency, say Δf, is considered or

$$f_m = \left(m - \frac{M_c + 1}{2}\right) \cdot \Delta f; \qquad m = 1, 2, \ldots, M_c, \tag{8.3}$$

but the analysis presented here is applicable to any other frequency plan.

The first step of the receiver includes a bank of receive filters to frequency-translate each carrier to baseband and to apply a filtering operation with impulse response $p_{m,R}(t)$ so that the noise is rejected in the nonsignal band, as shown in the lower portion of Figure 8.1. The input–output relationship of the mth receive filter bank is expressed as

$$x_m(t) = \int_{-\infty}^{\infty} r(t - \tau)\sqrt{M_c}e^{-j(2\pi f_m(t-\tau)+\theta_m)} \cdot p_{m,R}(\tau)d\tau. \tag{8.4}$$

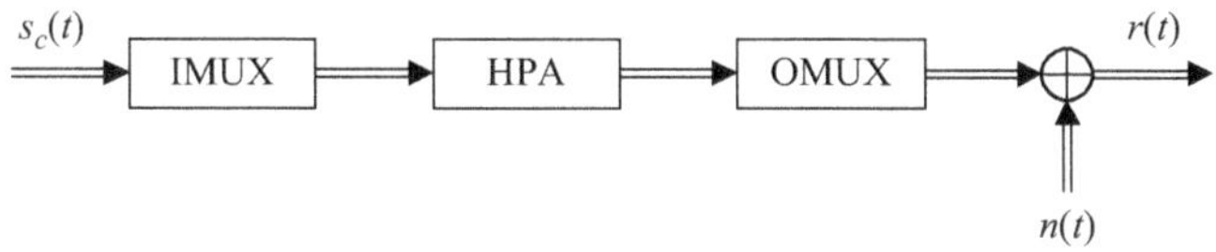

Figure 8.2 Block diagram of the satellite channel

The signals $x_m(t)$, $m = 1, 2, \ldots, M_c$, at the receive filter outputs in (8.4) are sampled at multiples of the symbol rate to allow for fractionally spaced (FS) group-delay (GD) equalization, which compensates for the linear phase distortion introduced by the multiplexing filters.

8.2.2 Satellite channel model

The satellite channel model, depicted in Figure 8.2, includes an input multiplexing (IMUX) filter, a nonlinear HPA and an output multiplexing (OMUX) filter. The IMUX filter selects the desired group of M_c signals, thereby limiting the impact of adjacent uplink signals. The OMUX filter, placed after the HPA, is used to limit nonlinear interference into adjacent transponders. The frequency responses associated with the IMUX and OMUX filters are obtained by using the scaling formula of the corresponding filter responses in [1] with a scale factor of M_c or

$$R'(f) = R\left(\frac{f}{M_c}\right); \tag{8.5}$$

$$G'(f) = \frac{1}{M_c} \cdot G\left(\frac{f}{M_c}\right). \tag{8.6}$$

An example of the frequency responses of the IMUX and OMUX filters is shown in Figure 8.3 for the case of $M_c = 3$. The amplitude and phase distortions introduced by the HPA, in terms of amplitude modulation (AM)/AM and AM/phase modulation (PM) conversions, respectively, are taken from [1] and shown in Figure 8.4.

The satellite channel adds downlink noise $n(t)$, assumed as additive white Gaussian noise (AWGN) with single-sided power spectral density (PSD) level of N_0 (W/Hz), which corrupts the signal at the OMUX output. The uplink noise is assumed negligible relative to the downlink noise, a situation achieved through proper satellite link parameters including the size of the transmit antenna.

It is insightful in nonlinear satellite channels to report performance results in terms of total degradation (TD) incurred when achieving a target error rate as a function of OBO. The OBO is defined as the loss in power, relative to saturation, of the signal at the output of HPA. The parameter TD, in decibel (dB), is defined as [5]

$$\mathrm{TD} = \mathrm{OBO} + \left.\frac{E_S}{N_0}\right|_{NL} - \left.\frac{E_S}{N_0}\right|_{L}, \tag{8.7}$$

where $\left.\frac{E_S}{N_0}\right|_{NL}$ is the average per-symbol signal-to-noise ratio (SNR), in dB, required to attain a target error rate in the presence of nonlinear distortion, and $\left.\frac{E_S}{N_0}\right|_{L}$ is its AWGN counterpart.

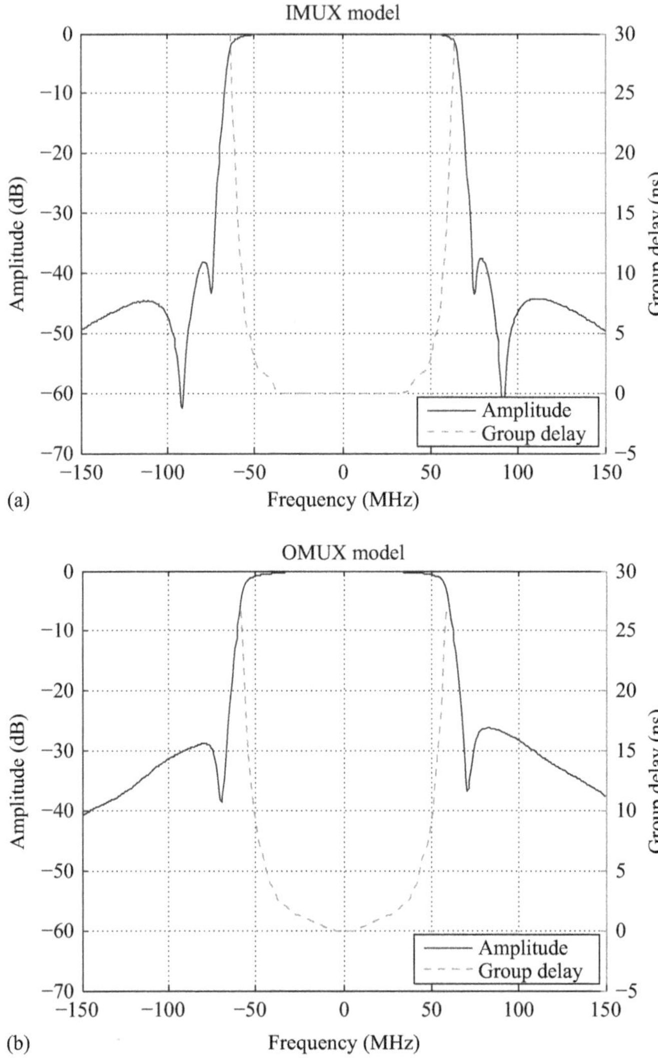

Figure 8.3 Characteristics for wideband satellite filters for (a) IMUX and (b) OMUX when $M_c = 3$

8.3 Multicarrier analysis of IMD

8.3.1 Multicarrier Volterra representation

The placement of the HPA between the multiplexing filters in the satellite transponder effectively creates a nonlinear channel with memory, which is well-known to be accurately modeled using Volterra series. This section analyzes third-order IMD at the

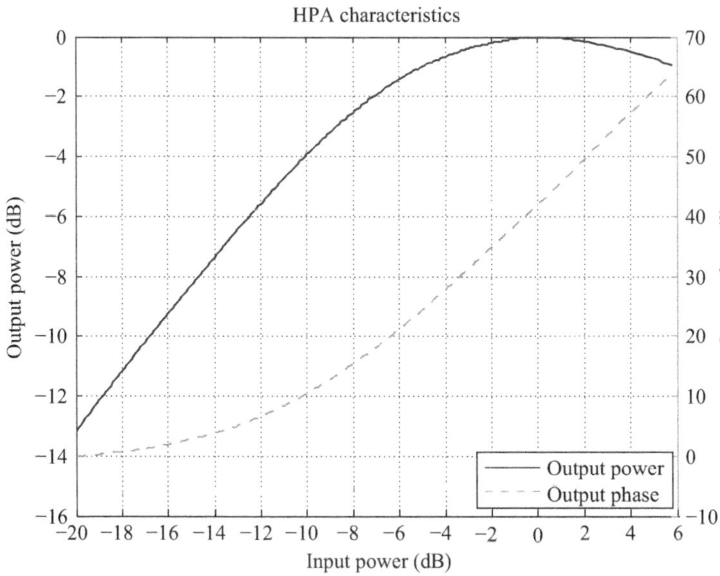

Figure 8.4 The AM/AM and AM/PM characteristics for nonlinear HPA

receive filter bank that results when a single satellite transponder HPA is utilized by multiple high-order modulation carriers based on multicarrier Volterra series representation. In addition, the amount of the resulting nonlinear distortion among carriers is analytically quantified.

In particular, let m_d be a specific branch of the filter bank and let $z_{m_d}^{(3)}(t)$ be its waveform prior to the receive filtering operation which is due to the third-order term in the power series representation of the nonlinearity. Then $z_{m_d}^{(3)}(t)$ is expressed as [3]

$$z_{m_d}^{(3)}(t) = \frac{\gamma^{(3)}}{M_c} \cdot \sum_{m_1=1}^{M_c} \sum_{m_2=m_1}^{M_c} \sum_{m_3=1}^{M_c} 2^{1-\delta_{m_1 m_2}} \cdot e^{j(2\pi(f_{m_1}+f_{m_2}-f_{m_3}-f_{m_d})t+(\theta_{m_1}+\theta_{m_2}-\theta_{m_3}-\theta_{m_d}))} \cdot s_{m_1}(t) \cdot s_{m_2}(t) \cdot s_{m_3}^*(t) \tag{8.8}$$

where δ_{ij} is the Kronecker Delta function, equal to unity if $i = j$ and zero otherwise. By substituting (8.8) into (8.4), and sampling at symbol rate, the third-order contribution of the nonlinearity to the receive filter bank, $\zeta_{m_d}^{(3)}((n+\varepsilon_{m_d})T_s)$, is expressed as

$$\zeta_{m_d}^{(3)}((n+\varepsilon_{m_d})T_s) = \int_{-\infty}^{\infty} z_{m_d}^{(3)}\left((n+\varepsilon_{m_d})T_s - \tau\right) \cdot p_{m_d,R}(\tau)d\tau. \tag{8.9}$$

The effective channel impulse response of IMD is expressed in terms of the generalized definition of the Volterra kernels, introduced in [5], as

$$h^{(3)}_{bcde}(t_1, t_2, t_3; f_o) \triangleq \int_{-\infty}^{\infty} p_{b,T}(t_1 - \tau) \cdot p_{c,T}(t_2 - \tau) \cdot p^*_{d,T}(t_3 - \tau) \cdot p_{e,R}(\tau) \cdot e^{-j2\pi f_o \tau} d\tau, \quad (8.10)$$

where $1 \leq b, c, d, e \leq M_c$. In this generalized definition, there are four subscripts: *bcd* refer to the three individual transmit filters involved in the third-order product, whereas *e* corresponds to the receive filter. The frequency parameter f_o, a function of the spacing between carriers, plays a significant role in the multicarrier analysis. To expose its significance, we express the generalized Volterra kernel of (8.10) in the frequency domain by using properties of Fourier transforms as

$$h^{(3)}_{bcde}(t_1, t_2, t_3; f_o) = \int_{-\infty}^{\infty} \Big[P_{b,T}(f - f_o) e^{j2\pi(f - f_o)t_1} \underset{f}{\star} P_{c,T}(f - f_o) e^{j2\pi(f - f_o)t_2} \underset{f}{\star} P^*_{d,T}(-(f - f_o)) e^{j2\pi(f - f_o)t_3} \Big] \cdot P_{e,R}(f) df, \quad (8.11)$$

where $\underset{f}{\star}$ denotes the convolution operator in the frequency domain. The bracketed term in (8.11) represents the IMD term resulting from the third-order product and is as wide in frequency as the sum of the widths of the individual transmit pulses, because of the convolution operation. In addition, this IMD term is centered at f_o relative to the receive filter, $P_{e,R}(f)$, which is centered at the origin. Fortunately, this contribution diminishes rapidly when $|f_o| > \Delta f$ for practical bandwidth-efficient choices of transmit and receive filters.

By examining (8.8)–(8.11), the following important observations can be made:

- Utilizing the same nonlinearity by multiple carriers creates multitude of third-order IMD terms, for a total of $M_c^2(M_c + 1)/2$ distinct terms. By using [5, Eq. (15)] to provide explicit expressions in terms of the interfering symbols, each of these third-order IMD terms appearing at the m_dth branch when sampled at the symbol rate is expressed as

$$\begin{aligned} \text{IMD}^{(3)}_{m_d}([m_1 m_2 m_3]) &= \frac{\gamma^{(3)}}{M_c} \cdot 2^{1-\delta_{m_1 m_2}} \cdot e^{j2\pi(f_{m_1} + f_{m_2} - f_{m_3} - f_{m_d})(n + \varepsilon_{m_d})T_s} \cdot e^{j(\theta_{m_1} + \theta_{m_2} - \theta_{m_3} - \theta_{m_d})} \\ &\cdot \sum_{k_1=-\infty}^{\infty} \sum_{k_2=-\infty}^{\infty} \sum_{k_3=-\infty}^{\infty} a_{m_1, n-k_1} \cdot a_{m_2, n-k_2} \cdot a^*_{m_3, n-k_3} \\ &\cdot h^{(3)}_{m_1 m_2 m_3 m_d}((k_1 - \delta\varepsilon_{m_1})T_s, (k_2 - \delta\varepsilon_{m_2})T_s, \\ &\quad (k_3 - \delta\varepsilon_{m_3})T_s; f_{m_1} + f_{m_2} - f_{m_3} - f_{m_d}), \end{aligned} \quad (8.12)$$

  where $\delta\varepsilon_{m_i} = \varepsilon_{m_i} - \varepsilon_{m_d}$, and $1 \leq m_1, m_2, m_3, m_d \leq M_c$. This suggests that the third-order IMD at the sampled receive filter output represents a discrete convolution of the interfering symbols with a three-dimensional impulse response quantified by the generalized Volterra kernel. These kernels are specified at a multiple factor of the carrier spacing, a factor that depends on the carrier combination $[m_1 m_2 m_3]$ at hand.

- The IMD terms can be classified on the basis of their frequency centers f_o in (8.11). There are $3(M_c - 1) + 1$ centers of IMD spanning the range

$$\begin{bmatrix} -\frac{3}{2}(M_c - 1)\cdot\Delta f - f_{m_d} \\ (-\frac{3}{2}(M_c - 1) + 1)\cdot\Delta f - f_{m_d} \\ \vdots \\ -\Delta f \\ 0 \\ +\Delta f \\ \vdots \\ \frac{3}{2}(M_c - 1)\cdot\Delta f - f_{m_d} \end{bmatrix}. \tag{8.13}$$

- The most significant terms are those with IMD that is frequency-centered at zero, achieved when the condition $f_{m_1} + f_{m_2} - f_{m_3} - f_{m_d} = 0$ is satisfied. The next set of significant IMD terms is centered at $\pm\Delta f$, achieved when the condition $f_{m_1} + f_{m_2} - f_{m_3} - f_{m_d} = \pm\Delta f$ is met. For equally spaced carriers, the number of such distinct terms can be shown to be

$$N_{m_d}(f_0) = \begin{cases} \frac{1}{4}(M_c^2 - (-1)^{m_d}\cdot M_c \pmod 2) \\ +\frac{1}{2}m_d(M_c - m_d + 1), & f_0 = 0 \\ \frac{1}{4}(M_c(M_c - 2) + (-1)^{m_d}\cdot M_c \pmod 2) \\ +\frac{1}{2}(M_c m_d - (m_d - 1)(m_d - 2)), & f_0 = -\Delta f \\ \frac{1}{4}(M_c(M_c + 2) + (-1)^{m_d}\cdot M_c \pmod 2) \\ +\frac{1}{2}m_d(M_c - m_d - 1), & f_0 = +\Delta f \end{cases}. \tag{8.14}$$

Table 8.1 contains the carrier combinations $[m_1 m_2 m_3]$ that produce these terms for IMD centered at zero when the number of carriers is 2, 3, 4, and 5, at each m_dth branch of the filter bank; $1 \le m_d \le M_c$. Carrier combinations for IMD centered at $\pm\Delta f$ are included in [3, Table II].

- Among the numerous terms in (8.8), there is one "special" summand that is associated with the condition $m_1 = m_2 = m_3 = m_d$. It can be alternatively described as nonlinear ISI and naturally has a frequency center of zero.

Table 8.1 Carrier combinations that produce IMD, frequency-centered at zero, for equally spaced multiple carriers

	Receive filter branch				
	$m_d = 1$	$m_d = 2$	$m_d = 3$	$m_d = 4$	$m_d = 5$
$M_c = 2$	[111]	[121]	–	–	–
	[122]	[222]	–	–	–
$M_c = 3$	[111]	[121]	[131]	–	–
	[122]	[132]	[221]	–	–
	[133]	[222]	[232]	–	–
	[223]	[233]	[333]	–	–
$M_c = 4$	[111]	[121]	[131]	[141]	–
	[122]	[132]	[142]	[231]	–
	[133]	[143]	[221]	[242]	–
	[144]	[222]	[232]	[332]	–
	[223]	[233]	[243]	[343]	–
	[234]	[244]	[333]	[444]	–
	–	[334]	[344]	–	–
$M_c = 5$	[111]	[121]	[131]	[141]	[151]
	[122]	[132]	[142]	[152]	[241]
	[133]	[143]	[153]	[231]	[252]
	[144]	[154]	[221]	[242]	[331]
	[155]	[222]	[232]	[253]	[342]
	[223]	[233]	[243]	[332]	[353]
	[234]	[244]	[254]	[343]	[443]
	[245]	[255]	[333]	[354]	[454]
	[335]	[334]	[344]	[444]	[555]
	–	[345]	[355]	[455]	–
	–	–	[445]	–	–

- The accompanying noise term $n_m(t)$, $m = 1, 2, \ldots, M_c$, at the receive filter output is zero-mean additive colored complex-valued Gaussian process with covariance

$$\mathbb{E}\{n_m^*(t) \cdot n_{m'}(t')\} = N_0 \cdot e^{-j(2\pi(f_{m'}-f_m)t' + (\theta_{m'}-\theta_m))} \left[\int_{-\infty}^{\infty} p_{m,R}^*(\alpha) \cdot p_{m',R}(\alpha + t' - t) \cdot e^{-j2\pi(f_{m'}-f_m)\alpha} d\alpha \right]. \quad (8.15)$$

We here provide analytical performance evaluation in terms of mean-square error (MSE) of distortion that results when sharing multiple equally spaced carriers through a nonlinearity. As an example, each carrier is modulated by 16APSK with frequency spacing Δf of values $1.25T_s^{-1}$, $1.13T_s^{-1}$ and $1.10T_s^{-1}$. The transmit and receive filters, $p_{m,T}(t)$ and $p_{m,R}(t)$, are a matched pair of root-raised cosine (RRC) filters with a rolloff factor of 0.25. Consider a nonlinearity as containing only third-order components,

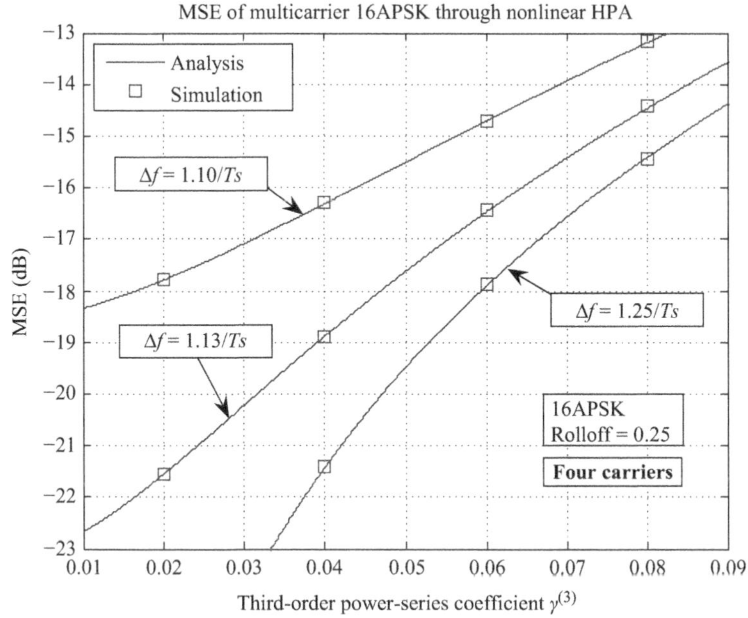

Figure 8.5 MSE evaluation of IMD for center carrier when multicarrier 16APSK pass through nonlinear HPA for four carriers

or $y = x + \gamma^{(3)} \cdot x \cdot |x|^2$. The MSE of distortion is displayed in Figure 8.5 versus the third-order parameter $\gamma^{(3)}$ for the center carrier when the number of carriers is four. The carrier combinations considered in the computation of MSE are those that produce IMD centered at zero and $\pm\Delta f$. Also displayed are Monte-Carlo simulations, marked by " □" at some selected values of $\gamma^{(3)}$. It can be seen that analysis and simulations are in complete agreement. This confirms the accuracy of the Volterra-based characterization of IMD among carriers and the sufficiency of including carrier combinations associated with IMD centered at zero and $\pm\Delta f$.

8.3.2 Multicarrier Volterra filter formulation

A formulation that models the nonlinear components derived in Section 8.3.1 is described here. This formulation is linear in terms of its input so it can be directly utilized for deriving compensation algorithms. We first start with an instantaneous version that only incorporates the receive filter bank at the current time instant but will be expanded at the end of the section to include successive time samples. By denoting

$$\underline{x}(n) \triangleq \left[x_1((n+\varepsilon_1)T_s), x_2((n+\varepsilon_2)T_s), \cdots, x_{M_c}((n+\varepsilon_{M_c})T_s)\right]^T, \tag{8.16}$$

the output of the bank of receive filters at the nth time instant is described in matrix-vector form as

$$\underline{x}(n) = \mathbf{H}^{(3)}(n) \cdot \underline{a}_{\mathrm{NL}}^{(3)}(n) + \underline{n}(n) \tag{8.17}$$

where $\underline{n}(n)$ is a zero-mean complex-valued Gaussian noise vector with covariance matrix $\mathbf{R}_{\mathbf{N}}(n)$, assembled using the component-wise relation (8.15). In (8.17), the matrix $\mathbf{H}^{(3)}(n)$ models the IMD products among carriers and $\underline{a}_{\mathrm{NL}}^{(3)}(n)$ is the corresponding vector of nonlinear combination of the symbols, both are detailed in [3]. The instantaneous multicarrier formulation can be further generalized to include successive time samples of the received filter bank spanning L' symbols around the nth time instant by stacking vectors $\underline{x}(n)$ of (8.16) for $n - ((L'-1)/2), \ldots, n + ((L'-1)/2)$

$$\tilde{\underline{x}}(n) \triangleq \begin{bmatrix} \underline{x}\left(n - \frac{L'-1}{2}\right) \\ \underline{x}\left(n - \frac{L'-1}{2} + 1\right) \\ \vdots \\ \underline{x}\left(n + \frac{L'-1}{2}\right) \end{bmatrix} \tag{8.18}$$

where $(\tilde{\cdot})$ indicates stacked construction. Then, $\tilde{\underline{x}}(n)$ in (8.18) is expressed as

$$\tilde{\underline{x}}(n) = \tilde{\mathbf{H}}^{(3)}(n) \cdot \tilde{\underline{a}}_{\mathrm{NL}}^{(3)}(n) + \tilde{\underline{n}}(n) \tag{8.19}$$

with its quantities regarding generalized Volterra kernels summarized in [3].

In practice, the Volterra matrix in (8.19) can be calculated adaptively by using stochastic gradient-based algorithms such as the superior recursive least-square (RLS) method [13]. This way, the formulation can be evaluated without prior knowledge of the nonlinearity characteristics and can be rapidly responsive to variations in the environment, including back-off level adjustment and long-term aging effects. Application of RLS to nonlinear channel identification is incorporated into the numerical examples of Section 8.4. Moreover, the formulation presented in this chapter is modular, where only the modules pertaining to carrier combinations with significant contribution need to be retained. Within each module of carrier combination $[m_1 m_2 m_3]$, the time combination with small contribution to the output can be discarded for more reduction in the size of matrices.

8.3.3 Reduced-complexity Volterra construction

One special reduction of the full Volterra representation is the generalized memory polynomial, proposed in [14], obtained by only retaining terms that form products of the input and its exponentiated envelope. For simplicity, we use this reduced Volterra model which can be expressed here as

$$\mathscr{H}_{\mathrm{rdcd}}^{(D)}(u[n], \underline{w}; L, L_b, L_c) = \sum_{d=1}^{D} \sum_{k=0}^{L} \sum_{m=-L_c}^{L_b} w_{d,k,m} \cdot u[n-k] \cdot |u[n-k-m]|^{d-1}, \tag{8.20}$$

where $\underline{w}$ is the vector of coefficients $w_{d,k,m}$, constructed below, L is the memory span of the nonlinearity, and L_b, L_c are associated with the memory length of the lagging and leading exponentiated envelope, respectively. If the parameters L_b and L_c in (8.20) are set to zero, the representation reduces further to the memory polynomial model [15,16]. The reduced representation (8.20) includes nonlinear terms with even and odd orders to gain more flexibility in the modeling of the nonlinearity. Also, the coefficients $w_{d,k,m}$ are related to the Volterra kernels through the expression

$$w_{d,k,m} = h^{(d)}[k, k+m, \ldots, k+m]; \quad \text{for odd } d, \tag{8.21}$$

and that the memory span in (8.20) has the same value for all nonlinearity orders d, set at L, L_b, and L_c. For good performance, the memory span of the model has to match the memory associated with the nonlinear system to be compensated. The number of terms associated with this reduced Volterra model grows linearly in the nonlinearity degree D and equals $D \cdot (L+1) \cdot (L_b + L_c + 1)$. A compact vector form of the reduced-complexity Volterra representation using the generalized memory polynomial is mathematically expressed as

$$\mathscr{H}^{(D)}_{\text{rdcd}}(u[n], \underline{w}; L, L_b, L_c) = \underline{w}^T \cdot \underline{u}^{(D)}_{\text{NL}}[n; L, L_b, L_c], \tag{8.22}$$

where the details of the vectors are described in [17].

8.4 Powerful nonlinear countermeasures

An instructive exposition is presented here of advanced compensation architectures that can efface the linear and nonlinear distortions, experienced when operating a multicarrier satellite system with high efficiency. These solutions utilize the analytical characterization and modeling topology of multicarrier IMD previously elucidated in Section 8.3.

Some of these solutions are applied at the receiver in the form of turbo Volterra equalization that iteratively exchanges soft information between equalizer and FEC decoders, detailed in Section 8.4.1. Other solutions are applied at the transmitter in the form of predistortion (PD), described in Sections 8.4.2–8.4.4.

Generally, there are two categories of approaches for PD: data versus signal. Data PD is employed before the transmit filter and is applied on the transmitted symbols at the symbol sampling rate. In contrast, signal PD is employed after the transmit filter and uses a sampling rate that is related to the signal bandwidth which is higher than the symbol rate. Because of these differences, data PD does not cause uplink spectral regrowth but can contribute to it at the output of the nonlinearity, on the downlink. However, this is suppressed by the OMUX filter on-board the satellite. On the other hand, digital signal PD provides control over spectral shaping after the nonlinearity but causes spectral regrowth at the transmitter output impacting uplink out-of-band (OOB) emissions.

Many of these advanced techniques are later utilized and compared in Section 8.5 when exploring OFDM for broadband satellite applications.

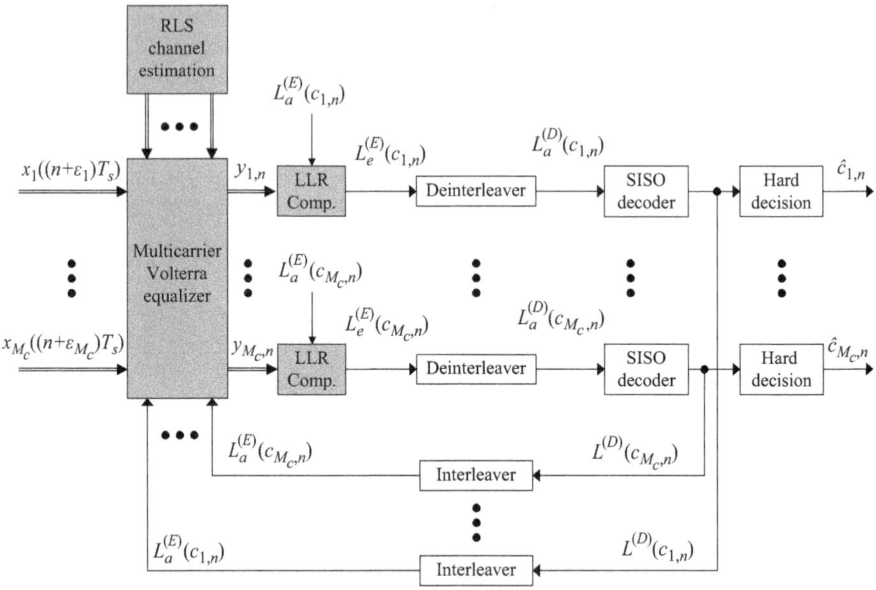

Figure 8.6 Block diagram of the multicarrier Turbo Volterra equalizer

8.4.1 Turbo Volterra equalization

The turbo Volterra equalizer employed at the receiver, developed in [3], is displayed in Figure 8.6. Its novel component is the multicarrier Volterra equalizer which is capable of reconstructing IMD among carriers by applying the analysis of Section 8.3. It operates on the bank of receive filters $x_m((n+\varepsilon_m)T_s)$ and uses *a priori* log-likelihood ratios (LLRs), $L_a^{(E)}(c_{m,n})$, of code bits from interfering carriers. This soft information is the set of *a posteriori* LLRs supplied by a bank of soft-input–soft-output (SISO) single-carrier FEC decoders, $L^{(D)}(c_{m,n})$, after interleaving. The multicarrier Volterra equalizer is also adaptive as it receives generalized Volterra kernel estimates, obtained by RLS channel estimation at the end of known-sequence training and utilizes a modular matrix-vector formulation, reviewed in Section 8.3.2. Compensation of the nonlinear interference is done through linear minimum MSE (MMSE) equalization. An LLR computer is needed to convert the equalizer output $y_{m,n}$ into extrinsic LLRs regarding the code bits, $L_e^{(E)}(c_{m,n})$, by using *a priori* LLRs from the previous decoding iteration. This updated set of soft information about the code bits is then deinterleaved and provided as *a priori* LLRs, $L_a^{(D)}(c_{m,n})$, for the next decoding iteration. Single-carrier solutions for turbo equalization have been applied for mitigating nonlinear ISI in [18–20] but unfortunately cannot handle the nonlinear interaction among the carriers.

Equalization here is done by applying linear MMSE filtering with both feed-forward and feedback coefficients on the matched filter output vector $\underline{\tilde{x}}(n)$ of (8.18) as

$$y_{m,n} = \underline{c}_f^T \cdot \underline{\tilde{x}}(n) + c_b. \tag{8.23}$$

Coefficients $\underline{c}_f$ and c_b, representing sum of feedback terms, are derived by minimizing the MSE between $y_{m,n}$ and the desired symbol $a_{m,n}$. It is straightforward to obtain the equalizer output as [21]

$$y_{m,n} = \underline{c}_f^T \cdot \left(\underline{\tilde{x}}(n) - \tilde{\mathbf{H}}_I^{(3)}(n) \cdot \mathbb{E}\left\{\underline{\tilde{a}}_I^{(3)}(n) \left| \mathbf{L}_a^{(E)} \right.\right\}\right) \tag{8.24}$$

where subscript I indicates that the component associated with desired symbol $a_{m,n}$ is set to zero and the feed-forward coefficients are detailed in [3].

By making a simplifying choice for $\underline{c}_f$, as all-zeros vector except for unity in the component $(m-1)\cdot L' + (L'+1)/2$, a lower complexity version of the equalizer (8.24) can be expressed as

$$\begin{aligned} y_{m,n} &= x_{m,n} - \left[\mathbf{H}^{(3)}(n)\Big|_{\texttt{mth row}}\right] \cdot \mathbb{E}\left\{\underline{a}_{\mathrm{NL}}^{(3)}(n) \left| \mathbf{L}_a^{(E)} \right.\right\} \\ &\quad - \mathbb{E}\left\{\mathscr{P}_m^{\mathrm{centroid}}(a_{m,n}) \left| \mathbf{L}_a^{(E)} \right.\right\}\right], \end{aligned} \tag{8.25}$$

where $\mathscr{P}_m^{\mathrm{centroid}}(a_l)$ is the centroid value associated with a_l, computed during training mode. In (8.25), we invoke the use of centroids [22] to account for the constellation warping caused by the presence of nonlinearity. This can be used when estimating the nonlinear interference for the sake of improving performance.

In (8.25), the expectations $\mathbb{E}\{\underline{a}^{(3)}(n) \left| \mathbf{L}_a^{(E)} \right.\}$ can be computed using component-wise relations of first- and third-order symbol products. Namely,

$$\begin{aligned} &\mathbb{E}\left\{a_{m_1,k_1} a_{m_2,k_2} \cdots a_{m_p,k_p} a^*_{m_{p+1},k_{p+1}} a^*_{m_{p+2},k_{p+2}} \cdots a^*_{m_q,k_q} \left| \mathbf{L}_a^{(E)} \right.\right\} \\ &= \prod_{m=1}^{M_c} \prod_{i=n-(L-1)/2}^{n+(L-1)/2} \mathbb{E}\left\{a_{m,i}^{\nu_{m,i}} \left(a_{m,i}^{\nu^*_{m,i}}\right)^* \left| \mathbf{L}_a^{(E)} \right.\right\}, \end{aligned} \tag{8.26}$$

where the product of expectations is possible due to the independence across carriers and across symbols, as provided by the interleaving operations. The parameter $\nu_{m,i}$ is defined as the number of indices of the mth data symbol stream a_{m,k_j} taking on the value i, whereas $\nu^*_{m,i}$ is the number of indices of the conjugate of the mth data symbol stream a^*_{m,k_j} when it takes on the value i. The individual terms in the product of (8.26) are then computed as

$$\mathbb{E}\left\{a_{m,i}^{\nu_{m,i}} \left(a_{m,i}^{\nu^*_{m,i}}\right)^* \left| \mathbf{L}_a^{(E)} \right.\right\} = \sum_{l=1}^{M} a_l^{\nu_{m,i}} \left(a_l^{\nu^*_{m,i}}\right)^* \cdot P\left\{a_{m,l} = a_l \left| \mathbf{L}_a^{(E)} \right.\right\}, \tag{8.27}$$

where conditional symbol probability $P\{a_{m,l} = a_l \left| \mathbf{L}_a^{(E)} \right.\}$ is formed on the basis of the *a priori* LLR of the corresponding code bits provided by the bank of SISO decoders at the previous iteration.

Figure 8.7 includes the bit error rate (BER) performance of single-carrier versus multicarrier Volterra equalizers for the center carrier when the number of carriers is four, modulation is 32APSK, encoded by low-density parity check (LDPC) code with rate of 11/15, and the HPA is operated at aggregate OBO level of 2.8 dB. The pair (u, v) in the performance curve is used to enumerate the iterations where u is the number of equalization iterations and v is the number of decoding iterations

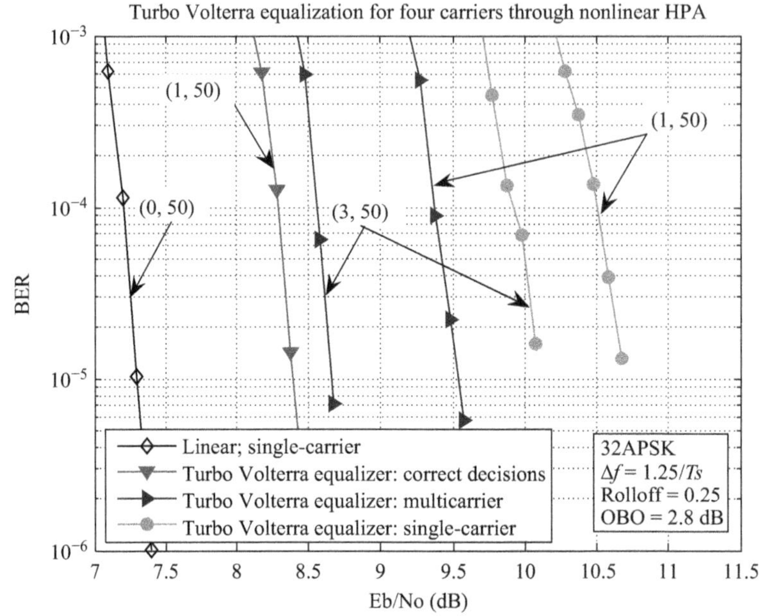

Figure 8.7 BER of multicarrier turbo Volterra equalization for center carrier when four carriers of LDPC-coded 32APSK pass through nonlinear HPA at OBO 2.8 dB

within the LDPC decoder. When using state of the art single-carrier methods that compensate for only ISI, the performance degradation relative to the ideal case of correct decisions at BER of 2×10^{-5} is 2.3 and 1.7 dB, without and with turbo processing, respectively. However, this degradation can be reduced to within 0.25 dB when using turbo processing and incorporating the multicarrier IMD analyzed in the chapter.

8.4.2 Volterra-based data predistortion

A multicarrier data PD scheme is introduced in [23] which modifies the transmitted symbols by a third-order multicarrier Volterra inverse of the nonlinearity, applied in a single stage and simplified using the memory polynomial approach of [16]. It is processed at the symbol rate and is placed prior to the pulse shaping filters. Single-carrier PD that uses Volterra inverse of the nonlinearity is described in [24,25].

This approach of PD benefits from the multicarrier Volterra analysis of [3] to approximate the inverse channel which is also nonlinear with memory. This is done by applying an inner-product between the nonlinear input combination $\underline{a}_{\mathrm{NL}}^{(3)}(n)$ and the PD coefficient vector $\underline{g}_{m_d}$ as

$$\tilde{a}_{m_d}(n) = \underline{g}_{m_d}^{T} \cdot \underline{a}_{\mathrm{NL}}^{(3)}(n). \tag{8.28}$$

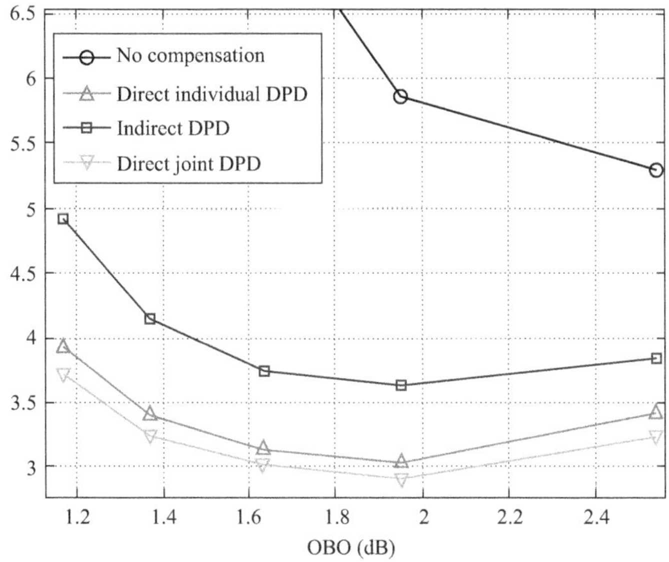

Figure 8.8 Coded total degradation versus OBO level for the inner carrier when three 16APSK carriers share nonlinear transponder. © 2014 IEEE. Reprinted, with permission, from Reference [23]

Two methods for estimating the PD coefficient vector $\underline{g}_{m_d}$ in (8.28) are available: indirect versus direct learning. For the former, the post-inverse is computed first and then copied to the predistorter during the second step. In contrast, the direct-learning method resembles the pre-inverse and computes the coefficients directly based on the nominal constellation and the output of the filter bank, minimizing

$$e_{m_d}(n) = a_{m_d,n} - x_{m_d}((n + \varepsilon_{m_d})T_s). \tag{8.29}$$

The direct-learning method is expected to outperform the indirect method as it applies a pre-inverse, instead of post-inverse, prior to the nonlinearity, and that the order of nonlinear operations is not commutative. It however has the disadvantage of requiring an estimate of the nonlinear system to be inverted.

Individual versus joint cost functions are considered in [23]. The individual estimation method reduces to M_c distinct optimization processes, minimizing $e_m(n)$, $m = 1, 2, \ldots, M_c$, and generating the PD coefficients for different carriers separately and is run in parallel. The joint estimation method achieves a global optimum of the PD coefficients, jointly minimizing $\sum_{m=1}^{M_c} |e_m(n)|^2$ which incurs higher complexity.

Figure 8.8 contains performance comparison for the inner carrier when three 16APSK carriers share the same nonlinearity. Multicarrier data PD provides significant gain with respect to a case of no PD. Direct estimation of the predistorter

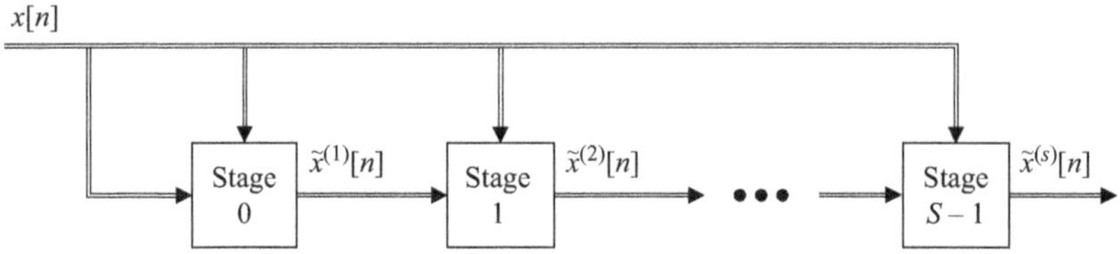

Figure 8.9 Block diagram of successive signal predistortion with S *stages*

coefficients provides further gain relative to the indirect method. The joint optimization yields the best performance, providing about 0.5 dB over the indirect method for the same predistorter complexity and training length.

8.4.3 Volterra-based successive signal predistortion

Signal PD is a digital compensator that is applied ahead of a nonlinear system so as to cancel the resulting distortion. The focus here is on methods that incorporate memory effects as they are becoming more prominent with the increasing demand for transmitting broadband signals. Techniques of nonlinear PD with memory for the case when the HPA is co-located with the predistorter include classical Volterra-based inverse [15,16,26] and more recent successive solution with significantly improved performance [17]. These methods can be applied directly on the signal composite because they are agnostic to the number of signals sharing the nonlinearity. A family of schemes is detailed in this section, capable of suppressing downlink spectral regrowth and in-band distortion simultaneously with an added tunability feature.

In particular, letting the original complex-valued input to the nonlinear system with memory be $x[n]$, the objective of the PD structure is to modify this input so the output when processed by the nonlinear system approximates a desired response that is free of nonlinear distortion. The sought solution is determined successively by using recursion which is of the stochastic approximation type [27], intended for finding zero-crossing of an unknown function when only its noisy measurements are available. Denoting the modified input at the sth-stage and nth discrete-time instant as $\tilde{x}^{(s)}[n]$, this recursion updates the predistorted signal in the following manner

$$\tilde{x}^{(s+1)}[n] = \tilde{x}^{(s)}[n] + \mu^{(s)} \cdot e[n] \tag{8.30}$$

where $e[n]$, described below, is an error signal that is driven toward zero in multiple stages, indexed $s = 0, 1, \ldots, S-1$. For initialization, the input to the zeroth-stage of PD uses the original undistorted input, or we set $\tilde{x}^{(0)}[n] = x[n]$. Figure 8.9 presents a block diagram of the successive solution with S stages.

The choice of sequence $\{\mu^{(s)}\}$ plays a central role in balancing the convergence speed and amount of residual error. Deterministic step-size rules, including numerical experimentation with step-size sequences, are examined later in the section.

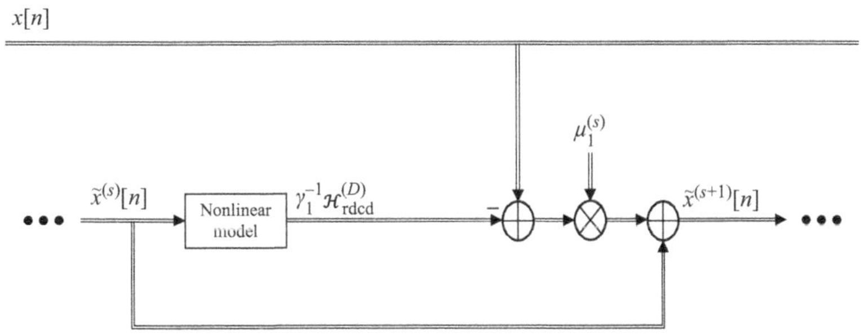

Figure 8.10 Block diagram of the sth-stage for the first scheme

Several PD schemes are described in this section to achieve different objectives in nonlinear distortion suppression. In one case, the scheme optimizes the suppression of spectral regrowth caused by the nonlinear system while simultaneously mitigating in-band distortion. For this, the error signal is formed based on the difference between the original input $x[n]$, selected to serve as the desired signal $d_1[n]$, and its estimate $\hat{d}_1^{(s)}[n]$ during the sth-stage, using reduced-complexity Volterra model $\mathscr{H}_{\mathrm{rdcd}}^{(D)}$, described in Section 8.3.3. This is mathematically expressed as

$$e_1[n] = d_1[n] - \hat{d}_1^{(s)}[n], \tag{8.31}$$

where

$$d_1[n] = x[n], \tag{8.32}$$

$$\hat{d}_1^{(s)}[n] = \gamma_1^{-1} \cdot \mathscr{H}_{\mathrm{rdcd}}^{(D)}\left(\tilde{x}^{(s)}[n], \underline{\tilde{w}}^{(s)}; L, L_b, L_c\right), \tag{8.33}$$

$\underline{\tilde{w}}^{(s)}$ is the vector of Volterra kernels associated with the sth-stage and γ_1 is a complex-valued gain correction aimed at removing nonlinearity-induced warping effects. The γ_1 may be obtained by

$$\gamma_1 = \frac{\sum_n \hat{d}_1^{(s)}[n] \cdot d_1^*[n]}{\sum_n |d_1[n]|^2}. \tag{8.34}$$

Figure 8.10 contains an illustration of an sth-stage application of the successive PD for the first scheme.

A second scheme is considered to optimize suppression of in-band distortion as experienced at the output of the receive filter $p_R[n]$. The receive filter $p_R[n]$ is designed to reject noise in nonsignal band and reduce spillage from adjacent carriers. This scheme forms an error signal that uses a model of the receive filter $p_R[n]$ when producing the desired response and its approximation as

$$e_2[n] = d_2[n] - \hat{d}_2^{(s)}[n] \tag{8.35}$$

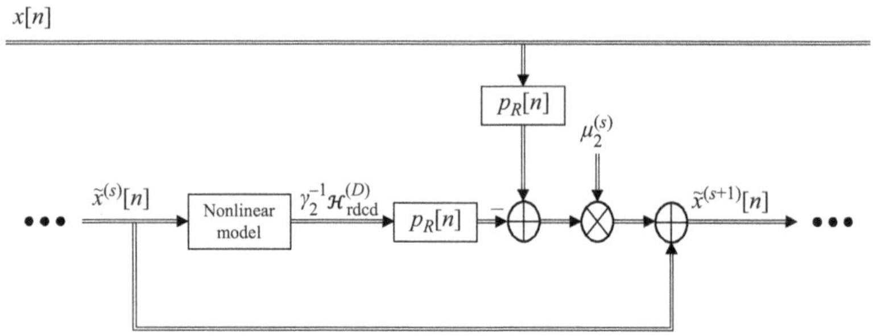

Figure 8.11 Block diagram of the sth-stage for the second scheme

where

$$d_2[n] = \sum_k d_1[n-k] \cdot p_R[k], \tag{8.36}$$

$$\hat{d}_2^{(s)}[n] = \left(\frac{\gamma_2}{\gamma_1}\right)^{-1} \cdot \sum_k \hat{d}_1^{(s)}[n-k] \cdot p_R[k], \tag{8.37}$$

and γ_2 is a complex-valued gain correction that is aimed at removing the warping effect caused by the nonlinearity after the filter. Figure 8.11 contains an illustration of an sth-stage application of the successive PD for the second scheme.

Furthermore, a third scheme is proposed in an effort to offer the designer tunability to trade-off suppression levels of spectral regrowth at the output of the nonlinearity with in-band distortion at the receive filter output. This scheme forms an error signal that is comprised of a linear combination, or weighted sum, of the error signals described above, $e_1[n]$ and $e_2[n]$, adjusted possibly at different rates. Namely, the recursion for the predistorted signal in this case is

$$\tilde{x}^{(s+1)}[n] = \tilde{x}^{(s)}[n] + \left(\alpha \cdot \mu_1^{(s)} \cdot e_1[n] + \beta \cdot \mu_2^{(s)} \cdot e_2[n]\right) \tag{8.38}$$

where α and β are nonnegative combining parameters which are designer-selected to balance levels of suppressing spectral regrowth versus in-band distortion.

Noteworthy in this regard is that all the schemes derived above require only one implementation of the nonlinear model per stage. Also, implementing a reduced-complexity Volterra model on-the-fly allows for coping with systems with high degree of nonlinearity and large memory span.

Figure 8.12(a) shows the levels of adjacent channel interference (ACI) for the system model when 64-ary quadrature amplitude modulation (QAM) passes through nonlinear Wiener–Hammerstein-based HPA model, with and without successive PD, for various levels of OBO. The associated results for normalized MSE (NMSE) are reported in Figure 8.12(b). The curves labeled with marker "■" in the figure are generated by the first scheme, which does not involve a receive filter. One observation

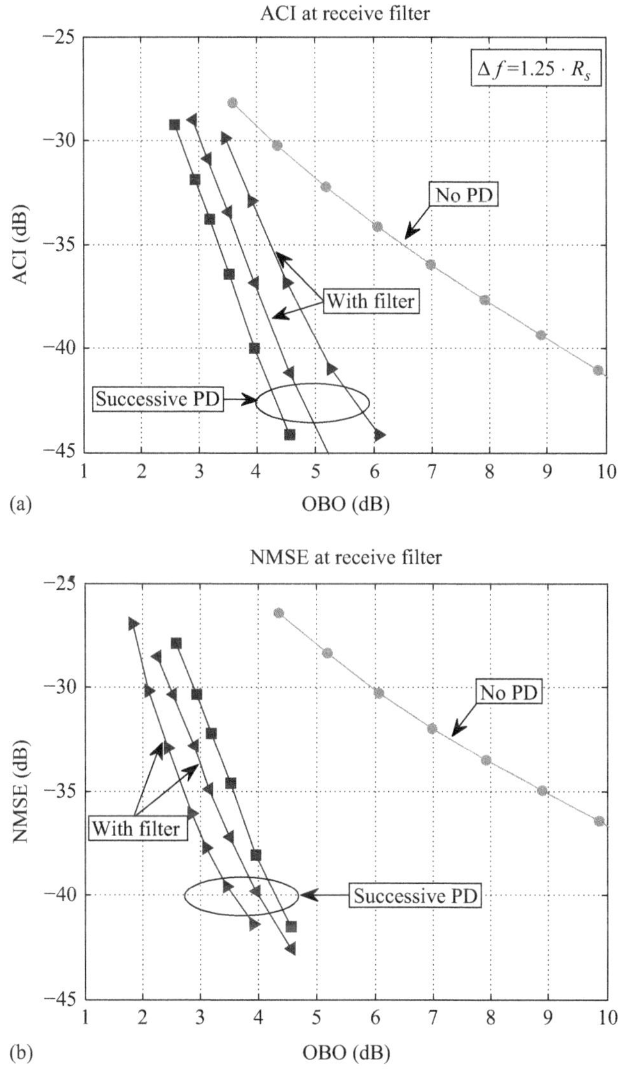

Figure 8.12 Levels of (a) ACI with $\Delta f = 1.25 \cdot R_s$ *and (b) NMSE versus OBO when 64-QAM is passed through nonlinear Wiener–Hammerstein HPA, with and without successive predistortion*

to be made is that the levels of ACI are drastically reduced by the considered technique, a strong indication of its ability in suppressing the spectral regrowth or OOB emission at the output of the nonlinearity. Second observation is that the successive signal PD is simultaneously effective at mitigating in-band distortion due to the nonlinear HPA behavior. Third observation is that incorporating the receive filter, which introduces

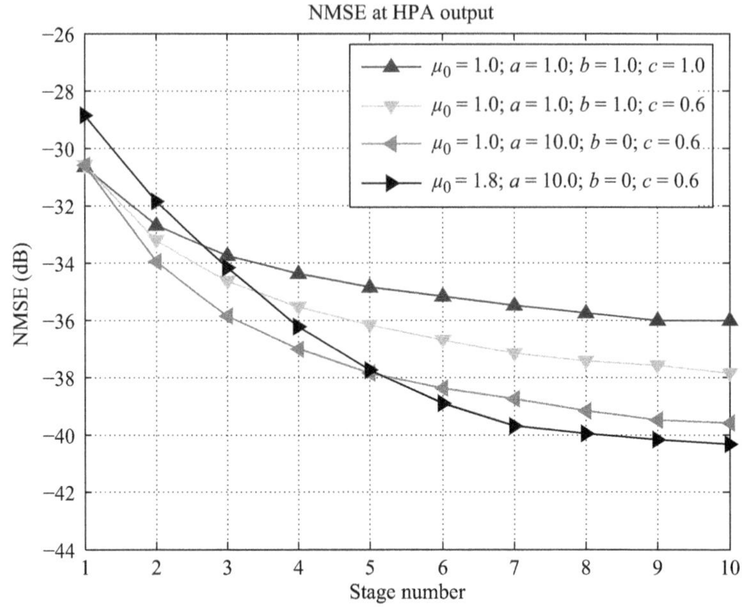

Figure 8.13 NMSE at the nonlinear Wiener–Hammerstein HPA output versus number of stages for the successive predistortion when using 64-QAM at OBO of 4.2 dB

more memory beyond that of the HPA, provides further suppression of in-band NMSE. An important feature to notice is the tunability that the considered signal PD offers to the designer in trading off levels of suppression of OOB distortion versus in-band distortion generated by the nonlinear HPA.

A general formula for deterministic step-size sequences that can satisfy the basic condition of convergence is given by

$$\mu^{(s)} = \mu_0 \cdot \frac{(a + (b/(s+1))}{(a + (b/(s+1)) + (s+1)^c - 1)} \tag{8.39}$$

for $s = 0, 1, \ldots$. Its evaluation is shown in Figure 8.13 for the case of 64-QAM at OBO of 4.2 dB, in terms of NMSE between the Wiener–Hammerstein HPA output, $s_{\text{NL}}(t)$, and its ideal counterpart. The figure displays the progression of NMSE versus the stage number of the successive PD method, using different values of μ_0, a, b, and c in (8.39). As visible, the performance is improved when more stages of distortion cancellation are utilized, and that only a small number of stages are needed.

Another evaluation is reported in Figure 8.14 when 256-QAM modulation is amplified by nonlinear Wiener–Hammerstein-based HPA model. The 256-QAM modulation has a constellation with more signal levels. As a result, it has higher PAPR and is expected to be more vulnerable to nonlinear distortion. As can be seen, the additional advantage of using successive signal PD compared with the one based on

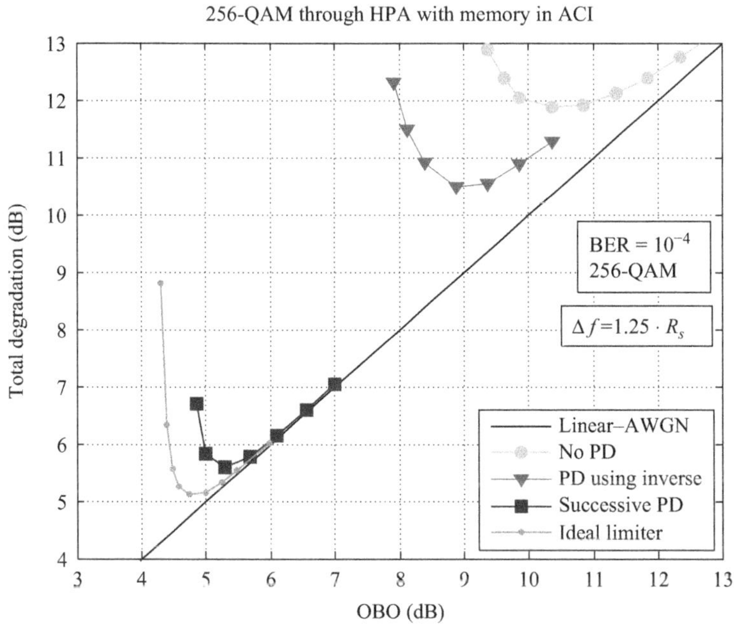

Figure 8.14 Total degradation versus OBO level for 256-QAM through nonlinear Wiener–Hammerstein HPA with adjacent interferes at $\Delta f = 1.25 \cdot R_s$*, with and without predistortion. (Results are at target BER of* 10^{-4}*.)*

inverse, commonly adopted in the literature [15,16,26], is as high as 5 dB reduction in TD and close to 3.6 dB improvement in OBO power-efficiency level. Last, the successive PD can approach the performance of the ideal limiter, representing the perfectly predistorted system, to within 0.5 dB.

8.4.4 Successive data predistortion

Successive data PD, introduced in [28] and illustrated in Figure 8.15, is placed at the transmitter or gateway and entails successively modifying the set of transmitted symbols to drive multicarrier distortion vector toward zero. This distortion vector results from passing the transmitted symbols from the multiple carriers, intrinsically accessible at the gateway, through the nonlinear satellite channel model. This method produces predistorted symbols for each carrier that contain the impact of past and future interfering data symbols, within a certain memory span, from all of the M_c carriers participating in the PD.

Dynamic data PD that is based on look-up tables (LUTs) for the single-carrier case, generated by minimizing MSE, is introduced in [29,30]. Unfortunately, the size of LUTs grows exponentially with the memory span of the predistorter, at a growth factor that equals the modulation order; both of these quantities need to be large in efficient satellite systems. An attractive feature of this considered method is

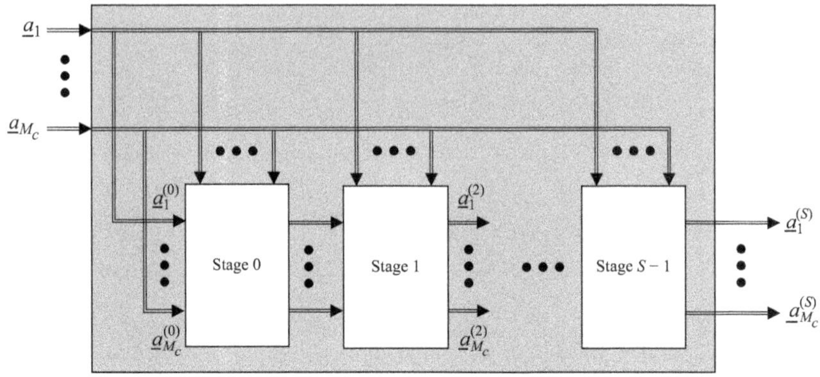

Figure 8.15 Block diagram depicting multicarrier successive data predistortion

that it uses on-the-fly computations, avoiding LUTs whose size in this case grows exponentially with the product of PD memory span and number of carriers. In one implementation of this method, the computational complexity increases only linearly with this product of parameters, allowing for the use of large values of the modulation order, number of carriers, and memory span of the predistorter.

In particular for the current method, let $\underline{a}_m^{(s)}$ be the vector of complex-valued data symbols associated with the mth-carrier at the sth-stage as

$$\underline{a}_m^{(s)} = \left[a_{m,0}^{(s)}, a_{m,1}^{(s)}, \ldots, a_{m,N-1}^{(s)}\right]^T, \tag{8.40}$$

where $s = 0, 1, \ldots, S-1$; $m = 1, 2, \ldots, M_c$; and N is the length of the data block which typically spans a codeblock of symbols. A special case of (8.40) is the first application of the PD for which the input is composed of the undistorted data symbols, or $\underline{a}_m^{(0)} = \underline{a}_m = \left[a_{m,0}, a_{m,1}, \ldots, a_{m,N-1}\right]^T$. We further define a vector of symbols $\underline{\alpha}^{(s)}$, whose size is $N \cdot M_c$, composed of the corresponding symbols at the output of the previous stage, associated with each one of the M_c carriers sharing the nonlinearity as

$$\underline{\alpha}^{(s)} = \begin{bmatrix} \underline{a}_1^{(s)} \\ \underline{a}_2^{(s)} \\ \vdots \\ \underline{a}_{M_c}^{(s)} \end{bmatrix}. \tag{8.41}$$

Next, we denote the vector $\underline{\mathscr{H}}_{m_d}(\underline{\alpha}^{(s)})$, of length N, as the estimate of received symbols at a specific m_dth carrier, where $m_d = 1, 2, \ldots, M_c$, and uses the vector $\underline{\alpha}^{(s)}$ defined in (8.41). Implementations of $\underline{\mathscr{H}}_{m_d}(\underline{\alpha}^{(s)})$ are detailed in [28], including computationally efficient polyphase structures for the interpolating and decimating filtering operations. The received symbol estimate is then utilized in generating distortion vector $\underline{e}_{m_d}(\underline{\alpha}^{(s)})$, also of length N, relative to undistorted constellation as

$$\underline{e}_{m_d}(\underline{\alpha}^{(s)}) = \underline{a}_{m_d} - \underline{\mathscr{H}}_{m_d}(\underline{\alpha}^{(s)}). \tag{8.42}$$

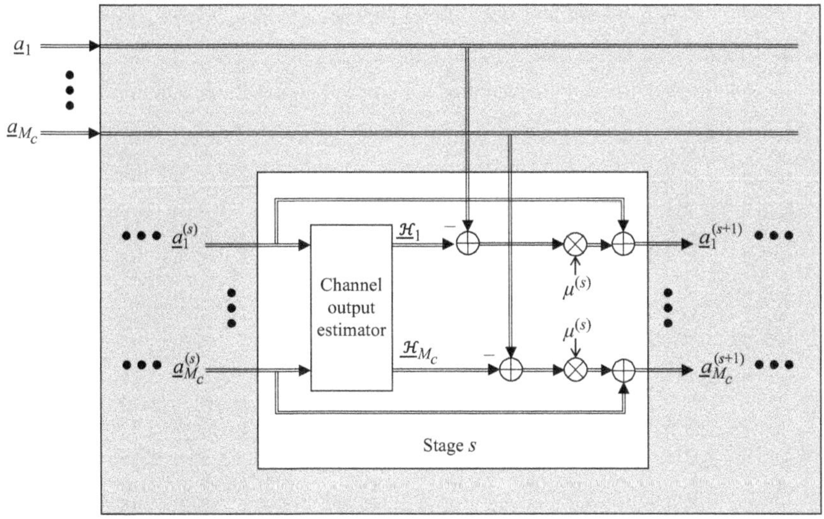

Figure 8.16 Block diagram of the s*th-stage of the multicarrier successive predistortion*

Through the estimate $\mathcal{H}_{m_d}(\underline{\alpha}^{(s)})$, the distortion vector in (8.42) contains the impact of past and future interfering data symbols, within the memory span of the predistorter, from all of the M_c carriers involved in the PD. Successive application of the PD is then used to drive the distortion vector in (8.42) toward zero. For this, the PD output is generated by modifying the predistorted symbols from the previous stage with a correction that is proportional to the distortion vector. Namely, the PD output at the sth stage for the m_dth carrier is mathematically expressed as

$$\underline{a}_{m_d}^{(s+1)} = \underline{a}_{m_d}^{(s)} + \mu^{(s)} \cdot \underline{e}_{m_d}(\underline{\alpha}^{(s)}), \tag{8.43}$$

where $\{\mu^{(s)}\}$ is a step-size sequence that is positive and decreasing to ensure progress toward a solution. Furthermore, this recursion is computationally simple in the sense of requiring only one evaluation of the distortion vector $\underline{e}_{m_d}(\underline{\alpha}^{(s)})$ at each stage. Figure 8.16 illustrates an sth stage application of the successive PD method.

Noteworthy in this regard is that the predistorted symbols are adjusted simultaneously for all of the M_c carriers during any sth stage. Also, the data symbols from the adjacent carriers are needed in the estimation of the channel output for deriving the distortion vector. However, the distortion vector is minimized separately for the individual carriers as in (8.42) and that the predistorted symbols are generated individually for each carrier as in (8.43).

Figure 8.17 reports the coded TD at a target packet error rate (PER) of 10^{-3} versus OBO level for the inner carrier when four 16APSK carriers, all using the rate 2/3 DVB-S2 LDPC code, share the nonlinear transponder. Two sets of curves are generated, one for a system employing the multicarrier successive PD and another for

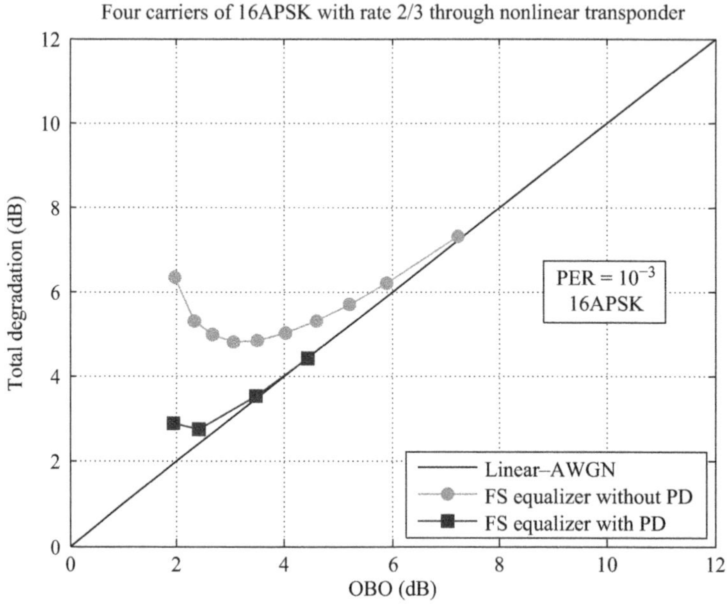

Figure 8.17 Coded total degradation versus OBO level for the inner carrier when four 16APSK carriers share nonlinear transponder. (DVB-S2 LDPC code rate is 2/3; target PER is 10^{-3}.)

a system without PD. In both scenarios, no additional receiver-based compensation beyond the linear equalizer is applied. Significant performance gains are seen over the no-PD case. Without PD, the TD attains its lowest value of 4.8 dB at an OBO of 3.1 dB. In comparison, the application of multicarrier successive PD makes it possible for the system to not only signal at a much lower TD of 2.7 dB but also improves the HPA power efficiency by allowing operation at a much lower OBO of 2.4 dB.

This multicarrier successive data PD technique is successfully adopted in [31] for the optimization of next-generation broadband medium-earth orbit (MEO) satellite systems employing Q/V-band (33–75 GHz) within the extremely high frequency (EHF) area of the radio spectrum. Detailed information-theoretic assessment reveals that successive data PD ensures significant gain when applied with single-user or multiuser detection.

8.5 OFDM-like signaling

In an effort to provide greater commonality with 5G terrestrial networks, this section endeavors to apply OFDM-like signaling for broadband satellite applications that are highly efficient in mass, power, energy, and bandwidth.

An OFDM-like system, introduced in [32], is considered here which invokes two layers of multicarrier operation. The first layer allows for multiple independent signals

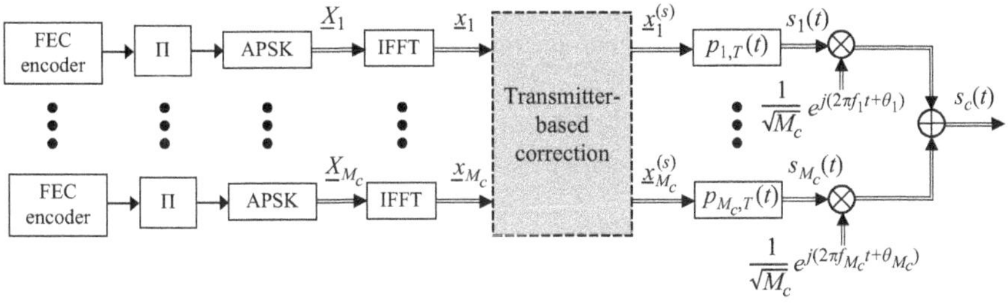

Figure 8.18 Block diagram of OFDM-like transmitter for broadband satellite applications

to share a single on-board HPA, maximizing payload mass efficiency. The second layer of multicarrier operation permits transmitted symbols from each individual signal to modulate multiple narrowband OFDM subcarriers. This is followed by interpolating filters to provide oversampling, suppress interference leaking into adjacent signals in the composite and to limit OOB emission levels to be compatible with satellite uplink transmission.

Several powerful countermeasure strategies from Section 8.4, originally developed for SCM, are generalized and their effectiveness is demonstrated in minimizing the distortion. This distortion includes not only linear and nonlinear interaction between the signals sharing the satellite transponder, but also the linear and nonlinear ICI amongst the OFDM subcarriers.

8.5.1 OFDM-like transmitter

Figure 8.18 displays the considered OFDM-like signaling that generates a composite of M_c frequency-multiplexed independent signals and has several features. The first feature includes the application of N-point inverse fast Fourier transform (IFFT) on blocks of the transmitted symbols, at the symbol rate T_s^{-1}, belonging to each of the individual signals $s_m(t)$ in the composite $s_c(t)$ so they modulate N narrowband OFDM subcarriers. The number of OFDM subcarriers or size of the IFFT, N, can be different for signals $s_m(t)$ to allow different OFDM numerology amongst them. A second feature is the application of each interpolating filter $p_{m,T}(t)$ on the aggregate of OFDM subcarriers, within $s_m(t)$, to provide oversampling, suppression of interference leaking into adjacent signals in the transmitted signal composite, as well as to limit the level of OOB emissions that is high for conventional OFDM. Conventional OFDM uses rectangular pulse shaping, exhibiting a slowly decaying $\sin(x)/x$ behavior in the frequency domain. A third feature is avoiding cyclic prefix (CP) that repeats the last part of an OFDM symbol. Using CP is advantageous in dispersive channels, not encountered here, but induces spectral efficiency loss due to redundancy, and causes energy loss as the CP symbols require additional energy to transmit, but are then discarded at the receiver. This energy loss is computed in dB as $10 \cdot \log_{10}((N + N_{CP})/N)$, where N_{CP} is the number of CP symbols. Also, using CP creates prominent ripples in the in-band region of a conventional OFDM spectrum, requiring power reduction to ensure regulatory compliance (as elaborated later in this section). However, the advanced techniques in this chapter are directly applicable to the case of including CP when needed to remove ISI encountered in frequency-selective multipath channels. Furthermore, in our OFDM-like signaling, no guard tones are inserted at the input of OFDM modulator to avoid reduction in throughput.

The input to the OFDM-like transmitter is complex-valued symbol sequences, at the symbol rate T_s^{-1}, $\{X_{m,n}; n = 0, 1, \ldots, N_s - 1; m = 1, 2, \ldots, M_c\}$, from M-ary APSK constellation, using a well-chosen bit-to-symbol mapping, of independent FEC-encoded, bit-interleaved bit stream for each signal. The parameter N_s is the length of the data block which spans a codeblock of symbols.

In particular, let $\underline{X}_m$ be the vector of complex-valued data symbols of size $N_f \times 1$, associated with the mth signal, that lie in the frequency domain, or

$$\underline{X}_m = \left[X_{m,0}, X_{m,1}, \ldots, X_{m,N_f-1}\right]^T. \tag{8.44}$$

The vector $\underline{X}_m$ is segmented into N_{OFDM} blocks to modulate N orthogonal subcarriers, which for ease of exposition is chosen to be the same for all signals, where $N_{OFDM} = N_f/N$. Padding of small number of extra symbols $(N_f - N_s)$ may be needed to make N_{OFDM} a whole integer. The padding symbols can be distributed into different blocks or introduced as one segment. The vectors relating to OFDM blocks are stacked to form $\underline{X}_m$ in (8.44), which can be equivalently represented as

$$\underline{X}_m = \left[\underline{\tilde{X}}_{m,0}^T, \underline{\tilde{X}}_{m,1}^T, \ldots, \underline{\tilde{X}}_{m,N_{OFDM}-1}^T\right]^T, \tag{8.45}$$

where

$$\underline{\tilde{X}}_{m,l} = \left[X_{m,l\cdot N}, X_{m,l\cdot N+1}, \ldots, X_{m,(l+1)\cdot N-1}\right]^T \tag{8.46}$$

is of size $N \times 1$, $l = 0, 1, \ldots, N_{OFDM} - 1$, and $m = 1, 2, \ldots, M_c$. Each vector $\underline{\tilde{X}}_{m,l}$ in (8.46) is further processed by an N-point IFFT to generate lth OFDM symbol for the mth signal as

$$\tilde{\tilde{x}}_{m,l,k} = \frac{1}{\sqrt{N}} \cdot \sum_{n=0}^{N-1} \tilde{X}_{m,l,n} \cdot e^{j2\pi kn/N}, \tag{8.47}$$

where $\tilde{X}_{m,l,n}$ is the nth component of vector $\underline{\tilde{X}}_{m,l}$ in (8.46) and $k = 0, 1, \ldots, N - 1$. The samples $\tilde{\tilde{x}}_{m,l,k}$ in (8.47) are stacked to form the input $\underline{x}_m$ in the time domain as

$$\underline{x}_m = \left[\underline{\tilde{x}}_{m,0}^T, \underline{\tilde{x}}_{m,1}^T, \ldots, \underline{\tilde{x}}_{m,N_{OFDM}-1}^T\right]^T, \tag{8.48}$$

where

$$\underline{\tilde{x}}_{m,l} = \left[\tilde{\tilde{x}}_{m,l,0}, \tilde{\tilde{x}}_{m,l,1}, \ldots, \tilde{\tilde{x}}_{m,l,N-1}\right]^T \tag{8.49}$$

is of size $N \times 1$.

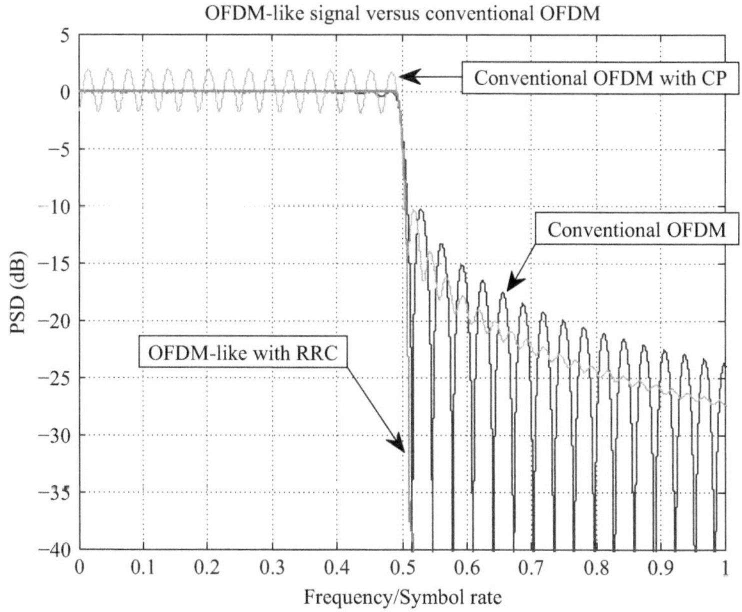

Figure 8.19 Uplink PSD of individual $s_m(t)$ *for conventional OFDM, conventional OFDM with cyclic prefix, and OFDM-like signal*

Alternatively, the OFDM block $\tilde{\underline{x}}_{m,l}$ in (8.49) can be generated by a matrix-vector multiplication as $\tilde{\underline{x}}_{m,l} = F^H \cdot \tilde{\underline{X}}_{m,l}$, where F is an $N \times N$ discrete Fourier transform (DFT) matrix and $l = 0, 1, \ldots, N_{OFDM} - 1$. As depicted in Figure 8.18, there is an optional time-domain successive compensator with S stages, which processes the resulting complex-valued symbol sequences, generating a modified set of symbols $\left\{x_{m,k}^{(S)}; k = 0, 1, \ldots, N_f - 1; m = 1, 2, \ldots, M_c\right\}$, also at rate T_s^{-1}.

The individual waveforms $s_m(t)$ are digitally modulated using the transmit pulse shaping filter $p_{m,T}(t)$ and given by

$$s_m(t) = \sum_{k=0}^{N_f-1} x_{m,k}^{(S)} \cdot p_{m,T}(t - kT_s). \tag{8.50}$$

They are then used to form the baseband composite signal $s_c(t)$ as in (8.1).

An exemplary uplink PSD for an individual waveform $s_m(t)$ is depicted in Figure 8.19, illustrating a comparison between conventional OFDM, conventional OFDM with CP, and OFDM-like signal when selecting $p_{m,T}(t)$ as RRC and 16APSK is employed. As clearly evident in the figure, conventional OFDM exhibits a slowly decaying $\sin(x)/x$ behavior in the frequency domain, and that about 3.6 dB of ripples are present in the in-band region when using CP [33]. Unfortunately, spectral ripples require reducing the transmit power so as not to violate strict emission limits, set by regulatory bodies based on the *peak* level of the spectrum.

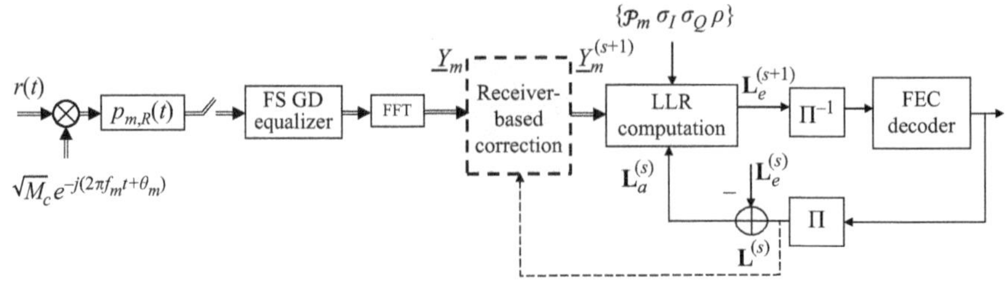

Figure 8.20 Block diagram of OFDM-like receiver for the mth signal

On the other hand, the spectrum associated with the considered OFDM-like signal does not suffer from in-band ripples and has excellent containment of its frequency content within the frequency band of interest. The latter provides minimum levels of interference leaking into adjacent signals $s_m(t)$ in the composite $s_c(t)$, even if orthogonality amongst them is compromised due to different OFDM numerology or synchronization offsets. This also ensures that the uplink OOB emission level is consistent with that of a traditional satellite signal using SCM.

8.5.2 OFDM-like receiver

Figure 8.20 shows the schematic of the OFDM-like receiver. It is noted that single-user detection only is applied, such that no information is exchanged with receivers of other users, as is typical in a satellite forward application. The signal $x_m(t)$ at the receive filter output in (8.4), sampled at multiples of the symbol rate, is processed through GD equalizer to compensate for the linear phase distortion introduced by the IMUX and OMUX filters. At its output, we have samples, at the symbol rate, $\{y_{m,k}; k = 0, 1, \ldots, N_f - 1; m = 1, 2, \ldots, M_c\}$. These samples are segmented into N_{OFDM} blocks, each containing N samples to be converted into the frequency domain via N-point fast Fourier transform (FFT), or

$$\overset{\approx}{Y}_{m,l,n} = \frac{1}{\sqrt{N}} \cdot \sum_{k=0}^{N-1} y_{m,l\cdot N+k} \cdot e^{-j2\pi kn/N}, \tag{8.51}$$

for $l = 0, 1, \ldots, N_{OFDM} - 1$, $n = 0, 1, \ldots, N - 1$, and assembled back into vector of size $N_f \times 1$, per individual mth signal, as

$$\underline{Y}_m = \left[\tilde{\underline{Y}}_{m,0}^T, \tilde{\underline{Y}}_{m,1}^T, \ldots, \tilde{\underline{Y}}_{m,N_{OFDM}-1}^T\right]^T, \tag{8.52}$$

where

$$\tilde{\underline{Y}}_{m,l} = \left[\overset{\approx}{Y}_{m,l,0}, \overset{\approx}{Y}_{m,l,1}, \ldots, \overset{\approx}{Y}_{m,l,N-1}\right]^T. \tag{8.53}$$

Alternatively, the frequency-domain block of symbols $\tilde{\underline{Y}}_{m,l}$ in (8.53) can be generated by a matrix-vector multiplication as $\tilde{\underline{Y}}_{m,l} = F \cdot [y_{m,l\cdot N}, y_{m,l\cdot N+1}, \ldots, y_{m,(l+1)\cdot N-1}]^T$.

Variables $\{Y_{m,n}; n = 0, 1, \ldots, N_s - 1; m = 1, 2, \ldots, M_c\}$, the nth components of $\underline{Y}_m$ in (8.52), are used to generate LLRs for the individual FEC decoders, after removal of the extra $(N_f - N_s)$ padded symbols. As shown in Figure 8.20, the receiver includes an option of implementing frequency-domain successive compensator using soft-information provided by the FEC decoder, over S iterations. In this case, a vector of frequency-domain samples at the output of the compensator during iteration $s+1$, denoted by $\underline{Y}_m^{(s+1)}$, is used to generate LLRs for the FEC decoder. In generating the required LLR, we apply a novel technique, first developed in [34], that takes into account the clustering and warping experienced by $Y_{m,n}$ due to the nonlinear distortion. This clustering can be different for symbols on different constellation rings and is also non-circular, with some rotation, prompting the use of bivariate Gaussian model for the evaluation of the LLRs. This is used in conjunction with the principle of bit-interleaved coded modulation with iterative decoding (BICM-ID) [35], involving exchange of soft information with the FEC decoder. More specifically, the LLR computation module in Figure 8.20 takes as input $Y_{m,n}$ and $\mathbf{L}_a^{(s)}$, the *a priori* information on the code bits provided by the FEC decoder during the sth iteration. The LLR computation module calculates the bit extrinsic information for the $\log_2 M$ bits that map to a particular symbol $X_{m,n}$ and can be expressed in terms of an LLR as

$$\mathbf{L}_e^{(s+1)}(b_{m,i}) = \log \frac{\sum_{\tilde{X}\varepsilon\chi_i^0} \exp\{f_{\mathrm{bi}}(Y_{m,n}|\tilde{X}) + \sum_{\substack{j=1\\j\neq i}}^{\log_2 M} g_j(\tilde{X})\mathbf{L}_a^{(s)}(b_{m,j})\}}{\sum_{\tilde{X}\varepsilon\chi_i^1} \exp\{f_{\mathrm{bi}}(Y_{m,n}|\tilde{X}) + \sum_{\substack{j=1\\j\neq i}}^{\log_2 M} g_j(\tilde{X})\mathbf{L}_a^{(s)}(b_{m,j})\}} \tag{8.54}$$

for the case of code bit $b_{m,i}$ corresponding to symbol $X_{m,n}$. In (8.54), we define $g_i(X_{m,n})$ as a function returning the ith bit used to label $X_{m,n}$ such that $i = 1, 2, \ldots, \log_2 M$ and $f_{\mathrm{bi}}(Y_{m,n}|\tilde{X})$ represents an improvement in evaluating the likelihood probability based on the bivariate Gaussian model and which is described in [34]. For the specific case of iteration $s = 0$, no soft-information is available from the FEC decoder, so $\mathbf{L}_a^{(0)} = \underline{0}$ is used. The vector of extrinsic information $\mathbf{L}_e^{(s+1)}$ is provided as an input to the FEC decoder, after deinterleaving. The decoder generates an estimate of the source bits after a maximum number of iterations is reached.

8.5.3 Successive transmitter- and receiver-based compensation

The advanced nonlinear countermeasures described in Section 8.4 are generalized, and their effectiveness is evaluated in minimizing the distortions caused by the linear and nonlinear ICI due to OFDM-like signaling. In particular, six different compensation strategies are evaluated:

1. Enhanced receiver architecture from [12] using FS linear equalizer whose taps are computed using the least-mean squares (LMS) adaptation algorithm
2. Iterative receiver-based nonlinear distortion cancellation [36–39] employing hard symbol decisions to recreate the distortion and a single gain correction applied at the output of the FFT operator
3. Successive signal PD applied on the signal composite $s_c(t)$ of (8.1)

4. Receiver-based successive soft cancellation that uses centroids and bivariate Gaussian statistics
5. Successive data PD applied at the symbol rate prior to transmit filters
6. Combined transmitter-based data PD with receiver-based soft cancellation, a concept similar to the disclosure in [40] for SCM

The iterative solutions implement $S = 10$ stages of distortion cancellation. Performance comparison is also made with a traditional system employing SCM-based signaling, with enhanced receiver architecture from [12], while also taking advantage of our centroid-based calculations of the bivariate Gaussian function.

Figure 8.21 provides a comparison of the uplink and downlink PSDs when using signal versus data PD for the case of single OFDM-like 16APSK signal, at their optimal OBO levels. The signal PD scheme is widely used in applications where the HPA and the predistorter are co-located. However, signal PD is implemented at the oversampled signal after the transmit filtering operation. Thus, it requires high sampling rate that is proportional to the product of the individual bandwidth of the signal, number of M_c signals, their frequency separation Δf, and the degree of the nonlinearity to be compensated. In addition, signal PD causes spectral regrowth prior to the HPA, making it less suitable for broadband satellite applications with their strict uplink emission requirements. In contrast, the data PD requires sampling rate that equals the symbol rate only, does not cause uplink spectral regrowth, and provides better performance. It can contribute to the spectral regrowth on the downlink, but this is suppressed by the OMUX filtering present on-board the satellite.

Figure 8.22 shows the coded TD at a target PER of 10^{-3} versus OBO performance of 16APSK using the rate 28/45 LDPC code, in a setup of single signal per transponder. Results indicate an improvement of 1.4 dB for the OFDM-based system with data PD and 1.2 dB improvement with receiver-based successive soft distortion cancellation. Note that the combined solution of using data PD and soft interference cancellation at the receiver provides additional gain of about 0.2 dB beyond PD alone.

Figure 8.23 shows performance when employing 64APSK with rate 7/9 LDPC code in a setup of single signal per transponder. The data PD technique reduces the degradation of the OFDM-based system by almost 4 dB. The receiver-based soft distortion cancellation technique also offers a comparable reduction in TD. Also, the described transmitter and receiver-based techniques provide a substantial reduction in the required OBO for an OFDM-based system. Further, combined successive compensation at the transmitter and at the receiver extracts an additional 0.35 dB improvement beyond PD alone.

Finally, Figure 8.24 shows the performance of the inner signal for the case when three signals share a transponder, each using 16APSK and LDPC code with rate 28/45. The per-signal symbol rate is 37 MBaud, with uniform carrier spacing of $\Delta f =$ 40 MHz. The receiver-based subtractive soft interference cancellation technique has limited effectiveness as the receiver does not have access to symbol estimates from the LDPC decoders of other users. In contrast, successive multicarrier data PD can accurately reconstruct the distortion experienced by the desired signal, leading to the successful mitigation of the resulting distortion. Our results indicate a reduction of close to 1.4 dB in the minimum TD and a 0.5 dB improvement in the required OBO.

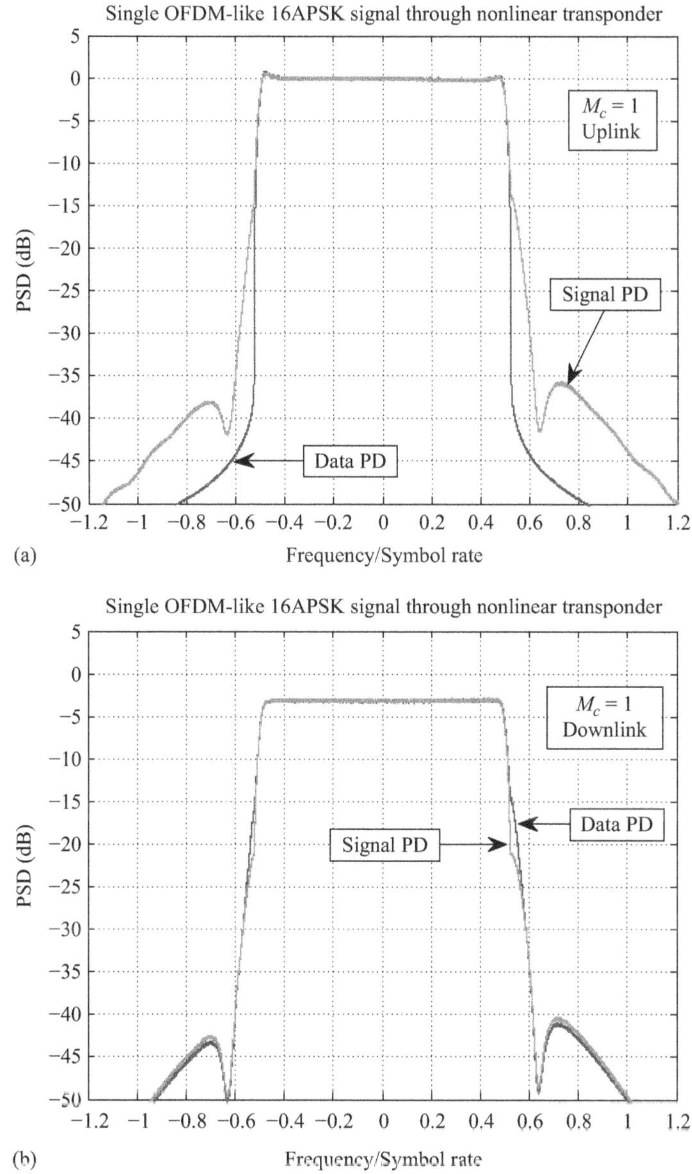

Figure 8.21 PSD when using signal versus data predistortion compensation schemes, for (a) uplink and (b) downlink when $M_c = 1$ *and OFDM-like 16APSK is used*

The successive signal PD technique is worse by 0.4 dB. It is interesting to note that the gap between the OFDM-based system and the SCM system with enhanced receivers is significantly smaller than what was previously observed in the results of single signal per transponder case. Also, the predistorted systems are within a mere 0.6 dB

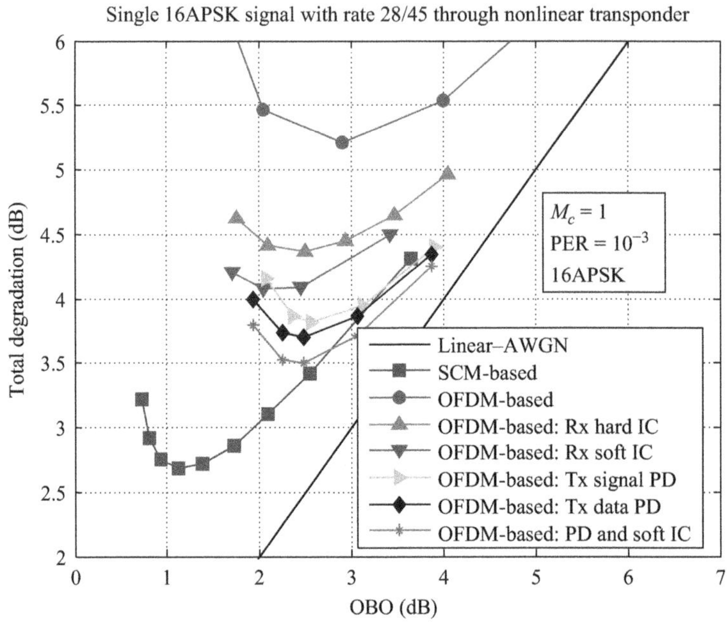

Figure 8.22 Coded total degradation versus OBO level for single 16APSK signal through nonlinear transponder, using OFDM-like signaling, with N = 32, and SCM-based signaling. (DVB-S2X LDPC code rate is 28/45; target PER is 10^{-3}.)

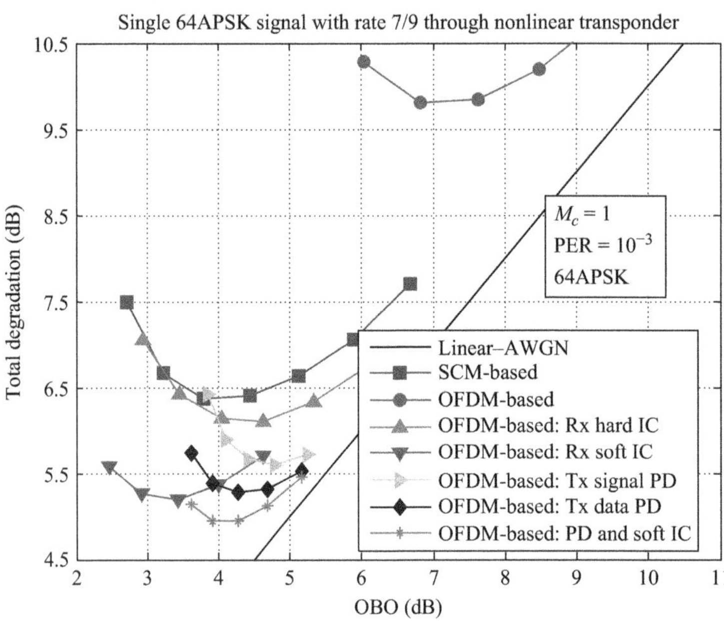

Figure 8.23 Coded total degradation versus OBO level for single 64APSK signal through nonlinear transponder, using OFDM-like signaling, with N = 32, and SCM-based signaling. (DVB-S2X LDPC code rate is 7/9; target PER is 10^{-3}.)

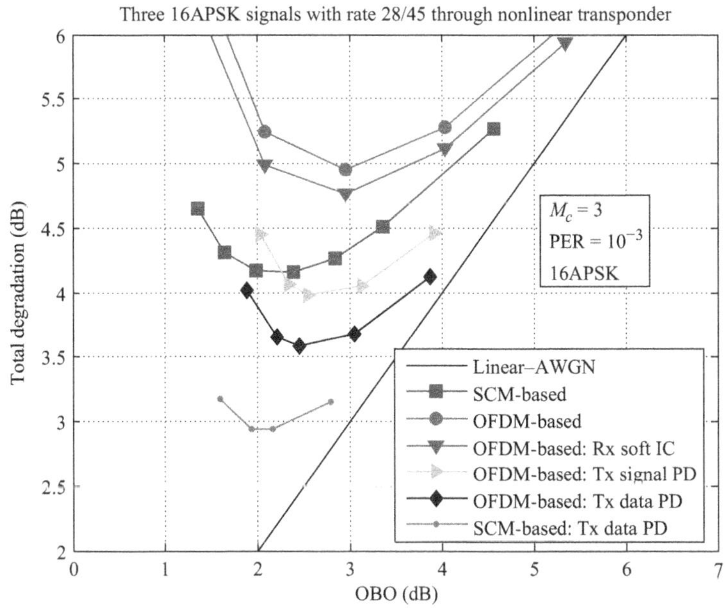

Figure 8.24 Coded total degradation versus OBO level for inner signal when three 16APSK signals share nonlinear transponder, using OFDM-like signaling, with $N = 32$, and SCM-based signaling. (DVB-S2X LDPC code rate is 28/45; target PER is 10^{-3}.)

of each other. This is because the PAPR of an SCM-based system and an OFDM-based system become more comparable when using high-order constellations and/or having multiple signals share the same transponder.

8.6 Conclusion

This chapter has described a satellite communications system for broadband and broadcasting applications that are highly efficient at many levels: mass, power, energy, and bandwidth. An important analytical characterization of the resulting nonlinear distortion has been presented using multicarrier Volterra series. This framework has been shown to be accurate and can be used to develop compensation methods, allowing for the successful operation of such systems while operating the HPA close to saturation. A cornucopia of powerful countermeasures has been investigated that minimizes linear and nonlinear distortion, applied at the transmitter in the form of PD, and at the receiver in the form of equalization, while exploiting the turbo processing principle of exchanging soft information with FEC decoders.

It is envisaged that satellites will play a vital role in the emerging 5G landscape which continues to include OFDM air interface. The second part of this chapter has endeavored to apply OFDM over satellite systems to establish greater commonality. The aforementioned countermeasures have been applied for OFDM-like signaling allowing OFDM-based satellite systems to be competitive with, and in some cases surpassing, traditional systems that use SCM, when employing high-order constellations and/or having multiple signals share the same transponder, consistent with the industry trend.

The presented analysis and techniques, with their attractive performance, deserve exploration in other important lines of research. For example, precoding for multi-beam satellite systems [41] is a promising technology that reduces the linear cochannel interference amongst the beams when aggressive frequency reuse is applied. Aspects of precoding include gateway-based solution requiring multiuser processing of the transmitted symbols. This creates synergy with multicarrier PD that can be utilized to combine precoding methods with mitigation of nonlinear distortions due to power-efficient operation of the satellite transponder.

Cognitive communications for satellite systems [42] is another promising technology that allows for spectral coexistence of satellite-terrestrial networks. The described characterization of nonlinear distortion can be computed by cognitive nodes to gain more awareness. Also, the considered OFDM-like signaling offers spectral flexibility, has excellent frequency containment, and is robust to distortion with the use of advanced techniques. Due to these properties, it can be exploited to enable further cognition.

References

[1] ETSI EN 302307-1. Second generation framing structure, channel coding and modulation systems for broadcasting, interactive services, news gathering and other broadband satellite applications; Part I: DVB-S2. Digital Video Broadcasting (DVB). 2005.

[2] ETSI EN 302307-2. Second generation framing structure, channel coding and modulation systems for Broadcasting, interactive services, news gathering and other broadband satellite applications; Part II: S2 Extensions (DVB-S2X). Digital Video Broadcasting (DVB). 2014.

[3] Beidas BF. Intermodulation Distortion in Multicarrier Satellite Systems: Analysis and Turbo Volterra Equalization. IEEE Trans Commun. 2011 June;59(6):1580–1590.

[4] Benedetto S, Biglieri E, Daffara R. Modeling and Performance Evaluation of Nonlinear Satellite Links—A Volterra Series Approach. IEEE Trans Aerosp Electron Syst. 1979 July;15(4):494–507.

[5] Beidas BF, Seshadri RI. Analysis and Compensation for Nonlinear Interference of Two High-Order Modulation Carriers over Satellite Link. IEEE Trans Commun. 2010 June;58(6):1824–1833.

[6] Mazo JE. Faster-Than-Nyquist Signaling. Bell Syst Tech J. 1975 October; 54(8):1451–1462.

[7] Liveris AD, Georghiades CN. Exploiting Faster-Than-Nyquist Signaling. IEEE Trans Commun. 2003 September;51(9):1502–1511.

[8] Piemontese A, Modenini A, Colavolpe G, *et al.* Improving the Spectral Efficiency of Nonlinear Satellite Systems through Time-Frequency Packing and Advanced Receiver Processing. IEEE Trans Commun. 2013 August;61(8): 3404–3412.

[9] Beidas BF, Seshadri RI, Eroz M, *et al.* Faster-Than-Nyquist Signaling and Optimized Signal Constellation for High Spectral Efficiency Communications in Nonlinear Satellite Systems. In: Proc. IEEE MILCOM Conference; 2014. p. 818–823.

[10] 3GPP TR 38 802 V14 1 0. Study on New Radio Access Technology Physical Layer Aspects (Release 14). 3rd Generation Partnership Project; Technical Specification Group Radio Access Network. 2017 June.

[11] Bingham JAC. Multicarrier Modulation for Data Transmission: An Idea Whose Time Has Come. IEEE Commun Mag. 1990 May;28(5):5–14.

[12] ETSI TR 102 376 V1 1 1. Implementation guidelines for the second generation system for Broadcasting, Interactive Services, News Gathering and other broadband satellite applications; Part 2: S2 Extensions DVB-S2X. Digital Video Broadcasting (DVB). 2015.

[13] Haykin S. Adaptive Filter Theory. 2nd ed. Englewood Cliffs, NJ, USA: Prentice-Hall; 1991.

[14] Morgan DR, Ma Z, Kim J, *et al.* A Generalized Memory Polynomial Model for Digital Predistortion of RF Power Amplifiers. IEEE Trans Signal Process. 2006 October;54(10):3852–3860.

[15] Kim J, Konstantinou K. Digital predistortion of wideband signals based on power amplifier model with memory. Electron Lett. 2001 November; 37(23):1417–1418.

[16] Ding L, Zhou GT, Morgan DR, *et al.* A Robust Digital Baseband Predistorter Constructed Using Memory Polynomials. IEEE Trans Commun. 2004 Januanry;52(1):159–165.

[17] Beidas BF. Adaptive Digital Signal Predistortion for Nonlinear Communication Systems Using Successive Methods. IEEE Trans Commun. 2016 May; 64(5):2166–2175.

[18] Heo SW, Gelfand SB, Krogmeier JV. Equalization Combined with Trellis Coded and Turbo Trellis Coded Modulation in the Nonlinear Satellite Channel. In: Proc IEEE MILCOM, *Los Angeles, CA*. 2000 October; p. 184–188.

[19] Pérez AL, Ryan WE. Iterative Detection and Decoding on Nonlinear ISI. In: Proc Int Conf Commun, *New York, NY*. 2002 May; p. 1501–1505.

[20] Burnet CE, Barbulescu SA, Cowley WG. Turbo equalization of the nonlinear satellite channel. In: Proc Int Symp Turbo Codes, *Brest, France*. 2003 September; p. 475–478.

[21] Beidas BF, El-Gamal H, Kay S. Iterative Interference Cancellation for High Spectral Efficiency Satellite Communications. IEEE Trans Commun. 2002 January;50(1):31–36.

[22] De Gaudenzi R, Luise M. Analysis and Design of an All-Digital Demodulator for Trellis Coded 16-QAM Transmission over a Nonlinear Satellite Channel. IEEE Trans Commun. 1995 February/March/April;43(2/3/4):659–668.

[23] Piazza R, Bhavani Shankar MR, Ottersten B. Data Predistortion for Multicarrier Satellite Channels Based on Direct Learning. IEEE Trans Signal Process. 2014 November;62(22):5868–5880.

[24] Biglieri E, Barberis S, Catena M. Analysis and Compensation of Nonlinearities in Digital Transmission Systems. IEEE J Select Areas Commun. 1988 January;6(1):42–51.

[25] Eun C, Powers E. A New Volterra Predistorter based on the Indirect Learning Architecture. IEEE Trans Signal Process. 1997 January;45(1):223–227.

[26] Zhou L, DeBrunner VE. Novel Adaptive Nonlinear Predistorters Based on the Direct Learning Algorithm. IEEE Trans Signal Process. 2007 January; 55(1):120–133.

[27] Kushner HJ, Yin GG. Stochastic Approximation and Recursive Algorithms and Applications. 2nd ed. New York, USA: Springer-Verlag; 2003.

[28] Beidas BF, Seshadri RI, Becker N. Multicarrier Successive Predistortion for Nonlinear Satellite Systems. IEEE Trans Commun. 2015 April;63(4): 1373–1382.

[29] Karam G, Sari H. A Data Predistortion Technique with Memory for QAM Radio Systems. IEEE Trans Commun. 1991 February;39(2):336–344.

[30] Casini E, De Gaudenzi R, Ginesi A. DVB-S2 Modem Algorithms Design and Performance Over Typical Satellite Channels. Intern J Satellite Commun Network. 2004;22(3):281–318.

[31] Kourogiorgas CI, Lyras N, Panagopoulos AD, *et al.* Capacity Statistics Evaluation for Next Generation Broadband MEO Satellite Systems. IEEE Trans Aerosp Electron Syst. 2017 October;53(5):2344–2358.

[32] Beidas BF, Seshadri RI. OFDM-Like Signaling for Broadband Satellite Applications: Analysis and Advanced Compensation. IEEE Trans Commun. 2017 October;65(10):4433–4445.

[33] van Waterschoot T, Nir VL, Duplicy J, *et al.* Analytical Expressions for the Power Spectral Density of CP-OFDM and ZP-OFDM Signals. IEEE Signal Process Lett. 2010 April;17(4):371–374.

[34] Beidas BF, Seshadri RI. Forward error correction decoder input computation in multi-carrier communications system. US Patent and Trademark Office, *Patent 9,203,680*. 2015 filed September 2012, granted December.

[35] Li X, Ritcey JA. Bit-Interleaved Coded Modulation with Iterative Decoding. IEEE Commun Lett. 1997 November;1(6):169–171.

[36] Kim D, Stuber GL. Residual ISI Cancellation for OFDM with Applications to HDTV Broadcasting. IEEE J Sel Areas Commun. 1998 October;16(8): 1590–1599.

[37] Kim D, Stuber GL. Clipping Noise Mitigation for OFDM by Decision-Aided Reconstruction. IEEE Commun Lett. 1999 January;3(1):4–6.

[38] Tellado J, Hoo LMC, Cioffi JM. Maximum-Likelihood Detection of Nonlinearly Distorted Multicarrier Symbols by Iterative Decoding. IEEE Trans Commun. 2003 February;51(2):218–228.

[39] Chen H, Haimovich AM. Iterative Estimation and Cancellation of Clipping Noise for OFDM Signals. IEEE Commun Lett. 2003 July;7(7):305–307.

[40] Beidas BF, Kay S, Becker N. System and Method for Combined Predistortion and Interference Cancellation in a Satellite Communications System. US Patent and Trademark Office, *Patent 8,355,462*. 2013 filed October 2009, granted January.

[41] Christopoulos D, Chatzinotas S, Ottersten B. Multicast Multigroup Precoding and User Scheduling for Frame-Based Satellite Communications. IEEE Trans Wireless Commun. 2015 September;14(9):4695–4707.

[42] Sharma SK, Chatzinotas S, Ottersten B. Cognitive Radio Techniques for Satellite Communication Systems. In: IEEE Vehicular Technology Conference; 2013. p. 1–5.

Chapter 9

Satellite multi-beam precoding software-defined radio demonstrator

Stefano Andrenacci[1], Juan Carlos Merlano Duncan[1], Jevgenij Krivochiza[1], and Symeon Chatzinotas[1]

Linear precoding, commonly known as multi-user multiple-input–multiple-output (MU-MIMO), exploits the spatial degrees of freedom offered by multi-antenna transmitters to manage interferences between multiple co-channel users. The technique relies on the channel state information (CSI) estimation by user terminals (UTs) and, as a consequence, a real-time test-bed implementation of the system is challenging and not straightforward. The estimation of the channel coefficient by UTs is a challenging operation, especially for satellite communications (SATCOM), since the estimation should also perform at very low signal-to-noise-plus-interference ratio (SNIR), while it is as well affected by a plethora of impairments given by different components and technologies, which goes to technologies at the gateway (GW) side, through the payload characteristics and till the UT impairments. Some of the most important and general impairments to be taken into account in the design of a precoding system are listed in the following:

- Frequency offsets, frequency instabilities and phase noise (PN) of the local oscillators (LOs) used at the transmitter, the channel emulator and all the receivers.
- Timing misalignment amongst different beams (for high throughput links).
- Differential phase/amplitude distortion on different payload chains due to on board non-linearity.
- Fixed point report of the CSI, fixed point computation of the precoder in field-programmable gate-array (FPGA).
- Limited computational capabilities, especially for high throughput links.
- Round trip time (RTT) delay of a satellite link.

The aim of this chapter is to demonstrate the ability of broadband multi-beam satellite systems to operate in aggressive frequency reuse modes, enabled by advanced signal processing methods, namely precoding, when practical constraints affects the implementation of signal processing techniques. To accomplish the objective,

[1]SnT-securityandtrust.lu, University of Luxembourg, Luxembourg

a specific hardware infrastructure composed by properly interconnected software-defined radios (SDRs) has been built. The infrastructure is able to emulate a satellite forward (FWD) link transmission using a GW emulator and a multi-beam satellite channel emulator, which includes, on top of the satellite impairments, the multiple-input–multiple-output (MIMO) user link channel and a set of independent UTs radio frequency (RF) impairments emulators. To enable real-time precoding implementation, a feedback channel from UTs to the GW is emulated accordingly. The general infrastructure includes a various number of SDR development platforms called universal software radio peripherals (USRPs), each of them connected to a central hub used for selecting the sub-infrastructure required for the specific test, while also providing control and monitoring functionalities. Each board is itself a single-antenna/multi-antenna system equipped with a RF module, digital-to-analog (DAC) and analog-to-digital converters (ADC) and a high performance FPGA for user-defined digital processing. The central hub is also supported by a high computational capabilities workstation equipped with a set of FPGAs, used for the centralized processing. The chapter is divided into the following sections. In Section 9.1, a general introduction on precoding techniques is included; in Section 9.2, an analysis of the practical constraints for precoding and possible solutions are investigated; in Section 9.3, a description of the precoding implementation is reported; in Section 9.4, the in-lab validation of the precoding techniques implemented is addressed; and finally, in Section 9.5, some conclusions and future work are reported.

9.1 Introduction on precoding

The new era of broadband internet and on-demand services challenges to come up with new approaches towards design of the SATCOM systems. The market importance of the broadband services and the limited frequency resources drive SATCOM industry and the academia towards development of novel smart and more efficient in terms of power and frequency wireless communication technologies.

Multi-beam satellites, on the one hand, are more power efficient and, on the other hand, have higher capacity in the satellite channel through the spatial multiplexing [1]. In addition, while conventional multi-beam systems employ a frequency reuse 4 (FR4) scheme or even higher, full frequency reuse (FFR) schemes are more attractive in spectrum limited scenarios. Therefore, application of MU-MIMO in SATCOM is highly challenging due to the practical constraints to be faced, but at the same time extremely rewarding academic task [2] from both literature and projects point of views.

9.1.1 Recent projects on precoding

In the framework of the European Space Agency (ESA) contract NGW 'Next Generation Waveforms for improved spectral efficiency' [3], the focus was mostly on the practical implementation aspects including degradations due to a multi-GW system, users grouping in multicast precoding and practical limitation in the CSI estimation accuracy and sensitivity. For example, in [4] it is stated that, depending on the relative scenario in which a FFR system is used jointly with precoding techniques at the GW, precoding gains in terms of overall capacity varies from 38% to 140% when compared with a four colour reuse system without the use of precoding.

In another ESA study, FoGBS 'future ground based beamforming techniques' [5] possible gains with respect to benchmark scenarios, given by the classical colour reuse scheme in a multi-beam satellite system, are evaluated from a system level point of view. Results showed potential gains in terms of system capacity for all the three scenarios, which were a regional broadband interactive service exploiting the Ka-Band spectrum, a high-throughput backhauling service over a continental region in C-Band and a mobile service in L-Band exploiting multi-satellite constellations, especially when proper scheduling approaches are assumed in the system.

The focus of the ESA project PreDem 'Precoding Demonstrator for Broadband System Forward Links' [6] is on the implementation of a software demonstrator for interference mitigation techniques at the GW side in a forward link SATCOM system, taking into account the recent Digital Video Broadcasting S2 extension (DVB-S2x), in particular for what concerns the Superframing option designed to support precoding techniques. The development of a full-fledged simulator that includes all system and physical layer aspects involved in precoding over satellite under practical impairments, is the main product to deliver, towards concluding this research activity.

The aim of the ESA project Optimus 'Optimized transmission techniques for satcom unicast interactive traffic' [7] is to go beyond the Direct-to-Home-like standards and applications in order to remove some constraints of broadcast-based satellite systems. According to the scope of the project, the design of a novel forward link air interface for broadband satellite systems which helps and favours the use of advanced interference mitigation techniques at the transmitter, namely precoding, is the main focus of the activity.

Based on the large experience gained through these research projects and activities, the first proof of concept project called SERENADE 'Satellite Precoding Hardware Demonstrator' [8], which is funded by the Luxembourg National Research Fund (FNR), has been kicked off in 2016. The aim is to designing and develop an in-lab software-defined radio-based multi-beam satellite precoding demonstrator, able to emulate and test the end-to-end link, taking into account all the practical impairments of the system. Differently from other MIMO-based test-beds [9,10], the current one is not based on LTE waveforms, it employs a programmable multi-beam satellite channel emulator and it performs CSI estimation at each receiver under practical constraints using DVB-S2x waveforms.

9.1.2 Related literature on precoding for SATCOMs

The future SATCOM will benefit from multi-beam satellites, which are capable of aggressive frequency reuse through advances signal processing techniques. Different precoding techniques were designed to enhance the SATCOM link in many ways: increase the physical layer security [11], optimize the system capacity and user scheduling [12] and manage interference via the spatial degrees of freedom offered by the multiple antennas [4,13–15]. These and numerous other works show the applications of such techniques in multi-beam satellites, which result in terms of increased system capacity, service availability, enhanced security and energy efficiency in SATCOM.

However, the non-linear nature of the HPA results in adjacent channel interference and increased peak-to-average power ratio [16], which limits the expected theoretical performance gains. In this context, studies are required on the energy efficient on-board pre-distortion techniques, which can optimize the performance of HPA by uniformly distributing the power load [17,18]. Furthermore, the MU-MIMO precoder at the transmitter utilizes closed-loop approach by employing the retrieved CSI from the UTs; hence, a feedback channel is required for the precoder to operate.

Generally, due to the inability of acquiring instantaneous CSI at the GW, precoding for mobile satellite systems can be very challenging. However, there is potential for specific types of applications such as aeronautical/maritime systems, where the channel is, in some cases, predictable, and there is no direct blockage of the line of sight component [19].

The problem of reporting accurate CSI is another not trivial aspect which has a considerable impact on the overall performance of a precoding-based system, especially in SATCOM where the channel is affected by non-ideal components and the UT is far from being impairments-free. Moreover, the frame-based nature of a SATCOM FWD link is based on long forward error correction (FEC) codewords, which vary their length in terms of symbols depending on the selected modulation and coding (Modcod). This variable FEC length has demanded for a novel air interface.

As a consequence, the DVB group, through its satellite technical module, has developed the superframe structure [20] as option of DVB-S2X [21,22], especially for interference mitigation techniques like precoding and beam-hopping. Two format specifications of the superframe have been implemented, which enable MU-MIMO technique for the FWD link of SATCOMs.

A novel UT synchronization and channel estimation procedure has been as well conceived to face the new scenario which is interference plus noise limited and in which the waveforms to be estimated can be extremely weak compared with the main and useful signal [4,22].

In the following, some challenges related to the implementation of precoding techniques in SATCOM are assessed and the corresponding results are reported.

9.2 Analysis of the practical constraints for precoding and possible solutions

9.2.1 System model

The general system model focuses on the forward link of a multi-beam satellite system, which aims at reusing the total available bandwidth among all beams of the coverage (the so-called FFR). We define N_t as the number of transmitting antenna elements and N_u as the total number of users in the coverage area. In the specified MIMO channel model, the received signal at the ith user is $y_i = \mathbf{h}_i^\dagger \mathbf{x} + n_i$, where $\mathbf{h}_i^\dagger$ is a $1 \times N_t$ vector representing the complex channel coefficients between the ith user and the N_t antennas of the transmitter, $\mathbf{x}$ is defined as the $N_t \times 1$ vector of the transmitted symbols at a certain symbol period and n_i is the independent complex circular symmetric (c.c.s.) independent identically distributed (i.i.d.) zero mean Additive White Gaussian Noise (AWGN) measured at the ith user's receive antenna.

Assuming a system having $N_t = N_u = N$, which is the present case, looking at the general formulation of the received signal, which includes the whole set of users, the linear signal model is

$$\mathbf{y} = \mathbf{H}\mathbf{x} + \mathbf{n} = \mathbf{H}\mathbf{W}\mathbf{s} + \mathbf{n} \tag{9.1}$$

where $\mathbf{y}$ and $\mathbf{n} \in \mathbb{C}^N$, $\mathbf{x} \in \mathbb{C}^N$ and $\mathbf{H} \in \mathbb{C}^{N \times N}$. In this scenario, we define the linear precoding matrix $\mathbf{W} \in \mathbb{C}^{N \times N}$ which maps information symbols $\mathbf{s}$ into precoded symbols $\mathbf{x}$. It is worth noting that in the following description, with the term 'waveform', we refer to the signal coming from the satellite feed.

9.2.2 Differential phase distortion for precoded waveforms

The issue of the instantaneous differential phase distortion for precoded waveforms caused by the non-zero AM/PM characteristics of the travelling-wave tube amplifiers (TWTAs) is a potential source of degradation for precoding, especially when non-linearized TWTAs are employed in the satellite payload chains.

In fact, since non-linearized TWTAs introduce different instantaneous phase offsets for different symbols in the transponders, a phase offset variation, that depends on the instantaneous symbol power at the input of the TWTA, can introduce mismatches between the precoding matrix and the channel. In fact, while the estimation of CSI accounts for the average amplitude and phase channel conditions, it is impossible to report instantaneous (i.e. per symbol) channel conditions, especially when the round trip delay of GEO and MEO orbits outdate the coefficient.

By looking at the phase distortion of a satellite power amplifier, the AM/PM characteristic affects a general transmitted waveform and, in particular, a precoded constellation, by introducing two main degradation effects:

- Differential phase distortion in space (DPhD). The effects of this instantaneous phase offset mismatch amongst transponders are a potential source of degradation for precoded waveforms. In this case, the instantaneous phase offset is a function of the instantaneous precoded symbol power.
- Phase distortion in time due to the non-linear AM/PM characteristic (PhC) of a single TWTA, which affects a generic constellation, both precoded and not precoded, irrespective to the phase mismatch amongst transponders. This effect has been intensively studied in the literature of terrestrial and satellite fields.

While payload technologies become more advanced and help the mitigation of undesired effects, the use of linearized TWTAs is reducing the impact of distortions due to the almost linear AM/AM, AM/PM characteristics of the amplifiers.

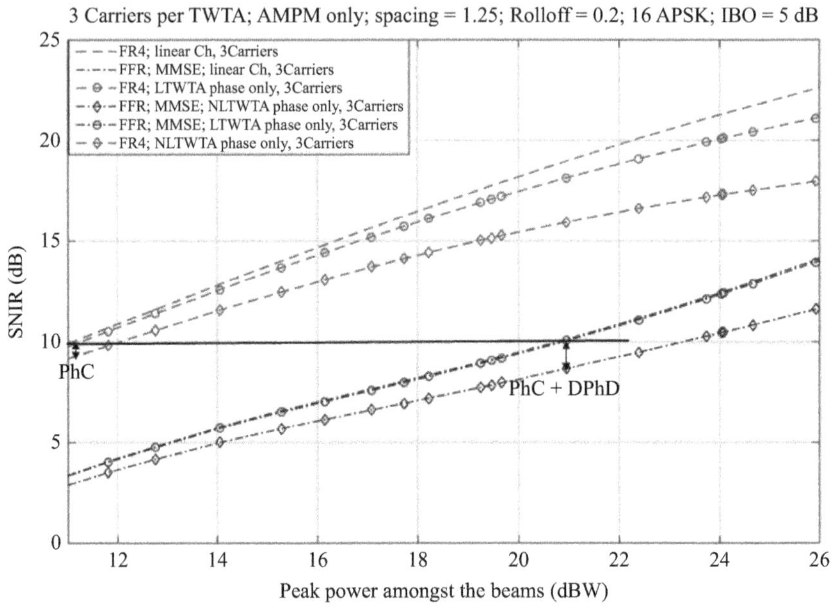

Figure 9.1 Evaluation of the impact of differential phase distortion over space in a multi-carrier scenario

In order to assess the impact of the differential phase shift over space for precoded waveforms, a comparison in terms of SNIR at UTs between precoded multi-carrier transmission in FFR and not-precoded multi-carrier transmission in FR4, affected by the AM/PM distortion of both linear and non-linearized TWTAs, is shown in Figure 9.1.

In the following simulations, $N_t \times N_c$ number of streams are independently and randomly generated at the transmitter side, where N_t is the number of feeds and N_c is the number of carrier per beam. Before the linear combination of the streams due to precoding (based on the perfect knowledge of the CSI), each stream is modulated with 16 APSK constellation. Assuming a multi-carrier transmission scheme, carriers of the same beam are precoded with the same coefficient; hence, the assumption is to have the same channel coefficient over different frequency bandwidths (non-frequency selective channel assumption).

After the application of precoding, carriers of the same beam are shaped and aggregated to generate the per-beam multi-carrier waveform. Each multi-carrier waveform is then distorted by separate AM/PM characteristics of TWTAs and then transmitted to the receivers. Finally, each receiver compute the SNIR for each carrier, and averaged results over the whole set of carriers are reported depending on the per-beam power used.

The general parameters used for the simulation are listed in Table 9.1.

Table 9.1 Simulation parameters for the multi-carrier transmission

Parameter	Value
Scenario	71 beams over Europe
Number of carriers per beam	3
Spacing between carriers	1.25
Roll-off	0.2
Modulation scheme	16APSK
Input back-off (IBO)	5 dB

In Figure 9.1, we report simulation results in terms of average received SNIR versus average per beam power, when both linear and non-linearized TWTAs are employed in the simulation chain.

Figure 9.1 reports the following curves:

- *FR4; Linear Ch, 3Carriers*: the benchmark curve (Linear Channel) for the FR4 case when 3 carriers per beam are used.
- *FFR; MMSE; Linear Ch, 3Carriers*: the benchmark curve (Linear Channel) for the FFR with Precoding (MMSE [15]) case when 3 carriers per beam are used.
- *FR4; LTWTA Phase Only, 3Carriers*: the AM/PM characteristic of a linearized TWTA is used for each beam of the FR4 multi-carrier case.
- *FFR; MMSE; NLTWTA Phase Only, 3Carriers*: the AM/PM characteristic of a non-linearized TWTA is used for each beam of the FFR precoding multi-carrier case.
- *FFR; MMSE; LTWTA Phase Only, 3Carriers*: the AM/PM characteristic of a linearized TWTA is used for each beam of the FFR precoding multi-carrier case.
- *FR4; NLTWTA Phase Only, 3Carriers*: the AM/PM characteristic of a non-linearized TWTA is used for each beam of the FR4 multi-carrier case.

Based on the results, a first outcome is that when linearized tubes are used (at least for low-medium SNIRs), differential phase does not introduce a significant degradation to both FFR and FR4 cases.

The focus is therefore on the distortion effects caused by non-linearized TWTA on the received SNIR. It is pretty evident how the SNIR is much more affected by degradation especially for high-peak power amongst beams; hence, there is a dependency between the suffered degradation and the level of interference.

In order to evaluate the impact of the differential phase distortion over space, we compare the curves (precoded and nor precoded) by fixing a common received SNIR. The current comparison uses a SNIR value of 10 dB as a threshold.

The discrepancy in the SNIR values of the FR4 (dashed) curves, which basically represents the level of the phase distortion over time, is about 0.8 dB. The discrepancy in the FFR (dashed-dotted) curves is due to both the phase degradation over space and time and it is estimated to be around 1.4 dB. Assuming that the two effects can be somehow decoupled, the differential phase distortion over space at SNIR = 10 dB for the chosen comparison is about 0.6 dB.

Possible solutions to the degradation effects are:

- The use of linearized TWTA, as shown in Figure 9.1
- The use of precoding techniques which limits the average power over beams (like per-antenna power constraint normalization)
- A joint precoding/pre-distortion technique

9.2.3 Timing misalignment on precoded waveforms

In order to generate the precoding matrix, it is foreseen that each terminal provides estimates of the channel parameters for all of the detectable received signals, i.e., the terminal specific waveform plus all of the detectable interferers [4,12]. This requires that a terminal synchronizes not only onto its own signal but also onto all the detectable interferers. DVB-S2X superframe structure, defined in Annex E of [21], is, by design, the framing structure which enables the possibility of estimating the CSI in a satellite system [22].

While DVB-S2X relies on the use of Walsh–Hadamard sequences in the SF-Pilots definition, the advantages provided by orthogonal cross correlated codes requires the transmitting waveforms to be perfectly synchronized in time, especially when a scrambling sequence is used as it happens in satellite links (see Figure 9.2). What

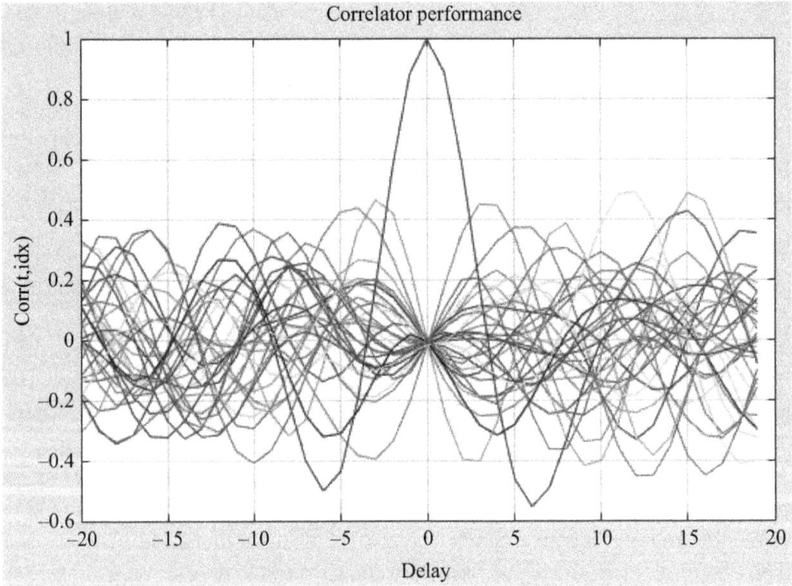

Figure 9.2 Superposition of all oversampled correlation functions amongst SF-pilots. Each curve defines a different pilot sequence. Delays are expressed in terms of kT_s/T_s where T_s is the symbol period

is more important, in some cases and depending on the per-carrier baudrate, the satellite payload introduces a timing misalignment amongst different transmitting antennas due to the different group delays and paths of the transponder filters.

As a consequence, the problem of the required waveforms alignment is important not only for CSI estimation accuracy but also for precoding techniques to avoid additional degradations due to ISI amongst waveforms of the superimposed signal at the UT, since precoding techniques assume the transmitting waveforms to be quasi-perfectly synchronized.

A procedure to pre-compensate the timing misalignments introduced by the payload, which is presented in the following, is fundamental to both increase the quality of the CSI estimations and to avoid performance degradation in the precoding process.

9.2.3.1 Analysis on the impact of timing misaligned waveform in precoded systems

The aim of the present section is to analyse the effect of timing misaligned carriers due to satellite payload impairments on a system which employs linear precoding techniques at the GW. The analysis and the validation sections are split into two parts:

1. Analysis of the effects on the CSI estimation accuracy
2. Analysis of the effects on the received SNIR for precoded symbols

9.2.3.2 Impact of time misalignment of bundled frames on channel estimation

Considering an interference limited scenario due to an aggressive frequency reuse amongst beams, known-symbols (Pilots)-aided algorithms provide advantages in distinguishing beam-specific waveforms from the superimposed signal at the UT. Pilot-aided algorithms are strictly influenced by the correlation properties of the set of sequences used. In case of time alignment, as in a downlink scenario of a wireless system, the use of a set of orthogonal sequences, as the Walsh–Hadamard set, is a preferable choice since it provides very good cross-correlation properties between sequences under some constraints. On the other hand, the autocorrelation properties are not suitable for detection purposes due to the presence of several sidelobes in the autocorrelation function. The use of a scrambling sequence on top which operates symbol by symbol, shared between all signals (as in DVB-S2x Annex E framing structure described in [21]) is very helpful in terms of autocorrelation function and in terms of frequency spectrum to avoid spectral lines.

The normalized correlation function between two discrete sequences is given by the well-known formula:

$$R_{ij}[n] = \frac{1}{N_{Pil}} \sum_{m=0}^{Npil-1} o_i[m]o_j^*[m-n] \tag{9.2}$$

where N_{pil} is the length of the sequences, o_i is the ith orthogonal Walsh–Hadamard sequence and $*$ represents the complex conjugate. This correlation function depends on the properties of the set of sequences which, for the case of Walsh–Hadamard set, is basically having a cross-correlation equals to 0 when $n = 0$ and $i \neq j$.

When the scrambling sequence $g[m]$ is considered, a new set of sequences should be considered which is given by $c_i[m] = g[m]o_i[m]$, where, according to the Super-framing description, $\mathbf{g}$ is the same for all the waveforms. Another property of the scrambling sequence is that $g[m]g^*[m] = 1$. If we substitute the new sequences in the equation, we obtain

$$R_{ij}[n] = \frac{1}{N_{Pil}} \sum_{m=0}^{Npil-1} c_i[m]c_j^*[m-n]\frac{1}{N_{Pil}} = \sum_{m=0}^{Npil-1} g[m]o_i[m]g^*[m-n]o_j^*[m-n] \tag{9.3}$$

From this formulation, it can be noticed that when $g[m]g^*[m-n] = 1$ for each m in the range $m = 0, \ldots, N_{pil}$, the correlation function is exactly the same as (9.2), and this happens when $n = 0$, hence, when the two sequences are time aligned.

In Figure 9.2, the auto and cross correlation functions in the oversampled domain of one selected sequence with respect to all other sequences are shown. It should be specified that to consider the effects of the payload data symbols on the correlation, which is a more realistic case, the correlation functions are calculated as follows (Format Specification 2 of the superframe is taken into account):

$$R_{ij}[n] = \frac{1}{N_{Pil}} \sum_{m=0}^{Npil-1} r_i'[m+n]c_j^*[m] \tag{9.4}$$

where $n = [-920, -920 + 1/ns, 0, \ldots, 955 - 1/ns, 955] \in \mathbb{R}$, $m \in \mathbb{N}$ and ns is the oversampling factor. r_i' is the ith received stream given by the successive concatenation of three vectors which are

$$r_i'[m] = \begin{cases} x1_i[m] & \text{when } -920 \le m < 0 \\ c_i[m] & \text{when } 0 \le m < N_{pil} - 1 \\ x2_i[m] & \text{when } N_{pil} \le m \le 955 \end{cases} \tag{9.5}$$

In the latter definition, both $x1$ and $x2$ are random data symbols having QPSK modulation for the sake of simplicity. The figure clearly shows that the orthogonality between sequences happens in case of perfect alignment only since all the correlation values in delay 0 are equal to 0. The curve which has a peak in delay 0 is of course the autocorrelation function R_{ii}. This justifies the importance of pre-compensating the timing misalignment due to the satellite payload.

9.2.4 *Numerical results on the quality of CSI with timing pre-compensated waveforms*

In the following, we report the numerical analysis and performance assessment considering the receiver algorithms (synchronization and channel estimation) described in [4,22]. The results can be divided into two parts: in the first part, the timing estimation performance in presence of timing misalignment amongst waveforms (no orthogonality) are shown, then, results obtained in terms of CSI estimation errors for the cases of pre-compensated and not pre-compensated waveforms are reported.

It is important to specify that the synchronization procedure, which happens before the CSI estimation in the receiver chain, is modelled according to residuals errors at the output of the synchronization chain.

For the numerical results, we consider, at the receiver side, a superposition of six waveforms given by the reference waveform (i.e. the only one with useful information to be decoded) and five interferer waveforms coming from five adjacent beams. Assuming that C is the power of the reference waveform and I is the power of the considered interferer waveform to be estimated (one CSI coefficient), the six waveforms have the following C/I distribution:

$$C/I_i = [0\ 4\ 8\ 12\ 16]\text{dB} \tag{9.6}$$

It is worth noting that the aim of this distribution is to highlight how CSI errors are affected by the other carriers when a timing misalignment is present. The algorithm used for the channel estimation procedure is a Pilot-aided algorithm described by the following formula:

$$\hat{h}_i = A_i e^{j\varphi_i} \tag{9.7}$$

$$A_i = \frac{1}{N_{Pilot} L_{Pilot}} \sum_{k=1}^{N_{Pilot}} \left| \sum_{j=1}^{L_{Pilot}} y_k^p[j] c_{ki}^*[j] \right| \tag{9.8}$$

$$\varphi_i = \frac{1}{N_{Pilot} L_{Pilot}} \sum_{k=1}^{N_{Pilot}} \angle \sum_{j=1}^{L_{Pilot}} y_k^p[j] c_{ki}^*[j] \tag{9.9}$$

where $\angle$ is the angle function, $\hat{h}_i$ is the estimate for the ith waveform, A_i and φ_i are respectively the amplitude and phase estimates for the ith waveform, N_{Pilot} and L_{Pilot} are, respectively, the number of pilot fields (the number of consecutive pilot blocks over which the estimate is averaged) and the length of the pilot fields, $y_k^p[j]$ is the portion of the received signal corresponding to the kth block of the transmitted pilots within the Superframe and $c_{ki}^*[j]$ is the beam specific sequence (composed by a beam specific Walsh–Hadamard sequence and a scrambling sequence).

A comparison in terms of CSI errors for time aligned and time misaligned waveforms is shown and described.

The simulation parameters used in the channel estimation procedure are reported in Table 9.2.

In Figure 9.3, results obtained in terms of mean and standard deviation of CSI amplitude errors in the case of both misaligned and aligned (hence pre-compensated) waveforms are shown versus the C/I value of the specific waveform. These values are calculated, starting from the amplitude errors obtained, by the following formula:

$$A_{errdB} = 20 \times \log_{10}\left(\frac{1}{N_{iter}} \sum_{i=1}^{N_{iter}} \frac{A_{lin} + \varepsilon_i}{A_{lin}}\right) \tag{9.10}$$

where A_{lin} is the value of the waveform amplitude in linear units, while ε is the error from the estimation of the amplitude. While the solid lines are the mean values,

Table 9.2 Simulation parameters for the CSI accuracy assessment

Parameter	Value
Symbol rate	500 MBaud
Roll-off	0.05
Oversampling factor	4
Time misalignments	$[-3\ T_s; +3\ T_s]$
Frequency misalignments	Negligible
Phase misalignments	$[-\pi/2; \pi/2]$
Residual from frequency estimation	Gaussian r.v. $\sigma = 0.0003\ R_s$
Residual from timing estimation	Gaussian r.v. $\sigma = 0.036\ T_s$
Residual from phase estimation	Gaussian r.v. $\sigma =$ Cramer Rao Bound
SNR (w.r.t. the reference waveform)	0–10 dB
L_{Pilot}	32 symbols
N_{Pilot}	639 consecutive pilot fields

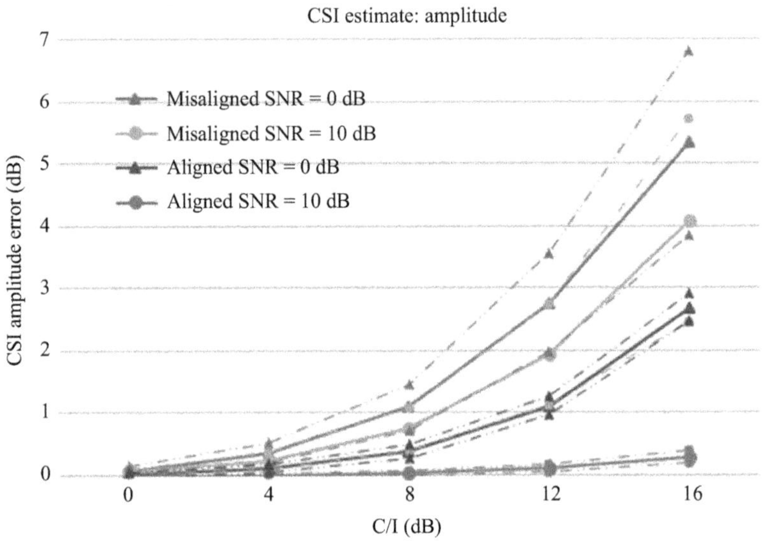

Figure 9.3 CSI amplitude errors (mean and standard deviation) [23] in case of timing misalignment and alignment for different SNR values. The dashed lines specify the window of the standard deviation w.r.t. the mean value

the respective (same colour) dashed dotted lines represent the window given by the standard deviation w.r.t. the mean value, meaning that those curves are the summation and the subtraction of the curve given by the mean values and the curve given by the standard deviation values. The first two curves of the legend are the results obtained

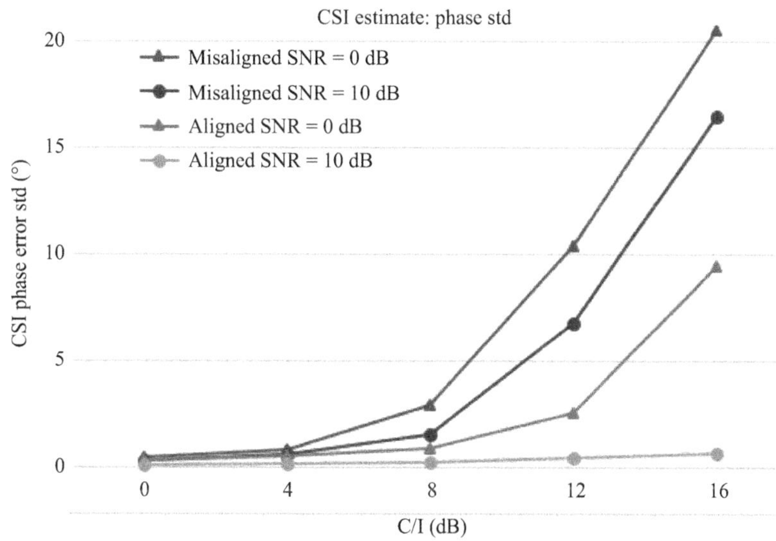

Figure 9.4 CSI phase errors (standard deviation) [23] in case of timing misalignment and alignment for different SNR values

respectively in case of SNR (reference user) equal to 0 and 10 dB for misaligned waveforms. On the other hand, the last two curves of the legend are the results obtained respectively in case of SNR equal to 0 and 10 dB for aligned waveforms.

It is quite straightforward to notice that for all curves, increasing the C/I the error is larger, which is quite obvious. By comparing the misaligned and the aligned cases, it is also very clear that the pre-compensation at the transmitter side provides huge gains in term of both CSI errors and reliability (i.e. standard deviation) of the estimation since both the mean and the standard deviation values are much larger in the misaligned case.

In addition, it is worth noting that considering a $C/I = 16$ dB, there are about 2 dB of difference between the mean values of the aligned and misaligned cases when the SNR of the reference waveform is 0 dB, while this difference increases when the considered SNR is 10 dB.

In Figure 9.4, results obtained in terms of mean and standard deviation of CSI phase errors in the case of both misaligned and aligned (hence pre-compensated) waveforms are shown versus the C/I value of the specific waveform. The mean values are here not reported since they are very small compared to the standard deviation. As it happened for the amplitude case, the advantages of using the pre-compensation [23] at the transmitter are large, especially considering that, for the weakest interferer, the standard deviation reaches values up to 20°, which degrades precoding performance significantly.

9.2.4.1 Precoded symbols analysis

In order to evaluate the effect of timing phase in a precoded system, a simplified 2×2 model is here assumed for the sake of clarity. The precoded symbols vector $\mathbf{x}$ can be formulated as follows:

$$\mathbf{x} = \begin{bmatrix} \widehat{w_{11}} & \widehat{w_{12}} \\ \widehat{w_{21}} & \widehat{w_{22}} \end{bmatrix} \begin{bmatrix} s^{(1)} \\ s^{(2)} \end{bmatrix} = \begin{bmatrix} x^{(1)} = \widehat{w_{11}} s^{(1)} + \widehat{w_{12}} s^{(2)} \\ x^{(2)} = \widehat{w_{21}} s^{(1)} + \widehat{w_{22}} s^{(2)} \end{bmatrix} \tag{9.11}$$

where $\widehat{w_{ij}}$ are the precoder coefficient (jth user and ith antenna) obtained using non-perfect CSI and $s^{(i)}$ is the symbol of the ith user. Assuming a noise-free transmission and, as a consequence, ZF precoding technique, the received precoded signal for the ith user $sy_i(t)$ can be expressed:

$$sy_1(t) = \sum_i x_i^{(1)} h'_{11}(t - iT) + \sum_i x_i^{(2)} h'_{12}(t - \tau - iT) \tag{9.12}$$

where h'_{mn} is the convolution of the channel coefficient h_{mn} for the shaping filter functions, T is the symbol period and τ is the timing misalignment between the first and the second transmitted signals. Sampling at kT instant:

$$sy_1(kT) = \sum_i x_i^{(1)} h'_{11}(kT - iT) + \sum_i x_i^{(2)} h'_{12}(kT - \tau - iT) \tag{9.13}$$

If we substitute the transmitted symbols $\mathbf{x}$ with the linear combination of the original not-precoded symbols, we obtain

$$\begin{aligned} sy_1(kT) &= \sum_i (\widehat{w_{11}} s_i^{(1)} + \widehat{w_{12}} s_i^{(2)}) h'_{11}(kT - iT) \\ &\quad + \sum_i (\widehat{w_{21}} s_i^{(1)} + \widehat{w_{22}} s_i^{(2)}) h'_{12}(kT - \tau - iT) \end{aligned} \tag{9.14}$$

$$\begin{aligned} sy_1(kT) &= \sum_i \widehat{w_{11}} s_i^{(1)} h'_{11}(kT - iT) + \sum_i \widehat{w_{12}} s_i^{(2)} h'_{11}(kT - iT) \\ &\quad + \sum_i \widehat{w_{21}} s_i^{(1)} h'_{12}(kT - \tau - iT) + \sum_i \widehat{w_{22}} s_i^{(2)} h'_{12}(kT - \tau - iT) \\ &= \left[\widehat{w_{11}} h'_{11}(0) + \widehat{w_{21}} h'_{12}(-\tau)\right] s_i^{(1)} + \left[\widehat{w_{12}} h'_{11}(0) + \widehat{w_{22}} h'_{12}(-\tau)\right] s_i^{(2)} \\ &\quad + \sum_{i \neq k} \left[\widehat{w_{11}} h'_{11}(kT - iT) + \widehat{w_{21}} h'_{12}(kT - \tau - iT)\right] s_i^{(1)} \\ &\quad + \sum_{i \neq k} \left[\widehat{w_{12}} h'_{11}(kT - iT) + \widehat{w_{22}} h'_{12}(kT - \tau - iT)\right] s_i^{(2)} \\ &= \left[\widehat{w_{11}} h'_{11}(0) + \widehat{w_{21}} h'_{12}(-\tau)\right] s_i^{(1)} + \left[\widehat{w_{12}} h'_{11}(0) + \widehat{w_{22}} h'_{12}(-\tau)\right] s_i^{(2)} \\ &\quad + \sum_{i \neq k} \widehat{w_{11}} h'_{11}(kT - iT) s_i^{(1)} + \sum_{i \neq k} \widehat{w_{21}} h'_{12}(kT - \tau - iT) s_i^{(1)} \\ &\quad + \sum_{i \neq k} \widehat{w_{12}} h'_{11}(kT - iT) s_i^{(2)} + \sum_{i \neq k} \widehat{w_{22}} h'_{12}(kT - \tau - iT) s_i^{(2)} \end{aligned} \tag{9.15}$$

which leads to the following final expression:

$$sy_1(kT) = \left[\widehat{w_{11}}h_{11} + \widehat{w_{21}}h'_{12}(-\tau)\right]s_i^{(1)} + \left[\widehat{w_{12}}h_{11} + \widehat{w_{22}}h'_{12}(-\tau)\right]s_i^{(2)} + \sum_{i\neq k}\widehat{w_{21}}h'_{12}(kT-\tau-iT)s_i^{(1)} + \sum_{i\neq k}\widehat{w_{22}}h'_{12}(kT-\tau-iT)s_i^{(2)} \quad (9.16)$$

Due to the fact that, at the right sampling instant, the inter-symbol interference (ISI) is 0

1. $$\sum_{i\neq k}\widehat{w_{11}}h'_{11}(kT-iT)s_i^{(1)} = 0 \quad (9.17)$$
2. $$\sum_{i\neq k}\widehat{w_{12}}h'_{11}(kT-iT)s_i^{(2)} = 0 \quad (9.18)$$

The same analysis can be replicated for the other signal, $sy_2(kT)$, in the same way.

We can highlight four main parts corresponding to the four factors in the final expression:

1. The received precoded symbols for the reference waveform (high-power useful signal for the user)
2. The received precoded symbols for the interferer waveform (low-power interference mitigated with precoding)
3. ISI on the precoded symbols for the useful information
4. ISI on the precoded symbols for the interferer information

The last two mentioned parts can be considered as a random variable with some characteristics that we will not investigate. These two components are sources of potential degradations for the SNIR performance at the receiver due to symbol by symbol detection.

It is also clear from the formulation that the performance of the received precoded symbols is affected by the selected channel.

9.2.5 Numerical results on precoding degradations with timing misaligned waveforms

In the following, simulations results for the timing misalignment effects on precoded waveforms under a realistic scenario are reported.

The simulation chain used for the results is the one showed in Figure 9.5. N_t number of streams are generated at the GW and modulated according to DVB-S2 waveforms. Precoding is then applied on the modulated symbols. Since in the present simulations, we are using perfect CSI, channel $\mathbf{H}$ is directly used to calculate the precoding vectors according to MMSE technique. The streams are then up-sampled and pulse shaped by a square root raised cosine filter. At this stage of the chain, the timing impairments are applied to the oversampled streams. An external block generates N_t instances of an uniformly distributed random variable. The values used can be found in the numerical results section. The channel matrix as well as the Gaussian noise is then applied. Before downsampling through matched filters, the received signal should be compensated according to the applied impairment as it

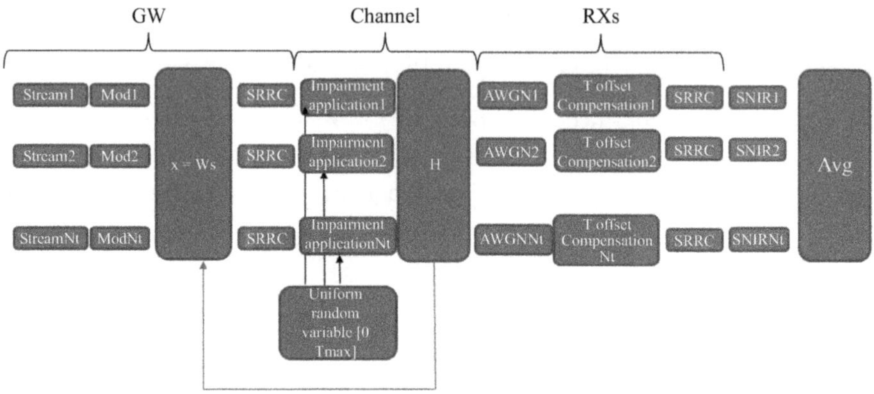

Figure 9.5 Simulation chain used for the results on the timing misalignment effects on precoded waveforms

Table 9.3 Simulation parameters for the precoding degradation under timing misaligned waveforms

Parameter	Value
Coverage	71 beams over Europe
User Link BW	500 MHz
Users' position	Beam centre
Carriers per Beam	1
Precoder	MMSE
CSI	Ideal CSI
Roll-off	0.05
Oversample	4
ModCod	QPSK 5/6

happens in a real receiver (timing offset of the reference waveform). This is needed in order to calculate the proper per-Rx SNIR but it does not remove the misalignment effects.

The simulation parameters used are the ones listed in Table 9.3.

Figure 9.6 shows the effects, in terms of SNIR versus per-beam peak power, of the timing misalignment on precoded waveform. Several values have been used for the maximum misalignment allowed, and random values up to this maximum are drawn in the simulation. The dashed curve is the benchmark curve which considers timing aligned waveforms. The other curves have been obtained using different maximum timing misalignments whose values are 1/20 T_s, 1/10 T_s, 1/8 T_s, T_s and T_s, where T_s is the symbol period.

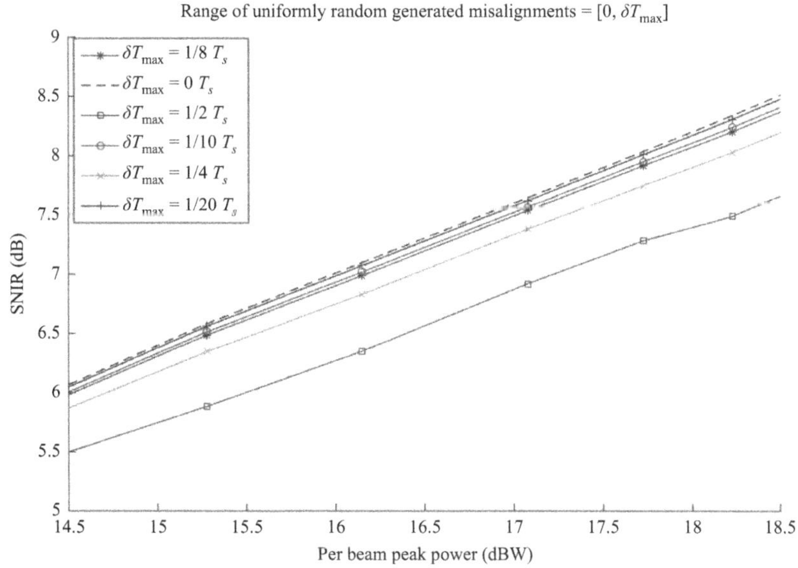

Figure 9.6 Effects of the timing misalignment on precoded waveforms considering different values for the maximum misalignment allowed

While the degradation is very limited if misalignments up to 1/10 of the symbol period are considered, the SNIR reduction is large when misalignments reach half of the symbol period. As a consequence, the use of a timing pre-compensation phase is recommended in order to avoid degradation in the precoding performances, especially when large bandwidth carriers are considered and, as a consequence, timing misalignments introduced by the payload are not negligible.

9.3 Description of the precoding implementation

9.3.1 Precoding technique

In addition to a ZF precoding we implemented a low-complexity symbol level precoding technique [24] into the demonstrator, which is the first step towards precoded waveform design where the precoder changes per symbol by taking as input of the optimization problem not only the CSI but also the symbols to be transmitted. We focus on the high multi-user interference regime, which can be generated by SATCOMs through FFR and large antenna size at the terminals (e.g. backhauling). In this regime, ZF and MMSE performance should converge. The SLP algorithm aims to minimize the total power of precoded symbols while sustaining a minimal SNR requirement for all received signals. The method optimally preserves constructive interference components to decrease the total consumed power at the transmitter side.

The essential difference from a linear precoding method is the optimization vector ($\mathbf{u} \in \mathbb{C}^N$), which is recalculated per every symbol set ($\mathbf{s}$) to construct optimal precoded signal

$$\mathbf{x} = \mathbf{W}_{\mathrm{ZF}}(\mathbf{s} + \mathbf{u}), \tag{9.19}$$

where $\mathbf{W}_{\mathrm{ZF}} = \hat{\mathbf{H}}^H \cdot (\hat{\mathbf{H}} \cdot \hat{\mathbf{H}}^H)^{-1}$ is zero-forcing precoding matrix, $\hat{\mathbf{H}}$ a channel matrix estimated from CSI. The precoding technique maintains the minimal SNR of the received symbols as

$$\begin{aligned} &\min_{\mathbf{x}} \|\mathbf{x}\|_2 \\ &\text{s. t.} |\mathbf{y}| \geq |\mathbf{s}|, \end{aligned} \tag{9.20}$$

for $\mathbf{HW}_{\mathrm{ZF}} = \mathbf{I}$ and $\mathbf{n} = \mathbf{0}$. It was shown in [24] that the problem (9.20) can be transformed into a non-negative least squares (NNLS) problem and solved for the vector $\mathbf{u}$. If a solution for a particular channel matrix cannot be found, then $\mathbf{u} = \mathbf{0}$ and (9.19) turns into conventional zero-forcing precoding [25]. Therefore, the minimal performance of the proposed precoding technique is expected at a level of zero-forcing using statistically averaged CSI data. We will refer to the proposed precoding technique as NNLS-SLP further in the paper.

9.3.2 Non-negative least squares algorithm

The low-complexity SLP design, described in the previous section, brings hardware and software implementation towards a unified solution. The key to this is an efficient algorithm to solve the NNLS optimization problem. It can be implemented using a software-defined radio and FPGA platforms with a reasonable level of complexity. We use a standard fast non-negativity-constrained least squares algorithm presented in [24]. The most time-consuming operation of this algorithm is the solving of unconstrained linear least squares sub-problems via QR decomposition. The asymptotic complexity of the QR decomposition of a square matrix ($\mathbb{R}^{n\times n}$) is $O(n^3)$. However, there are more efficient methods that can reduce considerably this level of complexity up to $O(n^2)$.

9.3.3 Impact of proposed SLP on constellation

Figure 9.7 shows how one symbol of the QPSK constellation can have an amplitude excursion in the horizontal or the vertical axis with the proposed NNLS-SLP algorithm, since the optimization problem of (9.20) impose an inequality constraint on the received symbols amplitude. Here, we obtain a theoretical BER expression assuming that the receiver recovers perfectly the phase of the reference symbols. This phase recovery may be approached in a realistic scenario by means of pilot symbols. In addition, an accurate synchronization can be maintained if modified symbols have in average the same phase of the mapping symbols. For the particular case of a symbol of a QPSK modulation, with an excursion ratio ε (see Figure 9.7) of the mapping symbol, which can be in the in-phase or quadrature axes, the BER is $p_{\mathbf{es}} = 0.5(\mathcal{Q}(\sqrt{\gamma}) + \mathcal{Q}\left(\sqrt{\gamma}(1+\varepsilon)\right))$ where $\mathcal{Q}(\cdot)$ is the standard

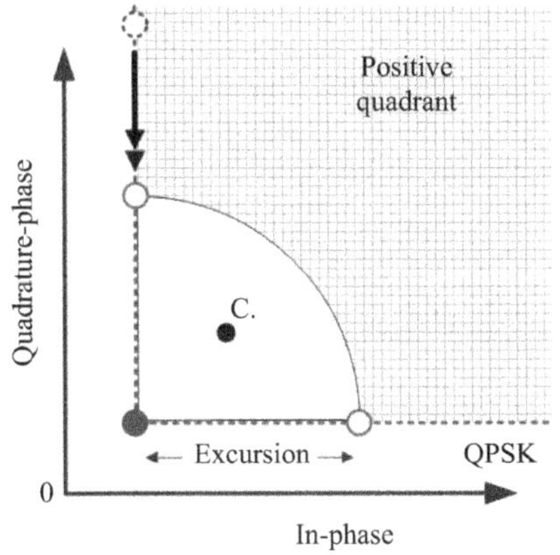

Figure 9.7 Symbol excursion in NNLS-SLP. The symbol excursion can be in the vertical or horizontal axis

Gaussian complementary cumulative distribution function, and γ is the SNR, where we assume that the received signal is affected by an additive zero-mean circulary-symetric complex Gaussian noise. The ensemble uncoded BER is computed to be $0.75\mathcal{Q}(\sqrt{\gamma}) + 0.25\mathcal{Q}(\sqrt{\gamma}(1+\varepsilon))$, under the assumption that all symbols have the same probability and half of the symbols have the same amplitude excursion in one dimension and the other half does not have any excursion.

9.4 In-lab validation of the precoding techniques

9.4.1 Experimental validation of a 2×2 sub-system

This subsection describes the experimental validation of a 2×2 precoder and demonstrates the feasibility of the proposed techniques. The demonstrator consists of the precoding transmitter (GW), a satellite MIMO channel emulator and two receivers using SDR platforms. The satellite channel emulator is able to generate the typical satellite impairments followed by the MIMO user link channel matrix. The channel matrix can be manually configured by the user according to specific scenarios. In order to perform a comparison of the two precoding techniques under different interference environments (representative of different user locations), the channel matrix is manually modified in both amplitude and phase A general block diagram of the precoding demonstrator is shown in Figure 9.8. The GW computes a precoding matrix employing the CSI obtained by means of a dedicated return channels from the set of receivers. Consequently, this precoding matrix is used to generate the transmitted symbols, which are translated into waveforms using a set of pulse-shaping filters.

The transmitted waveforms are sent to the MIMO channel emulator, which apply the channel matrix **H** and inject an additive white Gaussian noise with a controlled

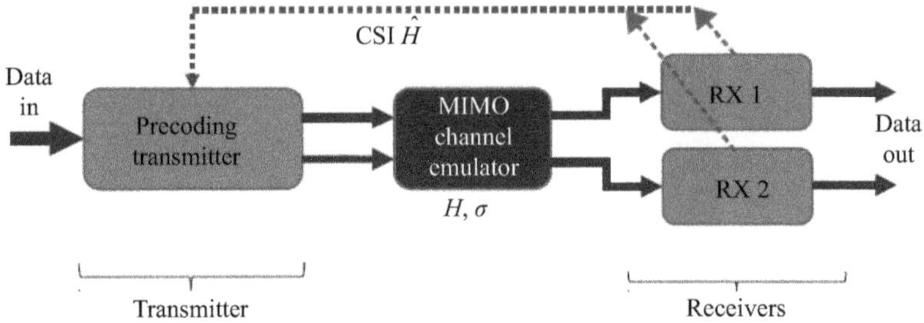

Figure 9.8 Precoding test bed diagram

Table 9.4 Experimental parameters of precoded transmission in 2×2 MIMO system

Parameter	Value
Modulation type	QPSK
TX to Emulator c. frequency	1210 MHz
Emulator to RXs c. frequency	960 MHz
Carrier bandwidth	250 kHz
Over-sampling factor	4
Pulse shaping filter	SRRC with 0.2 roll-off
Pilot duration	24 symbols
Data duration	896 symbols
Pilot repetition period	2,048 symbols

power. In addition, the MIMO channel emulator may have the capabilities of emulation of the satellite channel impairments. Some of these impairments are determined by the frequency response and the non-linearities of the satellite payload components, such as the OMUX and IMUX filters, and the high power amplifier (HPA). These channel functionalities are implemented in a FPGA which is integrated to the SDR platforms. Due to the baudrate of the used carriers (which assumes a multi-carrier transmission) and by selecting a back-off which allows the signal to work in the linear region of a linearized TWTA, the effect of the satellite is almost negligible in the experimentation. The RF inputs and outputs of the channel emulator operate at different carrier frequencies. Using this configuration we decrease mutual coupling between the transmission and reception links through the RF part of the channel emulator and therefore the accuracy in setting of the desired channel matrix. Table 9.4 shows a summary of the parameters of the precoded transmitted signals.

Before precoding, each of the input bit streams are XOR-scrambled with different gold sequences obtained from the combinations of the two maximum-length sequences with the characteristic polynomials $1 + x^3 + x^{20}$ and $1 + x^2 + x^{11} + x^{17} + x^{20}$. This scrambling is used in order to obtain a transmission in which all the symbols have the same probability of occurrence. The transmitted data is a set of two different

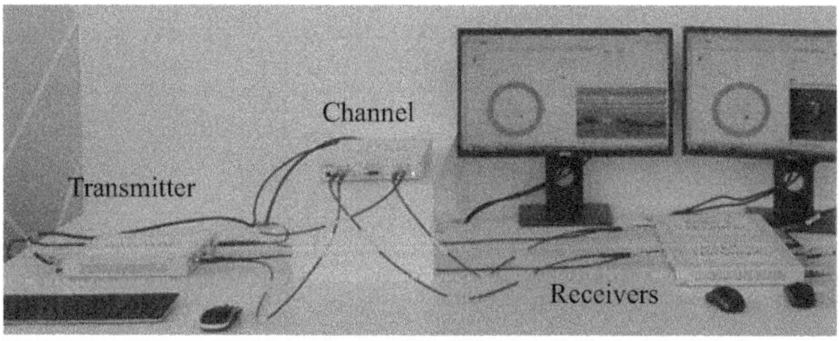

Figure 9.9 Precoding 2×2 experimental settings. The SDR platform used for transmitter, channel emulator and receivers is the NI-2944-R

video streams that are recovered at the receivers. Figure 9.9 shows the experimental set-up. We use National Instruments USRP-RIO NI-2944-R as SDR platform. Each of the SDR platforms is monitored and controlled by a dedicated PC host used for data collection, processing and visualization. The channel emulator can generate 2×2 complex channel matrices with a given condition number and set accurately the power of the AWGN. We as well use the channel emulator to measure actual transmission power on each port of the transmitter.

9.4.2 Symbol-level optimized precoding evaluation

We use the aforementioned experimental environment to benchmark the optimized symbol-level precoding technique. We generate a set of random channel matrices $\mathbf{H}$ with unitary matrix F-norm, defined by $||\mathbf{H}||_F = \text{trace}(\mathbf{H}^H\mathbf{H})$, and for different matrix conditioning numbers, defined by

$$\kappa_2(\mathbf{H}) = ||\mathbf{H}||_2 \cdot ||\mathbf{H}^{-1}||_2. \quad (9.21)$$

We apply the NNLS-SLP and compare the results to conventional channel-inversion ZF precoding. In both cases, we normalize the precoding matrix to have an unitary 2-norm, so that we obtain a constant value for the expectancy of transmitted power per antenna [to emulate automatic level control (ALC) on board of the satellite]. Under this constraint, we measure the power in the two receivers and compare the results for different channel realizations (representative of different user locations) for a set of channel matrix conditioning numbers between 2.5 and 4, as is shown in Figure 9.10. It is worth to note that in both cases ZF and NNLS-SLP we use the same channel inversion matrix. However, the difference for NNLS-SLP is the use of optimized symbols, which are limited to the unitary amplitude. From Figure 9.10, we can observe that the received power for ZF precoding is not constant for a given conditioning number as should be expected from the theory but it ranges within 1 dB. These variations come from the imperfections in the actual hardware implementation. Some of these imperfections are the limited accuracy in the CSI estimation (which happens real time, frame by frame), and its quantization error. Here, we can observe the gains

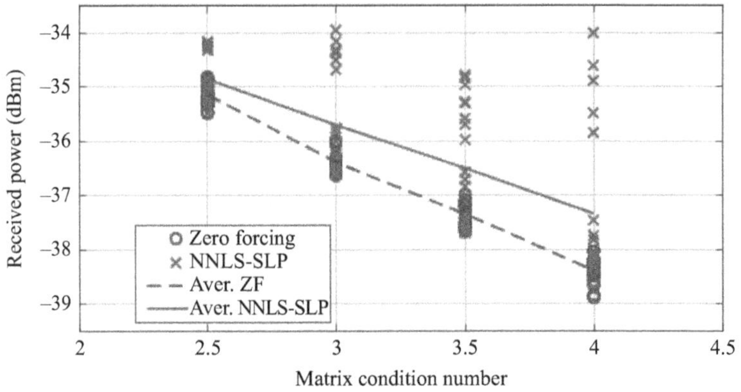

Figure 9.10 Different realizations of detected power, in 2 receivers indistinctly, for conventional ZF and NNLS-SLP [26]

in received power for the NNLS-SLP. These gains become more frequent as the matrix conditioning number is increased. There are particular channel realizations in which the NNLS-SLP result is the same of ZF for both receivers, and other realizations in which the optimized symbol is only produced for one of the receivers. Up to this point, we have observed the gains in received power for NNLS-SLP. In the following, we will observe how this gain is translated to BER performance in the receiver. It is worth noting that the power spread for the SLP case is related to the inequality constraint used in the optimization problem which allows the received constellation to move towards higher SNIR.

9.4.3 Un-coded bit error performance of NNLS-SLP

Figure 9.11 shows an example of received modified constellation with the NNLS-SLP algorithm with some AWGN already applied. This constellation will be difficult to demodulate by a conventional QPSK demodulator as the one used (LabVIEW Communications v.2.0), as the phase synchronization algorithm tracks for the mean square phase error, which is increased in the proposed received constellation due to the exploitation of the constructive interference in the used SLP. However, the symbol excursion will help for the cases in which the phase is correctly recovered and also for very low SNR conditions in which the received signal is very affected by additive noise. For this reason, we proceed to perform BER experiments for different SNR values. The SNR is set by means of the injection of artificial AWGN in the channel emulator. The noise power can be accurately controlled to adjust the desired SNR, knowing the exact value of the received signal power. First, we performed a single link BER measurement using an unmodified QPSK constellation. We use it as a reference to evaluate the effects of imperfect phase synchronization for low SNR values. Phase-locked loop of the demodulator is reset for every frame.

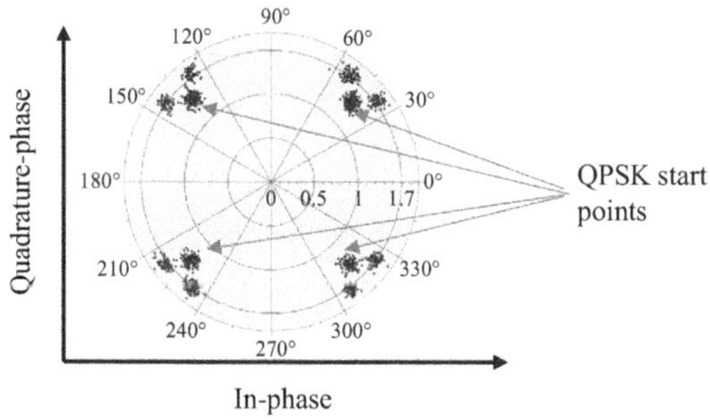

Figure 9.11 An example of NNLS-SLP modified received constellation

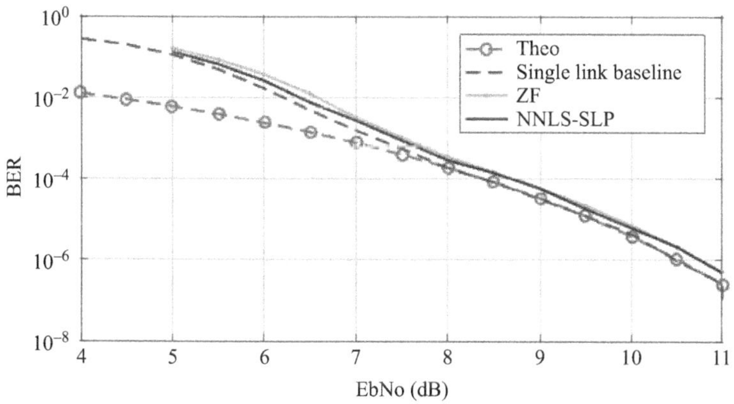

Figure 9.12 Experimental BER plots for ZF and NNLS-SLP compared to an experimental baseline non-interference QPSK BER and to the theoretical BER curve. The matrix condition for the precoded channel is 2.5. The NNLS-SLP in this case provides an excursion of 4%. Only a slight degradation is seen in the precoded system, which carries twice the data rate using the same frequency band [26]

For the case of NNLS-SLP, the precoded pilot symbols are not modified from the QPSK original mapping points. We performed measurements of BER for ZF and NNLS-SLP for different channel matrices, where the SNR was estimated solely using ZF precoding. This is a fair comparison, since, despite the average received power can increase while using NNLS-SLP, minimal received power can still match the one gained with ZF precoding.

Figure 9.12 shows the theoretical ideal QPSK BER values, the BER for a single non-interference link, and the BER for ZF and NNLS-SLP for a particular matrix with conditioning number 2.5 which gives and excursion (in horizontal and vertical axis)

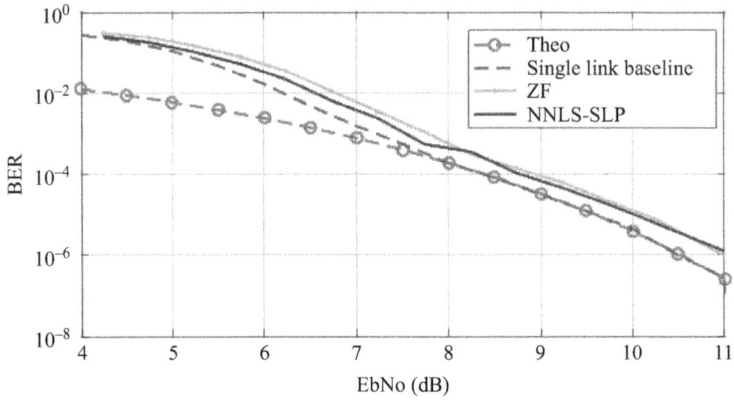

Figure 9.13 Experimental BER plots for ZF and NNLS-SLP compared to an experimental baseline QPSK BER and to the theoretical BER curve. The matrix condition number for the precoded channel is 3. The NNLS-SLP in this case provides an excursion of 20% [26]

of 4%. Here, we see a degradation of the BER plots which use precoding compared to the single link BER curve. This is attributed to the inaccuracies in the CSI estimation, which produces residual interference that affects the BER performance; however, we should remark that for the case of precoded signals we obtain twice spectral efficiency, since the system provide two separate streams using the same frequency band.

In the comparison between the ZF and NNLS-SLP, we have shown that the NNLS-SLP performs slightly better for low SNR values, and that the ZF performs better at some points of higher SNR values using conventional receiver. It is worth mentioning that the comparison is with the same Eb/No; hence, the comparison does not take into account the overall increase in the received SNIR due to the constructive interference exploitation of SLP. The experiment is repeated with some channel matrices with higher proposed excursion values which in some cases give a degradation in BER performance for high SNR values. Most of these errors are attributed to lack of phase synchronization and phase tracking. These effects can be observed as a rotated shaking in the constellation plots in the graphical user interface at the receiver. It is worth to clarify that at the transmitter QPSK modulator maps the transmitted symbols in correspondence to optimization excursion. However, at the receiver, QPSK demodulator normalizes the received symbols in correspondence to conventional QPSK symbol map only as receiver has no knowledge about optimized mapping. However, at the receiver, the QPSK start points are recovered in their original amplitude position, which is the same amplitude obtained when the ZF precoding is applied. Figure 9.13 shows the BER curves for ZF and NNLS-SLP for a particular matrix with conditioning number of 3 which gives and excursion (in horizontal and vertical axis) of 20%. Here we can observe how the NNLS-SLP performs slightly better than the conventional ZF for Eb/No values lower than 8 dB.

9.5 Conclusions and future works

In this chapter, we discussed practical challenges for a real implementation of MU-MIMO techniques, namely precoding, at the GW side of a multi-beam satellite system.

After a brief introduction to precoding concept and related works focussed on SATCOMs, in Section 9.2 we analysed some practical sources of degradation for precoding to work in a real satellite environment. In addition, some numerical assessments have been accomplished in order to quantify possible degradation with respect to the expected gains.

In Section 9.3, a general description of the precoding techniques adopted in the implementation phase has been described. A complexity estimation has been also carried out in order to motivate a real-time implementation.

Finally, Section 9.4 reported the in-lab experimental tests carried out using a small scale network composed by 2 transmitters and 2 UTs, having in between a multi-beam satellite channel emulator. Two precoding techniques have been studied and experimental results have been provided in comparison to a standard non-precoded system which employs a frequency division scheme to avoid interference.

Future foreseen works in the context of in-lab validation of techniques are listed below.

- use of carrier bandwidth up to 40 MHz;
- introduction of strong non-linear characteristics in the satellite payload chains as well as all satellite impairments and jointly precoding/pre-distortion techniques to deal with the degradations;
- performance evaluation using LDPC-based codewords taken from DVB-S2X set of modcods;
- statistical multiplexing performance with different group of users;
- performance evaluation of precoding techniques using large scale networks with beams clustering;
- performance evaluation of precoding techniques in Multi-GW environment with centralized and distributed techniques.

References

[1] Letzepis N, Grant AJ. Capacity of the multiple spot beam satellite channel with Rician fading. IEEE Transactions on Information Theory. 2008 Nov;54(11):5210–5222.

[2] Arapoglou PD, Liolis K, Bertinelli M, *et al.* MIMO over satellite: a review. IEEE Communications Surveys Tutorials. 2011 First;13(1):27–51.

[3] ESA Artes 1 Contract 4000106528/12/NL/NR, "Next Generation Waveform for Improved Spectral Efficiency". Final Report; 2015.

[4] Arapoglou PD, Ginesi A, Cioni S, *et al.* DVB-S2X-enabled precoding for high throughput satellite systems. International Journal of Satellite Communications and Networking. 2016;34(3):439–455. SAT-15-0019.R1. Available from: http://dx.doi.org/10.1002/sat.1122.

[5] ESA Artes 1 Contract AO/1-7723/13/NL/NR "Future Ground-Based Beam-Forming", Final Report; 2016.

[6] ESA Artes 5.1 Contract AO/1-7882/14/NL/US "Precoding Demonstrator for Broadband System"; 2015.

[7] ESA TRP Contract AO 1-8332/15/NL/FE "Optimized Transmission Techniques for SATCOM Unicast Interactive Traffic"; 2015.

[8] FNR Proof of Concept Project SERENADE "Satellite Precoding Hardware Demonstrator"; 2016.

[9] Malkowsky S, Vieira J, Liu L, *et al.* The World's first real-time testbed for massive MIMO: design, implementation, and validation. IEEE Access. 2017;5:9073–9088.

[10] http://www.cttc.es/project/cloud-architecture-for-standardization-development/.

[11] Kalantari A, Zheng G, Gao Z, *et al.* Secrecy analysis on network coding in bidirectional multibeam satellite communications. IEEE Transactions on Information Forensics and Security. 2015 September;10(9):1862–1874.

[12] Christopoulos D, Chatzinotas S, Ottersten B. Multicast multigroup precoding and user scheduling for frame-based satellite communications. IEEE Transactions on Wireless Communications. 2015 September;14(9):4695–4707.

[13] Christopoulos D, Arapoglou PD, Chatzinotas S. Linear precoding in multibeam SatComs: practical constraints. In: 31st AIAA International Communications Satellite Systems Conference. American Institute of Aeronautics and Astronautics; 2013. Available from: https://doi.org/10.2514/6.2013-5716.

[14] Vazquez MA, Perez-Neira A, Christopoulos D, *et al.* Precoding in multibeam satellite communications: present and future challenges. IEEE Wireless Communications. 2016 December;23(6):88–95.

[15] Cottatellucci L, Debbah M, Casini E, *et al.* Interference mitigation techniques for broadband satellite system. In: ICSSC 2006, 24th AIAA International Communications Satellite Systems Conference, 11–15 June 2006, San Diego, USA. San Diego, UNITED STATES; 2006. Available from: http://www.eurecom.fr/publication/1886.

[16] Chatzinotas S, Ottersten B, Gaudenzi RD. Cooperative and cognitive satellite systems. London: Academic Press is an imprint of Elsevier; 2015.

[17] Spano D, Alodeh M, Chatzinotas S, *et al.* Spatial PAPR reduction in symbol-level precoding for the multi-beam satellite downlink. In: IEEE SPAWC 2017; 2017.

[18] Alodeh M, Chatzinotas S, Ottersten B. Energy-efficient symbol-level precoding in multiuser MISO based on relaxed detection region. IEEE Transactions on Wireless Communications. 2016 May;15(5):3755–3767.

[19] Wang K, Sun Q, Tao X, *et al.* Partial precoding for integrated mobile satellite service system. In: 2013 IEEE 78th Vehicular Technology Conference (VTC Fall); 2013. p. 1–5.

[20] Rohde C, Alagha N, De Gaudenzi R, *et al.* Super-framing: a powerful physical layer frame structure for next generation satellite broadband systems. International Journal of Satellite Communications and Networking. 2016;34(3):413–438. SAT-15-0037.R1. Available from: http://dx.doi.org/10.1002/sat.1153.

[21] ETSI EN 302 307-2 Digital Video Broadcasting (DVB), "Second Generation Framing Structure, Channel Coding and Modulation Systems for Broadcasting, Interactive Services, News Gathering and other Broadband Satellite Applications, Part II: S2-Extensions (DVB-S2X)"; Available on ETSI web site (http://www.etsi.org); 2014.

[22] ETSI TR 102 376 V1.1.1 Digital Video Broadcasting (DVB), "Implementation Guidelines for the Second Generation System for Broadcasting, Interactive Services, News Gathering and other Broadband Satellite Applications; Part 2 – S2 Extensions (DVB-S2X)"; March 2015, Available on ETSI web site (http://www.etsi.org).

[23] Andrenacci S, Chatzinotas S, Vanelli-Coralli A, *et al.* Exploiting orthogonality in DVB-S2X through timing pre-compensation. In: 2016 8th Advanced Satellite Multimedia Systems Conference and the 14th Signal Processing for Space Communications Workshop (ASMS/SPSC); 2016. p. 1–8.

[24] Krivochiza J, Kalantari A, Chatzinotas S, *et al.* Low complexity symbol-level design for linear precoding systems. In: 2017 Symposium on Information Theory and Signal Processing in the Benelux. Delft University of Technology; 2017. p. 117.

[25] Peel CB, Hochwald BM, Swindlehurst AL. A vector-perturbation technique for near-capacity multiantenna multiuser communication – Part I: Channel inversion and regularization. IEEE Transactions on Communications. 2005 January;53(1):195–202.

[26] Merlano Duncan JC, Krivochiza J, Andrenacci S, Chatzinotas S, Ottersten B. Computationally efficient symbol-level precoding communications demonstrator. In: 2017 IEEE 28th Annual International Symposium on Personal, Indoor, and Mobile Radio Communications (PIMRC). 2017.

Chapter 10

Beam-hopping systems for next-generation satellite communication systems

Christian Rohde[1], Rainer Wansch[1], Sonya Amos[2], Hector Fenech[2], Nader Alagha[3], Stefano Cioni[3], Gerhard Mocker[4], and Achim Trutschel-Stefan[4]

10.1 Introduction

It is a global trend to have faster and more flexible communication all over the world. Terrestrial networks are well suited for serving densely populated areas. However, this trend will expand to include oceans, sky, diverse and sparsely populated areas. The classical approach to satellite communication service definitions has been to envelope all potential requirements for the lifetime of the satellite. This leads to inefficient use of the satellite resources, particularly in domains that might have fluctuating or uncertain markets or businesses that have requirements which change in time. In order to optimally adapt the system to changing traffic demands over time and location, the novel beam-hopping concept is introduced. Instead of static illumination, the satellite cycles in time through a set of coverages according to a schedule derived from the traffic demands. Thus, at any given time, only one coverage of the set is active with full power and bandwidth. Of course, there could be a number of such sets running in parallel on a given satellite.

The next generation of satellite communications aims at making more efficient use of the available system resources, the aim being to make services more cost effective. Beam-hopping is one avenue in this direction. One application is through high-throughput satellite (HTS) systems where matching the available capacity to the geographical distribution demand presents challenges, especially when considering the market evolution over the satellite lifetime which is typically 15 years.

Eutelsat Quantum is a development in commercial satellite communications where flexibility is available in all the major payload parameters and is an example satellite incorporating beam-hopping to extend its application. Beam-hopping on

[1]RF and SatCom Department, Fraunhofer Institute for Integrated Circuits (IIS), Germany
[2]Eutelsat SA, France
[3]ESA, The Netherlands
[4]WORK Microwave, Germany

Eutelsat Quantum allows capacity to be assigned anywhere on the earth as seen by the satellite with a smaller granularity in an efficient, agile and dynamic manner.

In both cases of HTS and Eutelsat Quantum, the key distinguishing factor is efficiency. In HTS, beam-hopping allows one to apportion a capacity between a number of spots such that the available capacity is geographically profiled in a dynamic fashion to match the demand. Eutelsat Quantum uses beam-hopping to provide services in geographically diverse areas while utilizing spacecraft resources efficiently.

In the first section, the idea, concepts and benefits of beam-hopping are discussed. The gains in user satisfaction and system throughput are shown with respect to a conventional broadband satellite system. Next, we discuss the physical layer transmission solutions. Based on the identified waveform key-requirements for applying beam-hopping, the super-framing specification of the already released DVB-S2X (Annex E) [1] standard is reviewed. As result, Format 2–4 are found to be ready to use for various beam-hopping system configurations.

Finally, actual and future technology for a beam-hopping system is discussed as well as implementation aspects and challenges. Specifically, the upcoming Eutelsat Quantum-Class Satellite designed for beam-hopping is presented along with its features like re-configurable beam-forming and highlights potential applications. The corresponding ground equipment is also discussed exploiting the advantages of wideband processing.

10.2 Beam-hopping system concepts

The ever-growing broadband satellite services in the past few years are often characterized by a time variable and in some cases uncertain traffic demands. Considering the non-uniform geographical distribution of capacity requests through the satellite coverage, the flexibility to adapt to different traffic demands is a key requirement for satellite systems in order to maintain their competitiveness. The need for supporting the traffic demand uncertainty throughout the satellite coverage area was recognized and studied in several activities including two ESA projects as reported in [2,3]. Among possible flexible payload and system architectures, beam-hopping systems have been investigated in these two ESA studies. In a beam-hopping system, at any given time, only a subset of the satellite beams is illuminated. A pre-configured illumination pattern determines the resource allocation to each beam.

In non-HTS applications, illumination of spatially diverse locations in time enables a greater geographical reach where each coverage is served by the smallest beam necessary. In HTS systems, beam-hopping enables the satellite resources to be shared over a large number of beams in the time domain. This means that only a portion of the beams are served at any one time, reducing the resources required and focusing them where needed. Very HTSs (VHTSs) system applications typically have a fixed coverage set of beams, and the fill factor is maximized within this definition. In both the VHTS and non-HTS applications, the capacity is optimized for the user requirements and geographical definition. Beam-hopping can additionally

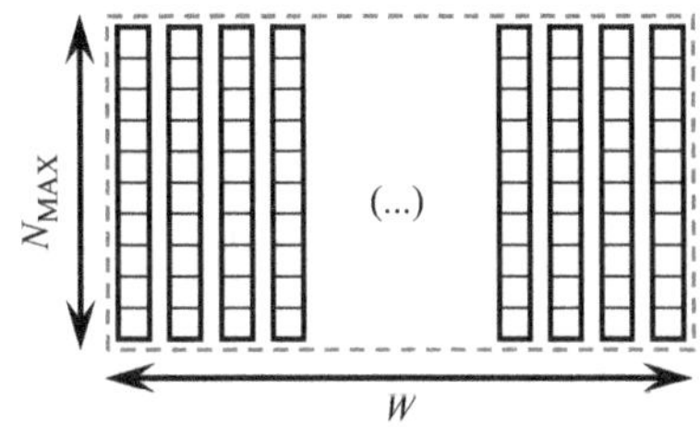

Figure 10.1 Beam-hopping window representations with no bandwidth segmentation

be particularly useful for areas of dense population and high capacity requirements where hopping of nearest neighbouring beams can improve the C/I.

As indicated by the study results in [4–6], the beam-hopping solution provides a high level of flexibility in accommodating irregular and time-variant traffic requests throughout the coverage area. Furthermore, in a beam-hopping system the whole available user link bandwidth can be assigned to a user beam in a single-carrier operation mode which in turn could improve the efficiency of the on-board high-power amplifiers. The set of illuminated beams changes in each time-slot based on a time–space transmission plan which is periodically repeated as illustrated in Figure 10.1. The time axis is divided into W time slots representing a beam-hopping window, which repeats following a regular pattern.

The window duration of W time slots is typically constant within a given operation cycle. The beam-switching pattern is optimized aiming at adapting to a different traffic distribution and traffic demands per beam. This process will be carried out in advance and communicated to the beam-hopping system. In each time slot, a different set of satellite beams is illuminated. In general, a maximum of N_{MAX} beams can be simultaneously illuminated. N_{MAX} is selected in order to limit payload architecture complexity. In Figure 10.1, each vertical column represents a vector of active beams in each time slot. The actual index of active beams can change from one time slot to the next. The time slot is the basic granularity for assigning resources to the satellite beams. The selection of the window length W (in time slots) is therefore carried out after a careful sensitivity assessment of system performance variations as a function of W.

In the more general case where bandwidth segmentation is assumed, each beam can be illuminated with a fraction of the total available bandwidth comprising N_f subbands. In this case, the beam-hopping matrix will have a three-dimensional representation, as shown in Figure 10.2.

A first example of beam-hopping implementation is the advanced communication technology satellite [7]. Another example of a satellite system with beam-hopping capability is the spaceway system from Hughes Networks [8], where beam-hopping techniques are applied to a large multi-beam satellite providing broadband services. An advanced beam-forming network (BFN) and a direct radiating array antenna allow for simultaneously illuminating 24 out of the 784 downlink beams [9].

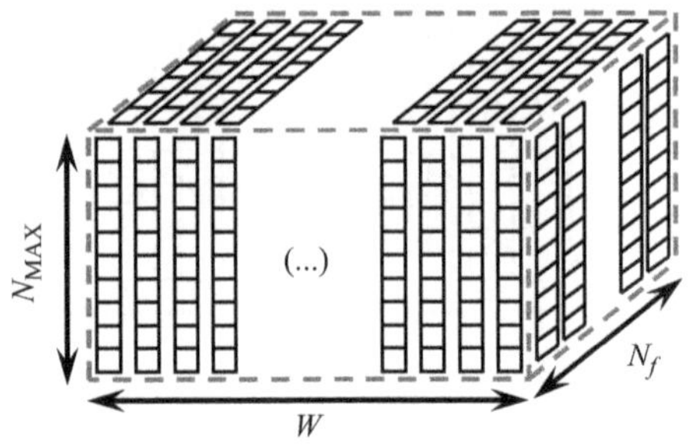

Figure 10.2 Beam-hopping window representations with bandwidth segmentation

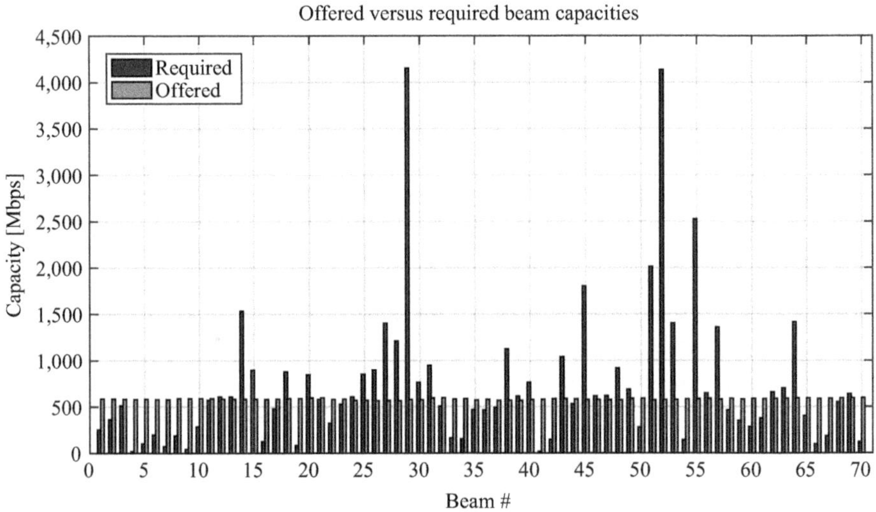

Figure 10.3 Required versus offered capacity per beam; conventional system

In recent ESA studies, the feasibilities of beam-hopping techniques in multi-beam satellite system were studied and identified their potential advantages compared to conventional system architectures from the payload and system perspective. Several payload architectures suitable to support beam-hopping are possible.

Figures 10.3 and 10.4 present examples of non-uniform traffic demand among 70 beams and of the offered capacity per beam in a non-beam-hopped and in a beam-hopped system, respectively. Both show the required (dark bars) and offered (light bars) capacity in Mbps versus the beam index. Obviously, the beam-hopped system can better meet the traffic demands, because the conventional system provides an equally distributed offered traffic of constant 580 Mbps per beam.

In theory, there are frequency and time–space duality principle to compare beam-hopping solutions and frequency flexible payload solutions. Examples of the

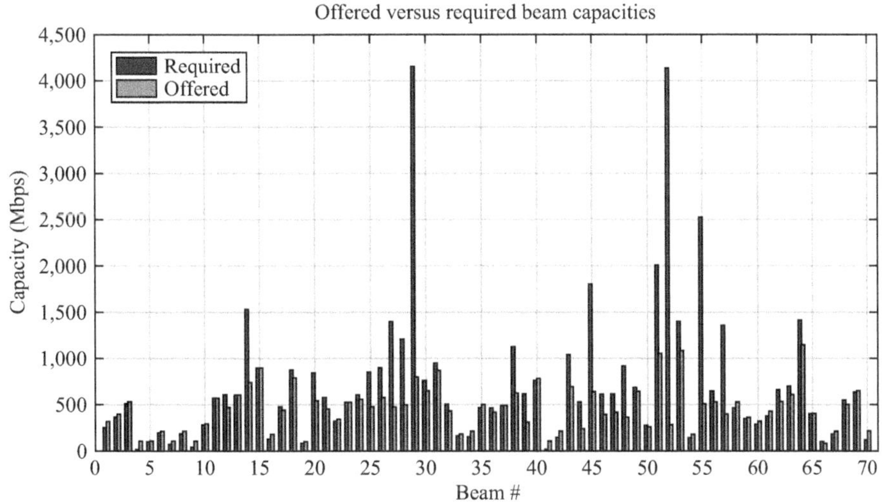

Figure 10.4 Required versus offered capacity per beam; beam-hopped system

theoretical framework for such comparison are reported in [10]. Despite the theoretical duality between the time- and frequency-domain resource sharing concepts, the implementation of the beam-hopping solutions as part of a satellite payload can offer some significant saving in component selection, accommodation, power consumption and power dissipation.

In some system scenarios, the use of beam-hopping can provide a considerable gain (15–20 per cent) in usable throughput in the presence of traffic uncertainty and non-uniform traffic demand distribution, or equivalently a significant saving in the DC power consumption in the case of a given realistic non-uniform traffic demand distribution over the coverage in a multi-beam payload application. Comparisons in the performance of conventional multi-beam systems and that of beam-hopping systems are discussed in [5,6], together with some numerical examples based on representative system scenarios.

10.3 Application of DVB-S2X waveform for beam-hopping

The principle setup of a beam-hopping system forward link transmission is illustrated in Figure 10.5. It refers to a transparent payload architecture without onboard data processing, which holds for all further considerations. The satellite changes the beams' directions to different areas on the ground, called coverages or service areas, according to a beam-switching time plan (BSTP). These service areas may include a varying number of remote terminals. In the example of Figure 10.5, only the forward link transmission is shown, i.e. in direction from the gateway (GW) to the terminals. Therefore, the user terminal situation corresponds to a point to point link with interrupted transmission.

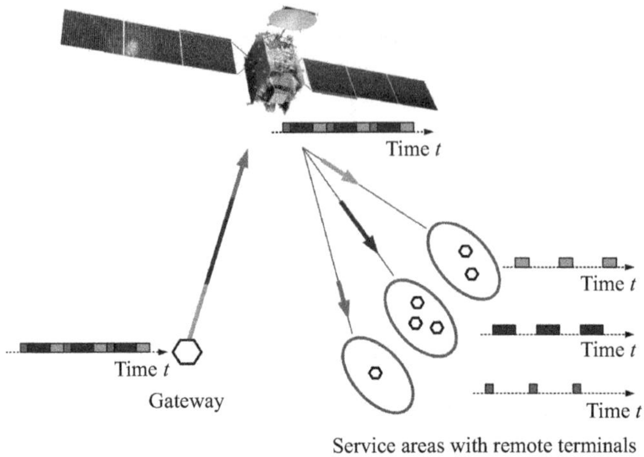

Figure 10.5 Example of data frames distributed to different service areas (Eutelsat Quantum satellite picture courtesy of Airbus Defence & Space)

As baseline, we assume that an BSTP update cylce like a change in illumination duration of each coverage happens within a system-defined granularity of large time frames compared to the illumination duration. And illumination durations are much bigger than a data symbol duration. These assumptions are needed to assure receivers with feasible complexity and to enable sleep mode functionality for power saving. At this general level, we identify three fundamental requirements for the waveform design:

1. Transmission with **guard times**, where the beam-switching takes place. In order to avoid corrupting the user data during the switching event, a guard time with do-not-care dummy data or even no data at all shall be transmitted.
2. **Regular framing structure** to align with the beam-switching schedule. A regular and clear framing structure significantly eases the GW-modulator design, since it has to coordinate and pre-calculate for the right transmission time of user data frames. Furthermore, it helps one to minimize the overhead in dummy data for the guard times.
3. **Anchor preamble sequence** to enable quick and reliable re-synchronization of the terminal working in burst-mode processing. Of course, the above-stated regular framing supports this terminal synchronization task as well.

Note that these requirements hold for feed-forward processing type of receivers. As described in Section 10.4.4 in more detail, it focuses on pipelined continuous processing. In comparison to that, the counter-part receiver type employs massive buffering in order to process data block by block, where requirements 2 and 3 become less significant. This is because detection algorithms can (iteratively) analyse more signal history stored in the buffer and check a variety of hypotheses, until a decision is made. As discussed in Section 10.4.4, this may lead to increased complexity and memory compared to a feed-forward processing receiver.

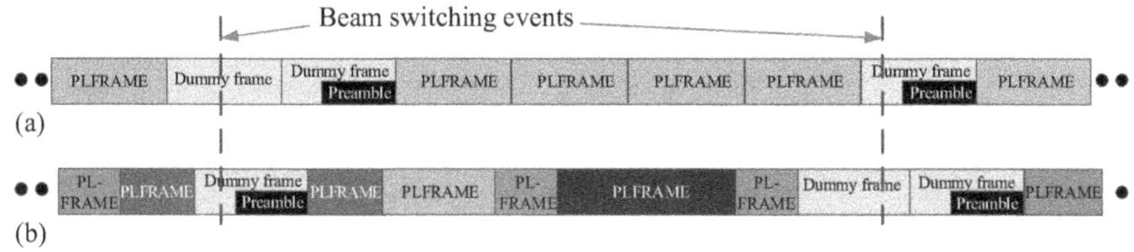

Figure 10.6 DVB-S2/S2X conventional framing uses dummy frames for hosting the preamble and providing guard time: (a) CCM-mode and (b) VCM/ACM-mode

Besides the requirements, there are also practical issues related to physical layer signalling in order to support all usage scenarios and terminal reception conditions. Some of these essential features are as follows:

- A **coverage- or beam-ID** helps a terminal with orientation and is the baseline for a handover management of mobile terminals. For this, the terminal has to feed-back to the GW which beams are received and under which signal-to-noise ratio (SNR).
- **Feedback about the envisaged traffic demand and type** represents valuable information to the GW in order to optimize the traffic scheduling.
- The GW may provide **side information on planned BSTP updates** to the terminals to enable efficient power-saving mode at the terminals.

In order to run the beam-hopping function of the satellite, a suitable waveform plays a major role. In the following, we analyse the latest DVB-S2X standard [1] with respect to these requirements. Also the potential application and flavours of the different waveform configurations are discussed. Both topics in this section extend and complement the discussion of [11].

10.3.1 DVB-S2X conventional framing

First, we discuss the conventional framing of DVB-S2X to distinguish from the later discussed super-framing. Since DVB-S2X [1] represents an extension to DVB-S2 [12], the term conventional framing may hold for both versions of the standard. One of the most fundamental characteristics of conventional framing is that the data frames vary in length with different modulation and code word length (normal or short) as well as when switching pilots on or off. Since the DVB-S2/S2X standard provides no explicit specification for beam-hopping, the following concepts represent potential extensions and ways to re-use the existing specifications as close as possible. Accordingly, continuous signal transmission of the GW is mandatory which excludes consideration of a guard gap. That is, it is not standard conform to switch the signal off during guard time or already after all data of the illumination is transmitted and resume transmission at the start of the next hop.

As shown in Figure 10.6, two cases are considered. The two different transmission modes of DVB-S2/S2X are considered: constant coding and modulation

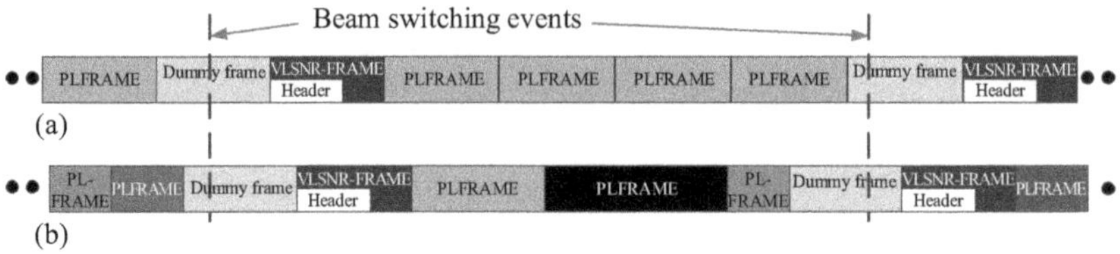

Figure 10.7 The VLSNR-frame of DVB-S2X conventional framing is used as preamble and dummy frames to provide guard time: (a) CCM-mode and (b) VCM/ACM-mode

(CCM) and variable/adaptive coding and modulation (VCM/ACM). Dummy frames are transmitted instead of data frames (physical layer frames, PLFRAMEs), where the beam-switching event is expected. Each PLFRAME consists of a physical layer header (PLH), where modulation and coding of this frame is signalled, and a modulated code word. A dummy frame may also host an anchor preamble sequence for data-aided start of illumination detection. The preamble may be placed at the end of the dummy frame to initialize the PLH/PLFRAME tracking. In doing so, there is a chance that one dummy frame is sufficient to accommodate both the beam-switching event and the preamble.

In both cases, the framing structure cannot be aligned to the beam-switching events in a regular way. Thanks to the constant frame sizes, CCM seems more suitable due to more regular structure. However, constant choice of modulation and coding would mean to take away flexibility from a beam-hopping system. Using VCM/ACM drops this constraint but results in a vast number of possible frame length combinations during each illumination.

Let us assume for a moment that a regular framing structure is not essential. The other two requirements are met according to Figure 10.6. An anchor preamble sequence can be placed at the end of the last dummy. A burst-mode receiver can detect this training sequence in order to initialize the frame tracker. Since these detections from different illuminations will not be equidistant because of non-aligned framing, neither a validation of the detection can be made with respect to a potential false alarm nor a prediction of the frame start in case the correlation peak is below the threshold. In consequence, either receiver reliability decreases or more effort is spent to correlate for a longer anchor training sequence.

An alternative framing structure is shown in Figure 10.7. It reflects usage of the so-called very low SNR (VLSNR) frame of DVB-S2X, whose header is meant for burst-mode detection down to an SNR of −10 dB. This structure can be combined either with CCM or VCM/ACM operation. Opposed to the approach of Figure 10.6, there is no chance to save one dummy frame, because the VLSNR-frame header is at the very beginning of this frame and its detection should not be impaired by the switching event. However, the well-protected data part of the VLSNR frame could be used for beam-hopping specific signalling.

From this discussion, we note that there are some possibilities and available waveform features for beam-hopping based on conventional DVB-S2/S2X framing

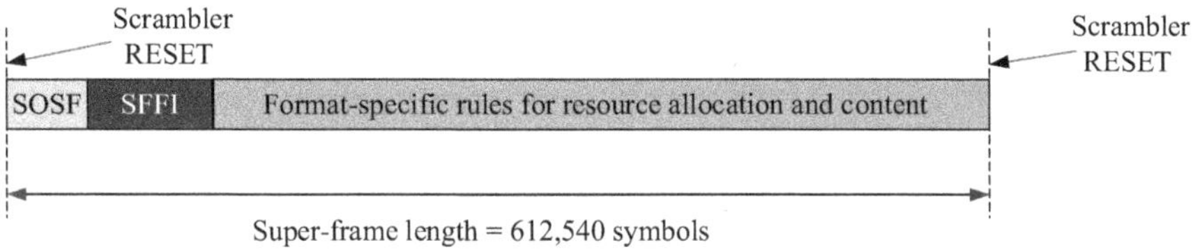

Figure 10.8 General structure of a super-frame according to DVB-S2X Annex E [1]

structure. However, conventional framing in its current form may not be efficient and practical to be utilized in beam-hopping systems. This is because the GW-side PLFRAME scheduling will be a very challenging task especially in VCM/ACM mode: Data-frame scheduling and time alignment with respect to all switching events of the BSTP have to be jointly solved and optimized. Figures 10.6 and 10.7 already demonstrate that each switching event is individually aligned to the framing so that it will be a moving target for the GW-side PLFRAME scheduler.

10.3.2 DVB-S2X Annex E super-framing

The super-frame (SF) structure is specified in Annex E of the DVB-S2X standard [1] as a container of different format-specific content. The general structure is shown in Figure 10.8. The start-of-SF (SOSF) represents a 270 symbols long preamble and the SF format indicator (SFFI) field of 450 symbols provides information on which format specification is valid in this SF. SOSF and SFFI together can be exploited as a 720 symbols long anchor sequence, which enables robust detection capability down to an SNR of −10 dB [13].

A regular framing was the original design criterion for the SF. This is accomplished by a predefined constant SF length and the fact that the SF size remains the same, independent whether SF-pilots are ON or OFF. Furthermore, a capacity unit (CU) of 90 symbols size is specified which can be used for resource allocation. In consequence, the above-mentioned GW-side scheduling dramatically simplifies with super-framing since it decouples the two tasks: One scheduling and network synchronization entity aligns the SFs to the switching events and calculates BSTP update requests. The second scheduling entity performs the resource allocation of the SFs with respect to the data PLFRAMEs being placed in the right SF for the target coverage.

Following this concept, two of the three fundamental requirements are already fulfilled by means of the general SF structure. Below, the guard-time requirement will be discussed format-specific along with some implications. An overview of the different SF formats and their purpose is given in Table 10.1.

From a general perspective, one could consider the static SF length of 612,540 symbols as a constraint, since it directly determines the illumination duration granularity. This SF-based granularity seems to be quite rough, but decreasing the SF-size significantly means increasing needed overhead for preamble and guard time. And when running the system with the beneficial wideband transmission (see

Table 10.1 Overview of the specified SF formats according to [1]

Format	Description and purpose
0	DVB-S2 conventional frames embedded in SFs for legacy support
1	DVB-S2X conventional frames including VLSNR frames embedded in SFs
2	Bundled PLFRAMEs of normal size for application of multiple-input–multiple-output (MIMO) techniques
3	Bundled PLFRAMEs of short size for application of MIMO techniques
4	Flexible format for wideband communication and VLSNR support
5 … 15	Reserved for future use

Section 10.4.2), the illumination durations become quite short in absolute time. Also the terminal synchronization benefits from the SF grid since it stays the same even in the case of BSTP updates. This allows a validation of illumination detection. Finally, the problem of resource wastage of serving a single user in a service area with a complete SF can be solved by changing coverage shape and size thanks to flexible beam-forming. Intentional beam side-lobes can be exploited as well.

10.3.2.1 Super-framing Format 0 and 1: S2 and S2X conventional frames

The SF Format 0 and 1 embed the conventional frames of DVB-S2 and -S2X into SFs. Thanks to the SF, the above-mentioned shortcomings of pure conventional framing are reduced. However, the needed guard time has still to be provided by dummy frame insertion at the end of the SF. Since these frames are aligned only to the CU grid but not to the end of the SF, the remaining part of the dummy frame spills over to the next SF. This has two consequences:

- At the beginning of each illumination, the terminals have to search for the location of the first PLH rather than just start PLH tracking at the first available CU. Although suitable search algorithms are already well established, this effort could be avoided.
- The remaining part of the dummy frame in the next SF is real overhead, because there is no purpose for it. Re-use for some application specific signalling may be complicated because of dynamic mapping to CUs.

Despite these implications, Format 0 and 1 could be used for beam-hopping. In the case of an update of the DVB-S2X standard, a suitable SF-padding scheme should be added to the Format 0 and 1 specification.

10.3.2.2 Super-framing Format 2 and 3: bundled frames structure

Format 2 and 3 specify bundled data frames in order to achieve constant length frames. As depicted in Figure 10.9, an SF of Format 2 hosts nine long bundled frames each of size 64,800 data symbols, which equals, e.g. two QPSK normal-size frames or

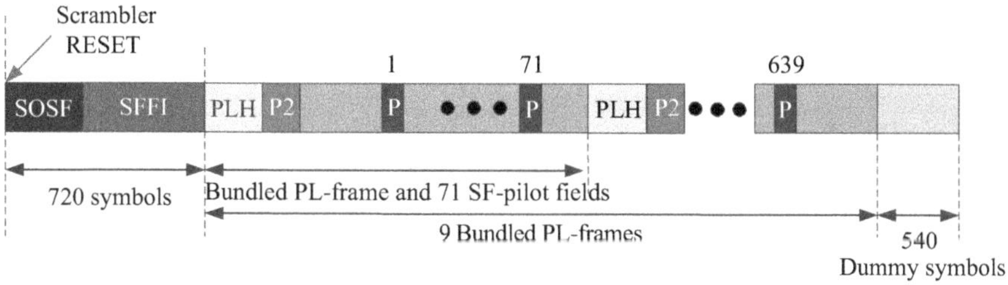

Figure 10.9 DVB-S2X Annex E SF structure according to Format 2

three 8PSK normal-size frames and so forth. Besides the PLH, each bundled frame comprises 71 normal pilots 'P' and a modulation-specific pilot field 'P2'. Correspondingly, an SF of Format 3 hosts 36 short bundled frames each of size 16,200 data symbols, where the same framing principle is applied to short-size frames. Thus, a modulation and coding selection as signalled by the PLH is always valid for a set of codewords, i.e. the bundled frame.

The framing structure of Format 2 is shown in Figure 10.9, where a static amount of 540 dummy symbols are inserted at the end of each SF – independent of end of illumination or not. If there are longer illumination durations, this padding, where no beam-switching event occurs, corresponds overhead because not exploited.

A similar structure holds for Format 3 specifying 396 dummy symbols. Note that the static configuration of dummy data insertion leads to different guard times for different symbol rates. For example, 396 dummy symbols mean a guard time of 1.98 μs @ 200 MHz symbol rate or 19.8 μs @ 20 MHz symbol rate. The consideration of whether this leads to a potential issue depends on the system application and design dependencies such as beam-switching event jitter and transition duration which could become unacceptable. Similar calculations are presented in [14] for the Eutelsat Quantum satellite, where it seems uncritical due to transition durations in the order of a few 100 ns. The long guard time at low symbol rates may result in a waste of capacity, whereas the short guard times at higher symbol rates may tend to be too small to cope with longer transition times and/or beam-switching jitter. Nevertheless, we can conclude that under typical conditions the achievable guard times should fit to the requirements.

In contrast to a pre-coded transmission system, the so-called P2 pilot field is not used. In order to serve the beam-hopping approach, the P2 fields could be redefined to carry physical layer signalling information such as coverage-ID or beam-hopping network status. However if beam-hopping shall be combined with multiple-input–multiple-output (MIMO) techniques like pre-coding, the original definition of the P2 pilot field will be needed.

10.3.2.3 Super-framing Format 4: flexible wideband approach

The SF Format 4 supports time-slicing as well as flexible ACM/VCM- and low-SNR support for wideband transmission and provides additional means for signalling. This is reflected in Figure 10.10, where the SF header (SFH) field and the SFH trailer (ST)

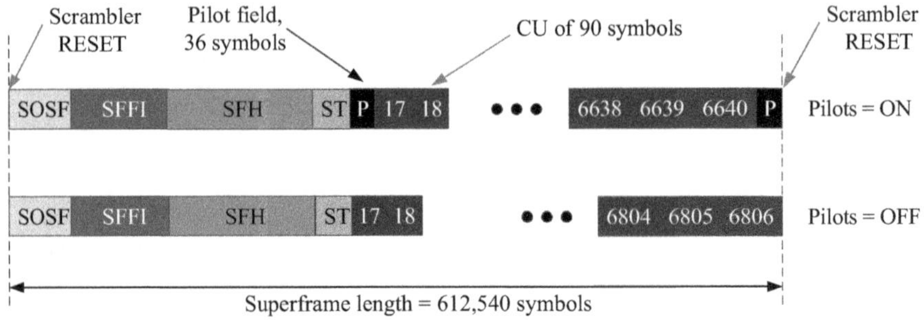

Figure 10.10 DVB-S2X Annex E SF structure according to Format 4

field are located directly after the SFFI and feature lengths of 630 and 90 symbols, respectively. The SFH signals whether SF-aligned pilots on or off, the PLH protection level as well as a pointer to the first complete PLH in the SF.

Although the ST field can be used as training data, it has no specific purpose so far. So, the ST field can be employed, e.g. to signal the actual target coverage-ID or beam-ID by choosing the corresponding Walsh Hadamard sequence index 0. . .63. Additionally, the pointer within the SFH can be used to signal up to 16 different states of the modulator or network status, etc. by means of using the yet-undefined PLH-pointer-values 0. . .15. This is already foreseen in the standard specification since these pointer values refer to all the signalling fields at the beginning of the SF. In addition, at start of illumination the first PLH will be located anyway directly after these signalling fields. This can be exploited for instance to signal an upcoming BSTP change in order to prepare the terminal not to go to sleep-mode but stay active to detect the new BSTP structure.

Apart from these potential features, the required guard time is supported in SF Format 4 by dynamic SF padding. This is accomplished by special dummy frames: Dummy frames of arbitrary content (time slicing number, TSN = 254, normal-size PLFRAME) or dummy frames of deterministic content (TSN = 255, normal-size PLFRAME) are used to terminate the SF at the end of the illumination with padding data. For longer illuminations over more than one SF, dummy frames for SF-padding are only inserted at the last SF for lower overhead.

This means that as much dummy data can be inserted at the end of the last SF as required to meet nearly any guard time requirement, e.g. depending on the network synchronization state. For overhead minimization and fine tuning, the number of dummy symbols can also be kept as small as possible. This dynamic padding length allows the modulator to reserve as much as guard time as needed depending on the actual symbol rate, beam-switching jitter and transition characteristics as well as network synchronization accuracy. Due to the automatic termination of the special dummy frames at the end of the SF, no useless dummy frame data spill over to the next SF, transmitted to another service area, happens, as already observed for Format 0 and 1. Although all SF formats satisfy the fundamental requirements, Format 4 provides already in its current specification very high flexibility for dummy data insertion to comply with practically all guard times and supports additional features like signalling of beam-ID or network status or BSTP update announcement.

10.3.3 Waveform conclusion

The DVB-S2X conventional framing as well as the super-framing according to Annex E have been discussed concerning application for beam-hopping. While the presented concepts how to use conventional framing for beam-hopping are not specified in the standard, the SF specification of formats is directly applicable to beam-hopping. The advantage of SF Format 2 and 3 is that it allows combining beam-hopping and MIMO techniques like pre-coding, while the guard time is static. SF Format 4 supports wideband transmission and is flexible enough to fully support beam-hopping (with dynamic guard time) and provides further features for practical application of beam-hopping.

This discussion has also shown that super-framing has higher relevance for practical feasibility than conventional framing, where the central aspect is the complexity of the GW-side scheduling and network synchronization. With super-framing, it dramatically simplifies compared to conventional framing. This is because super-framing decouples the two tasks of network synchronization with respect to alignment of framing to BSTPs and PLFRAMEs scheduling into separate entities, while it has to be joint optimization task for conventional framing.

10.4 Technology and implementation

10.4.1 Upcoming Eutelsat Quantum satellite for beam-hopping

The Eutelsat Quantum class of satellites is a software reconfigurable commercial telecommunication satellite that offers flexibility at its core. Manufactured by Airbus Defence and Space for the satellite operator Eutelsat, it provides reconfiguration in spectrum, power management and coverage definition and empowers the client to manage their resources in the most efficient and optimized way [15]. The incorporation of the beam-hopping functionality extends this flexibility by another significant step.

Traditionally, the approach to beam-hopping has been applied via static, small high performance spot beams in HTS systems. Through the apportion of capacity between a high number of pre-defined spots, the available capacity may be profiled in order to more efficiently use the resources. In a system in which requirements shift without significant change in geographical coverage, this is an efficient solution without significant additional complexity. Eutelsat Quantum solves the problem and provides increased efficiency of resources by enabling each spot, or hop, to have a varying shape, distributed over the visible Earth. Additionally, Eutelsat Quantum allows for the updating of the active coverage set so that the resources are always used efficiently. Coverages that are no longer useful can be deleted from the active set while new coverages that become required can be created.

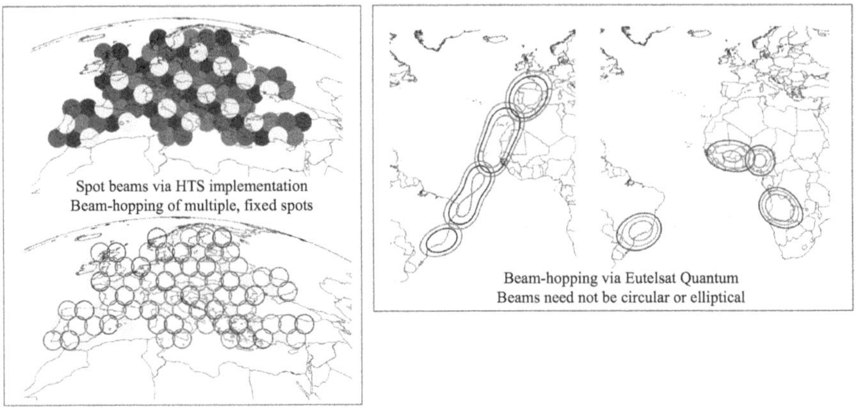

Figure 10.11 Example beam-hopping implementation via HTS systems and Eutelsat Quantum

Figure 10.11 highlights the difference in approach of the two systems. HTS systems offer beam-hopping from a high number of spots over a pre-defined coverage area typically through a single feed per beam or multi-feed per beam antenna approach. Eutelsat Quantum provides beam-hopping via individual reconfiguration of a beam in time. Since phased array antennas are utilized in conjunction with BFNs, the shape of the beam can be changed for each hop so that it is optimized geographically for every hop offering an optimized link budget at any given instant.

In addition, in combining the reconfiguration with the traffic profile controlled by the network control centre, it is possible to update the dwell time of each hop in response to the traffic demand, thus apportioning the available capacity amongst the beam-hopping set of coverages in the most appropriate way.

The ability to rapidly and seamlessly serve multiple regions in a time sharing basis is suited to a variety of rapidly evolving applications. We have already discussed how beam-hopping can be applied to spatially diverse coverages. When formed together over a flightpath or navigation route, beam-hopping can equally be applied to maritime and aeronautical industries where demand for capacity is ever increasing as service providers attempt to answer the need for 24/7 connectivity with evolving traffic profile over time. However, even though the need for capacity is increasing these are not followed by rising costs. So efficient solutions are critical.

Beam-hopping for mobility is not reserved for civil applications. Governmental and military entities could benefit from resourcing their routes in a much more controlled manner as they serve only their flightpath or region(s) of interest.

10.4.1.1 Beam-forming and beam-hopping

The implementation of beam-hopping can be carried out via a number of techniques. Eutelsat Quantum incorporates a BFN for each beam such that the amplitude and phase of each antenna element can be appropriately reconfigured to supply the desired beam-shape. The number of elements used by each antenna determines the shaping

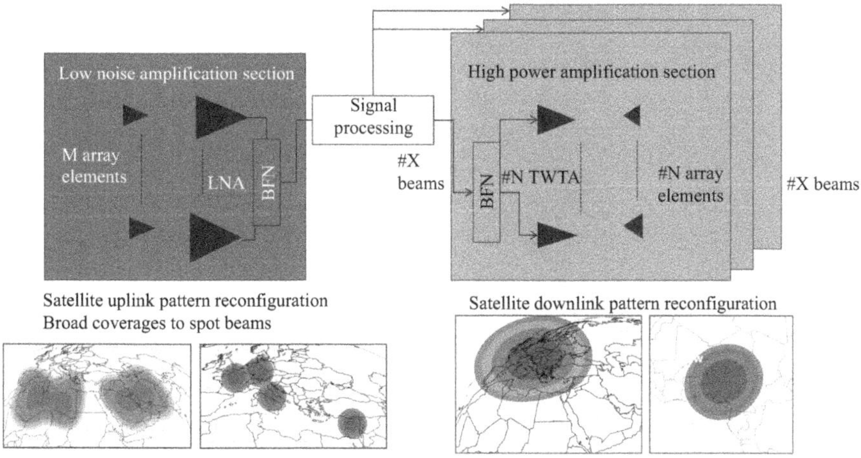

Figure 10.12 Example satellite reconfiguration in uplink and downlink

resolution of the resulting beam. Importantly when considering beam-hopping the number of elements also impacts the memory storage and time required to reconfigure each beam.

Furthermore, beam-forming enhances traffic shaping as an additional advantage. The optimization of an array of antenna elements allows coverage of spatially diverse locations simultaneously. This is visualized by Figure 10.12, where different coverage sizes and locations are configured independently for uplink and downlink. Thus, when previously a service area would be required to cover a single user with a complete SF, beam-forming allows covering the single user as part of a separated lobe or wider lobe of the main beam. With this optimized beam pattern, the single user is served jointly with the other users of said main beam. This enables for more efficient use of system resources.

10.4.1.2 Full duplex versus half duplex

Eutelsat Quantum is capable of operating in two modes of operation:

Mode 1: full duplex

- Uses a beam-hopping configuration (BHC) with one channel each. Therefore, there is a BHC for the forward link and a BHC for the return link. The BSTP, which applies to the forward downlink, and beam-hopping burst time plan (BHBTP), which applies to the return uplink, are executed in parallel with two different channels.
- Quantum supports 8 uplink and 8 downlink beams. Since full duplex assigns equal resource to the forward and return link, it is therefore capable of supporting four such networks.

As a single network example, Figure 10.13 depicts simultaneous forward link and return link transmission. The GW-side uplink and downlink beams are configured

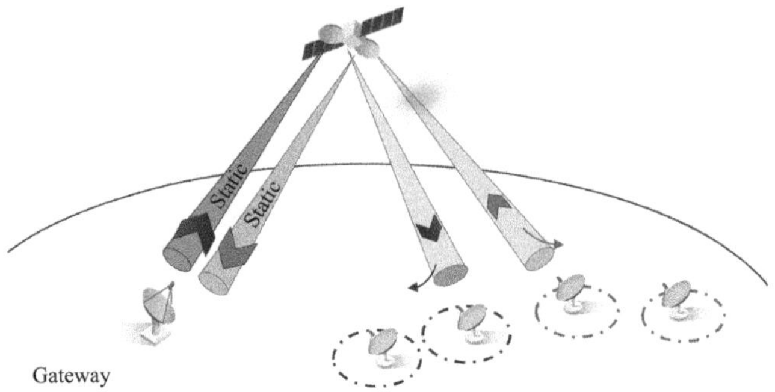

Figure 10.13 Full duplex

static, while the terminal-side uplink and downlink beams perform hopping according to BHBTP and BSTP, respectively.

Mode 2: half duplex

- Uses one BHC with one Quantum channel. That is, the BHC is shared in time between the forward and return-link communication. Consequently, the BSTP and the BHBTP are concatenated to share the same channel. Note that uplink and downlink frequencies are not the same due to frequency conversion at the satellite.
- Therefore each network employs only one BHC with a beam-hopping uplink beam and a beam-hopping downlink beam.
- The frame structure therefore covers both the forward link BSTP and return link BHBTP.
- Quantum supports 8 uplink and 8 downlink beams. Since half duplex shares the resource, Quantum is capable of supporting eight such networks.

Figure 10.14 visualizes one beam-switching frame of a half-duplex transmission, which is periodically executed. First in time, the forward link (FWD) communication uses beam-hopping according to the BSTP to distribute data to the terminals. Second, the return link (RTN) communication takes place following the BHBTP to gather the data from the terminals.

Each mode of operation has its advantages and disadvantages. Whilst full duplex is less complex in its implementation, it is also more limited and requires more resources. Operating in half duplex means that a beam-hopping network can be operated from a single pair of uplink and downlink beams, operating only one at any point in time. Given the different approaches, it is clear that both implementations suit a variety of applications. On consideration of implementation, both approaches may be feasible but the decision of which to utilize is likely to come down to efficiency of resources and the potential need for direct access.

Full duplex could be considered appropriate for scenarios in which the service is targeting a number of locations seamlessly. Since a channel is used for both uplink and downlink, it is capable of supporting larger bandwidths from the fixed GW

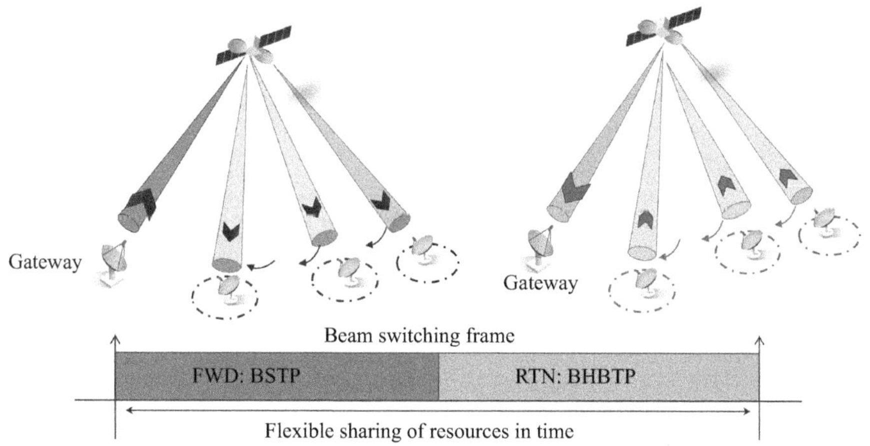

Figure 10.14 Half duplex

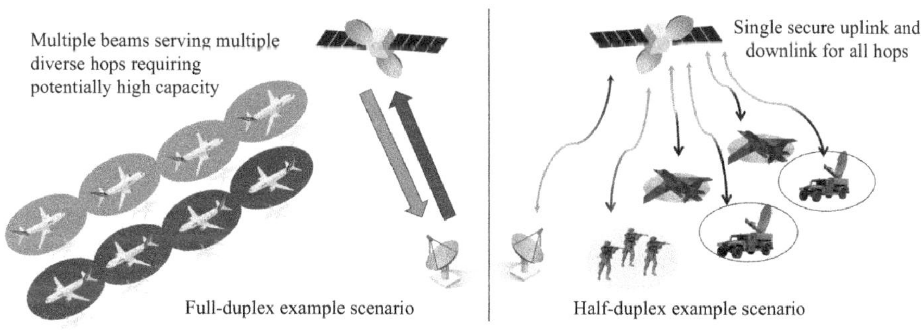

Figure 10.15 Full- and half-duplex scenario examples

locations. Typical examples could be considered as mobility applications in which maritime and aeronautical routes are required to be capable of supporting increasing capacity requirements as users demand continual access to video and data. In the mobility application, beam-hopping could be applied in order to track a single craft or seamlessly service a flight or busy maritime route, as depicted on the left of Figure 10.15.

Since half duplex employs a single uplink and single downlink beam, it is particularly suited to applications in which multiple clients each wish to operate or have access to their own beam-hopping network without additional resources. Such applications could be governmental or military missions which could be focussed to the smallest resource and coverage necessary and which potentially could be operated in a secure fashion. This is shown on the right of Figure 10.15. Half duplex can also be used to dynamically apportion the capacity between forward and return link.

10.4.2 Wideband transmission for beam-hopping

The satellite may support only a limited set of independent beam-hopping networks because of required independent beam-forming and beam-switching capabilities. Therefore, most efficiency will be achieved by means of wideband transmission instead of using standard signal bandwidths.

Assuming that a service area is illuminated in the conventional way by means of a standard-bandwidth transponder, e.g. 36 MHz, a symbol rate of 30 Msps may be used to serve the users. If a beam-hopping system shall serve, e.g. 10 of such service areas, a symbol rate of 300 Msps would be needed to provide the same average symbol rate per service area. This is due to the applied time-division multiplexing (TDM) approach.

Using a wideband transponder in the conventional way, i.e. in a frequency-division multiplexing (FDM)-like multi-carrier mode, exhibits several drawbacks even in non-beam-hopped systems compared to single-carrier mode:

- Waste of the rare frequency resources for guard bands (typically 10–16 per cent)
- Reduced gain of predistortion techniques
- Higher power back-off needed due to higher peak-to-average power ratio
- Inter-carrier interference due to intermodulation products
- No throughput enhancement due to wideband multiplexing gain in case of multi-service or multi-stream application scenarios

If beam-hopping comes into play as well, all carriers of a beam-hopped FDM transponder are forced to be switched jointly according to the actual BSTP. This seems an unfortunate condition in terms of maximizing flexibility and adaptivity.

So we can conclude that single-wideband-carrier mode would be most beneficial to choose, since a new system approach has to be implemented and no legacy system support has to be assured. Also the guard time overhead for the switching events can be kept very low $\ll$1 per cent. In the light of today's technology- and implementation-based constraints, it is therefore proposed that single wideband-carrier mode could be the most beneficial choice when considering that a new system is to be implemented and no legacy system support is required. This de-risks the impact of future changes to the system and whilst maintaining optimum flexibility where possible.

Note that wideband transmission can be used typically for forward links, where the terminals act as receivers. For the return links, the equivalent isotropically radiated power limits of the terminals may not allow one to easily use wideband transmission. Therefore, using more narrow-band transmission in a frequency-division/time-division multiple-access way may be more suitable for the return links. This is a common method since the data rate demands are usually asymmetric as well.

Also sharing the time resources among different operators can be achieved by reserving either different time slots for different operators or by sharing time slots. This is accomplished by using, e.g. different TSNs within the time slots with adjustable reservation of capacity per TSN at the data input processing for the different time slots. Such schemes are already common for quality of service processing. However, the operators have to run a joint resource allocation scheduler in a cooperative way.

10.4.3 Network synchronization aspects

The situation for transmission in direction to the terminals reflects either a broadcast situation or an outbound situation of a star-type two way system. In general, beam-hopping can be used also for the return links of a star-type two way system. The GW acts as part of a common uplink for the broadcast network or as part of a common hub of a star-type two way system. In addition, point-to-point links with interrupted transmission capabilities can also be considered as a special case of such generic network topologies.

A special challenge results from the requirement to synchronize the transmission of the GW to the BSTP of the satellite in such an order that the transmitted user data frames of the GW arrive correctly within the intended service areas.

The key parameters relevant for this are the following:

- The clock rate of the BSTP depends on the satellite high stability clock, which is accurate but not synchronized to any other time reference, which could also be made available to the GW on the ground.
- Standard satellite station keeping applies. So changes to the signal transmission latency between the GW and the satellite as well as the satellite and the remote terminals apply.
- The BSTP on the satellite is also known to the GW, but in the case of updates some uncertainty on the activation time applies.

All these uncertainties require that the GW shall be able to synchronize itself to the satellite by employing a special measurement loop. For this, a reference terminal is used, which receives the signal from the satellite similar to all other remote terminals but provides to the GW special information about required adjustments. This includes information about how to adjust the phase and clock rate of the BSTP of the GW in reference to the BSTP of the satellite.

Of course, other synchronization mechanisms are possible as well. For example, the satellite telemetry and control link could be used. However, time synchronization may not be accurate enough due to limited capacity of this link. In addition, BSTP update data has to be transmitted as well. Using a reference terminal on ground eliminates the need to install beam-hopping signal timing analysis at the satellite and transmission of the measurement results to the ground.

Also varying latency between the GW and the satellite as well as a possible small and also varying offset between the satellite clock and the GW clock need to be initially identified and corrected, as well as continuously tracked.

As already mentioned, the BSTP can also be updated from time to time, which is tentatively a process of satellite command and control, which includes significant and to some degree unknown latency. Nevertheless, the GW needs to switch over in timely coincidence with the satellite, which requires special procedures.

10.4.4 Signal synchronization at terminals

For a terminal, there are two fundamental burst-mode receiver concepts:

- **Decoupled block-based processing** of detection and data processing:
After detection of start of illumination, all samples are forwarded to a (large) buffer until end of illumination. The data processing like synchronization, demodulation and decoding works on the buffered data. This allows enhanced and fine synchronization with respect to timing, frequency and phase, because potentially all buffered data can be used for offset estimation and compensation (in various iterative steps), before demapping and decoding can be performed. However, this approach may need a very large buffer and exhibit throughput limitations with respect to support of different transmission scenarios and worst-case system configuration.
- **Feed-forward processing** of detection and data processing:
This approach has a pipelined processing architecture very similar to continuous signal processing. In contrary to the previous architecture, it has no large buffer and therefore less iterative methods. The synchronization modules perform permanent processing for detection as well as offset estimation and compensation. In the case of illumination, they forward the samples to the demodulation and decoding part. This architecture allows for a single receiver design supporting both, beam-hopped and conventional continuous signal reception. Therefore, high symbol rate and maximum throughput support are assured but at the price of some complexity overhead in the case of low illumination duty cycles.

Of course, these pro's and con's scale depending on the used symbol rate and the shortest illumination duty cycle supported by the receiver. Let us consider a few critical cases.

The shortest BSTP serves two coverage areas where one SF per coverage is scheduled. So a receiver would need to process every second SF. Furthermore, if all the data frames within this SF are assigned to one receiver, the processing speed has to cope with roughly half the system symbol rate. In this scenario, the feed-forward processing will clearly outperform the decoupled architecture in terms of complexity.

As a counter example, very long illumination duty cycles and only a few data frames per SF shift the rating in favour of the decoupled block-based processing architecture because of possible complexity scale down. Therefore, these two extremes tell us that there will be no globally optimum decision. Trade-offs and architecture mixture will be needed. Therefore, some tasks and algorithms are discussed below, which can be used in both architectures.

Nevertheless, for maximum flexibility in supporting various use-cases and system configurations, one will be on the safe side with the feed-forward processing approach despite some overhead in complexity in case of low duty cycles.

10.4.4.1 Reception scenarios

While general beam-hopping transmission scenarios are discussed in [14,15], three different exemplary reception scenarios are shown in Figure 10.16. In each of these

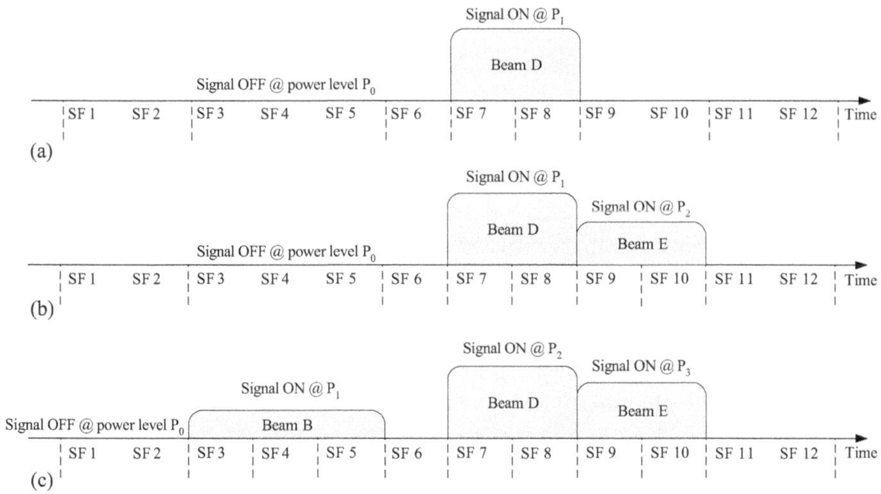

Figure 10.16 Terminal-side signal reception scenarios: (a) no neighbour signal, (b) one neighbour signal, and (c) two neighbour signals

cases, the horizontal time axis reflects the illumination time granularity with respect to SFs. The current received signal power is given in vertical direction. In case (a), the terminal receives only a single illumination of two SFs duration with respect to beam D at power level P_1 and otherwise pure noise at power level P_0. Therefore, this is the simple baseline scenario, where a terminal located in coverage area D observes only the target illumination by beam D. Since the BSTP is periodically performed, the terminal can exploit this repetitive character and synchronizes to the SF grid.

In the likely case of a terminal location near to the edge of coverage, reception scenarios as shown in cases (b) and (c) can occur. Now neighbouring beams B and E targeting adjacent coverages B and E, respectively, are received at different power levels than beam D. So a terminal cannot rely only on supporting baseline case (a). Note that scenarios (b) and (c) can also hold for terminals located in coverage area B or E. This means that the neighbouring beam signal has a higher power level than the own one. Thus, a proper terminal synchronization scheme has to cope with the more challenging cases as well. For this, a smart gain control and a high dynamic range is needed.

10.4.4.2 Power detection

The power detection represents the most essential part of the beam-hopping capable terminal. As a non-data-aided algorithm, it is independent of the used waveform. The power detection is also robust against timing and frequency offsets. Therefore, it is the backbone algorithm of the beam-hopping terminal in case other data-aided synchronization schemes fail. Furthermore, it is needed to implement a smart gain control and to trigger adaptation (re-)start of further synchronization algorithms.

Even in scenario (c) of Figure 10.16, a terminal located in coverage area B can exploit the signals of beam D and beam E for synchronization thanks to the common SF structure. But to do this, power detection is needed for identification. Of course data demodulation is still performed using the beam B SFs until the network controller schedules a hand-over to beam D or E. This requires feedback to the GW of power level estimates provided by the power detection of the terminals.

Since the instantaneous power values P_{act} are strongly fluctuating, an averaging has to be performed before applying detection techniques. In order to implement a low complexity moving average, one may either use an equal gain filter or a recursive filter. The equal gain filter offers linearly decreasing weighting of the memory, whereas the recursive filter exhibits exponentially decreasing weighting of the memory. This yields a linear increasing step response and an exponentially increasing step response, respectively. In order to immediately identify power variation related to the start and end of illumination, the quicker response of the recursive filter seems more advantageous over the equal gain averaging for such a detection task. The update equation of the average power P_{IIR} applies an infinite impulse response (IIR) recursive filter according to

$$P_{\text{IIR}}[i+1] = (1-\delta) \cdot P_{\text{IIR}}[i] + \delta \cdot P_{\text{act}}[i] \tag{10.1}$$

with sample time index i and where the forgetting factor δ is a small positive constant.

Three approaches for power detection are analysed in the following with different capabilities to find start and end of illumination as well as the illumination power level:

- Threshold-based power detector:
  From the averaged receive power signal, the minimum and maximum power is determined over an observation time. Thresholds are then calculated from these min/max power values for rising edge detection and falling edge detection. This procedure can be iterated to track slightly changing receive power over time.
- Slope-based power detector:
  The slope is calculated from the averaged receive power signal by means of a differential signal, i.e. subtracting power values of time lag Δ. Once the power changes significantly, there will be a peak in the differential signal, which can be checked against a threshold.
- Power level detector:
  While the two previous approaches search for identifying the start and end of illumination directly (by detection of rising/falling edge), this approach searches for power levels. According to a configurable snapshot distance, these snapshots of the averaged power are compared whether consecutive snapshots lie within a configurable margin. When storing detected power levels, they can be identified once they are recurring.

The detection principles of these approaches are provided for a single illumination at SNR $= -3$ dB. In Figures 10.17 and 10.18 the threshold-based detector and the slope-based detector are considered, respectively. Averaged power values of the IIR filters are shown versus sample time index i, where two configurations with respect to averaging depth are compared: IIR1 and IIR2 with forgetting factor $\delta = 2^{-10}$ and 2^{-17},

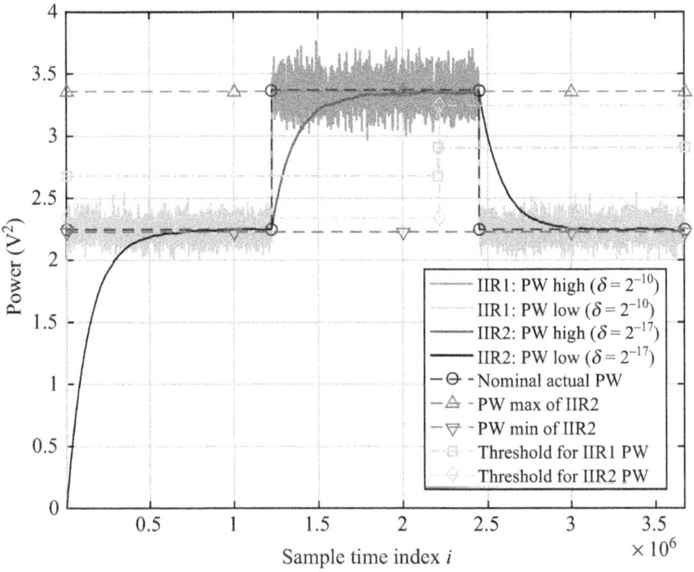

Figure 10.17 Start and end of illumination using a threshold-based detector evaluating min/max power

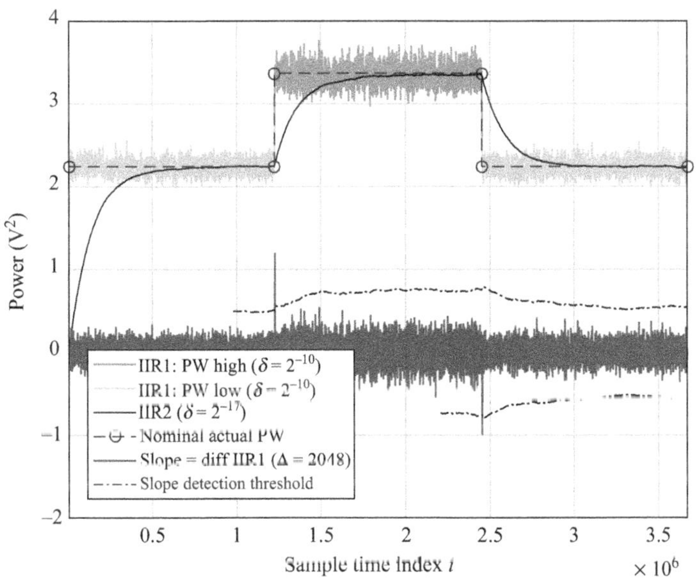

Figure 10.18 Start and end of illumination using a slope-based detector

respectively. The change in grey level of IIRx output values indicates detection of power high/low. A cosine-shaped transition between signal power on and off is used and random data symbols of 256 APSK constellation.

In Figure 10.17, maximum and minimum mean power values are determined from IIR2 because of more precision due to strong averaging ('PW max of IIR2', 'PW min of IIR2'). From these values, the threshold curves 'Threshold of IIR1' and 'Threshold of IIR2' are calculated, where the step in the threshold curve indicates switching from rising edge detection to falling edge detection. Obviously, this detection was successful for both evaluated IIR configurations since the grey level of the line changes when crossing the threshold. Note that here the scenario (a) (see Figure 10.16) of receiving only a single beam is considered. Further tests in scenarios (b) and (c) reveal that different neighbouring beam signals cannot be distinguished properly, which leads to missing rise or fall detections.

In Figure 10.18, the differential power signal is calculated based on IIR1 output values using $\Delta = 2{,}048$ samples. It is fluctuating around zero. Although the peaks in the differential signal can be observed and detected here, there is quite some chance under low SNR that the detection is not successful. This is due to the noise enhancing nature of differential signal calculation. This unreliable detection performance becomes even more severe in multiple beam scenarios as shown in Figure 10.16.

The principle of the power level detection approach is shown in Figure 10.19. Successful IIR1 power level detections are indicated by square markers also showing the detection interval, while star markers refer to successful IIR2 power level

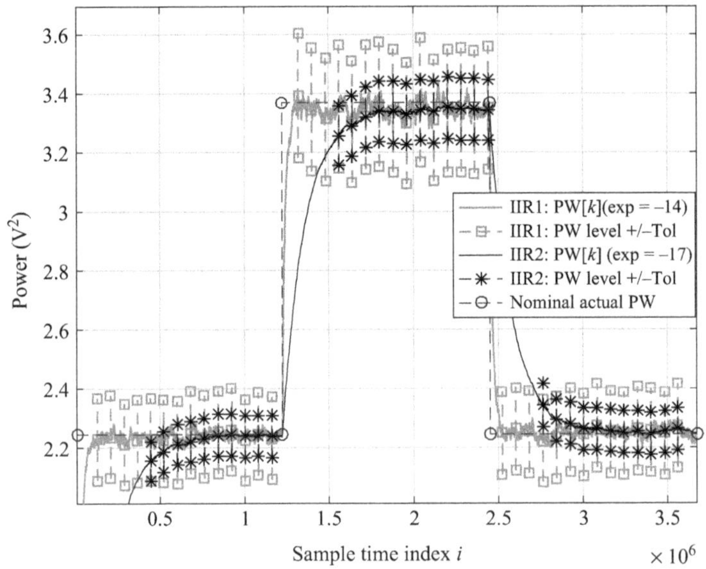

Figure 10.19 Start and end of illumination detection using power level detection

detections. Since a short history of (minimum two) snapshots is needed for power level detection, end of power level can be identified immediately, while start of power level decision is delayed by the used history length. For the example of Figure 10.19, a history of two snapshots is considered and compared against the actual snapshot. Note that a longer history allows to be more error tolerant, if a snapshot is by chance out of margin, but this may lead to further decision delay.

In essence, this algorithm may identify and distinguish the different power levels when receiving multiple beams, but the decision delay seems to be a drawback. Therefore, it will be most beneficial to combine the threshold-based approach with the power level detector. For example start of illumination can be securely identified by observing both events of 'leaving the low power level PW min' and 'PW above the rising PW threshold'.

10.4.4.3 Super-frame-related performance figures

The key performance figure of the SF is the detection probability of the 270 symbols long SOSF. It dictates the quality of the whole burst mode processing. In previous publications [16,17], detailed discussion and description of the considered subblock-based correlation algorithm are given. For this, a conventional full correlation of 270 symbols length is split in 18 subblock correlators each of 15 symbols length. Summation as combining of subblock correlator output values represents a full correlation. Another combining is pair-wise conjugate complex multiplication of neighbouring outputs and summation, which is termed cross-correlation algorithm (XCorr). A further combining method is to derive the absolute square of each output before summation, which is denoted as absolute square algorithm (Abs2).

In [16,17], perfect sampling with respect to detection of SOSF+SFFI was assumed. The more appropriate case of random sampling offset was considered in [13]. All considered a target false alarm probability of $\Pr(\mathrm{FA}) = 10^{-5}$, which is ok for continuous transmission. More reliable peak detection is needed for beam-hopping, which is why we focus on $\Pr(\mathrm{FA}) = 10^{-6}$. Since this worsens the probability of missed correlation peak Pr(MP), we analyse the simulation results in Figure 10.20. Correlation with respect to SOSF only is performed under relative frequency offsets of 0 and 0.01. XCorr seems robust enough to assure SOSF detection. In the case of tracking when carrier frequency offsets are compensated, one can switch to full correlation due to improvement of 2 dB in SNR.

As second performance indication applying SF Format 4, simulation results of ST-field-decoding are considered in Figure 10.21. To decide among the 64 different Walsh Hadamard sequences, the maximum correlation-based decoder is employed (with and without phase knowledge) as well as a low-complexity Hamming metric decoder. The impairment scenario is an additive white Gaussian noise channel with random complex phase (constant over each codeword). The code word error rates (CER) over SNR are shown in Figure 10.21.

Obviously, a CER of 10^{-6} is reached already at -4 dB SNR by the correlation decoder (without phase knowledge). Thus, robust signalling of the coverage-ID is guaranteed. The low complexity Hamming metric decoder using ideal phase knowledge is degraded by approximately 2.2 dB compared to the optimum correlation

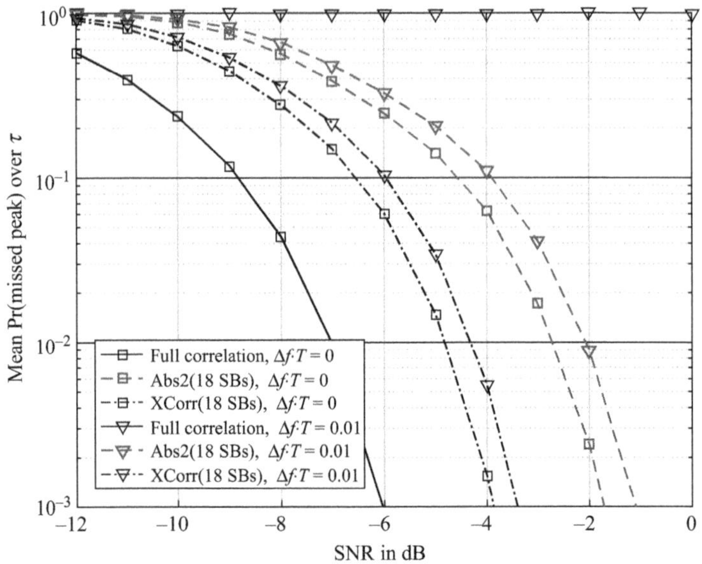

Figure 10.20 Mean probability of missed correlation peak averaged over sampling phase τ for relative carrier frequency offset $\Delta f \cdot T = 0$ and 0.01

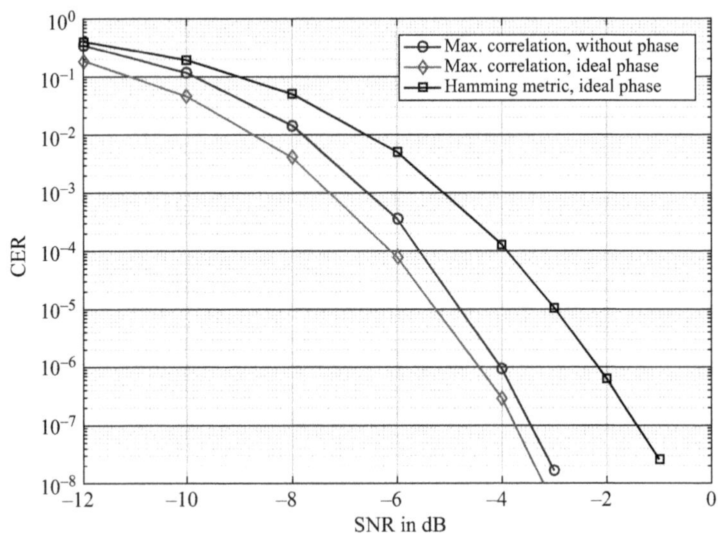

Figure 10.21 Code word error rates of a correlation-based decoder and a per-bit hard decision decoder for different knowledge on complex phase

decoder with ideal phase knowledge. For practical application, the Hamming metric decoder needs a phase estimate from the neighbouring pilot next to the ST field, cf. Figure 10.10.

10.5 Summary and conclusions

In this chapter, the concepts and benefits of beam-hopping were presented along with some detection performance considerations. In particular, the gains in user satisfaction and system usable throughput were compared to that of a conventional broadband satellite system with a static coverage. Complementing the beam-hopping system principles, we discussed the physical layer transmission solutions suitable for beam-hopping. Based on the identified waveform key-requirements for applying beam-hopping, the already released DVB-S2X standard was reviewed and analysed using practical and representative system examples. It was found that the super-framing specification offers high practical relevance compared to the conventional DVB-S2/S2X framing.

We also presented actual and future technology for a beam-hopping systems. Specifically, the upcoming Eutelsat Quantum-class satellite designed for beam-hopping was presented along with its features like re-configurable beam-forming and highlights potential applications. The corresponding ground equipment was also discussed exploiting the advantages of wideband processing. Furthermore, implementation feasibility has been demonstrated by means of detection performance results exploiting DVB-S2X SF Format 4.

Beam-hopping offers flexible system architecture to address changing traffic demands over time and geographical locations by sharing in time, power and frequency resources among multiple beams. Beam-hopping systems offer higher usable throughput by focusing the system resources where they are most needed at a time.

References

[1] "ETSI EN 302 307-2: Digital video broadcasting (DVB); second generation framing structure, channel coding and modulation systems for broadcasting, interactive services, news gathering and other broadband satellite applications; Part 2: DVB-S2 extensions (DVB S2X)," ETSI, European Telecommunications Standards Institute Std., Rev. 1.1.1, Oct. 2014.

[2] EADS Astrium Space Engineering, "ARTES-1 beam hopping techniques for multi-beam satellite systems – Final report," ESA, Tech. Rep., 2011. [Online]. Available: https://artes.esa.int/projects/beam-hopping-techniques-multi-beam-satellite-systems-eads-astrium.

[3] Indra Espacio, MDA, and Universitat Auttonoma de Barcelona, "ARTES-1 beam hopping techniques for multi-beam satellite systems – Final report," ESA, Tech. Rep., 2009. [Online]. Available: https://artes.esa.int/projects/beam-hopping-techniques-multibeam-satellite-systems-indra-espacio.

[4] P. Angeletti, D. Fernandez Prim, and R. Rinaldo, "Beam Hopping in Multi-Beam Broadband Satellite Systems: System Performance and Payload Architecture Analysis," in *Proc. of the AIAA*, San Diego, USA, Jun. 2006.

[5] J. Anzalchi, A. Couchman, P. Gabellini, *et al.*, "Beam Hopping in Multi-Beam Broadband Satellite Systems: System Simulation and Performance Comparison with Non-Hopped Systems," in *Proc. of 5th Advanced Satellite Multimedia Systems Conference (ASMS) and the 11th Signal Processing for Space Communications Workshop (SPSC)*, Cagliari, Italy, Sep. 2010, pp. 248–255.

[6] X. Alberti, J. M. Cebrian, A. Del Bianco, *et al.*, "System Capacity Optimization in Time and Frequency for Multibeam Multi-Media Satellite Systems," in *Proc. 5th Advanced Satellite Multimedia Systems Conference (ASMS) and the 11th Signal Processing for Space Communications Workshop (SPSC)*, Cagliari, Italy, Sep. 2010, pp. 226–233.

[7] R. T. Gedney and R. J. Schertler, "Advanced Communications Technology Satellite (ACTS)," in *Proc. ICC 1989, IEEE Int. Conf. on Communications*, vol. 3, Jun. 1989, pp. 1566–1577.

[8] D. Whitefield, R. Gopal, and S. Arnold, "Spaceway Now and in the Future: On-Board IP Packet Switching Satellite Communication Network," in *Proc. MILCOM 2006, IEEE Military Communications Conference*, Oct. 2006, pp. 1–7.

[9] R. J. F. Fang, "Broadband IP Transmission over SPACEWAY Satellite with On-Board Processing and Switching," in *Proc. GlobeCom 2011, IEEE Global Telecommunications Conference*, Dec. 2011, pp. 1–5.

[10] J. Lei and M. A. Vazquez Castro, "Frequency and Time-Space Duality Study for Multibeam Satellite Communications," in *Proc. ICC 2010, IEEE International Conference on Communications*, May 2010, pp. 1–5.

[11] C. Rohde, R. Wansch, G. Mocker, S. Amos, E. Feltrin, and H. Fenech, "Application of DVB-S2X Super-Framing for Beam-Hopping Systems," in *Proc. 23rd Ka and Broadband Communications Conference*, Trieste, Italy, Oct. 2017.

[12] "ETSI EN 302 307-1: Digital video broadcasting (DVB); second generation framing structure, channel coding and modulation systems for broadcasting, interactive services, news gathering and other broadband satellite applications; Part 1: DVB-S2," ETSI, European Telecommunications Standards Institute Std., Rev. 1.4.1, Nov. 2014.

[13] C. Rohde, H. Stadali, and S. Lipp, "Flexible Synchronization Concept for DVB-S2X Super-Framing in Very Low SNR Reception," in *Proc. 21st Ka and Broadband Communications Conference*, Bologna, Italy, Oct. 2015.

[14] E. Feltrin, S. Amos, H. Fenech, and E. Weller, "Eutelsat QUANTUM-Class Satellite: Beam Hopping," in *Proc. 3rd ESA Workshop on Advanced Flexible Telecom Payloads*, ESA/ESTEC Noordwijk, The Netherlands, Mar. 2016.

[15] H. Fenech and S. Amos, "Eutelsat Quantum-Class Satellites, Answering the Operator's Need for Flexibility," in *Proc. 3rd ESA Workshop on Advanced Flexible Telecom Payloads*, ESA/ESTEC Noordwijk, The Netherlands, Mar. 2016.

[16] C. Rohde, H. Stadali, J. Perez-Trufero, S. Watts, N. Alagha, and R. De Gaudenzi, "Implementation of DVB-S2X Super-Frame Format 4 for Wideband Transmission," in *Proc. WISATS 2015, 7th EAI International Conference on Wireless and Satellite Systems*, Bradford, United Kingdom, Jul. 2015.

[17] C. Rohde, N. Alagha, R. De Gaudenzi, H. Stadali, and G. Mocker, "Super-Framing: A Powerful Physical Layer Frame Structure for Next Generation Satellite Broadband Systems," *International Journal of Satellite Communications and Networking (IJSCN)*, vol. 34, no. 3, pp. 413–438, Nov. 2015, SAT-15-0037.R1. [Online]. Available: http://dx.doi.org/10.1002/sat.1153.

Chapter 11

Optical on–off keying data links for low Earth orbit downlink applications

Dirk Giggenbach[1], Florian Moll[1], Christopher Schmidt[1], Christian Fuchs[1], and Amita Shrestha[1]

Optical free-space links will shape the high-speed communications technology landscape for space missions substantially in the next years. The dramatically reduced signal spread – as compared to any radio frequency (RF) technology – provides a variety of advantages: increased power efficiency, the avoidance of interference and thus spectrum regulation issues, the inherent tap- and spoof-proofness and, most of all, the vastly increased data rates (DRs) will make this technology a 'game changer' comparable to the introduction of glass fibre instead of copper cables previously used in the global communication infrastructure.

As one use case of optical space links high-speed geostationary data-relays for the repatriation of low Earth orbit (LEO) observation satellite telemetry have been tested and are currently implemented operationally by various space agencies [1–4]. Deep space missions will also boost their DRs by several orders of magnitude by sending their data to large optical receiver telescopes, NASA is currently transforming its Deep Space Network to an optical DNS, and we also see European developments in optical deep space communications [5–7]. In order to connect very high-throughput communication satellite systems to the Tbps-regime (Terabit-per-second), optical uplinks can solve the spectrum bottleneck that RF links would otherwise encounter [8]. In the LEO regime (inter-satellite, as well as optical LEO downlinks – OLEODL), distances are way shorter, allowing very high data rates while, at the same time, reducing the requirement for high system sensitivity (where complexity and thus costs generally increase with sensitivity). Instead, components and technologies that are close to commercial-off-the-shelf (COTS) from terrestrial fibre communications can be used, allowing both very high throughputs and moderate-to-low system costs. Using COTS components in inter-satellite as well as downlinks is also supported by the shorter life time of LEO missions, implying less radiation exposure of these components. In the last years, several demonstrations of OLEODL have been performed by various agencies [9–15], and its commercialisation will be seen in near future.

[1]Satellite Networks Department, German Aerospace Center, Institute of Communications and Navigation, Germany

OLEODL serve for sensor data download from earth observation satellites, their link scenario is strongly asymmetric, since the data flow is mostly simplex or at least the downlink DR is orders of magnitude higher than the uplink (the later may only serve for tele-command and link protection). Therefore, the antenna gain can be distributed favourably: with small and lightweight transmitters in space and correspondingly moderately sized antennas (i.e., receiver telescope apertures) on the ground. The disturbing atmosphere only affects the lower end of the link close to the receiving ground station, which on the one hand allows for simple techniques for link stabilisation by aperture averaging but on the other hand complicates some of the advanced modulation and detection formats, since these may require sophisticated techniques like adaptive optics for coupling into single-mode fibres. Therefore, data format options for OLEODL focus mainly on rather low complex and robust direct detection (DD) techniques [16].

The following chapter introduces in its subsections:

- the implementation history of space terminals and optical ground stations (OGSs) and consequences of the link scenario geometry
- effects of the atmospheric transmission channel, link budget, modulation formats and link protection techniques
- system and component aspects, and an outlook to ongoing and future missions and systems.

11.1 The scenario and history of optical LEO data downlinks

11.1.1 Optical LEO downlink experiments overview

OLEODL – in contrast to their traditional RF counterparts – enable higher data throughput from earth observation satellites while avoiding spectrum regulation issues. This has attracted attention for several decades now and has resulted in multiple experimental or demonstration space missions, see Figure 11.1. One of the first were the downlink campaigns from the Japanese satellite Kirari (also named *OICETS*) to ground stations in Japan, Europe and the United States in 2006 and 2009 [17,18]. While this mission was compatible with the European geostationary Earth orbit (GEO)-relay terminals of semi-conductor inter-satellite link experiment [19] and thus used wavelengths in the semi-conductor laser domain, later on the OLEODL projects focused on 15xx nm as the carrier wavelength since this allows to build on component technology from terrestrial fibre communications like optical amplifiers and laser diodes. Furthermore, eye safety can be achieved more easily and solar background radiation causes less disturbance at longer wavelengths. These follow-on projects comprise SOTA by NICT (Small Optical Transponder, on-board SOCRATES Satellite) [20], OPALS (Optical Payload for Lasercomm Science, on-board the ISS) by the Jet Propulsion Lab (JPL) and the various development stages of DLR's OSIRIS (Optical Space InfraRed link System) [21]. Chinese and Russian experiments have also been reported. OLEODLs were also performed from the LCTs

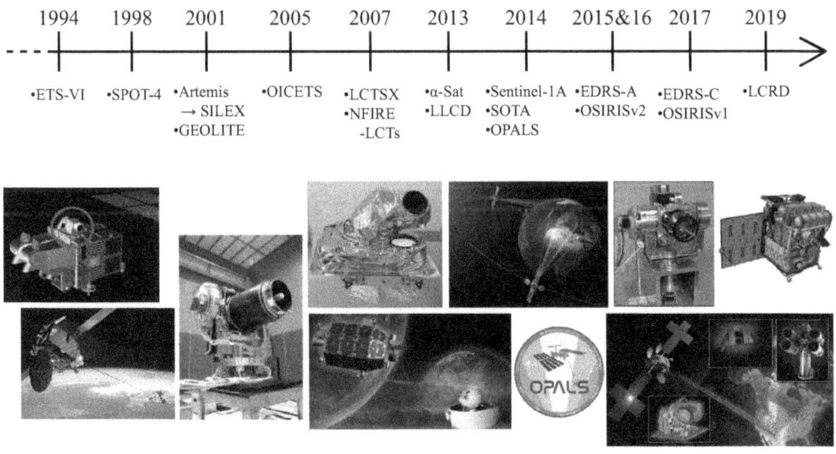

Figure 11.1 Recent timeline of space laser missions. Pictures: ESA, NASA, JAXA, NICT, DLR

on-board TerraSAR-X, testing sensitive and elaborate coherent BPSK-homodyne modulation [22]. Table 11.1 provides an overview of some project parameters.

Several institutions operate OGSs in order to carry out such downlink experiments not only from LEO but also from GEO and farther space probes. OGSs on Tenerife, in California and in Tokyo, have been established since the 1990s [23–26]. Newer – partly temporary – sites include Hawaii, White Sands in New Mexico, Oberpfaffenhofen near Munich, Observatoire de la Côte d'Azur (OCA) [27] and several more stations in Japan like Okinawa and Kashima. Other institutions also operate astronomical facilities in temporary use as ground stations in optical communication experiments. Table 11.2 illustrates the basic parameters of some OGS installations.

In the future, further developments in the domain of OLEODL are expected in Japan (VSOTA on RISESAT), Europe (OSIRIS-v3 [28], OPTEL-μ [29]) and the United States [30]. Standardisation efforts are ongoing in the *Consultative Committee for Space Data Systems* (CCSDS) to advance global cooperation in this domain [31].

11.1.2 Performance and geometrical restrictions

An OLEODL link includes the downlink signal from the satellite which is detected and tracked at the ground station telescope, and an uplink beacon signal from the ground station to the satellite which allows the space terminal precise tracking of the OGS location during the overflight. Optical links, just as any traditional RF LEO downlinks, encounter the same geometrical conditions here as depicted in Figure 11.2 and summarised in Table 11.3. A typical low altitude for Earth observation satellites is 400 km, whereas 900 km is typical for a satellite communication network. An optical downlink should generally start acquisition of the optical signals at around 5° elevation, and secure data transmission should work from 10° elevation upwards.

Table 11.1 Overview of optical LEO downlink projects (selection)

Optical terminal	LUCE	LCTSX	SOTA	OPALS	OSIRIS v2	OSIRIS v1	OSIRIS v3
Operator	JAXA	DLR	NICT	JPL	DLR	DLR	DLR
In orbit	2005	2006	2014	2014	2016	2017	2019
Satellite or platform	**OICETS/ Kirari**	**TerraSAR-X**	**SOCRATES**	**ISS**	**BIROS**	**Flying Laptop**	**TBC**
Orbit height (circular)	~600 km	515 km	600 km	~400 km	510 km	600 km	TBC
CPA type	az-el	periscope	az-el	az-el	Sat-Pointing w. 4QT	Sat-Point. open-loop	1-mirror
Tx wavelength	847 nm	1,064 nm	1,549 nm	1,550 nm	1,545 and 1,550 nm	1,550 nm	1,540 nm
Tx power (typical, mean)	0.1 W	0.7 W	35 mW	0.8 W	0.5 and 0.05 W	0.5 W	1 W
Tx divergence (FWHM)	5.5 μrad	–	223 μrad	940 μrad	200 and 1,200 μrad	200 μrad	TBC
Data rate, channel	50 Mbps	5.6 Gbps	1/10 Mbps	50 Mbps	1 Gbps	10/100 Mbps	10 Gbps
Uplink/beacon wavelength	820 nm	1,064 nm	1,064 nm	976 nm	1,560 nm	N.A.	1,590 nm
Uplink data rate	2 Mbps	5.6 Gbps	N.A.	N.A.	100 kbps	N.A.	TBD
Downlinks to OGSs	NICT-Tokyo OGS-OP ESA-Tenerife JPL-TMF	ESA-OGS OGS-OP Calar-Alto	NICT-Tokyo NICT–other OGS OGS-OP/TOGS CSA CNES-OCA	OCTL (TMF) OGS-OP/TOGS	OGS-OP/TOGS	OGS-OP/TOGS	OGS-OP/ TOGS–future
Mission status (Jan 2018)	Finished	Finished	Finished	Finished	Launched	Launched	In development

Table 11.2 Global installations of optical ground stations for OLEODL signal reception (selection)

OGS	Tenerife-Izana, Spain (ESA-OGS)	Tokyo–Koganei, Japan	Table Mountain, CA, USA (OCTL)	Oberpfaffenhofen, Germany (OGS-OP)	Worldwide (TOGS)
Operator	ESA	NICT	JPL-NASA	DLR	DLR
Operational since	1997	1994	2003	2006	2010
Location a.s.l.	2,400 m	70 m	2,288 m	600 m	Portable
Rx aperture diameter	100 cm	100 and 150 cm	100 cm	40 cm	60 cm
Telescope and mount type	Cassegrain and Coudé	Nasmyth and Coudé	Az.-El., Coudé	Cassegrain and Coudé	Ritchey Chretien
Employed in links from	OPALE (on Artemis) LUCE (on OICETS) LCTSX (on TerraSar-X) SOTA (on SOCRATES) OPALS (on ISS) LLCD (on LADEE)	ETS-VI LUCE SOTA	LUCE LLCD OPALS	LUCE OPALS SOTA OSIRIS	SOTA OSIRIS VABENE

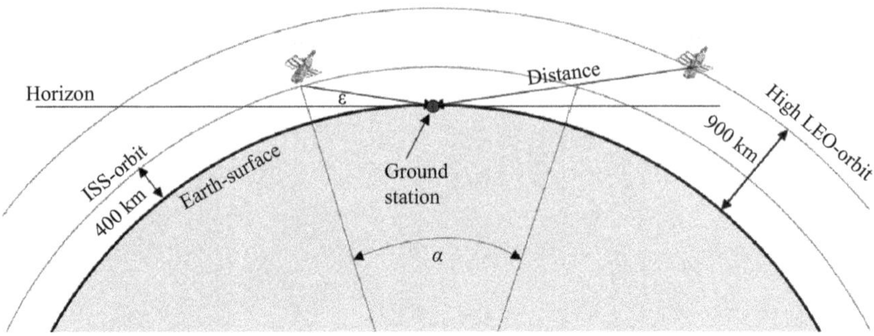

Figure 11.2 Link geometry of typical LEO satellite downlinks with circular orbits

Table 11.3 Parameters for the two satellite altitudes depicted in Figure 11.2. Absolute velocity and thus also the point-ahead angle of both orbits are nearly the same; however, their visibility time, distance, and maximum slew rate differ by ca. factor of two.

Orbit altitude (km)	Distance at 5° (km)	Max. link duration 5°→5° (s)	Angular slew rate at zenith (°/s)	Point-ahead at zenith (μrad)
400	1,804	475	1.1	51
900	2,992	831	0.48	49

A key parameter is the point-ahead angle (PAA) of the uplink versus the downlink direction, which originates from the fast orthogonal velocity of the satellite versus the ground station (the satellite moves several metres during the time of flight of the signals). Since optical signal divergence angles are small, they can be in the same order as this PAA, and as a result, the PAA offset must be taken into account for the alignment of the opto-mechanical systems.

When the LEO satellite is in the line of sight of an OGS, its viewing elevation is restricted to low elevations most of the time, as the simulation result in Figure 11.3 depicts for 500 km orbit height. When defining 5° as the minimum possible contact elevation, the satellite is seen between 5 and 20° for 64% of the total contact time. This has a major influence on the data format and link protection, since higher range loss is experienced at lower elevations and atmospheric disturbances have a greater impact.

11.1.2.1 Throughput advantage and spectrum issues

Optical link technology currently uses only one wavelength to achieve transmission rates of several Gbps; however, from terrestrial fibre communications, we see how this rate can increase into the Tbps-regime by multiple channels (dense wavelength

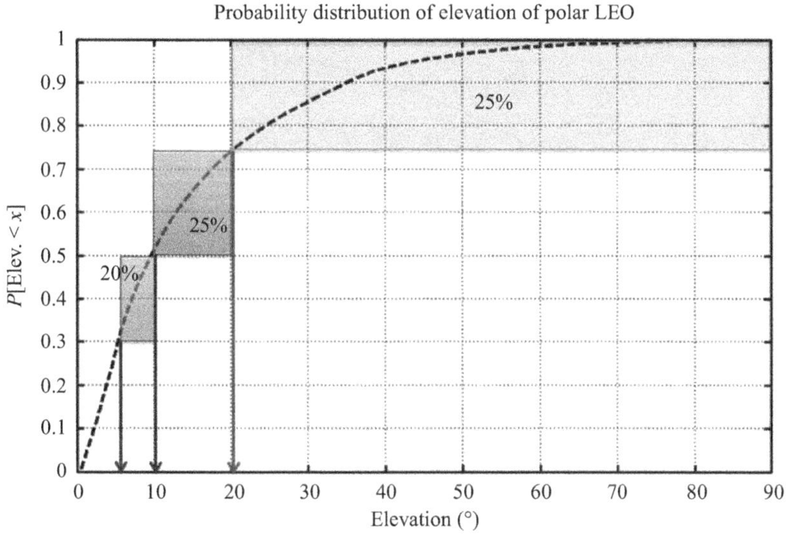

Figure 11.3 Typical distribution of the average viewing elevation for a polar LEO satellite (500 km orbit height). This relative distribution is qualitatively similar for any OGS location on earth, although of course the absolute overall visibility changes depending on orbit and OGS latitude [32]

division multiplexing) and higher order modulation formats. From available channel capacity, optical links offer several Terahertz of spectrum and thus according combined DRs, while RF links will always be strongly limited in spectrum and thus throughput. See [33] for an estimation of OLEODL system throughput taking into account realistic cloud blockage statistics.

Another motivation to move directly to optical links in LEO downlinks and avoid other higher frequency RF techniques is to avoid spectrum interference issues with future 5G mobile communications standards which are moving into the millimetre wave domain.

11.1.3 Data rates and rate change for a variable link budget

Targeted DRs in OLEODL range from a few megabits per second for very simple and low-cost satellites and terminals with limited pointing control and transmit power, to several gigabits per second for high throughput Earth observation sensor data downloads. Since the corresponding OGSs should not require adaptive optics for single mode fibre coupling in the first place, an upper channel rate limit of at least 10 Gbps is assumed – a rate at which multi-mode photo detectors can still be used. An optimised data throughput does, however, not only depend on the maximum possible DR, but also on the variation of the rate due to link constraints such as channel attenuation and

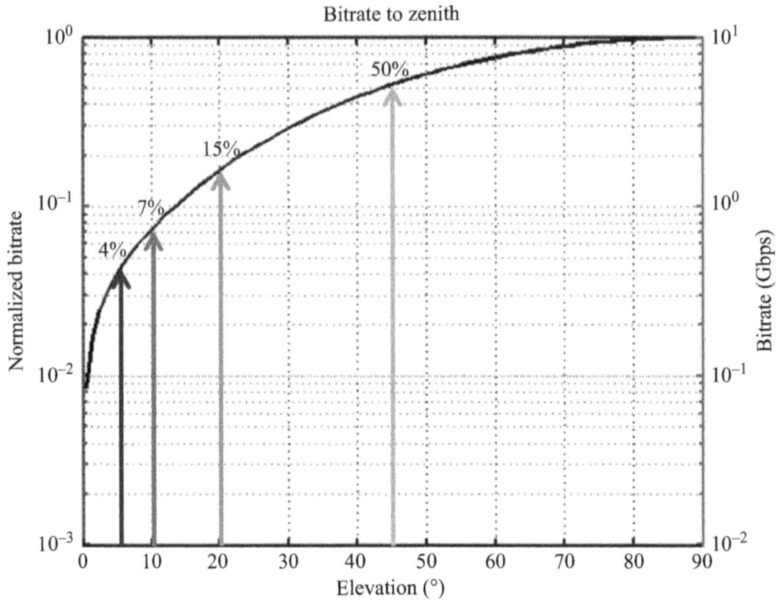

Figure 11.4 Downlink bitrate normalised to zenith, for constant energy per bit (i.e. sensitivity per bit is independent of data rate), including range and atmospheric losses

power variation as caused by the index-of-refraction turbulence (IRT) of the atmosphere. These two effects – varying link budget due to distance and atmospheric attenuation, and fast power scintillation due to atmospheric IRT – are the key challenges in OLEODL. The effect of atmospheric scintillation has been investigated in depth by various publications [34] and shall not be detailed in this text.

Directly connected to the link elevation is the maximum achievable DR. Assuming a receiver that performs with constant energy/bit, a link from 5° elevation to zenith allows a rate variation of around 25, as Figure 11.4 depicts. This plot includes the elevation-dependent atmospheric signal attenuation, but not the dynamic scintillation and fading effects caused by atmospheric turbulence, which will be explained later in this chapter. However, such an ideal receiver and the corresponding transmitter (one that can change its rate continuously) do not exist in practice. Therefore, few hard DR steps must be assumed or even just one fixed rate. The total throughput with a fixed rate would, even at best, be only one-third of the ideal maximum throughput with a continuously variable rate [35].

While the foregoing exemplification implies that the source DR equals the channel symbol rate, generally this is not the case since further mechanisms influence their relation (generally symbol rate is higher than DR), of which some are shown in Figure 11.5.

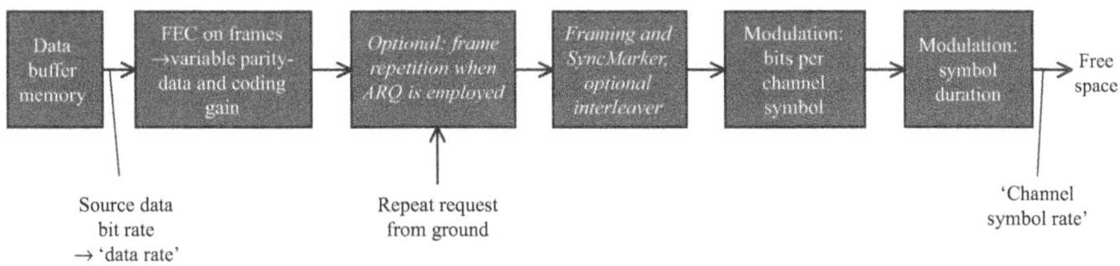

Figure 11.5 Steps in the transmitter data processing chain of the optical space terminal

FEC (forward error correction) is the standard technique used to protect data against bit errors in a simplex link, and its ratio of coding overhead versus total data payload, together with the according FEC-gain variation, allows for some rate variability. Automatic Repeat Request (ARQ) is an alternative – or additional – link protection mechanism, which, however, requires a return channel (uplink) which cannot be ensured. Other optional methods, such as burst transmission with pauses between data sections, frame repetition and inter-leaver techniques, partly proof advantageous in a fading channel. The variation of bits per channel symbol with on–off-Keying (OOK) modulation is, e.g. done with a pulse position modulation (PPM) or Amplitude-Shift-Keying modulation format, where one pulse transmits the information of more than one bit. Finally, the simplest way to vary the effective DR is to alter the length of one symbol time.

These mechanisms are used in different sophistication levels of rate variation modes, in order to maximise the overall downlink system throughput under varying link loss, while also securing a frequent access to the satellite. Note that variations in the effective source DR do not necessarily require a change in channel symbol rate.

Different modes of varying the DR in an OLEODL-system can be identified:

1. While a specific satellite terminal might only work at one DR, still an OGS may need to vary its Rx rate since it serves different types of satellite missions.
2. A constant rate during one downlink contact is chosen according to its pass geometry, e.g. to allow maximum throughput during this link.
3. Depending on the progression of the link elevation, the transmitter varies the effective DR on pre-programmed time steps, to adopt to the known elevation dependent link losses.
4. By exchanging channel state information between ground and satellite, the optimum rate is chosen dynamically, every time the link budget changes notably.

11.2 Link design

The basis for any system development is the preceding link design. In our approach, this comprises analysis of the propagation channel, definition of the transmission equation, calculation of the link budget, consideration of the pointing, acquisition

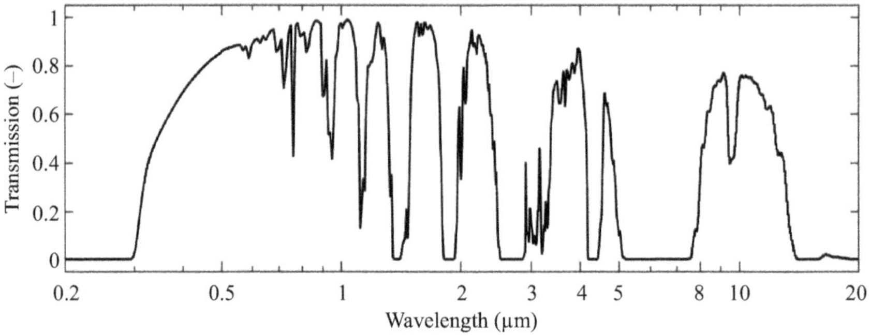

Figure 11.6 Clear sky atmospheric transmission spectrum from sea level to space in zenith direction, from 200 nm to 20 μm calculated with libRadtran using the LOWTRAN model. The transmission is the ratio I_{out}/I_{in} from (11.1) [36]

and tracking (PAT) process, modulation formats, receiver technology and impact of bit coding and higher layer coding and protocols.

11.2.1 Propagation channel model

With respect to propagation characterisation, we distinguish between two groups of effects: extinction of the atmosphere and turbulence effects. Extinction means a loss of energy of the propagating electro-magnetic wave by absorption and scattering processes. Presuming that the extinction is not dependent on the intensity of the wave, it can be described by Beer's law. It models the attenuation of a propagation path through a medium with an exponential law using the medium specific extinction coefficient $\alpha_{\text{ext}}(\lambda)$ (km^{-1}) and path length L (km), assuming a homogeneous medium and monochromatic light. Let I_{in} (W/m^2) be the input intensity to the medium and I_{out} (W/m^2) the output intensity, then

$$I_{\text{out}} = I_{\text{in}} \cdot \exp(-\alpha_{\text{ext}}(\lambda) \cdot L). \tag{11.1}$$

For the case of a non-homogeneous medium, the argument of the exponential function is defined by an integral over the path length.

The wavelength dependency of the extinction determines the atmospheric transmission spectrum. A calculation of the spectrum between 200 nm and 50 μm is given in Figure 11.6 (based on a clear sky atmosphere). The atmospheric windows are clearly visible.

While Figure 11.6 identifies the large spectral atmospheric transmission windows, when looking in detail at the situation around specific wavelengths and consider low link elevations, thin molecular absorption lines can become dominant. These lines are mostly produced by water vapour and carbon dioxide molecules and have a typical width of a few GHz, while their occurrence is roughly two lines per nm. As elucidated in Figure 11.7 (with atmospheric model *mid-latitude-summer*,

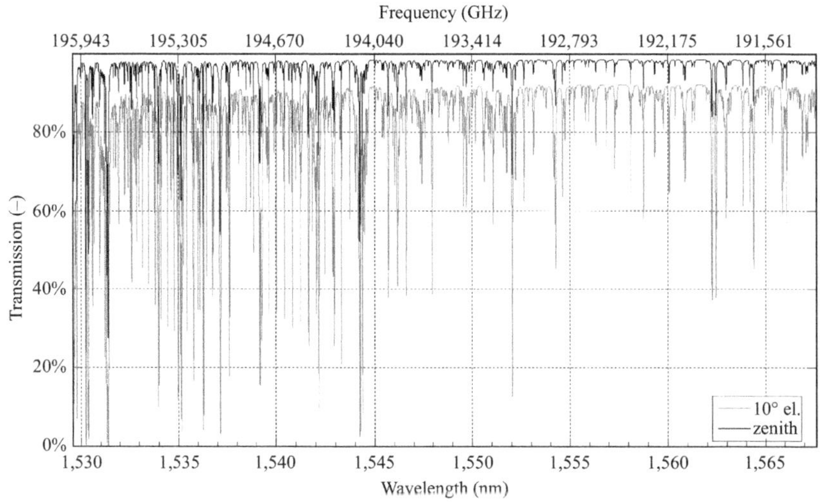

Figure 11.7 Molecular absorption lines mostly due to water vapour impact the atmospheric transmission in C-band (1,530–1,565 nm), especially at low link elevations. Simulated using the atmospheric constituent profiles and absorption coefficients derived from the HITRAN database [37]

continental-clean aerosol model and *volcanic activity two* out of four), it becomes obvious that for typical OLEODL elevations, the lower part of the commonly used C-band shows more of these absorption lines than the upper part. While the water content of the atmosphere reduces with altitude and thus ground stations on mountain tops will be less affected by these absorption effects, one must not limit the applicability of OLEODL technology to OGSs at favourable geographical locations. Rather, careful wavelength selection and stability control of up- and downlink sources can ensure reliable operation to any OGS site.

The second group of atmospheric effects relates to IRT. These effects cause phase distortions during propagation of the electro-magnetic wave from space to ground. The distorted phase front causes constructive and destructive self-interference of the wave which results in a stochastic intensity pattern of the beam changing spatially and temporally, called intensity scintillation. A variety of phase and intensity effects are created by the process of IRT which are isolated for the sake of easier modelling. The strength is governed by the strength of the turbulence, the length of the propagation path and, in the case of a slant path, the direction of propagation. Table 11.4 lists the most important effects.

These effects are usually modelled by means of a statistical description. Furthermore, different scenarios are categorised according to their fluctuation regime, which is used to select the appropriate model for the statistical description: weak, moderate and strong.

Table 11.4 Overview of effects on laser beam due to atmospheric turbulence

Effect	Type	Description
Wavefront distortions	Phase	Distortion of the spatial two dimensional wavefront
Beam tilt	Phase	Change of propagation direction as seen from the source
Angle of arrival fluctuations	Phase	Change of propagation direction as seen from the receiver
Intensity scintillation	Intensity	Spatial and temporal fluctuation of intensity
Beam broadening	Intensity	Causes increase of beam waist

11.2.2 Transmission equation

The transmission equation describes the link at the system level and is used to calculate the link budget for specific system designs. One particular form of the transmission equation which is suitable for optical satellite links and described in [38] reads

$$P_r = P_t \tau_t G_t L_{fs} G_r \tau_r \tau_{rp} \tag{11.2}$$

with P_r (W) being the received optical power, P_t (W) the average optical transmit power, τ_t [–] the optical loss in the transmitter, G_t [–] the transmit antenna gain, L_{fs} [–] the free-space loss, G_r [–] the receive antenna gain, τ_r [–] the optical loss in the receiver and τ_{rp} [–] the pointing loss of the receiver.

This equation does not contain the pointing loss of the transmitter τ_{tp} [–], the atmospheric extinction loss τ_{ext} [–], the atmospheric turbulence loss τ_{turb} [–], the loss due to background light τ_{bgl} [–] and coding gain G_c [–]. The extended transmission equation is

$$P_r = P_t \tau_t G_t \tau_{tp} L_{fs} \tau_{ext} \tau_{turb} \tau_{rp} \tau_{bgl} G_r G_c \tau_r \tag{11.3}$$

which is valid assuming independence of the individual loss and gain effects.

The peak antenna transmit gain in the case of a homogeneous intensity distribution is expressed by

$$G_t = \frac{16}{\theta_{div}^2} \tag{11.4}$$

with θ_{div} (rad) being the full divergence angle. It must be noted that the denominator in (11.4) is set to 32 in the case of a Gaussian intensity distribution since its peak is twice its mean intensity. The free-space loss is given by

$$L_{fs} = \left(\frac{\lambda}{4\pi z}\right)^2 \tag{11.5}$$

with the wavelength λ (m) and the propagation path length z (m). The receiving antenna gain is

$$G_r = \left(\frac{2\pi r_{Rx}}{\lambda}\right)^2 \tag{11.6}$$

with the radius of the receiving antenna r_{Rx} (m).

The extinction loss is defined by Beer's law [Equation (11.7)].

$$\tau_{ext} = \frac{I_{in}}{I_{out}}. \tag{11.7}$$

The loss due to background light τ_{bgl} [–] for incoherent systems can be written as

$$\tau_{bgl} = f(R_{atm}, \Delta\lambda_{bp}, r_{Rx}, \theta_{Rx}, P_{Rx}) \tag{11.8}$$

with the atmospheric radiance R_{atm} (W/m^2/nm/sr), the optical bandpass bandwidth $\Delta\lambda_{bp}$ (m) and the detector field of view θ_{Rx} (rad). The formalism is kept quite generic here since the background light loss strongly depends on the specific modulation and detection scheme. For a detailed analysis of background light loss, [39] can be consulted, for example, which contains a model for SNR degradation due to background light with a DD receiver using avalanche photo-diode (APD).

The optical losses in the transmitter and receiver depend on the material characteristics of the actual implementation, mainly on the quality of the anti-reflection and reflection coatings. Furthermore, a fraction of the energy may be split from the communication system to the PAT sensors, which is also considered an optical loss here. The losses due to miss-pointing of the transmitter and receiver are statistical losses and depend on the miss-pointing bias of the transmitter $\theta_{tp,bias}$ (rad), the miss-pointing jitter of the transmitter $\sigma_{tp,jit}$ (rad), the miss-pointing bias of the receiver $\theta_{rp,bias}$ (rad), the miss-pointing jitter of the receiver $\sigma_{rp,jit}$ (rad) and the probability p_{thr} [–] of the received signal dropping below a defined threshold F_{thr} (dB).

$$\tau_{tp} = f(\sigma_{tp,jit}, \theta_{tp,bias}, p_{thr}, F_{thr}) \tag{11.9}$$

and

$$\tau_{rp} = f(\sigma_{rp,jit}, \theta_{rp,bias}, p_{thr}, F_{thr}). \tag{11.10}$$

IRT of the air cause spatial and temporal intensity fluctuations which lead to fades and surges (scintillation) in the received power with millisecond timescale. The according dynamic signal quality loss depends on the specific transmission system and is defined similar to the pointing losses, i.e. it is a dynamic loss expressed through statistical parameters. The turbulence loss can be written in the very generic form as

$$\tau_{turb} = f(w_0, C_n^2(z), r_{Rx}, p_{thr}, F_{thr}). \tag{11.11}$$

This includes modelling of the turbulence channel with the path profile of the index of refraction constant $C_n^2(z)$ (m$^{-2/3}$) which describes the strength of the turbulence along the propagation path. In the special case of an incoherent system with OOK and DD, and assuming a turbulent channel with lognormal power fluctuation statistics, Giggenbach and Henniger [40] developed a model to assess turbulence loss for lognormal power distribution and a fixed loss threshold p_{thr}

$$\tau_{turb} = \frac{\exp\left\{\mathrm{erf}^{-1}(2p_{thr} - 1)\left[2\ln\left(\sigma_p^2 + 1\right)\right]^{1/2}\right\}}{\left(\sigma_p^2 + 1\right)^{1/2}}. \tag{11.12}$$

The power scintillation index σ_p^2 [–] covers the profile of the index of refraction structure parameter and the size of the receiver aperture.

The coding gain G_c [–] is defined according to [41] by

$$G_c = \frac{P_{\text{min,uncoded}}}{P_{\text{min,coded}}} \quad (11.13)$$

where $P_{\text{min,uncoded}}$ (W) is the necessary minimum power in the event that no coding is applied for a given target bit error rate BER_{tg} [–], and $P_{\text{min,coded}}$ (W) is the necessary minimum power in the event that a particular coding is applied. In the case of an atmospheric turbulent channel, the dependencies of the coding gain on channel parameters can be expressed with

$$G_c = f(\sigma_P^2, \tau_{\text{p,corr}}, \sigma_N^2, BER_{\text{tg}}) \quad (11.14)$$

where $\tau_{\text{p,corr}}$ (s) is the correlation time of the received power defined via the autocovariance function. Lognormal statistics of received power are once again assumed. The use of $\tau_{\text{p,corr}}$ assumes that the spectral shape of the fluctuations is known. However, since this is not necessarily the case, (11.14) may contain the power spectrum of the fluctuations $S_{\text{scint}}(f)$ instead of the correlation time. The parameter σ_N^2 denotes additional electrical noise.

11.2.3 Link budget

Based on the extended transmission equation (11.3), power budgets of the link can be calculated. It is customary to write the parameters of the link equation in dB and to present the link budget in a table. The influence of each parameter can thus easily be identified. In the following, we present example link budgets for satellite-to-ground downlinks as well as the beacon uplink.

We chose a satellite in a typical Earth observation orbit with an altitude of about 700 km. This results in a link distance of about 2,100 km at an elevation angle of 10°, which is considered the start elevation for the communication link, and about 2,500 km at an elevation angle of 5°, which is considered as start elevation for link acquisition. A wavelength of 1,550 nm is used as it is most common for OLEODL today.

Table 11.5 shows the resulting link budgets for downlink and uplink. Please note that several of the previously defined parameters are given in dB here. Satellite-to-ground links for Earth observation applications, in particular, can be designed highly asymmetric. A high throughput is only required to transmit mission data back to Earth, while a low-rate uplink is sufficient, e.g. for the exchange of channel status information. This allows for a small terminal in space – in the given example, a receiver aperture of only 25 mm is used for the satellite terminal.

Typical values of the downlink laser communication chain are assumed. The transmitter divergence angle is set to 100 μrad, receiver telescope size to 60 cm in diameter and transmit power is 1 W. The data rate is set to 10 Gbps, which results in a required Rx power of −29 dBm, assuming that the sensitivity of a state-of-the-art receiver front end (RFE) with an APD is about 1,000 Ph/bit at a BER of 1E–6, as a conservative value (better can be achieved in practice). The values for pointing loss, turbulence loss, extinction loss and background light loss are selected based on typical implementations. The coding gain is chosen as an example of a standard Reed–Solomon FEC implementation. When experiencing strong scintillation (e.g. when

Table 11.5 Example link budgets for data downlinks at 10 Gbps and beacon uplinks for tracking and tele-command, for a typical Earth observation satellite at 10° elevation, and the beacon at 5° elevation to start acquisition

Parameter	Unit	Data-downlink	Data-uplink	Beacon-uplink (5°)
P_t	dBm	30	40	40
τ_t	dB	−1.5	−1	−1
G_t	dB	92.0	78.1	78.1
τ_{tp}	dB	−3	−3	−3
L_{fs}	dB	−264.6	−264.6	−266.1
τ_{ext}	dB	−4	−4	−8
τ_{turb}	dB	−5	−5	−3
τ_{rp}	dB	−1	−1	0
τ_{bgl}	dB	−1	−1	−1
G_r	dB	127.7	100.1	100.1
G_c	dB	4	4	0
τ_r	dB	−2.5	−4.5	−4.5
$\boldsymbol{P_r}$	**dBm**	**−28.9**	**−61.9**	**−68.4**
$\boldsymbol{P_{req}}$	**dBm**	**−29**	**−69**	**−70**
Margin	**dB**	**+0.1**	**+7.1**	**+1.6**

Note: Bold values of the last three lines indicate the RESULT of the link budget calculation.

the receive aperture is small compared to scintillation pattern), standard interleaving techniques must be employed as mentioned below. Due to the high Rx-power required for 10 Gbps of data rate, the power split for the tracking sensor at the OGS is not critical and therefore not shown here.

A likewise approach has been taken with regard to the uplink direction; however, several parameters differ. For instance, a larger beam divergence is used in order to relax the requirement of OGS pointing and satellite orbit knowledge. Also, a larger Tx power can be used in uplink direction, since no strict power-efficiency limitations are apparent for the OGS. Again, typical values are chosen for the sensitivity of the data-receiver (1,000 Ph/bit for BER $= 10^{-6}$) and the tracking sensor (−70 dBm), which is a typical value to reach the required electrical SNR. Two laser beacons with 5W each are used to take advantage of transmitter diversity to reduce uplink beacon power variation. Eye-safety can be maintained at 1,550 nm even with such high powers when using moderately sized beacon collimators. It is assumed that the same laser is used for tracking (beacon-uplink) and data transmission (data-uplink) of a low-rate uplink with a rate of 1 Mbps that can be used for tele-command purposes or updates of on-board firmware.

It can be observed that in downlink direction, the link margin is small at the given DR of 10 Gbps. As the link shall be operated also at low elevation angles to maximise data throughput of any given mission, it becomes clear that scenarios with a high-link dynamic can benefit substantially from variable data rate techniques, since these allow a rate reduction at low elevation angles and thus a maximisation of the system throughput and link availability.

11.2.4 Pointing, acquisition and tracking

The process of PAT addresses the opto-mechanical system of a laser communication terminal. It is of high importance for any aerospace laser link to obtain line of sight. The first step, **pointing**, relates to the transmit beam steering towards the counter terminal based on a priori information of the position of the partner. In the case of a satellite link, this would be orbit data of the satellite and GPS location data of the ground station, for instance. Depending on the accuracy of the a priori data and the accuracy of the opto-mechanical system (gimbal accuracy, jitter, reference calibration [41]), an angular uncertainty area can be defined where the partner is expected to show up. If this uncertainty area exceeds the transmit beam cone, scanning algorithms must be applied. In the next step, **acquisition**, the beam is detected by the counter terminal using an acquisition sensor, and a control mechanism is activated that steers the beam into the tracking sensor's field of view. Finally, the **tracking** starts. The beam displacement measurement by the tracking sensor continuously creates an error signal used by the control loop to maintain the link lock.

The PAT process often uses a two-stage opto-mechanical system. A course pointing assembly (CPA) defines the field of regard of the satellite or ground terminal and corrects for low frequency, high amplitude bias and jitter. The opto-mechanical implementation is often a two-axes motorised lens/mirror system in combination with a static optical bench similar to a Coudé-path. Alternatively, turret systems that carry the entire electro-optical system are also an option. The precision of the CPA must be high enough to steer the beam into the field of regard of the fine pointing assembly (FPA). This subsystem corrects for high frequency, low amplitude bias and jitter. The sensor is often a four-quadrant diode, and the actuator a voice coil or piezo-driven mirror. For operation during day and night time, it is recommended to use modulated beacon lasers which enable the space segment to discriminate between the beacon laser and background light or earth albedo.

A block diagram for a ground segment that also shows the implemented PAT subsystem is shown in chapter 11.3. The PAT process is illustrated in Figure 11.8 for an exemplary LEO downlink system. The process comprises five steps. In step 1, the ground terminal illuminates the satellite with a high divergence beacon laser. The satellite acquires the signal and corrects its attitude in step 2. In step 3, the satellite points the transmit communications beam to the ground station. In step 4, the ground station acquires the satellite signal using it as a tracking beacon and corrects its pointing direction accordingly, thus both partners obtain line of sight. In step 5, communication is performed and line of sight is maintained via optical tracking.

11.2.5 Direct detection modulation formats and rate variation

Modulation formats considered for OLEODL are mostly based on OOK of the laser signal to encode the bit stream. Detection of such modulation is not hindered by atmospheric wave-front distortions and basically only requires power-in-the-bucket receiver technology which is offered by bulk APD receivers (Avalanche Photo Diodes). Still, if required for higher sensitivity or DRs, more sophisticated techniques, such as pre-amplification in conjunction with fibre coupling and adaptive optics, can be used.

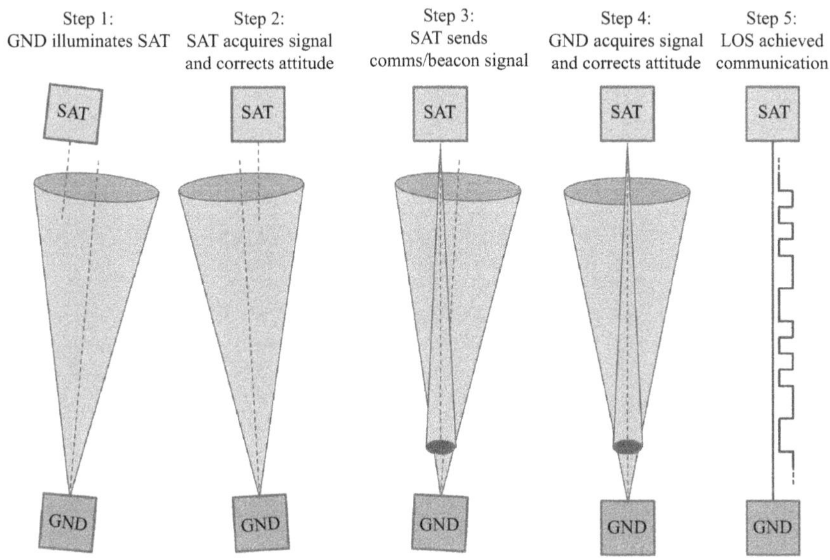

Figure 11.8 PAT process for LEO downlink: the cones qualitatively denote the laser beam divergence; the dashed line represents the optical axis of satellite and ground station

The overall process is therefore also called intensity modulation with DD (IM/DD). The phase of the optical signal does not contain any information, and thus deterioration of the phase does not degrade transmission sensitivity. However, sensitivities similar to coherent phase modulation can be achieved by IM/DD if the appropriate detection technology is used (theoretically 20 photons per on-bit for BER $= 10^{-9}$ when assuming Poisson noise statistics for photon arrival, versus 9 photons per any bit for coherent homodyne BPSK). Such OOK sensitivities could be achieved using the promising technology of single photon detection with superconducting nanowire detectors [42], while today's lower cost APDs reach sensitivities of a few hundred photons per bit and below. Different symbol-encoding schemes can also be applied with OOK, as described in the following, where we outline the most common waveforms.

OOK modulation can be considered the simplest modulation technique in which the intensity of an optical source is directly modulated by the information bit sequence. A bit '1' is represented by an optical pulse while a bit '0' is represented by the absence of an optical pulse. If the pulse occupies the whole bit duration, it is called Non-Return to-Zero (NRZ) OOK, and if the pulse occupies part of the bit duration depending on the duty cycle of the signal, it is called Return-to-Zero (RZ) modulation.

PPM is an orthogonal OOK modulation technique where information is encoded in the time slot when a pulse is transmitted [43]. It is more power efficient in comparison to NRZ and RZ but requires higher bandwidth, and additional complexity requirements must be met during synchronisation and post-processing. In M-ary

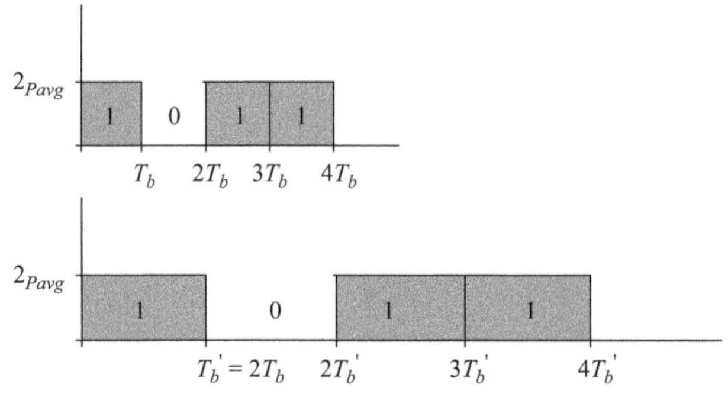

Figure 11.9 Data rate variation by increasing the pulse duration with NRZ-OOK. Top: high data rate (DR), bottom: half of DR

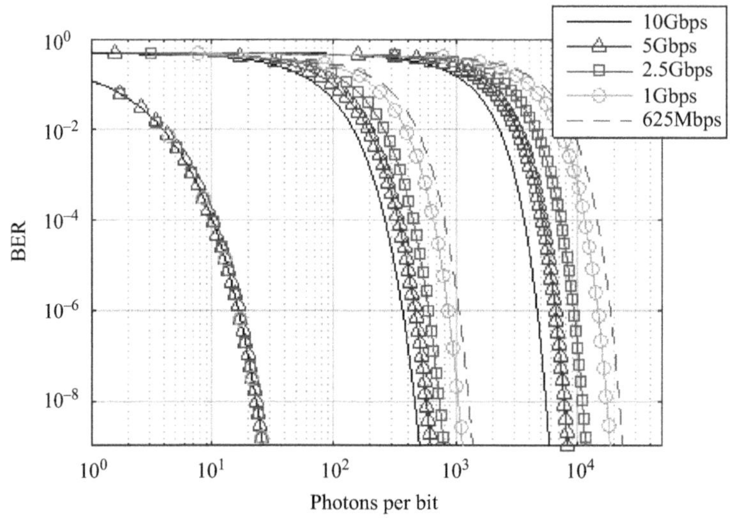

Figure 11.10 BER versus photons per user bit for different data rates and different receiver models (no FEC). Left: SNL, middle: APD, right: PIN

PPM, $M = 2^n$, where n is the number of bits in one symbol. The position of the pulse slot inside its symbol time (unless specified differently) corresponds to the decimal value of the n-bit input data. The symbol duration T_s is divided into L number of slots, each of duration T_b.

Options and effectiveness of data rate variation with different OOK modulation schemes: As explained above, the high channel variability in OLEODL (distance, attenuation and fading) requires variation of the system DR. With the NRZ modulation format, the DR can be lowered by simply increasing the pulse width. Figure 11.9 shows the signal waveform for transmitting NRZ-OOK signals at a high DR (top) and at half that rate (bottom) by doubling the pulse width. Figure 11.10 indicates the

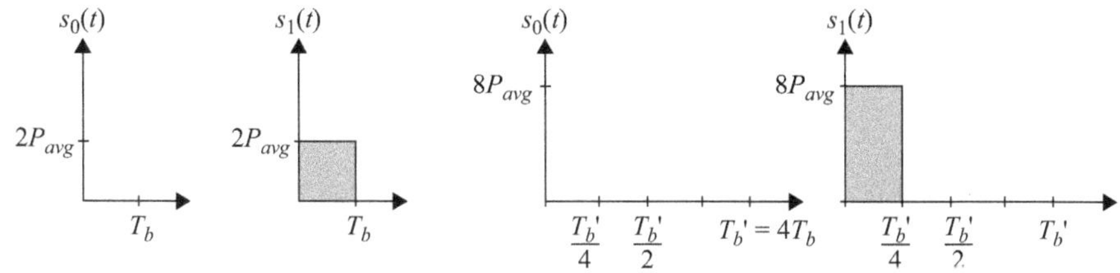

Figure 11.11 Data rate variation by reducing the duty cycle of RZ-OOK. Left: NRZ-OOK at high data rate (DR), right: RZ-OOK with 25% duty cycle and lower data rate (=DR/4)

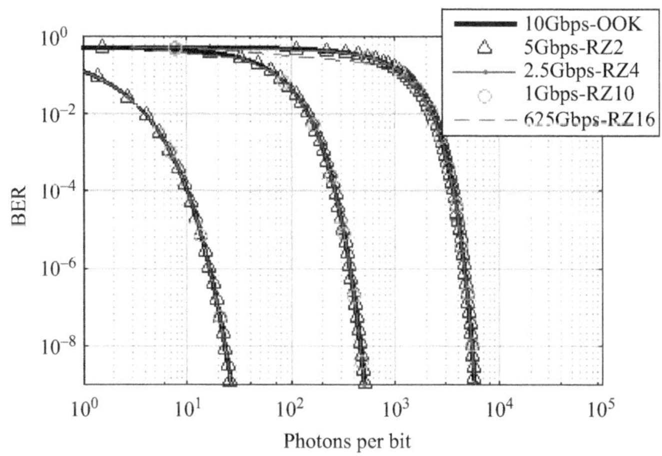

Figure 11.12 BER versus photons per user bit for different data rates and different receiver models. Left: SNL, middle: APD, right: PIN

performance of the system for different DRs for shot-noise-limited (SNL), practical APD and thermal limited positive-intrinsic-negative (PIN) receiver models (see next section for explanations of receiver sensitivity). With an ideal SNL receiver, the system sensitivity in terms of the number of photons per bit (thus energy per bit) required to achieve a certain BER remains constant for different DRs, whereas for APD and PIN, it degrades for higher DRs. For this rate variation scheme, the reception filter low pass in the RFE must be adapted according to channel rate.

With RZ-OOK, the variable pulse duty cycle enables an elegant way to keep the pulse width fixed (and thus also the RFE's reception filter), while the bit length is increased as shown in Figure 11.11. Figure on the left represents bit '0' and '1' at a high DR using NRZ-OOK modulation, while the right one represent bits '0' and '1' at a lower DR (=DR/4) using RZ-OOK modulation with 25% duty cycle. This method introduces longer pauses between the pulses, increasing the pulse amplitude accordingly in a transmitter with constant average power. As a result, system sensitivity in photons per bit for the different DRs stays constant for all types of receivers (SNL, APD, PIN) as seen in Figure 11.12 [44].

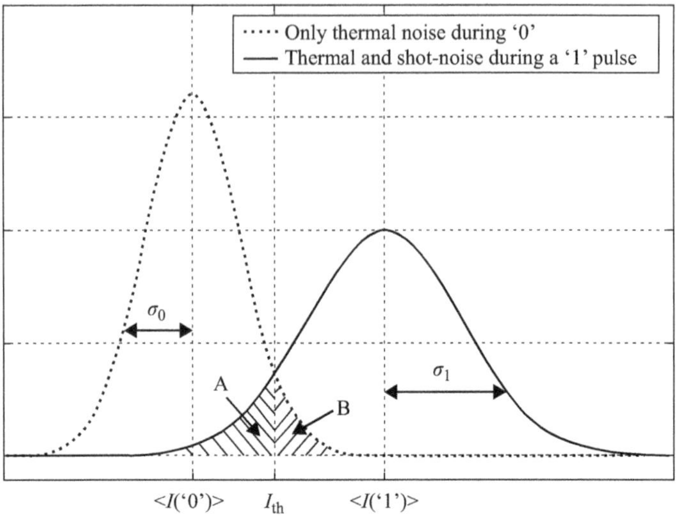

Figure 11.13 Probability distribution of received OOK signal with signal-dependent noise, when ‘0’ (left) and ‘1’ (right) is transmitted in presence of shot and thermal noise. Adopted from Reference [46]

Similarly, PPM also inherently lowers the DR with increasing order, if the pulse length is kept constant; therefore, variable PPM order can be used as a rate variation mechanism. However, the synchronisation effort increases, while the variability is limited due to the logarithmic relation between order and effective DR.

11.2.6 OOK RFE performance and impact on link budget

In OOK receivers, the receiver telescope collects the optical signal, filters the undesired background light and focusses onto the photodetector surface to convert it to an electrical signal current. This signal then has to be detected as pulse or no-pulse by a decision logic at the proper photocurrent threshold (I_{th}), which is derived, e.g. in [45]. If the detected signal is above the threshold, bit ‘1’ is detected, otherwise bit ‘0’ is detected. In addition to the modulated signal, shot noise (possibly signal-dependent) and thermal noise widen its level distribution, which may lead to false detection of the pulse or missed detection. Figure 11.13 shows the Gaussian probability distribution of the signal in addition to noise, and σ_0 and σ_1 are the noise variances, respectively. Areas A and B then indicate the probability of wrong decision leading to bit errors.

Considering all possibilities of errors explained above and assuming each symbol is equally likely, the bit error probability for NRZ-OOK is calculated as

$$BEP = \frac{1}{2} \cdot erfc\left(\frac{\sqrt{SNR}}{\sqrt{2}}\right); \quad \text{where } SNR = \left(\frac{\langle I(‘1’)\rangle}{\sigma_0 + \sigma_1}\right)^2 \tag{11.15}$$

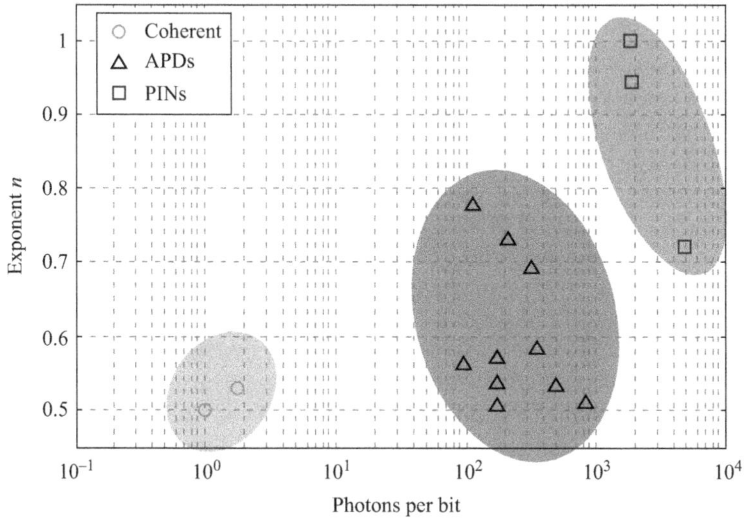

Figure 11.14 Performance ranges of different receiver implementations, derived from measured examples with COTS components. Abscissa indicates required photons per bit for $Q = 2$ (BER = 0.023), and ordinate shows the exponent n of the sensitivity run (measured RFE performances according to [47])

While theoretical derivation of the BER calculation from noise distributions is well understood, practical RFE performance depends on various parameters that often cannot be anticipated, especially in APD receiver realisations which are influenced both by thermal and shot noise. Instead, practically measured RFE performance in terms of BER(P_{Rx}) should be used to model system performance. One method is to use an absolute reference sensitivity (here the received power $P_{Q=2}$ for BER = 2.3% or quality factor $Q = 2$) and an exponent n defining the shape of the sensitivity slope [47].

$$BER = \frac{1}{2} \cdot erfc\left(\frac{Q}{\sqrt{2}}\right) = \frac{1}{2} \cdot erfc\left(\frac{f(\overline{P}_{\text{Rx}})}{\sqrt{2}}\right); \quad Q = \sqrt{SNR_{el}} \tag{11.16}$$

$$Q(\overline{P}_{\text{Rx}}) = 2\left(\frac{\overline{P}_{\text{Rx}}}{\overline{P}_{Q=2}}\right)^{n} \tag{11.17}$$

With this method, various RFE performances can be described sufficiently with their absolute sensitivity in photons per bit for $Q = 2$ and their sensitivity run. Measured examples are given in Figure 11.14. Here, the *coherent* SNL example is a BPSK homodyne receiver, while *APD* (Avalanche Photo Diode) and *PIN* are InGaAs-semiconductor DD receivers.

Channel sensitivities of 100 photons/bit can be achieved with APD-receivers, when the high BER of 2.3% is compensated with according FEC coding.

11.2.7 Error control techniques for Gaussian channels

The transmission of data bits from source to sink is always subject to noise, resulting in a certain probability of erroneous bit detection. To reduce this BER, either the SNR has to be increased (which reduces system efficiency) or techniques to reduce the final BER must be introduced (so-called error control algorithms). This can be ARQ, where bit errors in the received data packets are detected and corresponding repetition of these packets is requested. However, this technique does not apply to simplex links and does not work well for strongly delayed links, as can be the case with satellite downlinks. Alternatively, FEC techniques can be applied. Here, the source data and additional parity data – which are produced from the source data – are transmitted. This additional data allows correction of bit errors experienced during transmission over the noisy channel, and accordingly, the system sensitivity can be increased. Performance parameters of FEC techniques are on the one hand the required overhead (parity-bits) for the code, and on the other, the capability of the code to correct a certain number of erroneous bits in a transmission channel at a certain mean received power.

FEC has been the scope of intense scientific investigation and is indispensable in today's telecommunication field, see basic publications [48–50].

For space links, FEC is considered in-depth in the standardisation documents of CCSDS, e.g. [51]. Figure 11.15 shows a comparison of the basic forward correction codes applied in classical space links. The parameter E_b/N_0 denotes the ratio of received energy per source data bit to the noise power spectral density. This metric allows – amongst others – the comparison of the sensitivity (and thus efficiency) of different modulation formats and coding schemes.

Simple performance relations typically refer to a so-called Gaussian-noise channel, i.e. noise processes follow Gaussian statistics, and single error events are short (fast-fading channel). The picture changes when, e.g. the noise is no longer symmetric around its mean (as in single photon reception channels), or when the error rate changes with slow fading of the received signal. The later requires techniques that span the influence of a codeword over a longer time fraction, as described in the following section.

11.2.8 Interleaving in the atmospheric fading channel

Besides varying the channel symbol rate or symbol modulation order, other variation techniques are based on working directly on the data packets. Such techniques often combine the effective channel rate variation with variable error-control strength (FEC) [52]. Standard coding techniques improve sensitivity in a Gaussian channel but do not specifically compensate the long erasures caused by fading. Spreading the coded data through interleaving over timespans much larger than one fading event therefore helps in achieving an ergodic situation for subsequent FEC. This is of major concern when using small receive antennas which experience high scintillation dynamic, but it gets less important with larger antennas. Interleavers however lead to memory overhead and require additional processing which might be challenging in high DR transmission. Matrix and convolution interleaver are classical interleaver types.

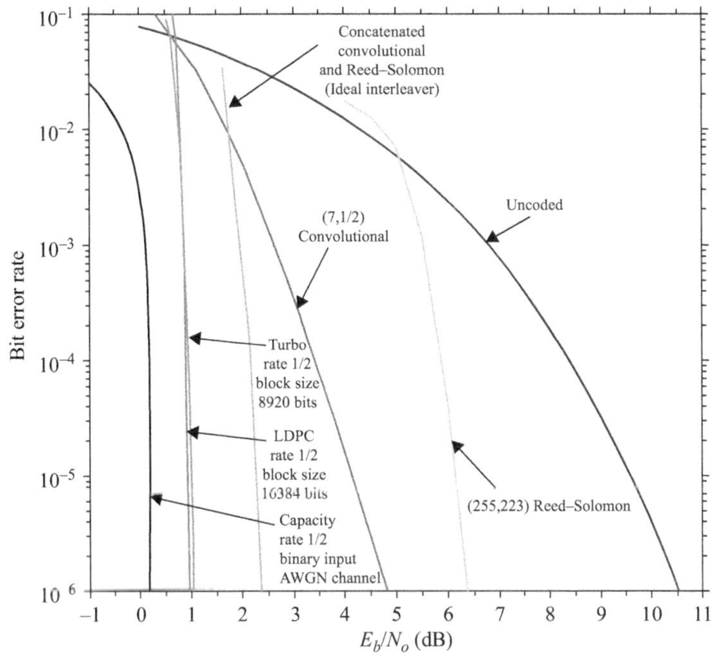

Figure 11.15 Comparison of various FEC coding schemes at different code rates, for a Gaussian (i.e. non-fading) channel and binary phase shift keying modulation. Reprint, with permission, from Figure 3-5 of [51]

In matrix interleaver, the input data is written in rows of a memory configured as a matrix, and then read out column-wise. In a convolutional interleaver, the data is multiplexed into and out of a fixed number of shift registers [53]. Such interleavers can be implemented at bit or codeword level. The basic idea of codeword interleaving is to resequence parts of a long codeword (instead of bits) before transmitting [54]. For optical fading channels, codeword interleavers might be more applicable than extremely large bitwise matrix interleavers (remember that with typical OLEODL DRs with fades in the order of several milliseconds, the memory requirement is in the order of Gigabit). Codeword interleaving can be done in different ways: A single codeword can be simply repeated after a delay longer than the channel correlation time [55]. More efficiently, a codeword may be split into several blocks, each affected from different fading states (so-called block-interleaving). This allows for further sophistication, e.g. sending the data- and parity sections of a systematic FEC-codeword separately [56], or applying a second level of FEC for the blocks. To summarise, the combination of FEC and interleaving works on both aspects – the fading compensation and effective DR variation – and they must be balanced for a specific channel situation with a particular scintillation strength and mean power.

11.3 Hardware

11.3.1 Space hardware

A key component of an optical communication system for LEO satellites in a direct-to-earth application is the satellite payload. The payload on board the satellite has to provide a laser signal, modulate it with the transmission data and keep the tracking of the ground station based on the received beacon laser while the whole payload needs to withstand the environmental influences during launch and in orbit.

Different system designs of the above-mentioned characteristics can be realised. These implementations largely depend on which of the following is selected:

- use of a beacon from the OGS
- active pointing assembly or body pointing of the satellite
- mono-static or bi-static system design.

The most simple and robust system design for an optical communication system on board a satellite is a pure laser source with transmission optics. For this design, the body pointing of the satellite is used together with a rather large divergence of the transmission system so that neither a beacon from the OGS nor a pointing device is required. This simple system design comes with the disadvantage of an inefficient link budget.

Adding a beacon laser on the ground station allows us to increase the efficiency of the system by reducing the transmitter divergence due to the improved tracking of either the body pointing of the satellite or the active pointing device. The use of a beacon from OGS also requires a receive path in the satellite payload in addition to the transmit path.

The tracking signal received from a tracking sensor, which receives the beacon signal from the OGS, can be used either for an improved body pointing of the satellite or for an active pointing stage. Using the body pointing of the satellite reduces the complexity of the optical communication terminal in the satellite but in turn increases the complexity of the attitude control of the satellite. If the attitude control accuracy is sensor-limited, the use of a beacon laser and tracking sensor can improve the attitude control accuracy via a sensor fusion of the satellite attitude sensor with the beacon detector. If the attitude control accuracy is limited by the actuators of the satellite, an active pointing device should be considered in the optical communication terminal. An active pointing device can either be a FPA, which delivers high precision and high speed but only in a small angular range, or a CPA, which covers a large angular range but offers less accuracy and speed – or a combination of both FPA and CPA.

Having both, a receive path for the beacon from OGS as well as a transmit path for the modulated data signal, means that either a mono-static or a bi-static system design needs to be implemented. A bi-static system design is characterised by two different apertures (one for the receive path and one for the transmit path) as shown in Figure 11.16, whereas only one aperture is used for both the receive and the transmit paths in a mono-static system design (compare Figure 11.17).

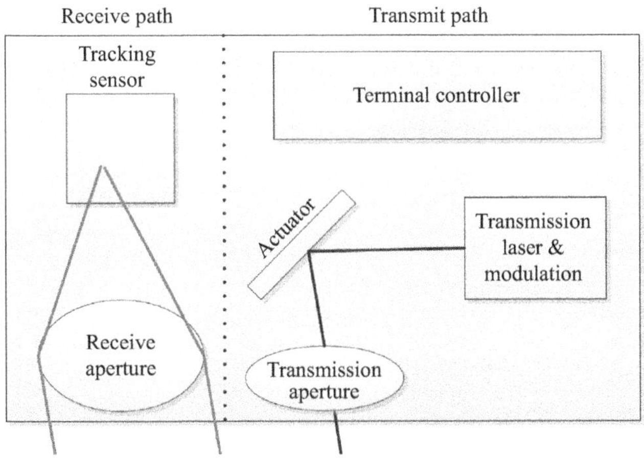

Figure 11.16 Bi-static system design with separate apertures for receive and transmit path

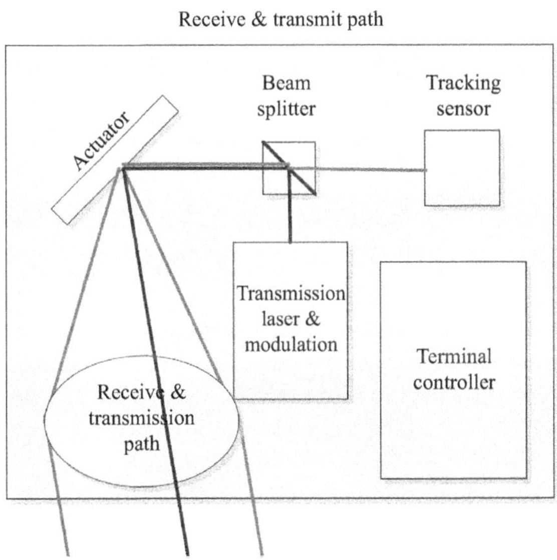

Figure 11.17 Mono-static system design with a combined aperture for receive and transmit path

Both system designs come with advantages and disadvantages. Table 11.6 summarises the advantages and disadvantages of both system designs.

All system designs share the requirement to withstand the environmental influences experienced during the launch of the system as well as during operation in orbit.

Table 11.6 Comparison of advantages and disadvantages of mono-static and bi-static system designs

	Advantages	Disadvantages
Bi-static design	• Simple and robust system design • No separation between transmit and receive path required	• Misalignment might occur due to different apertures • More space required for two separate apertures
Mono-static design	• Highly compact system design • Compensation of misalignment due to same aperture for Rx and Tx	• Separation between Rx and Tx required • More complex system design

While mechanical stress is the primary concern during the launch of the system (due to vibration loads of the launch vehicle), areas of concern during operation in orbit include thermal cycles as well as radiation effects. All of these effects influence the system design in one way or another. While vibration loads mainly have an influence on the mechanical structure and the optical design, radiation affects all electrical and optical components of the terminal. These effects may also lead to a degradation of the component performance on the electrical or optical level as well as a potential complete failure in the case of undetected latch-ups.

For both system designs – but especially for mono-static designs – the wavelengths selection for the receive and transmit path are essential. In the mono-static design, a beam splitter is used to separate the paths. The stray light and back reflections of the transmit path from the optical system need to be suppressed on the tracking sensor with a chromatic beam splitter in combination with filters to avoid self-blinding. The performance of these filters depends on the wavelength gap between receive and transmit signals. In addition to the separation of the receive and transmit path, the presence of absorption lines in the atmospheric spectrum plays a major role in the wavelength selection (compare Figure 11.7). Figure 11.18 shows exemplary band plans with different options for uplink beacon and downlink. For the selection of wavelengths, defining a spectral range free from absorption lines is a major driver. The absorption lines (resulting from water vapour and other molecules in the atmosphere) occur throughout the entire optical C- and L-bands and influence the transmittance of certain wavelengths, resulting in an attenuation that increases with lowering elevation. The band plan shows a favourable downlink wavelength range from 1,545 to 1,565 nm for multiple downlink channels to be selected within this window. Based on the wavelength gap between receive and transmit paths, which is ideally not less than 20 nm due to manufacturing complexity of the wavelength separation components, three options for beacon wavelengths are found: 1,064, 1,530 and 1,590 nm. Option 1

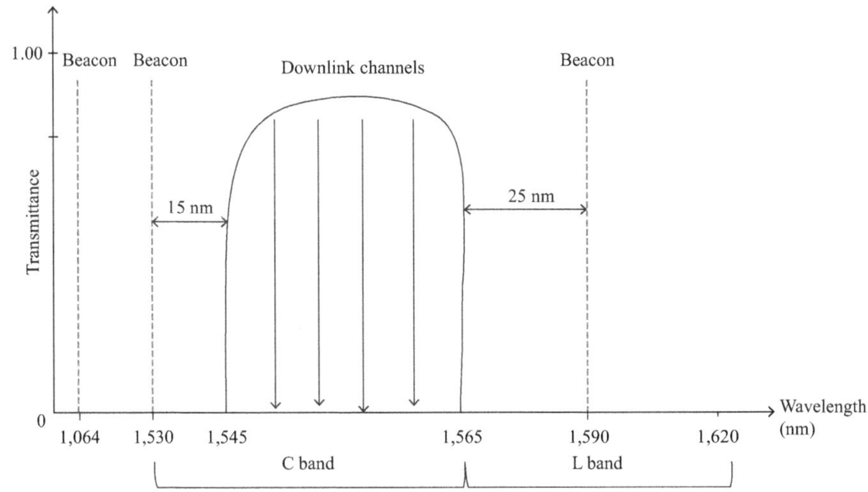

Figure 11.18 Band plan examples for beacon as well as transmission wavelengths

with 1,064 nm comes with the advantage of a large wavelength gap between receive and transmit paths together with a good availability of components but raises challenges regarding laser safety. Option 2 at 1,530 nm is the lowest wavelength within the optical C-band and allows to have both downlink and beacon in the same optical band, and the availability of components is good due to the use in fibre communication. However, this option comes with the disadvantage of a limited wavelength gap or reduced bandwidth for the downlink channels if the wavelength gap is increased, together with a high presence of absorption lines in this area and accordingly higher demand on beacon wavelength control. Option 3 at 1,590 nm (lower end of the optical L-band) allows for a wavelength gap of more than 25 nm while allowing to use the full downlink window in combination with lower presence of absorption lines in this area. Based on the requirements and characteristics of the scenario, an optimised beacon and downlink wavelength combination can be selected.

11.3.2 Ground hardware

LEO downlinks need an OGS as a receiver terminal. Setups with Cassegrain, Ritchey–Chretien and similar telescope configurations are common. Here, the data and tracking receivers would be installed in the Cassegrain focus or a conjugated plane. More experimental stations may deploy a Coudé focus. Then, more complex and experimental receivers and sensors can be set up on the Coudé focus optical bench. Most currently used ground telescopes have primary mirror diameters of about 20 cm–1.5 m, depending on the actual link distance and transmit antenna gain. A diameter of 40–60 cm is usually sufficient for receivers in LEO ground communications. The application of a wavefront correction system may also be needed if fibre coupling

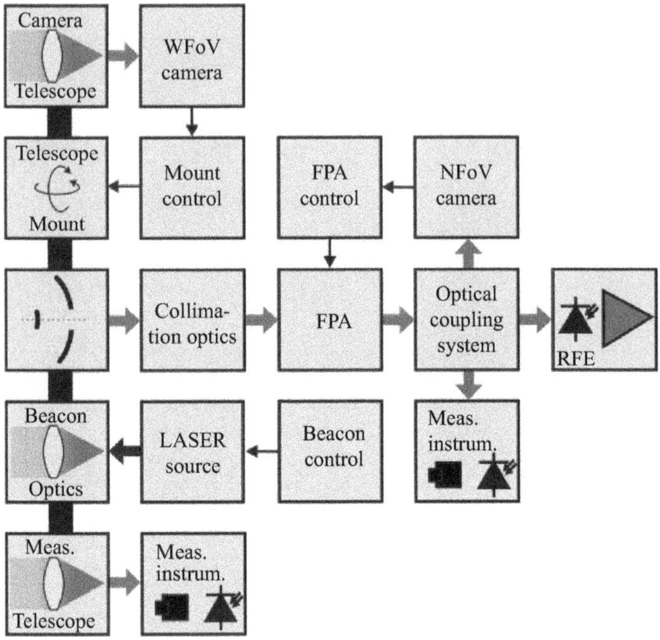

Figure 11.19 Basic block diagram of the OGS-OP optical system. The black bar indicates the mechanical connection of the functional blocks [57]

is necessary. The corresponding adaptive optics system is then often set up on a Coudé-bench.

To date, systems are mainly built for experiments and demonstrations, a list of known ground station installations is given in Table 11.2. Figure 11.19 shows exemplary the basic block diagram for DLR's Optical Ground Station Oberpfaffenhofen (OGS-OP). The black bar shall indicate that these elements are mechanically joint. The control software is steering the telescope mount to point towards the satellite. A wide field of view camera is installed to provide coarse optical tracking. A beacon laser telescope is co-aligned for the PAT process. Optional measurement telescopes are installed alongside for channel measurements. Behind the telescope, collimation optics together with the telescope form an afocal system. The FPA stabilises the beam to ensure that residual tracking errors are kept to a minimum. An optical coupling system (free-space) distributes beams to the near field of view camera for fine tracking, to the measurement instruments and to the data RFE.

The most important system aspects of a ground station are the antenna gain, the accompanying aperture averaging of scintillation, the tracking accuracy and beacon systems. The antenna gain is governed by the size of the primary mirror as described by (11.6). An increase in size not only increases the antenna gain but also reduces fading events due to the lower power scintillation seen on the data receiver. The aperture averaging factor *AAF*, i.e. the relation between scintillation with finite aperture size $\sigma_I^2(D)$ to infinitesimal small aperture size $\sigma_I^2(0)$, is a measure of how effectively a finite aperture can suppress fading events.

Another effect of aperture averaging is the transformation of intensity statistics from gamma–gamma or exponential distributions to lognormal distributions, i.e. from distributions in strong fluctuations to distributions in weak fluctuation conditions. See [58] for details on the transformation from intensity statistics to received-power statistics through aperture averaging under varying link elevation.

Tracking systems in the OGSs can be designed with almost arbitrarily high complexity. The minimum tracking capability requirement is to steer the whole telescope towards the satellite and keep line of sight. If this can be achieved with sufficient accuracy, no second stage tracking system, such as a fine tracking system for beam stabilisation or fibre coupling needs to be used. An example for a system with a one stage tracking system is DLR's transportable optical ground station (TOGS). This station achieved a residual peak tracking error in the demanding aircraft ground scenario of well below 100 μrad and is therefore precise enough to keep the signal spot on the RFE with field of view of 170 μrad. DLR's TOGS is also equipped with an uplink beacon laser system according to the band plan in Figure 11.18. Besides the wavelength of the beacon system, the optical output power, modulation frequency as well as divergence angle are to be considered in the system design.

11.4 Summary and outlook

Within this chapter, high-speed optical satellite data downlinks have been reviewed and the key characteristics of this application scenario have been described. The excellent properties of optical links, especially the high data rate, license free operation and favourable SWaP (size, weight and power) provide a game-changing technological alternative to RF-links for Earth observation satellite operators. Despite some drawbacks of the technology, industries and research organisations around the world are now developing optical communication systems that are suitable for downlink applications, demonstrating the potential of the technology and underlining its future importance for various applications.

Due to the fact that optical space-to-ground links suffer from limited availability due to clouds, OGS networks enable OGS diversity to ensure a reliable operation scenario. The availability of space-to-ground links is subject of current research [33,59]. It has been shown that, with a suitable OGS network design, the issue of limited link availability vanishes when a suitable buffer memory size is employed on the satellite to bridge weather-induced unavailability of OGSs. OGS networks are a key requirement for the future use of optical satellite downlinks and need to be established and operated.

A field which is gaining increasing attention is the installation of so-called LEO-Mega-Constellations for low-delay global communications, with orbital altitudes in the order of 1,000 km. These systems are not meant to serve for the transmission of remote sensing or Earth observation data to the ground. Rather, they are designed to enable Internet access in areas with limited terrestrial capabilities, as e.g. in developing countries. Several particularly rural regions in Europe also could benefit from Internet access through satellites. To avoid interference with terrestrial RF-communications and enable high DRs, the internetworking of these constellations will be favourably done with symmetric optical data links. Their link distances are similar to those of OLEODL, and accordingly terminal hardware will work in a likewise way. Thus, developments of optical LEO communications may see two use cases, allowing even compatible link technology between these transmission scenarios.

References

[1] T. Tolker-Nielsen, J-C. Guillen, "SILEX the first European optical communication terminal in orbit", ESA Bulletin 96, Nov. 1998.

[2] H. Zech, F. Heine, D. Tröndle, *et al.*, "LCT for EDRS: LEO to GEO optical communications at 1,8 Gbps between Alphasat and Sentinel 1a", Proc. SPIE 9647, 2015.

[3] Y. Chishiki, S. Yamakawa, Y. Takano, Y. Miyamoto, T. Araki, H. Kohata, "Overview of optical data relay system in JAXA", Proc. SPIE 9739, 2016.

[4] E. Luzhanskiy, B. Edwards, D. Israel, *et al.*, "Overview and status of the laser communication relay demonstration", Proc. SPIE 9739, 2016.

[5] D.M. Cornwell, "NASA's optical communications program for 2017 and beyond", IEEE International Conference on Space Optical Systems and Applications (ICSOS) 2017, 2017.

[6] A. Biswas, J.M. Kovalik, M. Srinivasan, *et al.*, "Deep space laser communications", Proc. SPIE 9739, 2016.

[7] ESA, RUAG, DLR, *et al.*, "DOCOMAS – Deep Space Optical Communications Architecture Study, Executive Summary", 2016, downloaded 25.Sep. 2017.

[8] D. Giggenbach, E. Lutz, J. Poliak, R. Mata-Calvo, C. Fuchs, "A high-throughput satellite system for serving whole Europe with fast internet service, employing optical feeder links", ITG-Conference "Breitbandversorgung in Deutschland", Berlin, Apr. 20–21, 2015.

[9] T. Jono, Y. Takayama, N. Kura, *et al.*, "OICETS on-orbit laser communication experiments", Proc. SPIE, 6105, 2006.

[10] T. Jono, Y. Takayama, N. Perlot, *et al.*, "Report on DLR-JAXA Joint Experiment: The Kirari Optical Downlink to Oberpfaffenhofen (KIODO)", JAXA and DLR, ISSN 1349-1121, 2007.

[11] D. Giggenbach, F. Moll, N. Perlot, "Optical communication experiments at DLR", NICT Journal Special Issue on the Optical Inter-orbit Communications Engineering Test Satellite (OICETS), vol. 59, pp. 125–134, 2012.

[12] A. Carrasco-Casado, H. Takenaka, D. Kolev, *et al.*, "LEO-to-ground optical communications using SOTA (Small Optical TrAnsponder) – Payload verification results and experiments on space quantum communications", Acta Astronautica, vol. 139, pp. 377–384, 2017.

[13] A. Biswas, B, Oaida, K. Andrews, *et al.*, "Optical Payload for Lasercomm Science (OPALS) link validation during operations from the ISS", SPIE Proceedings 9354, 2015.

[14] F. Moll, D. Kolev, M. Abrahamson, C. Schmidt, R. Mata Calvo, C. Fuchs, "LEO-ground scintillation measurements with the Optical Ground Station Oberpfaffenhofen and SOTA/OPALS space terminals", Proceedings of SPIE 9991 (Advanced Free-Space Optical Communication Techniques and Applications II), 2016, 999102-1–999102-8.

[15] C. Schmidt, M. Brechtelsbauer, F. Rein, C. Fuchs, "OSIRIS payload for DLR's BiROS satellite", International Conference on Space Optical Systems and Applications – ICSOS, Kobe, Japan, 2014.

[16] D. Giggenbach, A. Shrestha, C. Fuchs, C. Schmidt, F. Moll, "System aspects of optical LEO-to-ground links", International Conference on Space Optics, Biarritz, France, Oct. 2016.

[17] M. Toyoshima, K. Takizawa, T. Kuri, *et al.*, "Ground-to-OICETS laser communication experiments", Proc. of SPIE, 6304B, 2006, 1–8.

[18] N. Perlot, M. Knapek, D. Giggenbach, *et al.*, "Results of the optical downlink experiment KIODO from OICETS satellite to optical ground station Oberpfaffenhofen (OGS-OP)", Proc. of SPIE 6457, 2007.

[19] M.R. Garcia-Talavera, Z. Sodnik, P. Lopez, A. Alonso, T. Viera, G. Oppenhauser, "Preliminary results of the in-orbit test of ARTEMIS with the optical ground station", Proc. SPIE 4635, 2002.

[20] H. Takenaka, Y. Koyama, D. Kolev, *et al.*, "In-orbit verification of small optical transponder (SOTA) – Evaluation of satellite-to-ground laser communication links", Proc. of SPIE 9739, 2016.

[21] C. Schmidt, M. Brechtelsbauer, F. Rein, C. Fuchs, "OSIRIS payload for DLR's BiROS satellite", International Conference on Space Optical Systems and Applications 2014. ICSOS 2014, Kobe, Japan, 7.–9. Mai 2014.

[22] B. Smutny, H. Kaempfner, G. Muehlnikel, *et al.*, "5.6 Gbps optical intersatellite communication link", Proc. of SPIE 7199, 2009.

[23] M. Toyoshima, T. Kuri, W. Klaus, *et al.*, "4-2 Overview of the laser communication system for the NICT optical ground station and laser communication experiments in ground-to-satellite links", Special issue of the NICT Journal, vol. 59, no. 1/2, pp. 53–75, 2012.

[24] Z. Sodnik, B. Furch, H. Lutz, "The ESA optical ground station Ten years since first light", ESA Bulletin 132, Nov. 2007.

[25] K. Saucke, C. Seiter, F. Heine, *et al.*, "The Tesat transportable adaptive optical ground station", Proc. of SPIE 9739, 2016.

[26] K. Wilson, N. Page, J. Wu, M. Srinivasan, "The JPL optical communications telescope laboratory test bed for the future optical deep space network", IPN Progress Report 42-153, 2003.

[27] D.-H. Pung, E. Samain, N. Maurice, *et al.*, "Telecom & scintillation first data analysis for DOMINO – laser communication between SOTA onboard Socrates satellite and MEO OGS", Space Optical Systems and Applications (ICSOS), New Orleans, 2015.

[28] C. Schmidt, C. Fuchs, "The OSIRIS program – First results and outlook", IEEE International Conference on Space Optical Systems and Applications (ICSOS) 2017, 2017.

[29] T. Dreischer, B. Thieme, K. Buchheim, "Functional system verification of the OPTEL-μ laser downlink system for small satellites in LEO", International Conference on Space Optical Systems and Applications (ICSOS) 2014, 2014.

[30] T. Shih, O. Guldner, F. Khatri, *et al.*, "A modular, agile, scalable optical terminal architecture for space communications", IEEE International Conference on Space Optical Systems and Applications (ICSOS) 2017, 2017.

[31] B.L. Edwards, "An update on the CCSDS optical communications working group", IEEE International Conference on Space Optical Systems and Applications (ICSOS) 2017, 2017.

[32] D. Giggenbach, F. Moll, C. Fuchs, T. de Cola, R. Mata-Calvo, "Space communications protocols for future optical satellite-downlinks", 62nd International Astronautical Congress, 3.Okt–7.Okt 2011, Cape Town, South Africa, 2011.

[33] C. Fuchs, S. Poulenard, N. Perlot, J. Riedi, J. Perdigues, "Optimization and throughput estimation of optical ground networks for LEO-downlinks, GEO-feeder links and GEO-relays", Proc. SPIE 10096, Feb. 24, 2017.

[34] L.C. Andrews, R.L. Phillips, "Laser Beam Propagation through Random Media, 2nd Edition", SPIE-Press, Bellingham, WA, 2005.

[35] N. Perlot, T. De Cola, "Throughput maximization of optical LEO-ground links", Free-Space Laser Comm. Technologies XXIV, San Francisco, USA, 2012.

[36] F. Moll, M. Knapek, "Wavelength selection criteria and link availability due to cloud coverage statistics and attenuation affecting satellite, aerial, and downlink scenarios", Proceedings of SPIE 6709, 2007.

[37] I.E. Gordon, L.S. Rothman, C. Hill, *et al.*, "The HITRAN2016 molecular spectroscopic database", Journal of Quantitative Spectroscopy and Radiative Transfer, vol. 203, pp. 3–69, Dec. 2017.

[38] H. Hemmati, M. Toyoshima, R.G. Marshalek, *et al.* (Ed.) Near-Earth Laser Communications, CRC Press, Boca Raton, FL, 2009.

[39] W.R. Leeb, "Degradation of signal to noise ratio in optical free space data links due to background illumination", Applied Optics, vol. 28, pp. 3443–3449, 1989.

[40] D. Giggenbach, H. Henniger, "Fading-loss assessment in atmospheric free-space optical communication links with on-off keying", Optical Engineering, vol. 47, pp. 046001-1–046001-6, 2008.

[41] S.G. Lambert, W. Casey, Laser Communications in Space, Artech House, Norwood, MA, 1995.

[42] M.D. Eisaman, J. Fan, A. Migdall, S.V. Polyakov, "Single-photon sources and detectors", Review of Scientific Instruments, vol. 82, 2011.

[43] Z. Ghassemlooy, W. Popoola, S. Rajbhandari, "Optical Wireless Communications: System and Channel Modelling with MATLAB", 2013.

[44] A. Shrestha, D. Giggenbach, "Variable data rate for Optical Low-Earth-Orbit (LEO) Downlinks", ITG-Fachbericht 264: Photonische Netze, 12–13 May 2016, Leipzig, Germany, 2016.

[45] G.P. Agrawal, "Fiber-Optic Communication Systems, 3rd Edition", John Wiley & Sons, New York, 2002.

[46] D. Giggenbach, "Optimierung der optischen Freiraumkommunikation durch die turbulente Atmosphäre – Focal Array Receiver", PhD. Thesis, University of German Federal Armed Forces, Munich, 2004.

[47] D. Giggenbach, R. Mata-Calvo, "Sensitivity Modelling of Binary Optical Receivers", Applied Optics, vol. 54, no. 28, Oct. 2015.

[48] C.E. Shannon. "A Mathematical Theory of Communication", Bell System Technical Journal, vol. 27, no. 3, pp. 379–423, Jul. 1948.

[49] B. Sklar, "Digital Communications: Fundamentals and Applications, 2nd Edition", Prentice-Hall, Upper Saddle River, NJ, 2001.

[50] S. Lin, D.J. Costello, "Error Control Coding", Prentice Hall, Upper Saddle River, NJ, 2004.

[51] "TM Synchronization and Channel Coding – Summary and Rationale", CCSDS-130.1-G-2, Consultative Committee for Space Data Systems, Washington, DC, Nov. 2012.

[52] G. Gho, L. Klak, J.M. Kahn, "Rate-adaptive coding for optical fiber transmission systems", Journal of Lightwave Technology, vol. 29, no. 2, 2011.

[53] G.D. Forney, "Burst-correcting codes for the classic bursty channel", IEEE Transactions of Communications Technology, Vol. 19, no. 1971, pp. 772–781, 1971.

[54] A. Botta, A. Pescape, "IP packet interleaving: bridging the gap between theory and practice", IEEE, 2011.

[55] A. Shrestha, D. Giggenbach, N. Hanik, "Delayed frame repetition for free space optical communication (FSO) channel", ITG-Fachbericht 272: Photonische Netze, 11–12 May 2017, Leipzig, Germany, 2017.

[56] F. Xu, A. Khalighi, P. Caussé, S. Bourennane, "Channel coding and time-diversity for optical wireless links", Optics Express, vol. 17, no. 2, pp. 872, 2009.

[57] F. Moll, A. Shrestha, C. Fuchs, "Ground stations for aeronautical and space laser communications at German Aerospace Center", Proc. of SPIE 9647, 2015.

[58] D. Giggenbach, F. Moll, "Scintillation loss in optical low Earth orbit data downlinks with avalanche photodiode receivers", IEEE-Xplore, International Conference on Space Optical Systems 2017 (ICSOS), 13.–16. Nov. 2017, Naha, Japan, 2017.

[59] S. Poulenard, A. Mège, C. Fuchs, N. Perlot, J. Riedi, and J. Perdigues, "Digital optical feeder links system for broadband geostationary satellite", Proc. of SPIE, 10096, 2017, pp. 1009614–1.

Chapter 12

Ultra-high-speed data relay systems

Ricardo Barrios[1], Balazs Matuz[1], and Ramon Mata-Calvo[1]

The *SpaceDataHighway*, the first operational service of high-speed data relay system based on optical intersatellite links, has set a new milestone in space optical communications. Data relay systems are becoming crucial in applications such as Earth observation, where huge amounts of data need to be sent to Earth reliably and with low latency.

Optical communications plays a major role in such high-speed applications, since no regulations are needed, because of the lack of interference among users and the huge amount of available bandwidth. Since the end of the 1990s, several experiments have shown the feasibility of such technology with several demonstrations from low Earth orbit (LEO), geostationary equatorial orbit (GEO) and the Moon. The current state-of-the-art relay system architecture involves LEO and GEO satellites with optical intersatellite links, and direct Ka-band radiofrequency (RF) links from GEO to the Earth. Next-generation systems may involve also unmanned aerial vehicles (UAVs) and may rely only on optical communications to exploit the full potential of these frequencies.

The main challenges of using optical links are the turbulence effects, when the link traverses the Earth's atmosphere, and the degrading impact of platform microvibrations because of the inherently small divergence of the transmitted beam. Such aspects have to be taken into account when designing future systems.

Together with the modulation, the forward error correction (FEC) defines the communications performance of the system. Following Consultative Committee for Space Data Systems (CCSDS) coding recommendations, the performance of several coding schemes is analyzed; concretely Reed–Solomon (RS) codes, convolutional codes (CCs), turbo codes and low-density parity check codes are taken into account. One of the main characteristics of the atmospheric channel is the correlation of fading events, which requires further data protection to compensate for erasure events. Interleaving and packet (PKT) level coding in combination with FEC are compared through simulations.

Finally, different approaches for data correction are considered. The complexity on board the GEO satellite can specially limit the use of the most advanced decoding

[1]German Aerospace Center (DLR), Institute of Communications and Navigation, Satellite Networks Department, Germany

schemes and data-protection for the upcoming generations of relay systems. The trade-off between performance and complexity is crucial in order to allow further system enhancements in terms of capacity, without endangering the whole system availability.

12.1 Introduction

Data transfer from LEO or pseudo-satellites to the ground is crucial for several applications where security is fundamental and where large amounts of data need to be transmitted. Perhaps, the most prominent example is the Earth observation missions.

A relay system based on GEO satellites has two big advantages. First, it can provide coverage to the whole Earth surface with a few relay satellites. Second, it increases the data-transfer availability of the terminals at LEO or on pseudo-satellites. In addition, a system based on free-space optical communications satisfies both security and high-data-rates requirements. Data transmissions from hundreds of megabit per second to several terabits per second are possible, allowing expanding the optical network into space.

Since November 2016, the first operational high-speed data relay system is offering the *SpaceDataHighway* service, transferring data from LEO satellites to the ground via the European Data Relay System (EDRS) GEO satellites [1,2]. High-data-rate optical links are able to transfer data between satellites and a Ka-band link relays the data to the ground.

A further development of this relay system, or the development of new ones, requires a detailed analysis of the physical (PHY) layer, optimizing the system architecture by defining the optimal modulation formats, coding and data processing scheme, taking into account the platform limitations and channel impairments, while maximizing the data throughput.

The objective of this chapter is to define and analyze the key elements in the design of future ultra-high-speed relay systems based on optical technologies.

12.2 Relevant missions and demos

Since end of the 1990s, several optical communication terminals have been developed for LEO, GEO and Moon missions. Figure 12.1 summarizes the main missions related to optical communications, past and planed ones. The missions performing relay communications are highlighted. In the upper part of Figure 12.1, there are the missions for GEO [SILEX, AlphaSat-laser communication terminal (LCT), EDRS and Moon [Lunar laser communications demonstration (LLCD)]. The LCRD is currently in development and it is planned for lunch in 2018. All of them are commented in the following sections. In the lower part of Figure 12.1, there are the missions for LEO payloads. The OPALS, SOTA and OSIRIS projects are focused in direct downlinks to Earth and they will not be further commented hereafter. In the future, for the SOTA, mission is also planned links to aircraft and satellites [3].

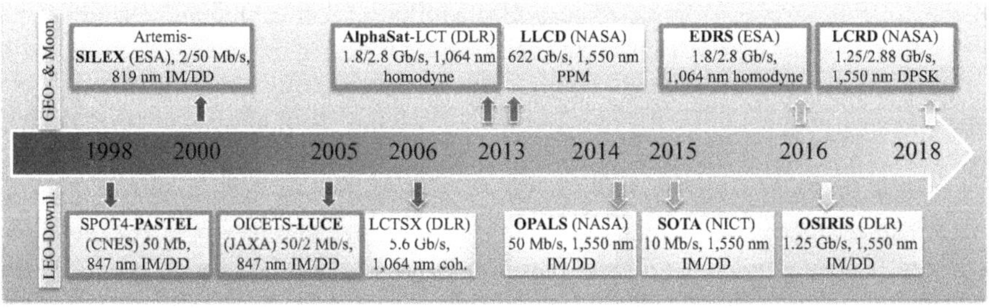

Figure 12.1 Timeline of optical terminals in space

Intersatellite links were the framework of the SILEX project with two main objectives: to demonstrate the feasibility and performance of intersatellite links and to relay video data from a LEO satellite to a ground station. The experiments involved two satellites that hosted the optical terminals: the ARTEMIS GEO satellite and the SPOT-4 LEO satellite. SPOT-4, developed by Matra Marconi Space for CNES, was successfully launched in 1998 and ARTEMIS, developed by Alenia for the European Space Agency (ESA), in 2001 [4]. The laser terminals were developed based on intensity modulation [50 Mb/s with on–off keying (OOK) with no return to zero for the forward link] and direct detection of laser beams in the 800-nm range, allowing 50 Mbps data rate transmission. Since November 2001, bidirectional links were performed between ARTEMIS and the ESA optical ground station (OGS) at Canary Islands, Spain [5].

Other intersatellite links were performed between ARTEMIS and OICETS satellites by JAXA and ESA since 2005, when the first bidirectional intersatellite link took place. OICETS performed the return link at 2 Mb/s with 2-pulse-position modulation (PPM). An avalanche photodiode was used as receiver (Rx) [6].

After the experience of SILEX, LEO intersatellite communications based on coherent communications were the next step. The TerraSAR-X hosted the first coherent communications terminal in LEO based on this communications technology. The terminal implements binary phase shift keying (BPSK) modulation and homodyne detection using an optical phase-locked loop (OPLL). The terminal was developed by TESAT-Spacecom under DLR funding [7]. The counter-partner was installed on the NFIRE satellite that was developed by the USA department of Defense. Homodyne BPSK at 5.625 Gbps was performed between both satellites over distances up to 4,900 km [8].

The EDRS—in operation since November 2015—relays data between LEO satellites to ground through a constellation of GEO satellites, and it will also support UAVs and aircrafts. After Alphasat and EDRS-A were launched, the first satellite constellation was already in orbit. On the GEOs, the LCT is part of a hybrid optical-RF payload for data relay [9]. The LCT is serving as input section for RF payloads that have different capabilities regarding the programs: In the Alphasat mission, the data output of the LCT is directly connected to a 600-Mbit/s Ka-band modulator (transparent connection), while in the EDRS mission, the data output of the LCT is subject to framing, encrypting and channel coding. Due to the resulting overhead, the data volume is increased and dumped through various Ka-band channels, each with 600 Mbps data rate. The ground segment performs decoding, decryption and deframing.

It is noteworthy that recently, NASA successfully demonstrated bidirectional links with the optical terminal—based on 1500 nm systems—on board the Lunar Atmospheric Dust and Environment Explorer (LADEE), with a series of ground-space optical links demonstrations [10–12]. In this demonstration, the Lunar Lasercom Space Terminal was used providing a maximum uplink and downlink rates of 20 and 622 Mbps, respectively [12]. The downlink operates with a 16-PPM modulation format, while the uplink does with a 4-PPM modulation.

The LCRD mission is currently under development by NASA, to serve as a testbed for different technologies and concepts required in a data relay system based on optical communications. LCRD will operate for a minimum of 2 years, with a terminal in GEO orbit hosting two optical communications modules, allowing for testing handover protocols between ground stations. The main goals to be demonstrated in this mission are high rate bidirectional communications between ground and GEO and to understand the feasibility of PPM for deep space communication—or other power-limited systems—or differential PSK (DPSK) for near Earth high-data-rate communications. Moreover, the LCRD will specifically target the study performance testing and demonstrations of coding, link layer and network layer protocols over optical links [13].

12.3 System architectures

At the top of Figure 12.2, there is a depiction of the relay scenarios considered hereafter. The data relay system architecture consists of a user (U) terminal node, a data relay (R) node and a ground (G) station node. In this system architecture, there are two links, namely, the user link and the feeder link. The U–R link is defined as the link between the user terminal node and the data relay node, while the feeder R–G link is defined between the data relay node and the ground station. The user terminal node can be a LEO (L) satellite or an UAV (X), and the data relay terminal is a GEO satellite. The ground station node can be, in principle, either optical or RF, accordingly to the desired feeder-link technology. At the bottom of Figure 12.2, an abstraction of such communication chain is also provided, where both the U–R and R–G channel are responsible for degrading the transmitted information, resulting in errors in the data transmission.

Among all considered schemes, full decoding on board of the relay offers the best trade-off between power, bandwidth and achievable error rate. From a channel coding perspective, the different communication links are considered independent, and the errors are recovered locally at the satellite, as well as on ground. This can be achieved by protecting the data stream over the U–R link via a forward error correcting code and decoding the data stream at the GEO relay. By doing so, the

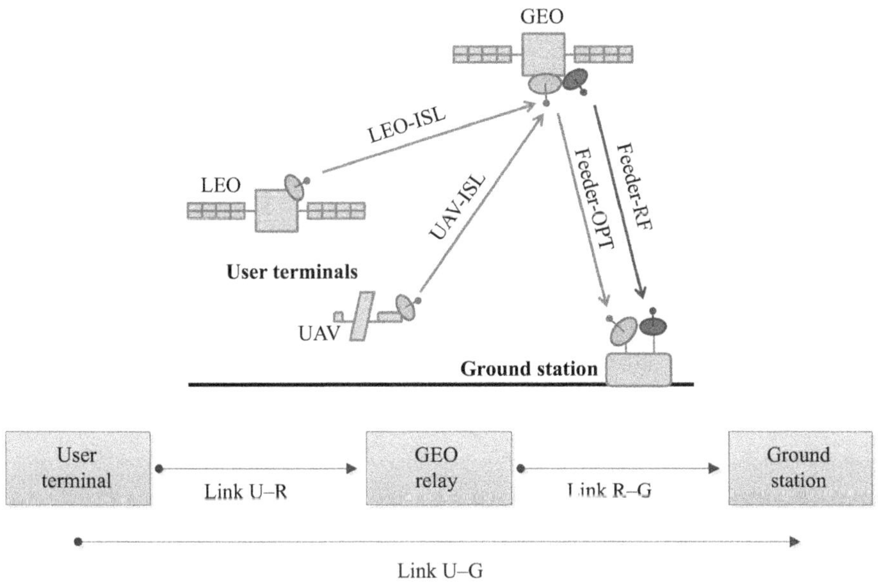

Figure 12.2 (Top) Scenarios for a GEO based relay system and (bottom) abstraction of the communication channel

redundancy introduced at the user terminal to cope with U–R link errors is removed at the relay, and upon proper dimensioning of the channel code virtually all errors are corrected. Therefore, the encoded information sent by the user terminal via U–R link is decoded and reconstructed at the GEO relay prior to transmission over the R–G link. The redundancy introduced over this link is exploited to cope with the errors affecting the R–G link. This approach, although optimal in the sense of minimizing the amount of redundancy over the two links—hence, maximizing the spectral efficiency of the system, has the major drawback of requiring decoding of at the GEO relay. The provision of a (quasi) error-free decoding of information at the relay may require, in fact, the use of a powerful error correcting code over the U–R link with a complex decoder at the relay.[1] Therefore, various other options are discussed shifting decoding complexity from the relay to ground. Note also that the channel codes for the U–R link need to be fixed in advance, making later changes difficult. Other schemes, such as layered decoding, offer more flexibility, since no decoding at the relay is performed.

To avoid implementing complex decoding algorithms at the relay, one may perform encoding at the user, route the data through the relay and perform all decoding

[1]Note that the definition of complexity is very vague and changes in time. At the time being, existing relay systems barely implement channel coding mechanisms (exceptions are simple repetition codes). Therefore, also with regard to the high user data rates in the order of Gbit/s, decoding of modern codes at the relay is assumed to be impractical in the mid-term.

operations on ground only. This scheme does not impose a strong complexity burden on the relay and provides some flexibility to change/update the PHY layer FEC scheme independent of the relay. In particular, for the U–R link—where medium/low code rates are required—this solution lacks spectral efficiency.

FEC coding termination options

- **Full decoding on board the GEO:** FEC coding is applied independently in each link (optical ISL and optical feeder link). The GEO has to correct errors in the ISL channel, and this may constraint the type and level of coding that can be applied, since resources on board the GEO are limited.
- **Decoding on ground only:** FEC coding is done, treating both the U–R and R–G links together. In this case, the GEO data relay does not perform any decoding. The ground station has then to correct errors occurred in both channels.
- **Partial decoding:** This scheme assumes that only some low-complexity decoding operations take place at the satellite. Another decoding step is done on ground where more decoding complexity is affordable.
  - **Layered coding:** This scheme implies that the user data is protected by an additional error correcting code on top of the PHY layer code. This code is not decoded on the relay, but only on ground, thereby, shifting the decoding complexity to the Rx.

An alternative is to allow some low-complexity decoding operations on board of the relay. To improve the spectral efficiency, one may recover as many errors as possible at the GEO relay, with the given complexity constraint. Hence, from a spectral efficiency viewpoint, the best possible error control scheme that fits with the complexity limitations at the relay shall be used to protect the U–R link. On ground, a further decoding attempt is made to correct the remaining errors. Hereafter this approach is referred as partial decoding.

This approach is used in EDRS, where the U–R link is protected by a line product code, while the end-to-end FEC is based on a (255, 239) RS code [14]. The repetition code works at the very low end of the complexity scale providing, however, no coding gain. Options to render the U–R link more reliable can be based on more complex but still very light in computational burden error correction mechanisms.

Another approach is based on layered decoding. In particular, when the U–R link is subject to severe error events, e.g., due to strong pointing jitter, an additional channel code can be added on top of the low-complexity PHY FEC scheme. In the following, this code is referred to as PKT level code or erasure code. In this case, at the GEO only the weak PHY code is decoded, correcting some errors on the U–R link. A mandatory error detection mechanism marks each PHY codeword either as correct or erroneous. All erroneous data is discarded. The remaining data, after some processing, is encoded and transmitted to the ground where after correcting the errors on the R–G link the PKT decoder attempts to recover the erroneous data from the U–R link—i.e., those not recovered at the GEO relay. Upon a proper design, high spectral efficiencies can be reached here with some penalty in performance with respect to full decoding on board of the satellite.

12.4 Optical channel model

12.4.1 Atmospheric channel

The atmospheric turbulence can be defined by the strength of the fluctuations in the refractive index, represented with the refractive-index structure parameter C_n^2 with units of $m^{-2/3}$. Hereafter, for all required calculations, the well-known Hufnagel-Valley vertical profile is used [15]. The intensity of the received signal (for both coherent and noncoherent modulations) is affected by fading, resulting in time-varying detected power, due to scintillation and beam wander. Scintillation is the result of self-interference processes due to phase distortions and beam wander are atmospheric-induced pointing errors. The former is defined by the scintillation index (SI)—i.e., the normalized variance of the received optical power—and the latter by the RMS value of the beam displacement. Expressions to calculate the SI are readily available elsewhere [15].

The intensity can be modeled as a random variable governed by a lognormal probability density function (PDF) in the case of weak turbulence regime for a point Rx and works well in all regimes of turbulence for aperture averaged data [16,17]. A process for the generation of lognormally correlated time samples has been presented elsewhere [18]. Additionally, the lognormal channel attenuation can be modeled as low pass process with a characteristic frequency—which depends on the atmospheric turbulence strength and the speed of the different turbulence layers—having −8/3 and −17/3 power law slope for low and high frequencies, respectively [19]. The cut off frequency characterizing the coherence time of the atmosphere is known as the Greenwood frequency [20]. The scale of the atmospheric coherence time, i.e., the inverse of the Greenwood frequency, is usually in the order of tens of milliseconds.

The block diagram, shown in Figure 12.3, describes a general channel model including all the channel impairment effects due to turbulence and pointing errors due to terminal microvibrations.

Once the basic scenarios are defined in Section 12.3, some calculation of the relevant parameters of the optical channel can be done in order to set the operational constrains of the different links. It is noteworthy that the L–R link is not affected by turbulence and, thus, the parameters related to atmospheric turbulence are calculated only for the X–R and R–G links.

The Fried parameter r_0 measures the integrated turbulence strength along a given propagation path and is given by

$$r_0 = \left(0.423 k^2 \sec\zeta \int_{h_0}^{H} C_n^2(h)dh\right)^{-3/5}, \tag{12.1}$$

where $k = 2\pi/\lambda$ is the wavenumber, with λ being the wavelength, and ζ is the elevation angle.

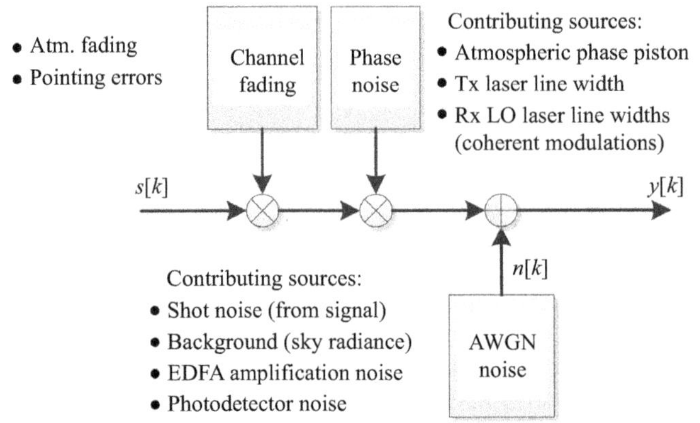

Figure 12.3 General block diagram of the channel model after the photodetector

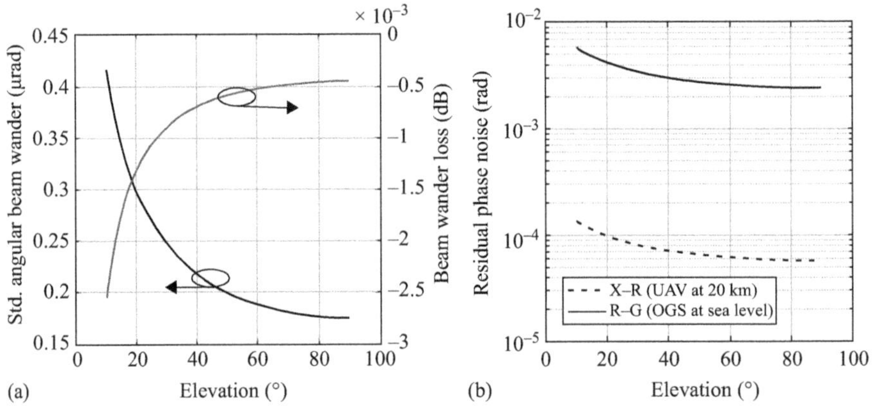

Figure 12.4 (a) Angular beam wander and beam wander loss for the X–R link and (b) residual phase noise error due to atmospheric piston of two different terminal altitudes and different elevation angles, for a direct link with GEO satellite

The higher the value of the Fried parameter is, the weaker the turbulence becomes. Typical values for weak turbulence are in the range of tens of centimeters. In the X–R link, the Fried parameter is about two orders of magnitude larger than the typical values for weak turbulence, indicating that little to no influence from turbulence should be present in such links.

Figure 12.4(a) presents an estimation of the beam wander effects over the X–R link. The angular beam wander, which represents the variance of the atmospheric induced pointing errors, can be calculated as [15]

$$\theta_{BW}^2 = 0.54\left(\frac{\lambda}{2W_0}\right)^2\left(\frac{2W_0}{r_0}\right)^{5/3}, \tag{12.2}$$

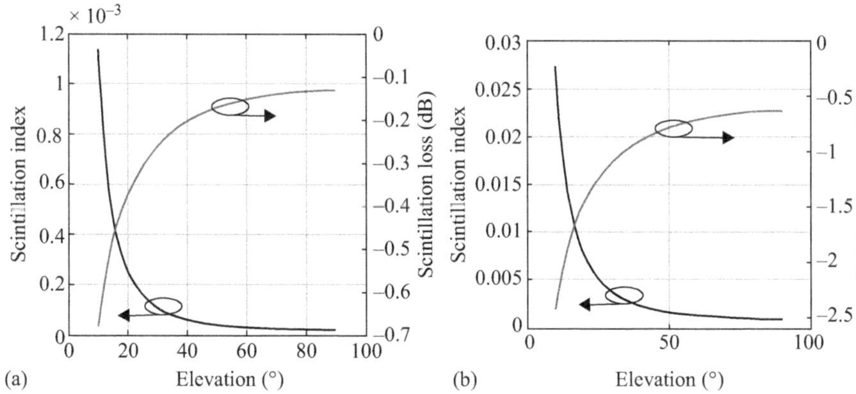

Figure 12.5 Scintillation index and scintillation loss for a direct link with a GEO satellite from (left) an UAV at 20 km, and to (right) an OGS (60 cm aperture) at sea level at different elevation angles. The target availability assumed was 99.6%

where W_0 is the beam radius at the transmitter (Tx) output plane. Moreover, it can be readily seen that the beam wander loss, which can be estimated as $L_{BW} = \exp(-G_T\theta_{BW}^2)$ for a Gaussian profile, is negligible for the UAV-to-relay link. This is so mainly due to the fact the standard deviation of their angular variations is about two orders of magnitude lower than the UAV Tx beam divergence, which is in the order of tens of microradians.

The atmospheric turbulence of the optical channel produces intensity and phase fluctuations. The phase distortions, induced by atmospheric turbulence, produce time-of-arrival jitter on the Rx signal, which is negligible for noncoherent modulation formats. In the case of coherent modulation formats, the influence of the atmospheric piston can be modeled through its effect on the residual phase noise as [21]

$$\sigma_\phi^2 = 1.328\left(\frac{v_\perp}{r_0}\right)^{5/3}\omega_n^{-5/3}, \tag{12.3}$$

where ω_n is the natural frequency of the Rx OPLL and $v_\perp$ is the wind speed vertical profile normalized with respect to the C_n^2 profile, which can be calculated as shown elsewhere [21]. Figure 12.4(b) shows the residual phase noise due to atmospheric piston for the X–R and R–G links, with $\omega_n = 50$ kHz, where it can be readily seen that values are always below 0.01 rad for all the analyzed conditions. It is already known that only values in the order of 0.1 rad or above can produce a significant deterioration of homodyne Rxs [21]. Therefore, it is determined that atmospheric piston does not play a significant role—when the OPLL is optimally designed [22]—in the reception of optical coherent modulation formats, for the relay scenarios analyzed here.

Figure 12.5 presents the SI value and the scintillation loss for the X–R and R–G links. The SI gives a measure of the normalized standard deviation of the received optical intensity and depends inversely on the link elevation angle, i.e., the lower the elevation the higher the SI as a longer atmospheric path is traversed. When estimating the scintillation loss, a target availability of 99.6% was assumed [23]. On the one hand, it can be seen that in the X–R link, for elevation angles above 15°, the SI loss is less than 0.5 dB, indicating very weak turbulence. On the other hand, for the R–G link, for a 60-cm receiving telescope the SI loss could go as high as 2.5 dB for low elevation angles. Nevertheless, typical elevation angles in a GEO-ground scenario are above 35°, where the SI loss would amount to approximately 1 dB or less. Because the SI value is always below 0.1, the atmospheric turbulence in all scenarios can be regarded to operate under a weak turbulence regime. The low values of SI are explained, as the propagation occurs only in the higher portion of the atmosphere for the X–R link case, where turbulence is the lowest. In the case of the R–G link, although the optical wave traverses the whole atmosphere, a fair amount of aperture averaging takes place effectively reducing the SIs. The SI expressions for the uplink and downlink have been given elsewhere [15].

12.4.2 Pointing errors and microvibrations

Microvibrations of the Tx platform contribute to the pointing errors and they can be modeled by a beta distribution, when the pointing bias is assumed to be zero. Therefore, the PDF of the received optical power, due to only pointing errors, is given by [24]

$$f_I(I) = \beta I^{\beta-1}, 0 \leq I \leq 1, 0 < \beta < \infty. \tag{12.4}$$

where the parameter $\beta = W_0^2/(4\sigma_e^2)$ characterizes the random microvibrations of the user terminal, W_0 being the laser bean radius at the Tx and $\sigma_e = [\int \theta_e^2(f)df]^{1/2}$ is the root-mean-square (RMS) of the random jitter [25]. To complete the model, a power spectral density (PSD) of the user terminal vibrations must be assumed, in order to take into account the temporal behavior of the transmitting telescope pointing errors. In the past, the ESA proposed a model for the microvibrations PSD, for the optical communication payload SILEX, given by [26]

$$\theta_e^2(f) = \frac{2\sigma_e^2}{\pi\left(1 + (f/f_{e0})^2\right)}, \tag{12.5}$$

where f_{e0} is the cut-off frequency of the PSD. For the case of the ESA model, this frequency was set to 1 Hz to model a LEO platform. For an UAV, it could be expected that the PSD to be spread over a larger bandwidth, taken into account the influence of wind gusts.

As an estimation of the order of magnitude of the RMS random jitter σ_e, a quick overview of available literature shows that for LEO terminals σ_e is in the range of 20–45 μrad [25], and for an aircraft terminal, a few hundreds of microradians has been reported [27]. These figures refer to total amounts due to vibrations of the space- or aircraft, which are more relevant in the initial pointing and acquisition stages of the link, and usually compensated through a coarse pointing assembly.

For the communication stage of link, a fine pointing assembly—usually a fast steering mirror—is most likely also part of the LCT that helps in further reducing the pointing errors. Nevertheless, there is always a residual pointing error, which is the relevant figure for the communications phase of the link. Values reported in the literature for the residual pointing errors (jitter) range from as low as 0.3 μrad [28] and 0.8–1.53 μrad [29] to as high as 2.6 μrad [30], for satellite platforms. In case of UAVs, for platforms flying altitudes of 10 km or above, reported residual pointing error jitters are in the order of some tens of micro radians [31].

To summarize, Table 12.1 presents a list of relevant parameter for all the scenarios defined, with some typical values for the sake of example. When the user terminal is a LEO satellite, the transmit power and aperture are changed to reflect the cases of a small and big LEO terminal. The small terminal has a 7-cm aperture with a 3-W power output, while for the big terminal a 15-cm aperture and a 5-W power output is assumed. In both cases, the Tx laser is assumed collimated. For the UAV, 50 W Tx power has been assumed, and 10 W for GEO platform.

On the one hand, it can be readily seen that for the X–R link, although it traverses the atmosphere, the atmospheric channel is quite benign due to the fact that only the upper part of the atmosphere plays a role. On the other, in case of the R–G link, although the whole atmosphere is within the propagation path of the downlink laser, a fair amount of aperture averaging takes place significantly reducing the effects of scintillation. This is due to the relatively large receiving aperture diameter at the ground station is 60 cm, when compared to typical values of the Fried parameter. Additionally, the atmospheric coherence time is about some tens of milliseconds, giving an indication of the interleaver size to cope with the correlated fading events. Finally, the residual jitter and its coherence time—for the U–R channel—are calculated by simulating the platform pointing errors using the model in (12.5). Next, the

Table 12.1 Relevant link parameters for all analyzed links for an optical GEO-based relay system. The links are LEO to relay (L–R) for a small and big platform, UAV to relay (X–R) and relay to ground (R–G)

Parameter	**Units**	**Small L–R**	**Big L–R**	**X–R**	**R–G**
Elevation	°	–	–	40.00	35.00
Link distance	km	40,000.00	45,000.00	35,980.00	38,394.12
Wavelength	nm	1,550.00	1,550.00	1,550.00	1,550.00
Fried parameter	cm			1,612.71	13.94
Greenwood frequency	Hz			0.34	34.46
Atmospheric coherence	ms	–	–	2,983.71	29.02
Res. pointing jitter	μrad	0.63	0.32	11.79	–
Coherence atm+jitter	μs	89.42	89.23	89.35	29,020.00
Scintillation index		–	–	1.74E−04	1.36E−2
Tx altitude	km	500.00	500.00	20.00	36,000.00
Atmospheric attenuation	dB	–	–	−0.01	−0.50
Tx power	W	3.00	5.00	50.00	10.00
Tx telescope diameter	cm	7.00	15.00	12.00	25.00
Tx divergence	μrad	19.94	9.30	50.00	5.58
Rx telescope diameter	cm	25.00	25.00	25.00	60.00

half-width-half-maximum point of the channel state autocorrelation at the Rx plane was measured, assuming that the Tx pointing mechanism can effectively reject vibration up to about 500 Hz, in the communication tracking phase of the link [32]. The total amount of initial jitter σ_e assumed was 20 and 45 μrad for the small and large LEO platform, with a PSD cut-off frequency f_{e0} of 1 Hz, following the ESA model [26]. For the UAV case, a σ_e of 100 μrad was assumed with $f_{e0} = 50$ Hz, to reflect the higher vibration regime due to the wind gust affecting the aircraft.

A special consideration is made for the X–R link, where the user is an UAV platform. Due to the strong residual pointing jitter, the divergence of the Tx telescope is optimized to counter the pointing loss effects. The resulting optimum divergence is about 50 μrad. The UAV's telescope is selected to be 12 cm as this size falls within the requirements of the tracking system [33,34]. Nevertheless, this aperture diameter has no impact in the link budget calculation on the X–R link, as the Tx is assumed noncollimated and its gain is obtained through the divergence value.

12.4.3 Light coupling efficiency

In every Rx chain, collected light by the telescope must be coupled into a photoelectric converter device, which might be preceded by fiber waveguide stage as in an erbium-doped fiber amplifier (EDFA) preamplified Rx chain case. When light needs to be coupled into a single mode fiber (SMF), the coupling efficiency under the presence of atmospheric turbulence is [35]

$$\eta_C = 8a^2 \int_0^1 \int_0^1 \exp\left[-\left(a^2 + \frac{D_R^2}{4\rho_0^2}\right)\right] I_0\left(\frac{D_R^2}{4\rho_0^2} x_1 x_2\right) dx_1 dx_2, \tag{12.6}$$

where $a = \pi D_R W_m/(2\lambda F)$, W_m is the field radius of the fundamental mode that propagates through the SMF (usually about 5 μm), F is the focal length of the receiving telescope and $\rho_0 = 0.48 r_0$ is the atmospheric coherence radius—which is directly related to the Fried parameter give in (12.1).

In the uplink direction—i.e., for the UAV to GEO relay—the turbulent structures defined by ρ_0 are much larger than the probable size of the GEO satellite receiving aperture and thus $\rho_0 \gg D_R$. Consequently, the maximum fiber coupling efficiency $\eta_C = 0.815$ can be obtained, provided that the Rx telescope has optimize the ratio D_R/F such that $a = 1.12$ [36].

In the downlink direction, for the GEO relay to ground link, the D_R/ρ_0 ratio is larger than unity, indicating that some amount of wavefront distortion is capture by the receiving aperture. Therefore, the shape of the focused light can differ greatly from an Airy pattern, effectively producing additional coupling losses. In order to counterattack this phenomena, adaptive optics (AO) is often used to correct the incoming distorted wave, which can be decomposed into several orthogonal modes described by the Zernike polynomials [37]. To estimate the possible gain when applying AO techniques, a generalized Fried parameter $r_{0,N}$ can be estimated in terms of the number of Zernike modes N corrected as [38]

$$r_{0,N} = 0.286 r_0 \left(\frac{3.44}{C_N}\right) N^{-0.362}, \tag{12.7}$$

where C_N is the corresponding coefficient for the number of modes N being corrected as given by Noll [37].

Finally, in cases where the light is directly coupled over the photodetector, the diameter of the time-averaged (long-term) focal spot can be larger than the detector diameter. If the Fried-parameter r_0 is smaller than the aperture diameter D_R, the long-term intensity distribution $I(r)$ can be modeled as a Gaussian distribution with standard deviation $\sigma \approx 0.42\lambda F/r_0$. Integrating the intensity distribution over the area of the detector yields the encircled—i.e., the detectable—power.

12.5 Noise model

The calculation of the available signal-to-noise ratio (SNR) is essential when assessing a link performance. The symbol-level SNR is defined as

$$\mathrm{SNR} = \frac{I_R^2}{\sigma_{\mathrm{s}}^2 + \sigma_{\mathrm{B}}^2 + \sigma_{\mathrm{ASE}}^2 + \sigma_{\mathrm{s-ASE}}^2 + \sigma_{\mathrm{ASE-ASE}}^2 + \sigma_{\mathrm{LO\text{-}ASE}}^2 + R_I^2 \mathrm{NEP}^2 B_e^2}, \qquad (12.8)$$

where R_I is the responsivity of the photodetector and $I_R = R_I P_R$ is the generated signal photocurrent, for a certain received optical power P_R. When the received signal is coherently modulated, and a local oscillator (LO) with optical power P_{LO} is used, then $I_R = 2R_I\sqrt{P_{\mathrm{LO}} P_R}$. Moreover, B_e is the electrical bandwidth of the photodetector or the subsequent low-pass electrical filter, which is chosen to match the required bandwidth for the specific symbol rate of the modulated received signal. In addition, the noise equivalent power (NEP) characterizes the noise figure of the photodetection process, which includes the effects of thermal and dark current noise.

The shot noise variance—product of the intrinsic quantum nature of the light—can be approximated by $\sigma_{\mathrm{s}}^2 = 2qI_R MFB_e$, where q represents the elementary charge, M is the mean avalanche gain (higher than unity for APD photodiodes) and F is the excess noise factor. Similarly, the noise due to optical background power is calculated in the same manner.

The total background radiation can be characterized by the spectral radiance of the sky that depends on the elevation angle and changes for day and night operation. In nighttime, the sky emissivity for a nearly horizontal path through the atmosphere is essentially that of a blackbody at the temperature of the lower atmosphere. The behavior for daytime conditions will be very similar to that of nighttime, with the corresponding change due to higher temperatures, and the addition of scattered sun radiation below 3 μm [39]. The background noise can be modeled as $\sigma_{\mathrm{B}}^2 = 2qR_I P_B MFB_e$, where $P_B = N_B B_o(\pi D_R \mathrm{FoV}/4)^2$ is the background optical power, which depends on the spectral radiance of the sky N_B, the Rx aperture D_R, optical filter bandwidth B_o and detector's field of view FoV.

Amplified spontaneous emission (ASE) noise is inherent property of the used optical amplifiers [40]. The PSD noise is assumed bilateral and for each component the complex noise variance can be written as $N_{0,\mathrm{ASE}} = h\nu(G-1)n_{sp}/2$, where h is the Plank's constant, G is the amplifier gain and n_{sp} is the spontaneous emission factor, which is always greater than one. It is noteworthy that the variance depends on the frequency ν, showing that ASE is not really white because of this dependence with ν. However, for the normal bandwidth values required by data transmission systems, the ASE noise is considered flat and thus can be assumed as an AWGN process.

At the optical-to-electrical conversion stage, an ASE shot noise and two beat components are generated, along with the beating noise between the signal and the ASE $\sigma^2_{\text{s-ASE}}$ and between the ASE with itself $\sigma^2_{\text{ASE-ASE}}$. Assuming only one polarization, all are assumed AWGN and are given by [41]

$$\begin{aligned} \sigma^2_{\text{ASE}} &= 2qN_{0,\text{ASE}}B_oR_IB_e, \\ \sigma^2_{\text{s-ASE}} &= 4I_RMFN_{0,\text{ASE}}R_IB_e, \\ \sigma^2_{\text{ASE-ASE}} &= R_I^2N_{0,\text{ASE}}^2B_e(2B_o - B_e). \end{aligned} \tag{12.9}$$

In the case of coherent detection, an extra beating noise term appears due to the interaction of the LO power P_{LO} with the ASE component from the EDFA preamplifier in the Rx chain, which is given by $\sigma^2_{\text{LO-ASE}} = 2R_I^2P_{\text{LO}}N_{0,\text{ASE}}B_e$ [42].

For the cases when an EDFA booster amplifier is used in the Tx side, its ASE noise can be referred to the Rx chain as part of the background noise, in the form of an additional background optical power given by $P_{ASE-T_x} = 0.2hcG_TF_TD_T^2D_R^2/\left(R^2\lambda^3\right)$, where G_T and F_T refer to the booster amplifier gain and noise factor, respectively [41].

12.6 Link budget

The channel model includes several effects: the transmission losses, the atmospheric turbulence effects and the platform microvibrations. A simple way to see the different phenomenon affecting the optical link is through the expression of the received optical power P_R detected at distance L, which is given by

$$P_R = P_TG_T\eta_T\eta_{\text{ATM}}L_{\text{FS}}L_pL_{\text{SI}}G_R\eta_R\eta_C, \tag{12.10}$$

where P_T is the transmitted average optical power with wavelength λ; $G_T = (\pi D_T/\lambda)^2$ and $G_R = (\pi D_R/\lambda)^2$ are the Tx and Rx gains, respectively; η_T and η_R are Tx and Rx efficiencies, respectively; while η_{ATM} is the atmospheric attenuation; $L_{\text{FS}} = (\lambda/4\pi L)^2$ is the free-space loss. From the terms in (12.10), G_T, G_R, η_T, η_R, η_{ATM} and L_{FS} are considered either static or slow-varying losses—respect to the time scale of the communication process—and do not have an impact on the statistical behavior of the fading process. Moreover, $L_p = \exp\left(-G_T\theta^2_{BW}\right)$ corresponds to the pointing errors. Finally L_{SI} is the SI loss, respectively. The former can be calculated with a method from [43], and the latter with an expression from [23]

$$L_{\text{SI}} = \left(3.3 - 5.77\sqrt{\ln 1/p}\right)\sigma_I^{4/5}, \tag{12.11}$$

where σ_I^2 is the SI, $p = 1 - \text{av}$ is the fractional outage time, and av is the target availability, which in the scenario analyzed here is set to 99.6%.

Finally, the telescope collected light must be coupled into a photodetector, which will exhibit a certain coupling efficiency η_C. When light needs to be coupled into a SMF—as in an EDFA preamplified Rx chain—the coupling efficiency under the presence of atmospheric turbulence, for the R–G link, is a function of the ratio of the Rx aperture diameter to the Fried's parameter D_R/r_0 [35]. For the downlink case, i.e., in the R–G link, it is assumed that 50 Zernike modes are corrected by applying AO correction. This represents an improvement of about 13 dB with respect to a system without AO, and about 7 dB with a system that compensates for the tip-tilt Zernike modes, i.e., corrects for angle-of-arrival fluctuations.

In the uplink direction, i.e., for the U–R link, the transversal coherence of the wave is much larger than the probable size of the GEO receiving telescope aperture. Consequently, the maximum fiber coupling efficiency $\eta_C = 0.815$ can be obtained.

In order to carry out the link budget calculations, some assumptions on the Tx and Rx chain have to be made. In the Tx, the booster amplifier is assumed to work on a regime with a 45-dB gain and a 6-dB noise figure. These parameters are necessary in calculating the effect on the Tx booster ASE noise, which is effectively included as an extra background power level.

For the Rx chain, it is assumed that an optical filter of 0.8 nm—i.e., corresponding to a dense wavelength division multiplexing (DWDM) grid of 100 GHz—is present, which is a well-known standard assumption for 1,550 nm. The Rx optical chain has a preamplifier with 30 dB of gain, with 4 dB noise figure. The preamplifier is assumed to be used in both coherent and noncoherent reception, thus, the light coupling power loss always refers to fiber coupling efficiency. The photodetector is a PIN diode with a maximum 20 GHz electrical bandwidth, 0.75 A/W responsivity, and a NEP $\text{NEP} = 2.5\text{pW}/\sqrt{\text{Hz}}$. In addition, for the case of coherent detection a 10-dBm LO laser is considered.

It is noteworthy that in the optical domain, data rates up to 40 Gbit/s are achievable with current technology using a single optical channel. Modulator and Rxs for 40 Gbit/s are also available. However, currently for data rates beyond 25 Gbit/s usually wavelength division multiplexing (WDM) techniques are taken into consideration. In fiber communications, it is a well-known technique that leads to the ITU recommendations G.694.1 and G694.2 for DWDM and coarse WDM (CWDM) spectral grids, respectively. Such recommendations fix the central frequencies of the Tx laser and the optical channels for the multiplexers and demultiplexers. This technology is however usually limited to wavelengths in the range of optical C- and L-bands for DWDM and in the range between 1,270 and 1,610 nm for CWDM.

Table 12.2 presents the link budget calculation for all the scenarios selected. In the U–R link, the user can be either an UAV or a LEO satellite—which can be a small or big platform. The bottom row gives the equivalent background noise power seen by the Rx photodetector, i.e., after the preamplifier, and includes the Tx booster ASE noise and the sky irradiance background noise.

Based on the total received power calculated in the link budget presented in Table 12.2, a calculation of the photons per bit (PPB) at different bit rates can be performed.

$$\text{PPB} = \frac{P_R}{E_\lambda R_b}, \tag{12.12}$$

Table 12.2 Link budget calculation for all link scenarios defined by Table 12.1

Parameter	Units	Small L–R	Big L–R	X–R	R–G
Tx power	dBm	34.77	36.99	47.00	40.00
Tx antenna gain	dB	102.15	108.77	95.05	113.20
Tx antenna efficiency	dB	−3.01	−3.01	−3.01	−3.01
Tx pointing loss	dB	−0.06	−0.15	−1.68	0.00
Free-space loss	dB	−290.22	−291.24	−289.30	−289.86
Atmospheric attenuation	dB	0.00	0.00	−0.01	−0.50
Scintillation loss	dB	0.00	0.00	−0.32	−1.84
Link margin	dB	−1.00	−1.00	−1.00	−3.00
Rx antenna gain	dB	114.10	114.10	114.10	120.88
Rx antenna efficiency	dB	−3.01	−3.01	−3.01	−3.01
Rx light coupling loss	dB	−0.89	−0.89	−0.89	−14.72
Total link loss	dB	−82.02	−76.44	−90.07	−81.00
Total Rx power	dBm	−47.24	−39.45	−43.08	−41.00
Total equivalent background power	dBm	−80.92	−74.75	−74.65	−69.24

Table 12.3 Average received photons per bit, for all link scenarios defined by Tables 12.1 and 12.2. The received average power is taken from Table 12.2

Bit rate	Small L–R	Big L–R	X–R	R–G
100 Mbps	1,470.09	8,837.77	3,831.27	6,182.63
1 Gbps	147.01	883.78	383.13	618.26
5 Gbps	29.40	176.76	76.63	123.65
10 Gbps	14.70	88.38	38.31	61.83
20 Gbps	7.35	44.19	19.16	30.91

where R_b is the uncoded data bit rate and $E_\lambda = hc/\lambda$ is the photon energy, with h being the Planck's constant and c is the speed of light in vacuum.

The PPB metric is useful for providing a first idea on the maximum bit rates that in principle could be achieved with an optically preamplified Rx. In [44], a rather complete table presents a list of high-sensitivity optical Rx demonstrations. There, previously reported sensitivities for uncoded transmission are in order of 147 PPB for OOK at 10 Gbps, 45 PPB for DPSK at 12.5 Gbps, and some 100 PPB for BPSK at 10 Gbps [44]. Hereafter, the assumption is made that for data rates in the order of few tens of Gbit/s—in a time frame of about 10 years from now—on-going developments could potentially allow for Rx sensitivities close to 50 PPB, for coherent modulations and DPSK, and about 100 PPB for OOK.

The estimation of the PPB for each link at 0.1, 1, 5, 10 and 20 Gbps is presented in Table 12.3. By inspecting the calculated values, it is readily seen that for a small LEO platform to the GEO relay data transmission using OOK would be possible for

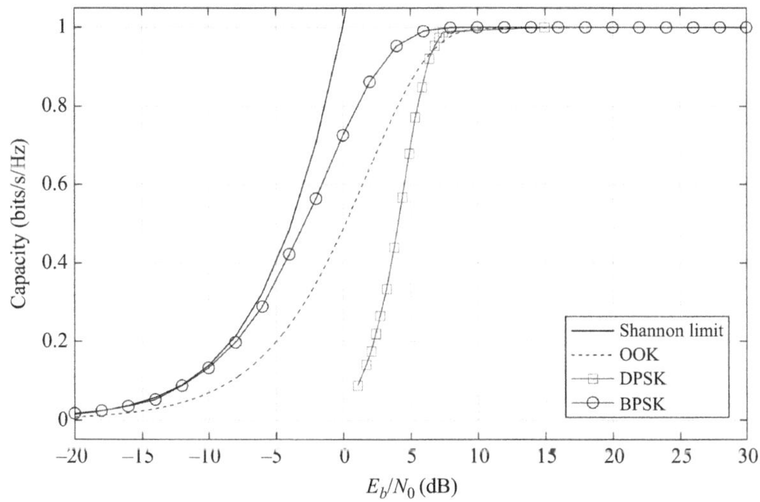

Figure 12.6 Maximum capacity as a function of the symbol-level SNR, for different modulation formats. The plot for DPSK corresponds to the performance when the observation window comprises two symbols. The curve for DPSK was taken from [45]

data rate below the Gbit/s regime and to transmit about 1 Gbps or more then DPSK or BPSK modulation would be required. In the case of a big LEO platform, transmission up to 10 Gbps seems to be possible. When the user communicating with the GEO relay is an UAV, data rates up to 5 Gbps would be feasible using either DPSK or BPSK, while OOK could work up to a few Gbit/s. Finally, in the downlink from the GEO relay to the OGS, data rates up to 10 Gbps could be possible, while for higher rates, it would be advisable to split the total throughput into various channels using WDM techniques.

Finally, the channel capacity for a given symbol-level SNR, for different modulation formats, is presented in Figure 12.6. The plot for DPSK corresponds to the performance when the observation window comprises two symbols [45]. Note that multisymbol detectors may close the gap with respect to the BPSK capacity curve. To make use of this information, the calculation of the available SNR for all links is presented in Table 12.4 in the case of direct detections. Values given are for OOK modulation format. In addition, values for DPSK and BPSK are given in square brackets and parenthesis, respectively. Moreover, only values for which reliable communication is possible—in terms of the Rx sensitivity discussion presented above—are given. It is noteworthy to mention that SNR values presented in Table 12.4 are based on the link budgets give in Table 12.2, where the Tx is assumed to be average power limited. Thus, the transmitted peak power for OOK is twice the average, while for DPSK and BPSK, the peak and average power are the same. Note that, although they

Table 12.4 Average symbol-level SNR in decibels, for all link scenarios defined by Tables 12.1 and 12.2. Values presented are for direct detection, i.e., OOK and DPSK (in square brackets), and for coherent detection, i.e., BPSK (in parenthesis). SNR values are not given for bit rates at which preamplified receiver sensitivity is not enough to allow reliable communication

Bit rate	Small L–R	Big L–R	X–R	R–G
100 Mbps	22.3 [22.0] (24.6)	33.1 [31.9] (32.3)	28.4 [27.6] (28.8)	31.2 [30.1] (30.7)
1 Gbps	12.3 [12.0] (14.6)	23.2 [21.9] (22.3)	18.4 [17.6] (18.8)	21.2 [20.1] (20.7)
5 Gbps	–	16.2 [15.0] (15.3)	11.4 [10.6] (11.7)	14.2 [13.1] (13.8)
10 Gbps	–	13.2 [12.0] (12.3)	–	11.2 [10.1] (10.8)
20 Gbps	–	–	–	–

assume the same average power, the SNR value for BPSK is larger than for DPSK, reflecting the fact that former uses coherent detection using a laser LO.

When compared with the maximum achievable capacity curve in Figure 12.6, the expected SNR values indicate that, in principle, maximum profit of the channel usage could be obtained. In this scenario, error correction with high code rates can be applied in order to maximize the bandwidth occupancy for the transmission of information bits.

Up to this point, all analysis has been performed considering uncoded transmission only. Nevertheless, a communications system will always be protected with an error correction code. In the following section, the implementation of FEC codes is presented, while taking into account the particularities of the user and feeder optical channels in a GEO relay scenario as well as the type of processing.

12.7 Forward error correction

An overview of different FEC codes defined in the framework of the CCSDS for near earth and deep space communications is provided in the following. These codes that were intended for point-to-point links, i.e., without relay, can be used as building blocks for data relay systems. Complete solutions for data relay systems will be discussed in the next subsections. Amongst others, the following channel codes are defined in CCSDS [46–48].

- **RS codes.** Hard decision decoding is done for short/medium sized blocks [49]. Due to the limited block length and the fact that soft information is not exploited at the decoder, coding gain is limited, in particular w.r.t. modern, iterative codes.
- **RS and CCs (RS+CC).** This serially concatenated scheme consists of an inner CC processing soft information and an outer RS code fixing residual (bursty) symbol errors of the inner code. Due to the lack of iterating between the blocks

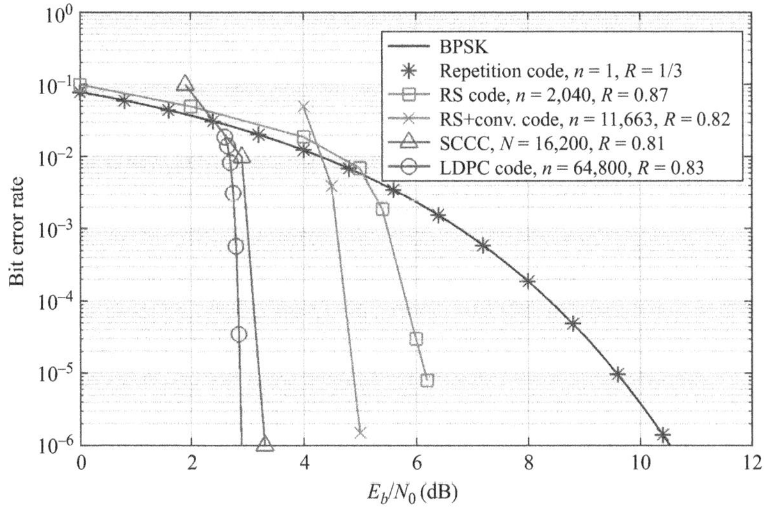

Figure 12.7 Comparison of different FEC schemes in terms of BER versus E_b/N_0 for a binary input AWGN channel. Additionally the bit error probability for uncoded BPSK is shown

and relying on soft-input–hard-output inner decoders, this option is inferior to modern codes in terms of performance.

- **Turbo codes.** Both serial concatenated CCs (SCCCs) and parallel concatenated convolution codes are proposed in the CCSDS standard and belong to the class of iteratively decodable turbo codes with excellent performance.
- **Low-density parity check (LDPC) codes.** Different types of LDPC codes are part of CCSDS. Some of them and originate from the Digital Video Broadcasting-Satellite 2 standard. Thanks to the large block lengths and soft decoders, LDPC codes belong to one of the most powerful coding schemes.

In Figure 12.7, bit error rate (BER) simulation results versus E_b/N_0—i.e., the energy per information bit to noise PSD ratio—for various CCSDS channel codes on an AWGN channel with BPSK are exemplified. One can observe from the figure that LDPC codes show the best performance among the considered channel codes. Notable gains in the order of a few dB are visible with respect to RS codes, concatenated RS and CC. Small gains in the order of a few tenths of dB are present with respect to SCCCs for the setup in the figure. From a BER performance point of view, LDPC and SCCCs are a natural choice, whenever complexity constraints are not stringent. As a complement, also the BER versus E_b/N_0 of a repetition code with rate 1/3 is depicted under soft decoding. Observe that there is a gap of around 5.6 dB with respect to the LDPC code at a BER of 10^{-4}. Despite this gap, repetition codes might be a reasonable choice if decoding complexity is a bottleneck.

For correlated fading channels, the following additions to the channel coding options above can be made:

- **Long PHY interleaver** is usually placed after the channel encoder. Thereby code symbols of several codewords are interleaved among each other before modulation and transmission over the channel. In this setting 'long' means that the interleaver duration shall exceed the coherence time of the channel. This way after deinterleaving at the Rx side, errors introduced by the fading are spread over several code words. If the interleaver is chosen long enough, there is virtually no degraded code performance compared to an uncorrelated channel [50]. However, the interleaver length is often limited by practical constraints (e.g., memory, delay, etc.).
- **PKT code** is placed as an additional layer of error protection as a complement to the PHY code. To this end, the user data is first portioned into PKTs and encoded by the PKT code where a code symbol in an entire PKT. The data is then further encoded by a PHY code. The duration of a PKT codeword shall be longer than the channel coherence time.

Next various FEC schemes for data relay systems are discussed.

12.7.1 Full decoding on board of the relay

Consider the U–R link. In fact, without complexity constraints on the relay, the best solution in terms of bandwidth efficiency/error rate performance is the following: encode the data on the user side and decode it completely on board of the relay. This way, upon a proper choice of the modulation and coding scheme, nearly all errors are corrected on board of the relay and all redundancy data for U–R link is removed at the relay. Then, a further encoding of the recovered user data (not containing any redundancy) takes place to protect the data from errors on the R–G link. A modern channel code with high coding gain, such as an LDPC code would be the natural choice here. Soft decoding of a modern code on board of the relay is problematic from a complexity point of view, at least nowadays. Therefore, given stringent complexity constraints, simpler codes might be used paired with simple, preferably hard decoders. This yields performance losses that can be mitigated by considering alternative FEC schemes (see, e.g., partial coding).

12.7.2 Decoding on ground only

As an alternative to full decoding on board of the relay, one may shift decoding complexity to the ground station where computation resources are plentiful. This scheme is called decoding on ground. On the one hand, this solution has the disadvantage that the bandwidth occupation increases, at least when the quality of the U–R link requires the use of medium/low code rates. For optical links, typically, power is plentiful, but fading events may require medium/lower code rates. On the other hand, decoding on ground only is a simple, low-complexity scheme with good performance and certain flexibility to modify the PHY FEC. In fact, among all considered schemes, it imposes the lowest computational burden on the relay.

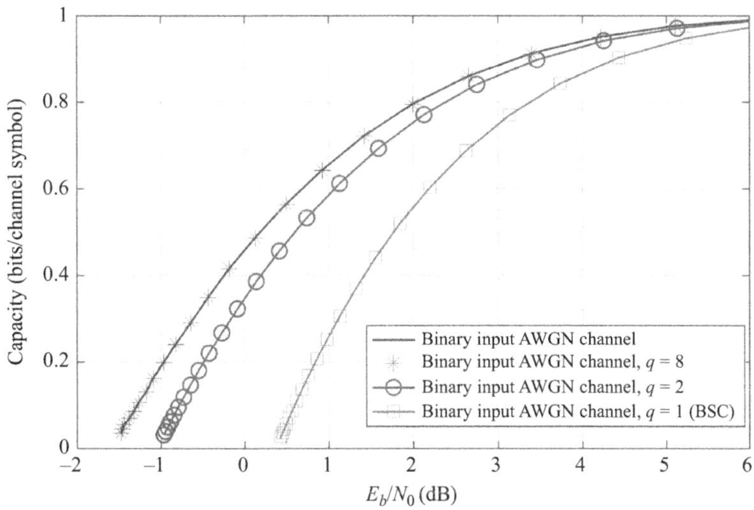

Figure 12.8 Channel capacity versus E_b/N_0 for binary-input AWGN channel with different quantization levels at the demodulator, where the variable q is the number of bits per quantization level

Due to technological constraints, often demodulation of the waveform at the relay takes place. This is followed by a modulation step. We call this type of relay systems semitransparent. To exploit the full capabilities of modern codes and their soft iterative decoders (on ground), they need to have access to soft channel information. Consequently, soft demodulation of the U–R link signal at the relay is desirable. Denote by 2^q the number of quantization levels at the demodulator output, i.e., each value is represented by q bits. This implies that q times more bandwidth and power (aggregated) are required to transmit soft information after demodulation compared to hard demodulation.

A special case consists for $q = 1$. Hard demodulation at the relay can be done in order to improve the bandwidth efficiency on the R–G link and to reduce complexity at the relay. Usually, the use of hard demodulators is paired with a performance loss of around 2 dB compared to soft demodulation (see Figure 12.8). However, in the current setup, the required bandwidth is reduced by a factor of q with respect to soft-demodulation and so does the required power (since only one and not q bits per demodulated code symbol need to be transmitted). Therefore, decoding on ground only with $q = 1$ is the preferred option.

There exist several flavors of decoding on ground only:

- **Two step encoding**: One may perform encoding at the user side to protect the data only from errors on the U–R link. Then, a second encoding step at the relay takes place adding additional complexity (and reducing flexibility of the scheme). Let us denote by R_{ur} the code rate for the U–R link and by R_{rg} the code rate for the

R–G link. The number of quantization levels is chosen to be $q = 1$. To transmit the k_{ur} bits, the U–R link needs to carry $k_{ur} \cdot 1/R_{ur}$ bits. After reencoding on board of the relay, the R–G link needs to carry $k_{ur} \cdot 1/R_{ur} \cdot 1/R_{rg}$ bits. By contrast, when decoding on board of the satellite is allowed, the amount of bits is at most $k_{ur} \cdot 1/R_{rg}$ for the R–G link. This means a factor of $1/R_{ur}$ increase in required data rate (bandwidth) w.r.t. decoding at the relay.

- **One step encoding**: Another alternative is to perform encoding only at the user with a rate R_{ug} in order to protect the data against impairments on both U–R and R–G links. Then, no encoding at the relay needs to be done. We have that $\min(R_{ur}, R_{rg}) \geq R_{ug} \geq R_{ur}R_{rg}$. The semitransparent relay performs demodulation and modulation. As sketched previously hard demodulation is the better choice from bandwidth/power consumption/complexity point of view. To transmit k_{ur} bits, now on both links $k_{ur} \cdot 1/R_{ug}$ bits need to be sent. For the same performance, higher bandwidth is required compared to decoding on board of the satellite. This option is a good choice when both communication channels allow high rate codes, i.e., when $R_{ur} \cdot R_{rg}$ is close to one or when bandwidth is plentiful.

Decoding on ground may not be the best choice in terms of bandwidth usage (or power usage for a fixed bandwidth). It is the most commonly employed scheme for relaying since it is flexible, highly performant and simple, i.e., it requires least processing capabilities on board of the relay.

12.7.3 Partial decoding scheme

There exist several options for partial decoding. The main idea is to decode parts of the data at the relay using simple codes and decoders and to decode the rest on ground where more computational power is available. This way some errors might be directly corrected at the relay and unnecessary redundancy on the R–G link is avoided.

One may consider the following approach. At the user side, encoding takes place and the data is transmitted to the relay, where a first low-complexity decoding attempt is done. If decoding is successful, the redundancy added at the user side can be removed and the user data is forwarded to the encoder at the relay. If decoding is not successful, the entire erroneous codeword is forwarded to the encoder. Additional redundancy is added at the encoder and the data is forwarded to ground for decoding. If decoding on board succeeds, this scheme is similar to decoding on board of the relay as sketched before in terms of bandwidth constraints. If decoding does not succeed, the scheme is similar to decoding on ground only as sketched before. Clearly, the success of decoding on board of the relay is strongly related to the U–R link quality and the complexity constraints on the PHY decoder. The setup is sketched in Figure 12.9(a).

Candidates for the PHY codes are for instance:

- **Low-memory CCs with interleaving**. Here, CCs with different memory can be considered, with coding gains with respect to the repetition code ranging from 4 dB (for the memory-2, rate 1/2 case) up to 7 dB (for the memory-6 case) under soft decision decoding.

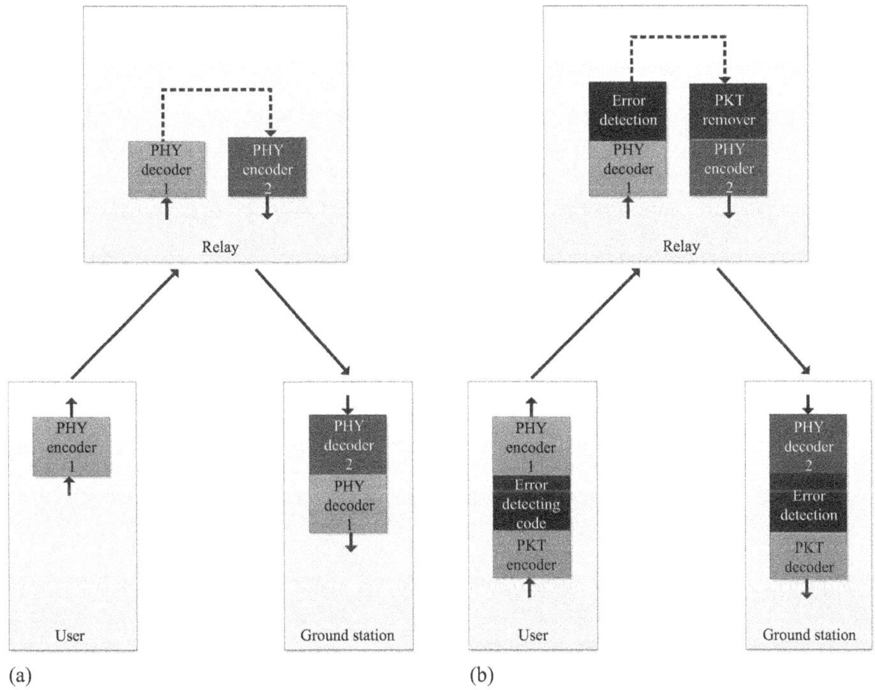

Figure 12.9 (a) Partial decoding scheme and (b) layered FEC scheme

- **Algebraic codes**, such as Bose–Chaudhuri–Hocquenghem (BCH) codes (or RS codes). Efficient syndrome decoders based on look-up tables available and are, for example, used in terrestrial fiber optical communications for 100 Gbps links [51].
- **Concatenated schemes.** Concatenations of the above codes may yield a more powerful channel code. An example are BCH product codes as also used in terrestrial fiber optical communications [51] whose component codes might decoded at the relay, while (upon decoding failure at the relay) the product code is decoded on ground. Modern codes, such as LDPC or turbo codes, also belong to the class of concatenated schemes [52,53]. Similarly, their component codes might be decoded at the relay (eventually using simple decoders), while the concatenated scheme is decoded on ground.

12.7.4 Layered coding scheme

A promising alternative in case of U–R (fading) links lies in the use of an additional PKT code. To this end, user data is split into K PKTs, each of them having L bits, and encoded by means of a PKT code yielding N PKTs, each of them having L bits. Each PKT is further subject to an error detection mechanism (usually a CRC code or inherent error detection capability of the PHY decoder) in order to ensure its integrity after transmission. The PKTs are forwarded to lower layers. At PHY, usually a simple error correcting code is additionally used to protect the PKTs against sporadic bit errors due to noise, since a single bit error may corrupt an entire PKT. The aim of using a PKT level code is to protect the data against sequences of errors introduced by the (correlated) communication channel.

After transmission on the U–R link, PHY decoding at the relay takes place to correct sporadic bit errors. Note that the PHY code is a simple code here, which can be tailored to the complexity limitation of the relay. In a next step, error detection takes place to check the integrity of all PKTs. All corrupted PKTs are discarded at the relay.

In order to spare bandwidth on the R–G link further PKTs at the relay can be discarded by a PKT remover as follows. Denote by K' the number of correctly received PKTs at the relay. $K \leq K'$ is a necessary condition for successful decoding. For many codes, $K \leq K'$ is not sufficient to ensure decoding success. Therefore, let us require $K + \Delta \leq K'$, where Δ is a design parameter (also referred to as overhead) that is usually much smaller than K (e.g., in the order of a few percent of K). Else, decoding will fail with a high probability and one may discard all PKTs already at the relay. Assume that at the relay K' PKTs are correctly received. Then, a PKT remover at the satellite discards PKTs until only $K + \Delta$ PKTs remain. The choice of the overhead gives a trade-off between the code performance and bandwidth occupation on the R–G link.

After the PKT remover, no decoding of the PKT code takes place at the relay. Instead, the remaining $K + \Delta$ PKTs are forwarded to lower layers, encoded again and transmitted over the R–G link. On ground, decoding of the code for the R–G link takes place. Then, again, error detection for each of the PKTs takes place. Finally, a PKT decoder attempts to correct the missing PKTs. The setup is sketched in Figure 12.9(b).

An advantage of the layered scheme lies in the fact that on board of relay no complex decoding operations take place. Only PHY decoding of a simple code needs to be done, followed by an error detection and PKT removal step. The code used on PHY can be an algebraic code or a low memory CC. Its purpose is to correct sporadic bit errors. Another advantage of the layered scheme is that the relay forwards only $K + \Delta$ PKTs to the lower layers, where K is the number of information PKTs. For sake of comparison with the former schemes assume that $K \cdot L = k_{ur}$. To transmit a file of k_{ur} bits, one has to send $(K + \Delta) \cdot L \cdot 1/R_{rg} = (k_{ur} + \Delta \cdot L) \cdot 1/R_{rg}$ bits on the user link. The parameter Δ is chosen to be a small fraction of K, typically in the order of a few percent.

Layered coding can be seen as a special case of partial decoding. Both schemes may implement a similar PHY code, complemented by a PKT code for layered decoding. While at the relay a low complex decoding attempt of the PHY code is done, on ground the PKT layer code is decoded in order to resolve residual errors on the U–R link. We point out that PKT level codes perform best on correlated communication channels. They work well if the PKT codeword duration is much longer than the coherence time of the channel.

12.7.5 Interleaving options

12.7.5.1 Long PHY interleaver

A typical strategy for correlated channels is the employment of long PHY interleavers. These interleavers spread over a multitude of code words. The goal is to let every codeword experience good and bad channel states. In this way, the number of errors in every codeword shall be similar after deinterleaving. Upon a proper choice of the code parameters, the number of errors in a codeword shall not exceed its error correcting capabilities and successful decoding is possible. On the contrary, without interleaving some code words would contain too many errors others maybe none. For a given date rate D, the length of the interleaver (interleaver depth d) is usually chosen such that d/D is much larger than the coherence time t of the channel. More formally $d = D \cdot t \cdot c$, where $c \gg 1$. The value of c determines the code performance and needs to be carefully chosen for the targeted communication channel.

To assess the effect of the interleaver, consider performance of a rate 2/3 LDPC code of length 64,800 on a lognormal block-fading channel with AWGN assuming BPSK. For the lognormal fading, let us choose the parameter $s = 0.5$ (standard deviation of the underlying Gaussian process). Further, choose m (mean of the underlying Gaussian process) such that the average power of the lognormal process is one.

The block-fading channel is implemented as follows. Based on a Markov process with average state duration $1/p_{ij}$ a channel state is selected. Each channel state is associated to a fading amplitude, sampled from a lognormal distribution. For the experiments different $1/p_{ij}$ were considered, where high values of $1/p_{ij}$ mimic a strongly correlated communication channel. The results are summarized in Figure 12.10.

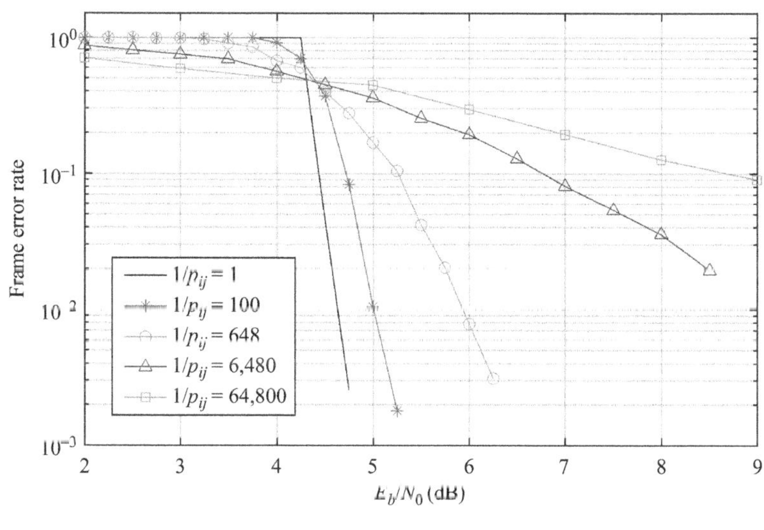

Figure 12.10 FER versus E_b/N_0 of an (64,800, 43,200) LDPC code on a lognormal block fading channel for different average state durations

In Figure 12.10, the curve with $1/p_{ij} = 1$ represents the frame error rate (FER) versus E_b/N_0 for a rate-2/3 LDPC code of length $n = 64{,}800$ symbols on a lognormal fading channel with no correlation. Significant losses in performance are visible if $1/p_{ij}$ is comparable to the codeword length n, i.e., for $1/p_{ij} = 64{,}800$. This is owing to the fact that code symbols in a codeword often experience similar level of fading and the channel code is not capable of compensating for it. For $1/p_{ij} = n/100 = 648$ the loss compared to the uncorrelated case ($1/p_{ij} = 1$) is within 1.4 dB at a FER of 10^{-2}. These observations suggest that for the example lognormal fading channel the interleaver depth shall be at least 100 times larger than the channel coherence time times the data rate to avoid significant losses in performance.

Regarding the interleaver dimensioning the procedure is as follows:

- Determine the channel coherence time t and the required data rate D.
- Fix a value of c, based on simulations and/or constraints on available memory and/or delay constraints.
- Compute the interleaver depth $d = D \cdot t \cdot c$.

12.7.5.2 PKT code with interleaved code symbols (packets)

Whenever PKT codes are used, the length of a PKT code word has to be chosen such that $NL/R_{PHY} = c \cdot D \cdot t_{ur}$, where L denotes the PKT size in bits, c a constant usually larger than one, D the data rate, t_{ur} the coherence time of the U–R link and R_{PHY} the code rate of the PHY code on the U–R link. The constraints here are as for the long PHY interleaver. In case of structured LDPC PKT codes on correlated channels, it is required that the code symbols (PKTs) are interleaved among each other before transmission to avoid performance losses.

12.7.6 Comparison of coding schemes

Consider a simplified setup where the R–G link is assumed ideal. This assumption can be justified if the PHY code is dimensioned in both cases such that it can correct quasi all error events on the R–G link. For the U–R link, consider a block lognormal fading channel with Gaussian noise and BPSK modulation. The mean state duration $1/p_{ij}$ was set to 64,800 BPSK modulated channel symbols, while the parameter s of the lognormal distribution[2] was varied from 0.5 (considerable fading) to 0.015 (weak fading). The following is analyzed:

- **Layered coding (i.e., partial decoding complemented by a PKT code).** Here, a RS PHY code is selected, complemented by a maximum distance separable PKT level code. The length of the RS-encoded PKT codeword was chosen to be approximately 6,480,000 bits with an overall rate of 1/2. The RS codeword is selected such that the information length corresponds to the packet size L.
- **Full decoding on board of the satellite.** Assume a SCCC with rate 1/2 that is interleaved with a long PHY interleaver of length 6,480,000 symbols.

[2]As before s is the standard deviation of the underlying Gaussian process.

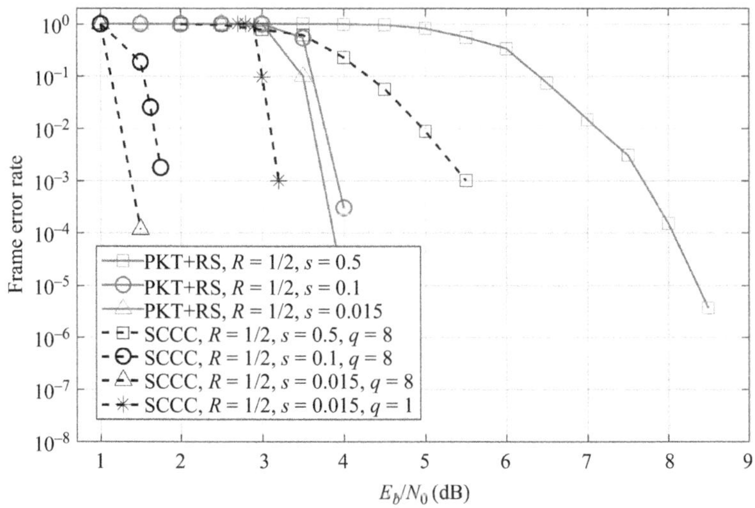

Figure 12.11 Comparison of layered coding and decoding on board of the satellite

Further, at the semitransparent satellite demodulation is done with q bits per bit reliability, with $q = 1$ or $q = 8$ for the SCCC scheme, while for layered coding always hard demodulation is done.

Figure 12.11 shows FER performances of both schemes. Note that if for demodulation with $q = 8$ (quantized soft demodulator) of the SCCC-coded modulation symbols there is a gain of around 2–2.4 dB w.r.t. the layered coding (at the price of a q times higher data rate on the R–G link). If q is chosen to be one (hard demodulator), the gain is around 0.5 dB for this specific setup.

The results in Figure 12.11 suggest layered coding is a suitable option for the considered correlated lognormal fading channel while for other channels dedicated simulations have to be done. In fact, assuming hard demodulation at the satellite, it provides similar performance (0.5 dB gap) to the best decoding strategy, full decoding at the relay, but with much lower complexity burden at the relay, while having similar spectral efficiency on both links. Note that (quantized) soft demodulation on board of the relay is often not desired. In particular, consider decoding on ground only: as discussed in Section 12.7.2, for $q > 1$ the data rate/bandwidth requirements are increased, but also the overall power requirements (since q symbols instead of one symbol need to be transmitted). This is clearly not desirable.

12.8 Summary

The analysis presented gives a general overview on different aspects for the communication chain in a relay-based system for high-speed data rates. The user sends its data down to Earth through a GEO satellite, and therefore, there are two main links, namely, the U–R and R–G links. A distinction of the user has been made, where either LEO satellites or UAVs have been considered. In case of the LEO platform user a small—e.g., CubeSat—and a large satellite have been taken into account. A channel model has been defined assuming that transmission through the U–R and R–G links is done optically. Special attention was taken into modeling the effects of the pointing errors, due platform microvibrations, for the user terminal, and dimensioning of the corresponding link has been accordingly. Next, based on the channel model, link budget calculations were performed in order to give an idea of the possibilities of future ultra-high-speed data relay systems. In addition, a Rx sensitivity analysis was done, based on extrapolation of previously reported experiments on the sensitivities for uncoded transmission. From this, possibly achievable maximum data rates were estimated for each link in the relay scenarios considered here, taking into account whether the Rx is set to work with direct or coherent detection.

Code design for relay systems depends on several constraints. Under strong complexity constraints on the relay and high powers on the U–R link (thus high code rates) decoding on ground only is the preferred option. Whenever the U–R link requires the use of a medium/low rate code, partial coding schemes and layered schemes might be a good choice depending on the communication channel. For correlated fading channels, layered coding schemes exploit their full capabilities. If complexity constraints on the relay are not stringent, full decoding at the relay is the best choice.

Abbreviations

AO	adaptive optics
ASE	amplified spontaneous emission
AWGN	additive white Gaussian noise
BCH	Bose–Chaudhuri–Hocquenghem
BER	bit error rate
BPSK	binary phase shift keying
CC	convolutional codes
CCSDS	consultative committee for space data systems
CWDM	coarse wavelength division multiplexing
DPSK	differential phase-shift keying
DWDM	dense wavelength division multiplexing
EDFA	erbium-doped fiber amplifier
ESA	European Space Agency
FEC	forward error correction
FER	frame error rate
Gbit/s, Gbps	gigabits per second
GEO	geostationary equatorial orbit
LCT	laser communications terminal
LDPC	low-density parity check
LEO	low Earth orbit
LO	local oscillator

L–R	LEO to relay
Mbit/s, Mbps	megabits per second
NEP	noise equivalent power
OGS	optical ground station
OOK	on–off Keying
OPLL	optical phase-locked loop
PDF	probability density function
PHY	physical
PKT	packet
PPB	photons per bit
PPM	pulse-position modulation
PSD	power spectral density
RF	radiofrequency
R–G	relay to ground
RMS	root-mean-square
RS	Reed–Solomon
Rx	receiver
SCCC	serial concatenated convolutional codes
SI	scintillation Index
SMF	single mode fiber
SNR	signal-to-noise ratio
Tx	transmitter
UAV	unmanned aerial vehicle
U–R	user to relay
WDM	wavelength division multiplexing
X–R	UAV to relay

References

[1] D. C. Troendle, C. Rochow, K. Saucke, *et al.*, "Alphasat TDP1 three years optical GEO data relay operations," in *34th AIAA International Communications Satellite Systems Conference*, 2016, pp. 2016–5700.

[2] M. Agnew, L. Renouard, and A. Hegyi, "EDRS-SpaceDataHighway: Near-real-time data relay services for LEO satellites and HAPs," in *30th AIAA International Communications Satellite System Conference, ICSCC*, 2012.

[3] Y. Koyama, M. Toyoshima, Y. Takayama, *et al.*, "SOTA: Small optical transponder for micro-satellite," in *Space Optical Systems and Applications (ICSOS), 2011 International Conference on*, 2011, pp. 97–101.

[4] G. D. Fletcher, T. R. Hicks, and B. Laurent, "The SILEX optical interorbit link experiment," *Electronics & Communication Engineering Journal*, vol. 3, no. 6, pp. 273–279, 1991.

[5] A. Alonso, M. Reyes, and Z. Sodnik, "Performance of satellite-to-ground communications link between ARTEMIS and the optical ground station," in

Optics in Atmospheric Propagation and Adaptive Systems VII, 2004, vol. 5572, no. 1, pp. 372–383.

[6] Y. Takayama, T. Jono, Y. Koyama, *et al.*, "Observation of atmospheric influence on OICETS inter-orbit laser communication demonstrations," in *Free-Space Laser Communications VII*, 2007, vol. 6709, no. 1, p. 67091B.

[7] M. Gregory, F. Heine, H. Kämpfner, R. Meyer, R. Fields, and C. Lunde, "Tesat laser communication terminal performance results on 5.6 Gbit coherent inter satellite and satellite to ground links," in *International Conference on Space Optics ICSO 2010*, 2010.

[8] Z. Sodnik, B. Furch, and H. Lutz, "Optical intersatellite communication," *IEEE Journal of Selected Topics in Quantum Electronics*, vol. 16, no. 5, pp. 1051–1057, 2010.

[9] G. Muehlnikel, H. Kämpfner, F. Heine, *et al.*, "The Alphasat GEO laser communication terminal flight acceptance tests," in *Proc. International Conference on Space Optical Systems and Applications (ICSOS)*, 2012.

[10] D. Giggenbach, P. Becker, R. Mata-Calvo, C. Fuchs, Z. Sodnik, and I. Zayer, "Lunar optical communications link (LOCL): Measurements of received power fluctuations and wavefront quality," in *Proc. International Conference on Space Optical Systems and Applications (ICSOS)*, 2014.

[11] Z. Sodnik, H. Smit, M. Sans, *et al.*, "Results from a Lunar Laser Communication experiment between NASA's LADEE satellite and ESA's optical ground station," in *Proc. International Conference on Space Optical Systems and Applications (ICSOS)*, 2014.

[12] D. M. Boroson, B. S. Robinson, D. V. Murphy, *et al.*, "Overview and results of the Lunar laser Communication Demonstration," in *Free-Space Laser Communication and Atmospheric Propagation XXVI*, 2014, vol. 8971, p. 89710S.

[13] B. L. Edwards, D. Israel, K. Wilson, J. Moores, and A. Fletcher, "Overview of the laser communications relay demonstration project," in *Proceedings of SpaceOps*, 2012, vol. 1261897.

[14] B. Friedrichs, "Data processing for broadband relay satellite networks–digital architecture and performance evaluation," in *Proc. Int. Commun. Satell. Syst. Conf.*, 2013.

[15] L. C. Andrews and R. L. Philips, *Laser Beam Propagation through Random Media*, 2nd ed. Bellingham: SPIE Press, 2005.

[16] N. Perlot and D. Fritzsche, "Aperture-averaging – Theory and measurements," in *Free-Space Laser Communication Technologies XVI*, 2004, vol. 5338, pp. 233–242.

[17] F. Vetelino, *Fade Statistics for a Lasercom System and the Joint PDF of a Gamma–Gamma Distributed Irradiance and its Time Derivative*, Department of Mathematics, University of Central Florida, Orlando, FL, 2006.

[18] B. Epple, "Simplified channel model for simulation of free-space optical communications," *Journal of Optical Communications and Networking*, vol. 2, no. 5, pp. 293–304, 2010.

[19] J.-M. Conan, G. Rousset, and P.-Y. Madec, "Wave-front temporal spectra in high-resolution imaging through turbulence," *Journal of the Optical Society of America A*, vol. 12, no. 7, pp. 1559–1570, 1995.

[20] G. A. Tyler, "Bandwidth considerations for tracking through turbulence," *Journal of the Optical Society of America A*, vol. 11, no. 1, pp. 358–367, 1994.

[21] J. Horwath, F. David, M. K. Knapek, and N. Perlot, "Coherent transmission feasibility analysis," in *Free-Space Laser Communication Technologies XVII*, 2005, vol. 5712, pp. 13–23.

[22] S. Schaefer, M. Gregory, and W. Rosenkranz, "Coherent receiver design based on digital signal processing in optical high-speed intersatellite links with M-phase-shift keying," *Optical Engineering*, vol. 55, no. 11, p. 111614, 2016.

[23] D. Giggenbach and H. Henniger, "Fading-loss assessment in atmospheric free-space optical communication links with on–off keying," *Optical Engineering*, vol. 47, no. 4, p. 046001, 2008.

[24] K. Kiasaleh, "On the probability density function of signal intensity in free-space optical communications systems impaired by pointing jitter and turbulence," *Optical Engineering*, vol. 33, no. 11, pp. 3748–3757, 1994.

[25] M. Toyoshima, Y. Takayama, H. Kunimori, T. Jono, and S. Yamakawa, "In-orbit measurements of spacecraft microvibrations for satellite laser communication links," *Optical Engineering*, vol. 49, no. 8, pp. 83604–83604, 2010.

[26] M. E. Wittig, L. Van Holtz, D. E. L. Tunbridge, and H. C. Vermeulen, "In-orbit measurements of microaccelerations of ESA's communication satellite OLYMPUS," in *Free-Space Laser Communication Technologies II*, 1990, vol. 1218, pp. 205–214.

[27] J. Horwath and C. Fuchs, "Aircraft to ground unidirectional laser-communications terminal for high-resolution sensors," in *Lasers and Applications in Science and Engineering*, 2009, vol. 7199, pp. 7199-09.

[28] R. Nelson, T. Ebben, and R. Marshalek, "Experimental verification of the pointing error distribution of an optical intersatellite link," in *1988 Los Angeles Symposium–OE/LASE'88*, 1988, pp. 132–142.

[29] G. Planche and V. Chorvalli, "SILEX in-orbit performances," in *5th International Conference on Space Optics*, 2004, vol. 554, pp. 403–410.

[30] M. Toyoshima, "Near-Earth Laser Communications," H. Hemmati, Ed. Boca Raton, FL: CRC Press, 2009.

[31] F. David, D. Giggenbach, H. Henniger, J. Horwath, R. Landrock, and N. Perlot, "Design considerations for optical inter-HAP links," in *Proceedings ICSSC, 22nd AIAA International Communications Satellite Systems Conference & Exhibit, Monterey, CA*, 2004.

[32] R. G. Marshalek, "Near-Earth Laser Communications," H. Hemmati, Ed. Boca Raton, FL: CRC Press, 2009, p. 59.

[33] F. Fidler, "Optical Communications for High-Altitude Platforms," Wien: Technische Universität Wien, 2007.

[34] D. Giggenbach, R. Purvinskis, M. Werner, and M. Holzbock, "Stratospheric optical inter-platform links for high altitude platforms," in *Proceedings of the 20th International Communications Satellite Systems Conference, Montreal*, 2002.

[35] H. Takenaka, M. Toyoshima, and Y. Takayama, "Experimental verification of fiber-coupling efficiency for satellite-to-ground atmospheric laser downlinks," *Optical Express*, vol. 20, no. 4, pp. 15301–15308, 2012.

[36] Y. Dikmelik and F. M. Davidson, "Fiber-coupling efficiency for free-space optical communication through atmospheric turbulence," *Applied Optics*, vol. 44, no. 23, pp. 4946–4952, 2005.

[37] R. J. Noll, "Zernike polynomials and atmospheric turbulence," *Journal of the Optical Society of America*, vol. 66, no. 3, pp. 207–211, 1976.

[38] M. P. Cagigal and V. F. Canales, "Generalized Fried parameter after adaptive optics partial wave-front compensation," *Journal of the Optical Society of America A*, vol. 17, no. 5, pp. 903–909, 2000.

[39] R. D. Hudson, *Infrared System Engineering*. New York: John Wiley & Sons, 1969.

[40] L. N. Binh, *Optical Fiber Communication Systems with Matlab and Simulink Models; 2nd ed.* Hoboken, NJ: CRC Press, 2014.

[41] P. J. Winzer, A. Kalmar, and W. R. Leeb, "Role of amplified spontaneous emission in optical free-space communication links with optical amplification: impact on isolation and data transmission and utilization for pointing, acquisition, and tracking," in *Proc. SPIE*, 1999, vol. 3615, pp. 134–141.

[42] M. Seimetz, *High-Order Modulation for Optical Fiber Transmission*, vol. 143. Springer, 2009.

[43] B. Moision, J. Wu, and S. Shambayati, "An optical communications link design tool for long-term mission planning for deep-space missions," in *Aerospace Conference, 2012 IEEE*, 2012, pp. 1–12.

[44] D. O. Caplan, "Laser communication transmitter and receiver design," *Journal of Optical and Fiber Communications Reports*, vol. 4, no. 4–5, pp. 225–362, 2007.

[45] M. Peleg and S. Shamai, "On the capacity of the blockwise incoherent MPSK channel," *IEEE Transactions on Communications*, vol. 46, no. 5, pp. 603–609, 1998.

[46] Huawei, "Polar codes: A 5G enabling FEC scheme," *HIRP Journal*, vol. 1, no. 1, pp. 2044–2047, Jun. 2015.

[47] CCSDS, "Flexible Advanced Coding and Modulation Scheme for High Rate Telemetry Applications," no. 131.2.B.1. Consultative Committee for Space Data Systems (CCSDS), Washington, DC, Mar. 2012.

[48] ETSI, *Digital Video Broadcasting (DVB); Second Generation Framing Structure, Channel Coding and Modulation Systems for Broadcasting, Interactive Services, News Gathering and Other Broadband Satellite Applications (DVB-S2)*. ETSI, 2009.

[49] A. K. Majumdar, *Advanced Free Space Optics (FSO)*. New York, NY, USA: Springer, 2015.

[50] S. J. Zhang, X. Blow, and K. Fowler Liu, "Low-complexity decoding for non-binary LDPC codes in high order fields," *IET Communications*, vol. 11, no. 13, pp. 2042–2048, Nov. 2017.
[51] B. Smith, A. Farhood, A. Hunt, F. R. Kschischang, and J. Lodge, "Staircase codes: FEC for 100 Gb/s OTN," *Journal of Lightwave Technology*, vol. 30, no. 1, pp. 110–117, Jan. 2012.
[52] R. Gallager, *Low-Density Parity-Check Codes*. Cambridge, MA, USA: MIT Press, 1963.
[53] C. Berrou, A. Glavieux, and P. Thitimajshima, "Near Shannon limit error-correcting coding and decoding: Turbo-codes," in *Proc. IEEE Int. Conf. Commun. (ICC)*, 1993.

Chapter 13
On-board processing for satellite-terrestrial integration

Rainer Wansch[1], Alexander Hofmann[1], Christopher Stender[1], and Robért Glein[1]

Flexibility in satellites is one of the major requirements to make use of them in the 5G environment. A key method to achieve this goal is on-board processing. Satellites have become more and more flexible in recent days, started by the development years ago with the development of digital transparent processors to gain flexibility in frequency and channel allocation.

First, this chapter gives a brief history of on-board processers (OBPs) followed by a classification of OBPs. For illustration, the current design of the Fraunhofer OBP (FOBP) is described followed by an exemplary 5G use case for OBP using low-earth orbiting (LEO) satellites. The chapter closes with a short summary.

13.1 Brief history of on-board processing

In the following, selected examples are shown for digital transparent and digital regenerative processors operated in satellite communication systems. A detailed description of this classification is given in Section 13.2.

13.1.1 Airbus Inmarsat processor

The Inmarsat satellite communication system is operating in L-band providing "global" coverage by building global beams, 19 wide spot beams and more than 200 narrow spot beams. This system uses GEO satellites and its upgrade to the Inmarsat 6 generation is at the horizon. They use OBPs on their satellites since many years, especially since the introduction of the broadband global area network (BGAN) system, which can aggregate a number of channels and provide data rates up to about 1 Mbit/s.

Narrow spot beams are formed digitally by the OBP shown in Figure 13.1, which can also direct the needed power and spectrum to an individual area. The digital

[1] Fraunhofer Institute for Integrated Circuits IIS, RF and SatCom Systems Department, Erlangen, Germany

Figure 13.1 Photo of the Inmarsat processor as built by Airbus DS, United Kingdom. © Airbus DS. Extracted, with permission, from Artes Webpage [1,2]

transparent processor switches any gateway uplink to any mobile user downlink beam and vice versa.

The main features of this OBP are displayed in the following list:

- L-band interfaces
- Configurable digital filters between 200 kHz and some MHz
- 200 spot beam interfaces incl. digital beamforming capability
- 600 channels in total

Currently, the development and manufacturing of the Inmarsat 6 is also conducted by Airbus DS. This next generation will have a much larger number of beams and therefore needs a much more advanced/powerful OBP. More than 60,000 channels have to be switched and routed. This is done by signal fractioning into small channels and recombining them after routing to set up the required bandwidth. Technology base are application specific integrated circuits (ASICs), which provide the basic functionalities.

Overall, this kind of processor is digital transparent (filter and switch) based on ASICs.

13.1.2 Thales Alenia Space Spaceflex processor

Thales Alenia Space France also provides a digital transparent processor focusing on broadband solutions for high-throughput satellites (HTS) based on ASICs. One of the current versions of this processor (SpaceFlex4, depicted in Figure 13.2) has the following main properties:

- Input and output ports: Up to 20 × 20 (16 × 16 active)
- Port bandwidth: 250 MHz
- Channel bandwidth: programable from 312.5 kHz up to 125 MHz
- Channel spacing: down to 312.5 kHz

Figure 13.2 Photo of SpaceFlex4 processor EM. © Thales Alenia Space. Taken, with permission, from Thales Alenia brochure [3]

Table 13.1 Table of main performance parameters of the SpaceFlex processors [4]

Products	SpaceFlex 2	SpaceFlex 4	SpaceFlex 24	SpaceFlex 64
Number of I/O	Up to 8/8	Up to 16/16	Up to 48/48	>128 scalable product
I/O Bandwidth	Up to 250 MHz	Up to 250 MHz	500–600 MHz	500–600 MHz
Capacity (GHz) useful bandwidth	Up to 2	Up to 4	24 minimum	64 minimum
Embedded digital BFN	No	No	BFN ready	Yes
Dynamic management	Option	Option	Baseline	Baseline
			Spaceflex the new generation	

- Port dynamic range: higher than 40 dB
- Channel dynamic range: 30 dB
- Power of 540 W (for 16 × 16 × 250 MHz active matrix)
- Mass of 50 kg (for 20 × 20 × 250 MHz matrix)

In Table 13.1, it can be seen that SpaceFlex can provide a number of different configurations from a minimum of 2 GHz useful bandwidth up to at least 64 GHz. The input bandwidth has been increased from 250 to 500/600 MHz which allows the usage within the next generations of HTS that can provide wide channel bandwidths. SES uses an even more powerful version of this processor, SpaceFlex VHTS, on its SES17 satellite which is expected to be launched in 2020/21 and shall be used for

Figure 13.3 Main parts of the Redsat OBP, from left to right: Thales L-band Tx processor, L-band Rx-processor, a Mier Comunicaciones Ku/L-band down and up-converter and in the front Thales filters. © Thales Alenia Space. Reprinted from ESA Website [5]

more than 15 years. The processor adds the required flexibility for this satellite and enables to connect almost 200 spot beams.

Enabling technology for this OBP is an ASIC developed by TAS using ATMEL AT65RHA technology based on the ST C65Space process together with fast signal converters provided by Teledyne-e2v [9].

13.1.3 Thales Alenia Space Redsat

Thales Alenia Space Spain has developed a regenerative processor for the recently launched Hispasat 36W-1 satellite (based on the SmallGEO platform from OHB) operating in Ku-band. The satellite will provide Spain, Portugal, the Canary Islands and the South America with multimedia services using the reconfigurable Redsat payload, which offers better signal quality and flexible land coverage.

This processor enables broadband Internet data services using the DVB/MPEG-2 standard as transport layer. To make effectively use of this OBP, a dedicated ground control and management system has been developed to connect satellite terminals and gateways in a flexible manner. Next to this regenerative part, the satellite also provides transparent channels. The satellite comprises a direct radiating array receive antenna which incorporates beam-steering capabilities for four beams.

The Redsat OBP has the following main features:

- 4 channels
- L-band interfaces
- DVB-S2, 36 MHz, up to 118 Mbps
- DVB-RCS, up to 8 Mbps

It consists of a number of components (shown in Figure 13.3) such as the Ku to L-band down- and up-converter, the processor units covering the core functionality and filters to shape the signal.

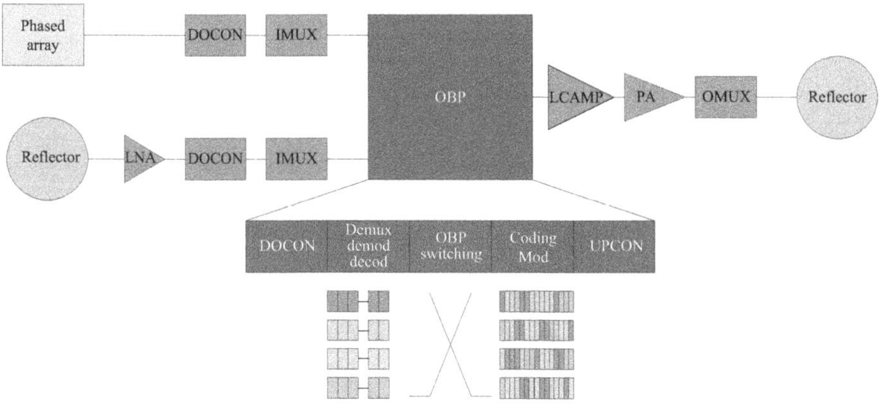

Figure 13.4 Main functional blocks of the regenerative part on the Hispasat satellite

Figure 13.4 shows the main processing functions of the satellite and the OBP. There is also an active array antenna embarked which provides optimum G/T for selected beams and users. It can be seen that Redsat demultiplexes, demodulates and decodes the received DVB-RCS signals and multiplexes the received data to build 4 DVB-S2 down-links. There are also means to configure the OBP for supporting Quality of Service (best effort, high priority and jitter sensitive traffic) demands.

13.2 Classification and applications of OBPs

13.2.1 Satellite payload architectures

Satellite payloads can be grouped in three different architecture categories:

- Bent-pipe
- Digital transparent
- Regenerative

These different architectures are briefly described in the following chapters.

13.2.1.1 Bent-pipe

The bent pipe architecture (Figure 13.5) is the classical satellite architecture as it provides the minimum needed functionality coming with the lowest complexity. It can be described as filter and forward architecture which receives the signal, amplifies it, converts it to the respective downlink frequency and amplifies it with the power amplifier to downlink the signal to earth.

Since the signals are untouched inside the satellite, all higher layer (e.g., network and connectivity) functions have to be addressed by complex gateways which provide the necessary connectivity to the ground system and provides less configurability. This configuration is mainly used in broadcast systems but also in communication

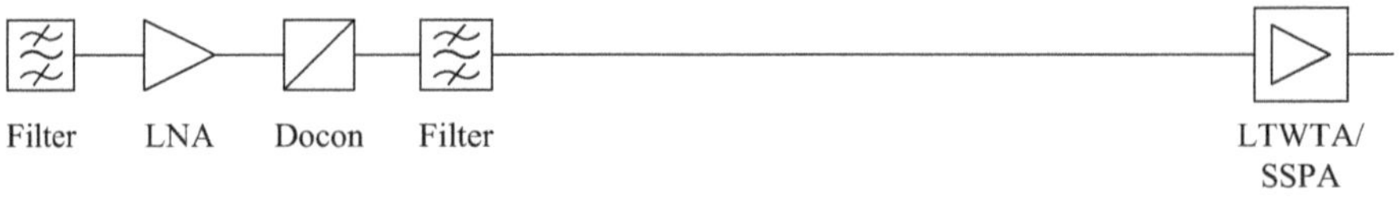

Figure 13.5 Block diagram of bent-pipe architecture with the main building blocks input filter (or multiplexer), low-noise amplifier (LNA), down-converter (DOCON), channel filter and high power amplifier [linearized travelling wavetube amplifier (LTWTA) or solid state power amplifier (SSPA)]

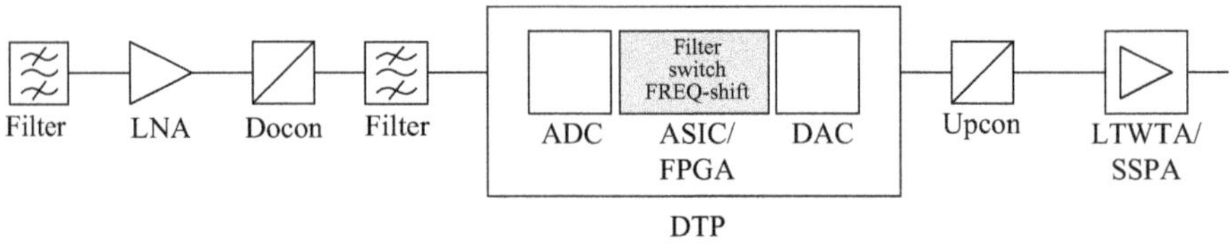

Figure 13.6 Architecture of DTP. Additional to the bent pipe architecture the DTP consisting of analog-to-digital converter (ADC), ASIC/FPGA and digital-to-analog converter (DAC) together with an upconverter (Upcon) is needed

systems, where a fixed gateway to beam connectivity is sufficient. By adding additional switches for routing purposes, some flexibility can be introduced in these systems.

13.2.1.2 Digital transparent processor

Digital transparent processors are used in satellites to introduce a higher level of flexibility. This architecture (Figure 13.6) adds a processor in the signal path to filter and switch/route the signals to the required beams. It does not regenerate the signal so it could also be used for analog signals. Many different implementations are possible for different applications. If a large number of channels is needed, an ASIC may be the most resource efficient variant.

This architecture brings a lot of additional flexibility in terms of signal processing capabilities such as filtering and routing. It is more complex and more power consuming as the bent-pipe architecture.

13.2.1.3 Regenerative processor

Architectures based on a regenerative processor add additional features as the received signals can be regenerated, and therefore, some additional gain can be achieved within the overall link-budget. The architecture (Figure 13.7) is very similar to the digital transparent case but offers different or additional signal processing features. Demodulation and decoding as well as modulation and encoding may be needed to make full use of this architecture.

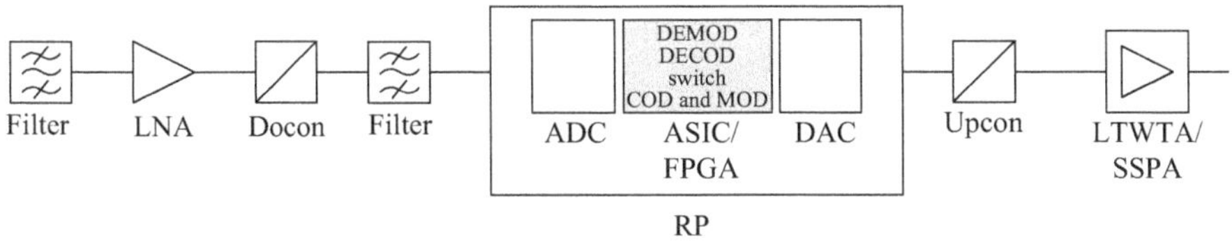

Figure 13.7 Block diagram of regenerative architecture

The benefit from demodulation and decoding lies in additional gain in the link budget as all the distortions and impairments in the signal are removed. A regenerative OBP is also capable of handling different terminal sizes with different received power at the satellite and therefore with different signal-to-noise ratios (SNR). After regeneration of the signals, these can easily be linked to the ground station or other user terminals.

This architecture leads to higher power consumption and introduces additional complexity in designing the signal processing. For very high data rates and very large bandwidths, current technology leads to boundaries which are hard to overcome.

A way to cope with this challenge is to only demodulate the signal without decoding. The drawback is that the decoding gain is lost and errors in the signal propagate through the processor. Different links with different encoding schemes can be handled by this approach as the decoders need not to be implemented. This can especially be used if user, gateway and intersatellite links (ISLs) have to be operated at the same time.

13.2.2 Digital payload technology matrix

As already discussed in the previous section, we distinguish between digital transparent and digital regenerative processors. This relates to the signal as it is transported through the processor. We look at the implementation of the OBP and differentiate it in terms of reconfigurability in the following way:

- Not reconfigurable—the functionality of the OBP is fixed
- Partly reconfigurable—parts of the OBP can change the functionality (e.g., adding additional signal paths, reconfiguring filter by changing of the coefficients, etc.)
- Fully reconfigurable—the entire function of the OBP can be changed in terms of updating firmware and software (e.g., by upload from ground)

Technologies to implement the digital processing, which may be used in OBPs, are:

- ASIC
- Anti-fuse field programmable gate array (FPGA)
- Reconfigurable FPGA (SRAM or flash-based)

The matrix in Figure 13.8 shows how these technologies can be used in the different transparent/regenerative and reconfigurability approaches.

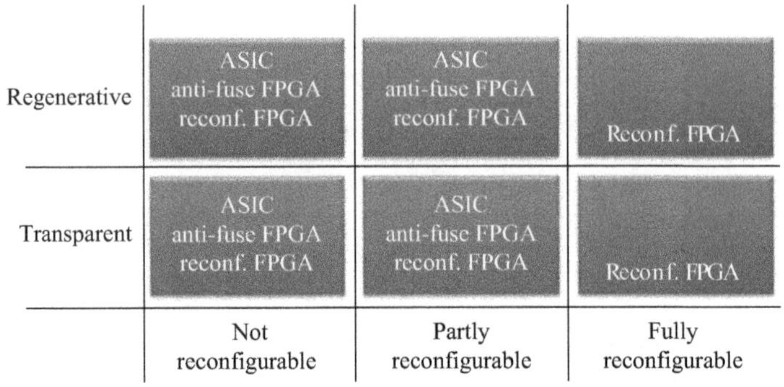

Figure 13.8 Technology matrix for OBPs matching circuit technologies to signal architecture and reconfiguration grades

Table 13.2 Comparison of Xilinx FPGAs that may suit space applications

FPGA	**Virtex-5QV XQR5VFX130**	**Kintex-7 XC7K325T**	**Kintex-Ultrascale XQRKU060**	**Zync-Ultrascale+ ZU19EG**
Process node	65 nm	28 nm	20 nm	16 nm FinFET
Grade	RHBD	COTS	Radiation tolerant	COTS
Slices	20,480	50,950	82,920	130,625
Emb. mem. (Mbit)	11.0	16.4	39.8	70.6
Multiplier	320	840	2,760	1,968
Emb. processors	0	0	0	6
SEEs (a^{-1})	6,008	12,516	25,592	N/A

Chip technologies available for space applications currently rely on 65 nm radiation hardened processes. Even ESA has developed its own FPGA, which should be available in the near future based on this process technology.

Current Complementary Metal-Oxide-Semiconductor (CMOS) technologies such as 20 nm are used in the current generation of Xilinx devices and may offer some radiation tolerance so that they could be used in space. Another technology which promises to offer good radiation properties seems to be 28 nm FD-SOI or 22 nm FD-SOI.

The smaller the processing nodes, the higher are the costs to develop and produce chips. Therefore, a clear choice has to be done which functionality shall be implemented in which technology.

Possible FPGA solutions

Table 13.2 shows *Xilinx* FPGAs of the latest generations with technology, resources and single-event effect (SEE) rates according to the assumptions of [6] (GEO and 7 mm aluminum shielding). The SEE rates do not take mitigation, except of error-correcting code in the block random-access memories (BRAMs), into account [10].

Table 13.3 Comparison of FPGAs from different vendors that may suit space applications

FPGA	*Microsemi* RTG4	*ESA* BRAVE medium	*Altera* 5SGSMD5-H3F35I4
Conf. memory	Flash	SRAM	SRAM
Process node (nm)	65	65	65
Grade	RHBD	RHBD	COTS
Slices	9,489	2,188	10,788
Emb. mem. (Mbit)	5.2	2.8	39.0
Multiplier	462	112	3,180
SEL performance	Immune	Immune	Vulnerable

The radiation hardening by design (RHBD) and the radiation tolerant FPGA can be used for any *Earth* orbit. The commercial off-the-shelf (COTS) FPGAs may be used in the LEO, but an assessment has to be done especially for destructive effects such as the single-event latch-up (SEL). The FinFET CMOS technology tends to be less vulnerable to SEEs by a few orders of magnitude compared with the planar CMOS technology, but the SEL problem seems to be still present.

Besides these SRAM-based FPGAs from *Xilinx*, designers may take FPGAs from *Microsemi*, *ESA* and *Altera* into account. Table 13.3 summarizes these FPGAs based on Flash and SRAM configuration memory. In contrast to the SRAM-based FPGAs, the RTG4 FPGA is specified only for 200 configuration write cycles and is not reconfigurable in space. The advantage of this component is that no additional external boot device is necessary, since the Flash stores the configuration nonvolatile. The RTG4 may be used as flight computer because of its outstanding SEE performance. The BRAVE FPGA from *ESA* can be used as an alternative for the Virtex-5QV and is planned in a small, medium and large version. The *Altera* 5SGSMD5 is an alternative for a COTS component.

13.2.3 Advantages of reconfigurable OBPs

Reconfigurable OBPs offer a set of advantages:

- Flexibility for future communication system techniques
- Routing flexibility
- Ability of reencoding
- Cost reduction when using generic processor approach
- Time-to-market reduction
- Flexibility for future Business Models
- Application changes (e.g., frequency, bandwidth, modulation, coding)
- Beamforming and phased array control
- IP-routing and ISL routing capability
- Possibility for using adaptive redundancy concepts

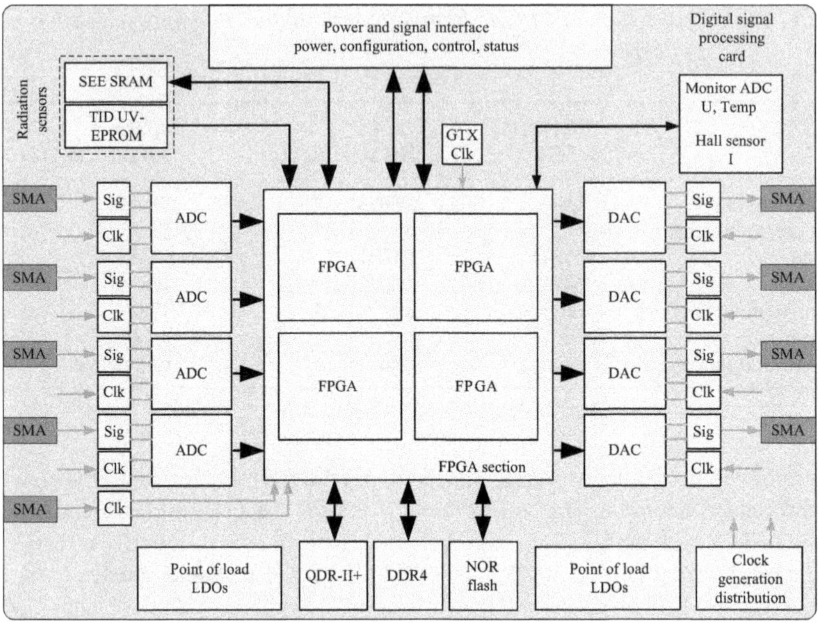

Figure 13.9 Block diagram of future high-performance signal processing module for reconfigurable OBP

13.2.3.1 Technical flexibility

As reconfigurable, OBPs offer reconfiguration during their usage in space; they offer great opportunities to adapt to future communication standards and flexibility in using them in different application scenarios.

They can be adapted to more efficient communication standards (advanced modulation and coding technologies such as DVB-S2 or DVB-S2X) during lifetime (subject to their processing capabilities). This may lead to better link qualities (lower bit error rates) and therefore enable the use of smaller terminals at the user side. This is achieved by demodulation and reencoding, de-multiplexing, error detection and correction, flexible switching, buffering, remultiplexing, modulation and network synchronization.

The current Fraunhofer IIS approach (FOBP) is based on reconfigurable FPGAs in the signal-processing modules. With these modules, we can achieve an analog channel bandwidth of up to 750 MHz, which is provided by the ADCs and DACs. This leads to possible 15 GHz/kW with 20 channels and 20 digital signal processing (DSP) modules.

Future signal-processing modules based on FPGAs may achieve an analogue channel bandwidth capability of roughly 1,500 MHz. This would lead to 120 GHz/kW with 80 channels and 20 DSP modules (assuming comparable power consumption of the DSP modules). A principle block diagram is shown in Figure 13.9.

With this approach a highly flexible and future-proof processor can be built.

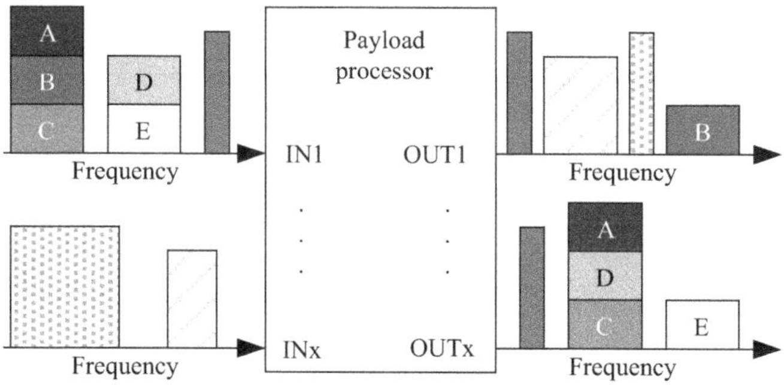

Figure 13.10 Routing possibilities provided by a payload processor

13.2.3.2 Flexibility in routing

As can be seen in Figure 13.10, a payload processor can flexibly route signals from the input to the output ports. Uplink frequency blocks may be switched to different downlink beams, or can be connected to more than one output port and may also be remodulated and reencoded to achieve a higher spectral efficiency.

13.2.3.3 Time-to-market reduction

Using FPGAs as processing cores within an OBP can lead to tremendous time-to-market reduction compared to ASIC developments because these FPGAs already exists and have been developed by external companies. A comparison of project time lines for both technologies is shown in Figure 13.11.

The design of an ASIC for an OBP has to start very early with technologies available for space applications and with a set of requirements which may change during the development phase as the satellite operator may also change parts of the application. Therefore, these ASIC developments will have to be partly based on abstracted requirements which reflect also additional flexibility to cover future applications.

The design of the FPGA firmware can start much later and the milestone of design freeze is much closer to the completion of the hardware and to the satellite integration of the processor. An estimation of the reduction in development time results is 12–18 months depending on the design complexity. Even during the integration, phase testing and debugging of the code are possible and updating the firmware just before start could be done. When using SRAM-based FPGAs also design updates of the FPGA firmware are possible during the satellite operation phase. This enables the operators to adaptively react on new business models and application scenarios.

13.2.3.4 Beamforming and phased array antenna control

Digital beamforming would open a new dimension in flexibility for satellites as a high number of beams could be served in parallel. The most challenging point is currently the high number of antenna elements which would be needed to achieve the required antenna gain. This is directly reflected in the number of ports for the beamforming processor—which are likely to be in thousands. To solve this, one possible solution

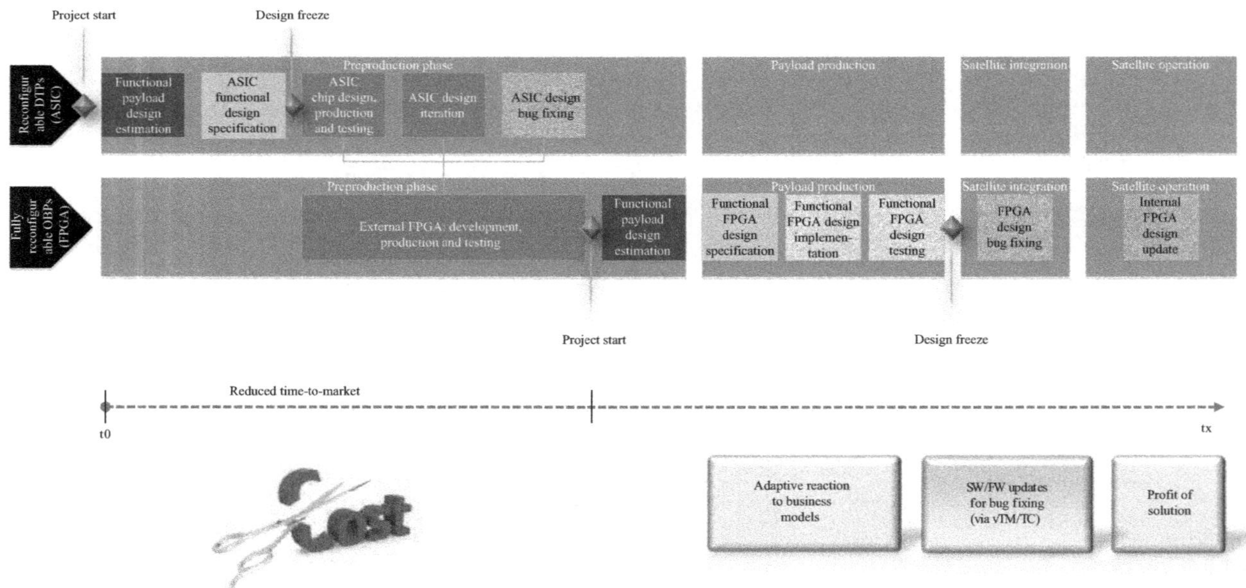

Figure 13.11 Project planning for OBP realization based on ASIC and on FPGA showing significant time-to-market reduction

is to place ADCs (and for transmit DACs) directly in the antenna aperture introducing a complicated assembly, power dissipation issues and the operation over a large temperature range. The interface to the processor may then be realized by high speed serial lanes. Also, processing high bandwidths and many beams will end in a huge number of operations to be handled in parallel. With state-of-the-art architectures and components, this remains a dream for the next years.

Controlling phased array antennas instead is a relatively easy thing to do, as only an interface to load new antenna weights is needed. This can easily be applied using modern technology. Next generation satellites like Eutelsat Quantum will use this approach by controlling the antenna through the OBP.

13.2.3.5 IP-routing and inter satellite links

Most of the announced LEO constellations rely on ISLs. These can be seen as mandatory to make full use of the constellation as most of the satellites will be above the oceans and do not have a direct gateway connection. An OBP can connect the user beams with the ISLs in a flexible way (although the flight of the satellites will follow a deterministic path). This can be used to route the signals to different gateways and also to establish an end-to-end connection only using the constellation satellites. These ISLs should be capable of routing about 5 Gbps between the satellites.

Higher layer routing could also be introduced by using an OBP. This function is not necessarily bound to FPGAs as it is a well-established technology and may not need reconfiguration on gate level. Therefore, existing chip-sets, ASICs or even the available CPU cores of next generation FPGAs could serve this.

13.3 The Fraunhofer OBP as an example

13.3.1 Payload architecture

The FOBP was designed for the German Heinrich Hertz-Mission (H2Sat) which is scheduled to be launched in late 2021. The mission is based on the SmallGEO platform of OHB which was developed with the help of ESA and is now operating in space as Hispasat 36W-1. H2Sat will cover another OBP (based on COTS) and a number of additional devices (filters, switches and amplifiers) which will be tested in space. The mission is twofold as it is divided into a commercial part and a scientific part which contains an experimental payload. Up- and downlinks for the latter part operate in Ka-band.

As Figure 13.12 shows, the FOBP can be switched in the signal path to achieve the flexibility in the scientific payload (the other experimental blocks are not shown). It supports two different bandwidths of 36 and 450 MHz, which can be provided by the satellite. Thus, it is possible to perform experiments with wideband signals on the satellite.

13.3.2 Main building blocks

Figure 13.13 shows a block diagram of the FOBP covering the main building blocks. The FOBP has two input and two output ports in L-band supporting 36 and 450 MHz

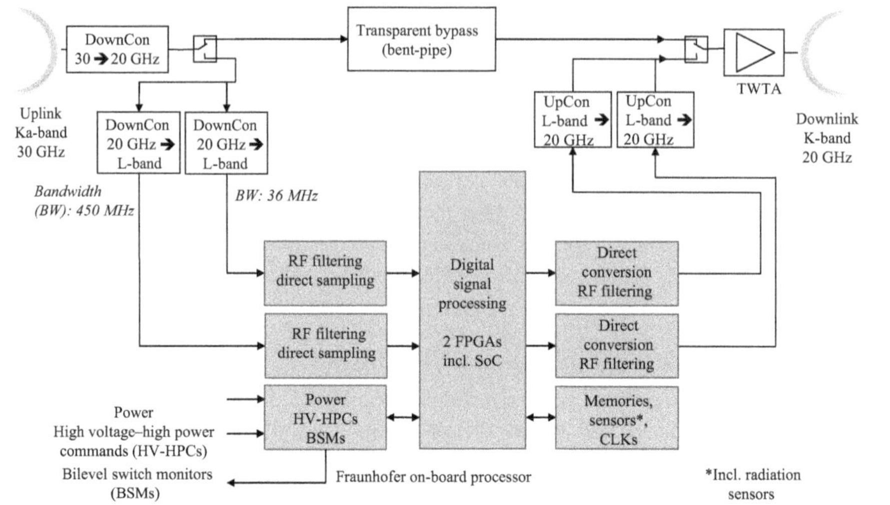

Figure 13.12 Principal integration diagram of FOBP in H2Sat

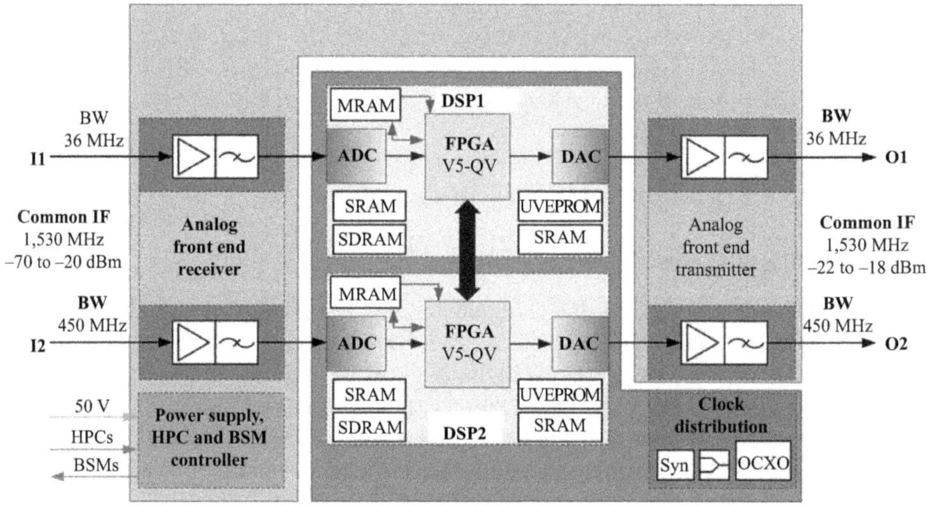

Figure 13.13 Block diagram of the FOBP showing the main building blocks

wide channels. It is supplied with a 50-V bus voltage and interfaces with satellite compatible command and monitoring lines.

The main building blocks of the FOBP are:

- Power supply, high power command (HPC) and bistatic monitor (BSM) controller
- Analog frontend receiver and transmitter for two ports each

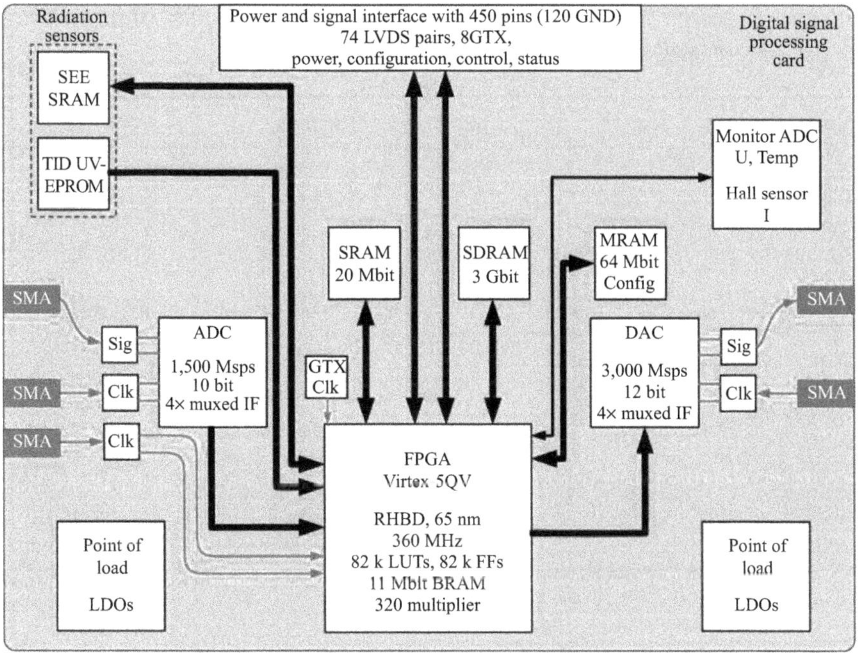

Figure 13.14 Block diagram of the signal-processing module

- Clock generation and distribution
- Two DSPs with ADC, FPGA, DAC and according memories

These building blocks have been implemented on four cards. Two DSP cards, one power supply unit and one RF and clock card. To reduce costs, an approach has been chosen which is based on two qualification categories—called scopes A and B.

Scope A (shown in light gray in Figure 13.13) covers all interface to the satellite (power, RF, HV-HPC and BSMs) and is designed for 15 years operation and only contains components which fulfill ECSS-Q-ST-60C Rev.2, Class 1. These interfaces are most critical as they are directly connected to the rest of satellite and shall not introduce any shortcomings.

Scope B parts were designed in a way that critical components are space grade. The rest of the components are at least packaged in space grade packages. As the DSP module is very complex, the PCB design rules cannot comply with ECSS standards.

13.3.3 Digital signal processing

The current model of the signal-processing module is equipped with space-grade devices, e.g., the *Xilinx Virtex-5QV* FPGA. Figure 13.14 depicts the block diagram of the proposed module. We set up the signal processing chain with the SEL immune reconfigurable FPGA, a high-speed ADC and DAC. These devices are supplied with separate clocks to enable different sampling and processing rates.

Table 13.4 Software-defined radio capabilities of the signal-processing module, limitations and FOBP setup [11]

Parameter	Range	Limited by	Remark	FOBP setup
Analog f_{in}	1–2,250 MHz	ADC	1.-3. (-5.) Nyquist	1,530 MHz
Noise power ratio in	43 dB	ADC	10 Bit ADC	10 Bit
Input bandwidth	5–750 MHz	Synthesizer, ADC		612 MHz
Interface data rate per I/O	0–720 Mbit/s	FPGA	Double data rate	306 Mbit/s
Processing speed	0–360 MHz	FPGA		51; 102; 306 MHz
Output bandwidth	5–1,500 MHz	Synthesizer, DAC		612 MHz
Noise power ratio out	45 dB	DAC	12 Bit DAC	12 Bit
Analog f_{out}	1–6,000 MHz	DAC	1.-5. Nyquist	1,530 MHz
Power consumption	15–40 W	–	Worst case	Typical 25 W (per module)

Besides the SRAM for data buffering and the SDRAM as working memory, the module is equipped with radiation sensor memories. We use a nonvolatile magneto-resistive RAM (MRAM) to store the initial FPGA configuration (bit file). The power supply, consisting of point of loads (POLs) and low-dropout (LDO) regulators, as well as the monitoring complement the signal-processing module. A power and a signal interface enables power supply and high data rate connections to other modules.

Since the signal-processing module is versatile, Table 13.4 shows the general software-defined radio (SDR) capabilities, its limitations and the setup of the FOBP. We recommend direct sampling (band-pass under-sampling) of the in- and output to save one analog mixer stage. The FOBP does so, by sampling a 450-MHz-wide band with a center frequency of 1,530 MHz in the third Nyquist band with a sample rate of 1,224 MS/s. Note that, we implemented a synthesizer for the ADC and DAC clock to change the sampling rate on demand. This synthesizer is not part of the signal-processing module, to allow for application specific adaptions. It is located at the radio frequency module of the FOBP.

The most challenging parts of the system design are the digital interfaces of ADC (40 LVDS pairs) and DAC (48 LVDS pairs), the power consumption and the heat dissipation of the FPGA. The interfaces of the ADC and DAC are fourfold multiplexed to transmit and receive the digital data. We solved the power consumption and heat dissipation issues by using POLs and a customized thermal concept, based on a cooling finger.

13.3.4 Virtual TM/TC

To monitor and control the payload a dedicated link is needed. As the FOBP is reconfigurable, it requires the upload of configuration data. Therefore, we need higher data

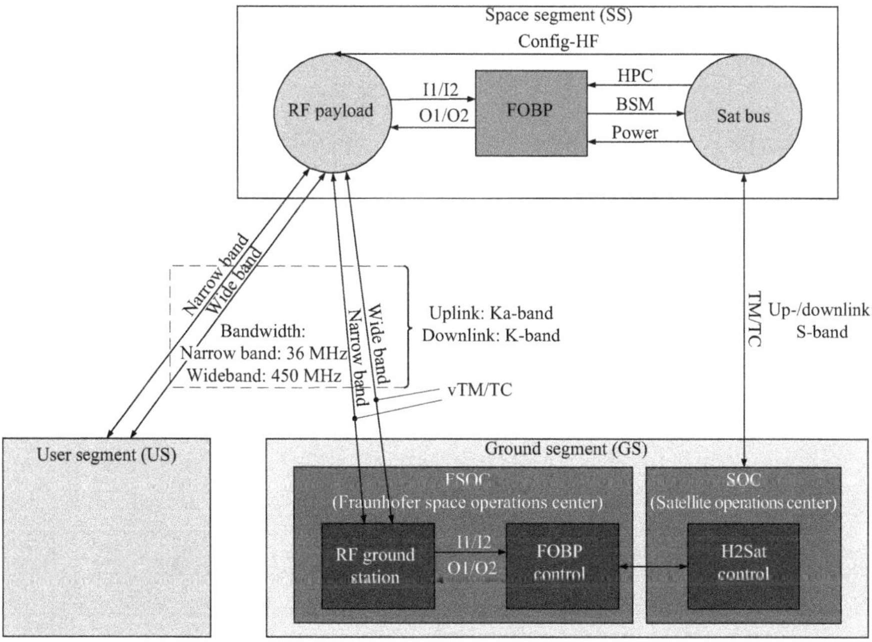

Figure 13.15 Block diagram of system showing both vTM/TC link on the left and "standard" satellite TM/TC link on the right

rates compared to conventional TeleMetry and TeleCommand (TM/TC) systems—which are usually shared between all modules of the satellite—to arrange the upload (e.g., bit file) in an adequate time. Typical TM/TC links are limited to a few kbit/s for one payload module and the access has to be coordinated with the satellite ground control.

We designed the so-called virtual TM/TC (vTM/TC) to overcome these issues. Additionally, we avoid the cost intensive interface, in terms of qualification between the payload and the satellite bus (e.g., MIL-STD-1553B), by using this in-band vTM/TC. To secure the vTM/TC, an encryption has to be implemented.

The in-band vTM/TC link is colocated in the K/Ka-band communication links and has a channel bandwidth of 2 MHz, which is able to provide a data rate of 1 Mbit/s using a differential quadrature phase-shift keying (DQPSK) modulation. Figure 13.15 shows the different links—vTM/TC on the left side integrated in the user links and the satellite TM/TC on the right side as classical TM/TC link in lower frequency bands. The optimization goal of the vTM/TC is not linked to spectral efficiency but rather to the link availability of 99.9% and highest reliability. This gives us the ability to upload FPGA bit files of about 6 MB each in less than a minute. Additionally, control and status information such as temperatures, voltages, current, total ionizing dose (TID) and single event effect (SEE) upset rates are provided via our vTM/TC as well. For channel coding a combination of Reed–Solomon encoding together with convolutional coding was used as recommended in CCSDS standards [7,8].

The receiver chain as shown in Figure 13.16 consists of a digital down conversion, synchronization, demodulation and channel decoding. Data output of the channel decoding are 188-bytes MPEG-TS packets which are fed into the next layer.

For the higher layers we used the internet protocol (IP) for the network layer. Therefore, the FOBP inside the satellite and the ground control gets IP addresses as shown in Figure 13.17. On top of the IP, we use the transport layer protocols UDP and TCP. TCP guaranties reliable data transmission even if packets get lost or corrupted. This is perfectly suitable for transmitting TCs or FPGA bit files. Above the transport layer is the application layer where we can use also standard protocols, e.g., FTP for bit file uploads or telnet for TCs. In essence, it is now possible to write on-board software using very well-known network programing techniques. If an application requires some special controlling or monitoring, the necessary flight and ground control software can be written in a few hours or days instead of weeks or months.

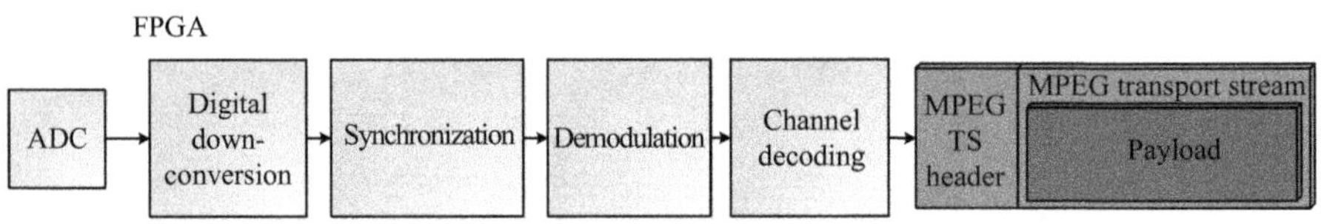

Figure 13.16 Block diagram of vTM/TC receiver chain implemented in the FPGA

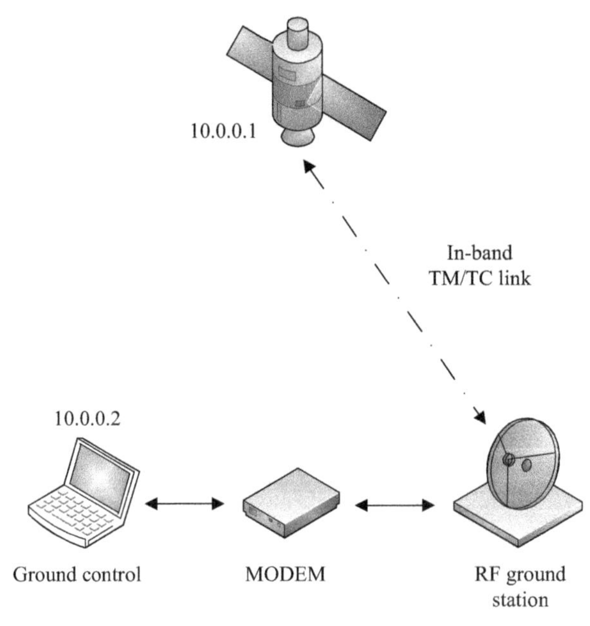

Figure 13.17 Communication chain to address the FOBP on the satellite

With this, we enable the control of the flexible payload in a way that it can easily be embedded in future 5G systems. It also provides means to upload new applications, which have been designed on ground thus increasing the flexibility of the satellite payload.

13.4 Exemplary 5G use case for OBP using LEO satellites

To highlight the demands in future satellite communication systems assisting 5G or fully integrated in 5G, we display the example of a LEO constellation where satellites may have to play different roles at different locations. These LEO systems inherently demand high flexibility of the satellites and the ground equipment as all of these may be on the move—at least the satellites as they fly on low orbits [12].

One of these systems is announced by OneWeb with 648 Ku-band satellites in 18 orbital planes each of them up to 200 kg and an orbital height of 1,200 km. LEOSAT wants to operate 108 satellites in 9 orbital planes and additionally uses optical ISLs. Telesat is intending to use 117 satellites in two orbits at 1,000 km and at 1,248 km. The latter two constellations use Ka-band for connecting the users. LEO-constellations promise a 15-dB advantage compared to GEO satellites because of the closer distance although only smaller antennas can be accommodated on the satellites. To make full use of these systems, ISLs are mandatory to also have connections when not flying over land or if gateway stations cannot be reached.

The following figures display a use case in which LEO satellites are integrated in the 5G landscape and fulfill different tasks at different positions during their orbit flight. In Figure 13.18, the satellite operates directly as 5G base station to connect ships or airplanes. There, reencoding between the user terminal and ISL or gateway connections is needed to deliver the relevant data and internet connection. A direct gateway connection is often not possible as ground stations are mainly deployed on the continents. So, ISLs are of major importance to make use of LEO systems in the 5G landscape.

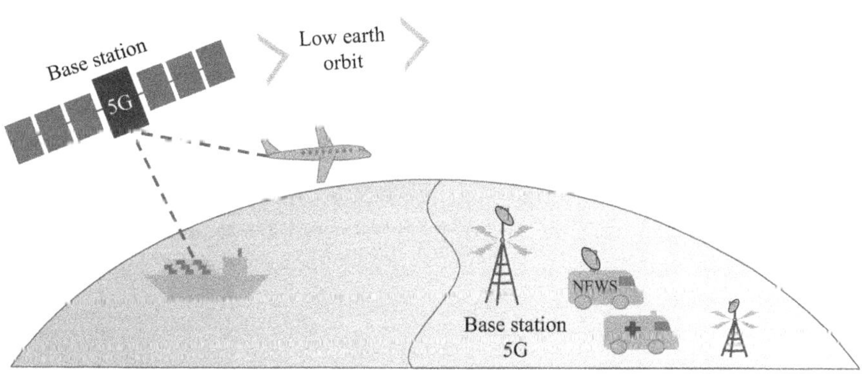

Figure 13.18 LEO satellite operating as 5G base station to serve ships and airplanes over the sea

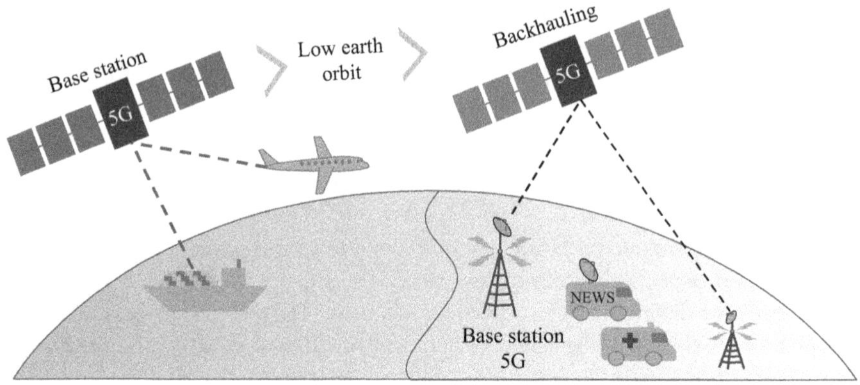

Figure 13.19 Change of behavior of LEO satellite during its orbital flight as different service can be provided for different regions

During the orbital flight of the LEO satellite, it may act differently when flying over ground (Figure 13.19). In the displayed example, the LEO satellite now operates as backhaul connection to bring capacity to rural areas where no high-speed connections are available as fiber connections are very costly when deployed over large distances. Another possible application may be additional data rate enhancement of urban areas at almost no additional costs.

It can be seen that a high demand for flexibility is needed on board the satellite.

13.5 Summary

This chapter provided a short overview on on-board processing and its necessity to integrate satellites in 5G. A highly flexible satellite is needed to account for the dynamics inside the future telecommunications infrastructure. This can only be achieved when using satellites which provide on-board processing to flexibly route the data where it is needed—to different users, different ground stations and to other satellites. We propose to use OBPs based on FPGAs and a dedicated monitoring and control channel to make full use of the flexibility. It offers a means to provide a software defined payload (SDP) which may be controlled from the network it is supposed to work in.

Combining the flexibility and strengths of OBP with clever network integration of the satellite payload can pave the way to operate satellites in future 5G networks.

Acronyms

ADC	analog-to-digital converter
ASIC	application specific integrated circuit
BGAN	broadband global area network

BSM	bistatic state monitor
COTS	components of the shell
DAC	digital-to-analog-converter
DOCON	down-converter
DQPSK	differential quadrature phase shift keying
DSP	digital signal processing
DTP	digital transparent processor
DVB	digital video broadcasting
EM	engineering model
FDSOI	fully depleted silicon on insulator
FOBP	Fraunhofer on-board processor
FPGA	field programmable gate array
GEO	GEOstationary satellite
H2Sat	Heinrich hertz-satellite mission
HPC	high power command
HTS	high-throughput satellite
I/O	input/output interface
ISL	intersatellite link
LDO	low dropout regulator
LEO	low-earth orbiting satellite
LNA	low-noise amplifier
LTWTA	linearized travelling wavetube amplifier
LVDS	low voltage differential signaling
MPEG	moving pictures expert group
MPEG-TS	MPEG transport stream
MRAM	magnetoresistive RAM
OBP	on-board processor
POL	point of load
QoS	quality of service
RP	Reconfigurable Processor
SDP	software-defined payload
SDR	software-defined radio
SEE	single event effects
SNR	signal-to-noise ratio
SSPA	solid-state power amplifier
TID	total ionizing dose
vTMTC	virtual TeleMetry TeleCommand

References

[1] https://artes.esa.int/news/astrium-team-completes-web-system, August 2006, accessed 29.11.2017.
[2] https://artes.esa.int/sites/default/files/hiresimage/OPB_hi-res.jpg, August 2006, downloaded 29.11.2017.

[3] https://www.thalesgroup.com/sites/default/files/asset/document/Digital_Transparent_Processor_april2012.pdf, March 2012, accessed 29.11.2017.

[4] P. Voisin, A. Barthere, O. Maillet, *et al.*, Flexible Payloads for Telecommunication Satellites – A Thales Alenia Space perspective, 3rd ESA WS on advanced flexible payloads, March, 21–24, 2016, Noordwijk, The Netherlands.

[5] http://www.esa.int/spaceinimages/Images/2016/11/Redsat, 2016, accessed 20.12.2017.

[6] A. Hofmann, R. Glein, L. Frank, R. Wansch, and A. Heuberger, "Reconfigurable on-board processing for flexible satellite communication systems using FPGAs," in *2017 Topical Workshop on Internet of Space (TWIOS)*, 2017, pp. 1–4.

[7] CCSDS Green Book, CCSDS Protocols over DVB-S2 – Summary of Definition, Implementation, and Performance, Informational Report, 130.12-G-1, November 2016.

[8] CCSDS Green Book, Overview of Space Communications Protocols, Informational Report, CCSDS 130.0-G-3, July 2014.

[9] H. Gachon, V. Enjolras, P. Voisin, and G. Lesthievent, Spaceflex Digital transparent processor for advanced flexible payloads, 3rd ESA WS on advanced flexible payloads, March, 21–24, 2016, Noordwijk, The Netherlands.

[10] R. Glein, P. Mengs, F. Rittner, R. Wansch, and A. Heuberger, BRAM Implementation of a Single-Event Upset Sensor for Adaptive Single-Event Effect Mitigation in Reconfigurable FPGAs, 11th NASA/ESA Conference on Adaptive Hardware and Systems (AHS2017), July, 24–27, 2017, Pasadena, CA, USA.

[11] Robért Glein, Scalable Signal Processing based on Reconfigurable FPGAs for Satellite Payload Applications, 3rd ESA WS on advanced flexible payloads, March, 21–24, 2016, Noordwijk, The Netherlands.

[12] M. Russ and A, Hofmann, Architectural considerations on Software Defined Payloads (SDP) of interests to 5G Community, EuCNC 2017, June, 12–15, 2017, Oulu, Finland.

Chapter 14
On-board interference detection and localization for satellite communication

Christos Politis[1], *Ashkan Kalantari*[1], *Sina Maleki*[1], *and Symeon Chatzinotas*[1]

Interference is identified as a critical issue for satellite communication (SATCOM) systems and services. There is a growing concern in the satellite industry to manage and mitigate interference efficiently. In this context, an on-board spectrum monitoring and localization unit can be used to detect and localize the interference reliably. Current satellite spectrum monitoring and localization units are deployed on the ground, and the introduction of an in-orbit spectrum monitoring and localization unit can bring several benefits, e.g., simplifying the ground-based station in multibeam systems. This chapter presents the interference detection and localization techniques which take place on-board the satellite within a digital transparent processor (DTP) satellite payload or in a partially regenerative satellite. First, the conventional energy detector (CED) is presented, which is an efficient technique to monitor strong interference in SATCOMs. However, weak interference is not so easily detectable because of its low interference-to-signal-plus-noise ratio (ISNR). To address this issue, a second detector is discussed, which exploits the frame structure and pilot symbols of the SATCOM standards. Assuming that the pilot signal is known at the receiver, it can be removed from the total received signal, and then, an ED technique can be applied on the remaining signal to decide on the presence or absence of interference. Nevertheless, the detection at low values of ISNR may require more samples than the number of pilots supported by the standards. For this reason, a third detector is introduced by demodulating the desired signal, removing it from the total received signal and applying an ED in the remaining signal for the detection of interference. After detecting the interference, the interferer needs to be localized and, hence, this chapter describes the current techniques for on-ground interference localization and presents an on-board interference localization technique using frequency of arrival (FoA) via a single satellite.

[1]Interdisciplinary Centre for Security, Reliability and Trust, University of Luxembourg, Luxembourg

14.1 Introduction

Interference issues have been identified as a major threat for the commercial satellite telecommunication systems and services [1]. Interference has a financial impact on the satellite operators that can run into several million dollars, ranging from the revenue loss because of the throughput degradation to the increase of the expenses from the buying of interference monitoring, detection, localization and mitigation equipment. Except the satellite operators, their users also suffer from the interference due to the decreased quality of service [2]. The situation is likely to become worse over the next years, as there is a steady increase of satellites in-orbits and congestion of already crowded frequency bands. Hence, a strategy for the management of interference appears essential for the commercial satellite industry.

Effectively tackling interference is a complex task to be performed at various levels: interference monitoring, interference detection and isolation, interference classification, interference localization and interference mitigation [1]. In this chapter, we focus on two of them, the detection and localization of interference. Interference detection could be performed either in-space or on-ground. Referring to on-ground implementations, the satellite acts as a transparent transponder, and all the processing is performed on-ground, possibly combined with other functionalities. On the other hand, the introduction of an in-orbit spectrum monitoring unit (SMU) would bring several benefits, e.g., allowing faster reaction to resolve interference before the downlink impairment, simplifying the ground-based stations in multibeam satellites by avoiding monitoring equipment replication in multiple Earth Stations (ESs) and offering the capability to process uplink signals which are not affected by additional downlink impairments and possible distortions related to the transponder [1,3,4]. Here, we should mention that a single-monitoring equipment can be used on-board for the detection of strong interference, such the wideband detectors. However, for the detection of weak interference, we need more complex algorithms, where an interference detector is used in each channel. To overcome this issue, these detectors may be applied only in the channels which the satellite operators have characterized more suspicious for the appearance of interference or a higher level of protection is necessary. Finally, it is worth mentioning that the on-board implementation faces some technical challenges, which have to be taken into account, with the most important one being the minimization of the complexity and power consumption.

In this chapter, we also aim to localize an interference with unknown location using FoA technique by only relying on the measurements obtained through a single satellite. This can be done both on-ground and on-board. The on-ground method acts as the benchmark for the on-board method to understand the on-board interference localization advantages. In on-ground approach, the satellite samples the interference in each time instance and forwards it to the gateway to estimate its frequency. Since the satellite moves, each estimated frequency includes a Doppler shift, which is related to the location of the unknown interferer. The satellite's position, velocity, oscillator frequency and the interference frequency are used at the gateway to build a location-related equation between the estimated frequency and the location of the unknown interference. Simultaneously with the interference signal, the satellite samples a reference signal to calibrate the estimated frequency and compensate for the mismatches between the available and real values of the satellite's position, velocity, and oscillator frequency. Multiple location-related equations obtained based on the FoA measurements (at least two), along with the equation of the earth surface are used to localize the unknown interference. In the on-board method, the satellite performs the localization algorithm on-board the satellite, and hence, it avoids the frequency error of the down conversion oscillator as well as the errors of the estimated velocity and the position of the satellite in the downlink transmission. Furthermore, the on-board localization approach can improve the localization accuracy and, hence, results in substantial reduction in the localization error.

In this part, we clarify with more detail the motivations of using on-board satellite localization [1,5]. Following motivations can be considered for performing on-board localization:

1. The collected interference signal does not go through the downlink channel from the satellite to the GW. Hence, it does not get distortion, attenuation and noise due to the channel and the rain effect. Therefore, better estimation of the location can be obtained.
2. Only professional uplink stations will be equipped by carrier ID[1] [2], by 2018. However, very small aperture terminal (VSATs) will not be equipped by carrier ID due to being cost sensitive.
3. Mobile users such as airplanes are not equipped by carrier ID.
4. Illegal uplink stations (bandwidth piracy) do not use carrier ID.

On-board interference detection and localization clearly can help the satellite industry, however, before continuing any further with the on-board current techniques, we first provide an overview of the digitalization of the satellite.

14.2 On-board digitization

Almost all the commercial satellites consist of a number of building blocks that allow them to receive and redistribute signals from Earth. The satellite filters the received signal, amplifies it through a low noise amplifier (LNA), converts it to another frequency of interest, multiplexes it, amplifies it again with a high power amplifier (HPA) and, then, sends it back to the ground for further processing [6]. This process is depicted in Figure 14.1. Therefore, the satellite contains a large number of analog hardware, such as filters, switches, multiplexers, converters and amplifiers.

However, nowadays, there is a revolution on the digitization of the satellites [7]. The introduction of digital signal processors (DSPs) totally changes the way that

[1] Carrier ID is a simple concept—every transmitted carrier will have a unique ID which can be decoded by satellite operators. If a carrier is causing interference, the unique ID will be decoded to identify who is transmitting the interference. A satellite operator will be able to decode the unique ID in the carrier, contact the uplinker causing the interference and reduce the duration of service interruptions caused by interference.

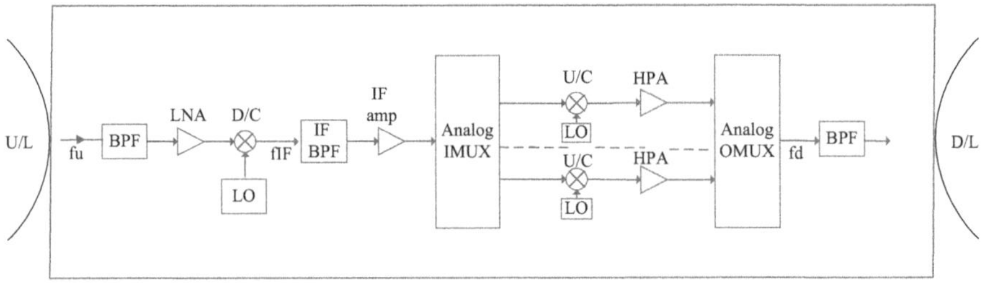

Figure 14.1 Transparent satellite payload

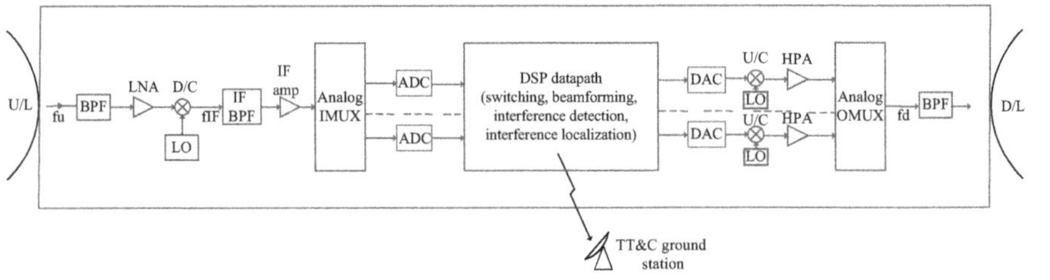

Figure 14.2 Digital transparent processor satellite payload (where U/L: uplink, BPF: bandpass filter, LNA: low noise amplifier, D/C: down-conversion, LO: local oscillator, IF: intermediate frequency, ADC: analog-to digital-converter, IMUX: input multiplexer, OMUX: output multiplexer, DAC: digital-to-analog converter, U/C: up-conversion, HPA: high-power amplifier and D/L: downlink)

satellites operate, interact and serve customers [8]. The satellite design greatly changes, allowing a large portion of the aforementioned on-board analog hardware to be replaced, where the signal will be passing from the DSPs for conversion, transformation and digital amplification.

Furthermore, the digitization of the satellite enables the design and operation of flexible and adaptive payloads, offering several benefits to the satellite operators and their customers. The flexibility of a satellite payload optimizes the resource management, offering the capability to adapt the satellite use according to demands and based on the real traffic conditions in a given zone. To this extent, the DTP [9,10] is a promising technology to offer the flexibility. The DTP is designed to provide a nonregenerative DSP on uplink signals, as shown in Figure 14.2. The DTP is the first step in the direction of a more advanced vision: the full payload digitization [7]. A full digital payload is designed to enable regenerative DSP such as demodulation, decoding, coding and modulation on-board the satellite.

In the next sections, the sources of interference on-board the satellite are described, and also, the current techniques for interference detection and localization are presented.

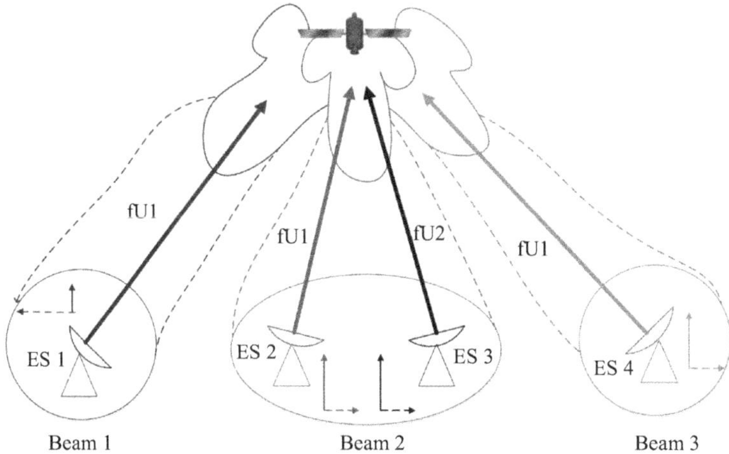

Figure 14.3 Sources of intrasystem interference

14.3 Satellite interference

By definition, interference is the undesired power contribution of other carriers in the frequency band occupied by the wanted carrier [6]. There is a large number of scenarios where interference can occur which are described in this subsection, focused on the uplink satellite interference. The latter can be classified into two categories: intrasystem and external interferences [11].

14.3.1 Intrasystem interference

The intrasystem interference is produced over carriers transmitted by ESs belonging to the same system [12,13]. Some potential sources of intrasystem interference in the satellite network are cochannel interference, adjacent channel interference and crosspoll interference [6,11–15] as they are depicted in Figure 14.3. This figure presents three beams assuming that the ES 2 of beam 2 is the useful ES.

- **Cochannel interference** is generated due to imperfect isolation between different beams. In Figure 14.3, the ES 2 transmits a signal which is received by the antenna which defines the beam 2, in the main lobe with the maximum antenna gain. Moreover, the ES 4 of beam 3 transmits a signal at the same frequency and polarization as the ES 2 and the signal is received by the side lobes of the antenna defining the beam 2, with low but nonzero gain. Therefore, the carrier of beam 3 appears as interference noise in the spectrum of the carrier of beam 2, producing cochannel interference.
- **Crosspoll interference** is the result of the opposite polarization field of the carriers. In Figure 14.3, if the ES 1 of beam 1 transmits at the same frequency but opposite polarization as the carrier of the useful ES, crosspoll interference is produced.

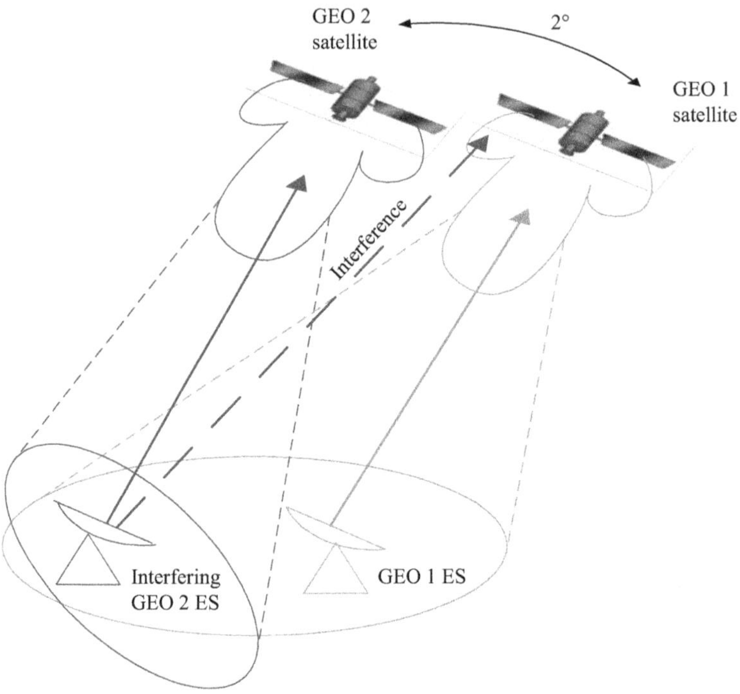

Figure 14.4 Adjacent system interference

- **Adjacent channel interference** is produced due to the fact that part of the power of the adjacent carrier at frequency f_{U2} is captured by the satellite tuned to the carrier at frequency f_{U1}. In Figure 14.3, we see that part of the power of the signal transmitted by the ES 3 of beam 2, at the same polarization but different frequency as the ES 2, is introduced as a result of imperfect filtering in the channel occupied by the carrier of ES 2, generating with this way adjacent channel interference.

14.3.2 External interference

The external interference is produced by carriers from ESs belonging to a different system [12,13]. Some examples of potential external interference sources are: adjacent system interference, in-line interference, terrestrial interference and intentional interference.

- **Adjacent system interference** is generated by an ES into an adjacent satellite. This type of interference is typically accidental, due to operator errors, poor inter-system coordination or poor equipment setup. A scenario of adjacent system interference is presented in Figure 14.4, where the interfering source is transmitting toward the operational satellite.

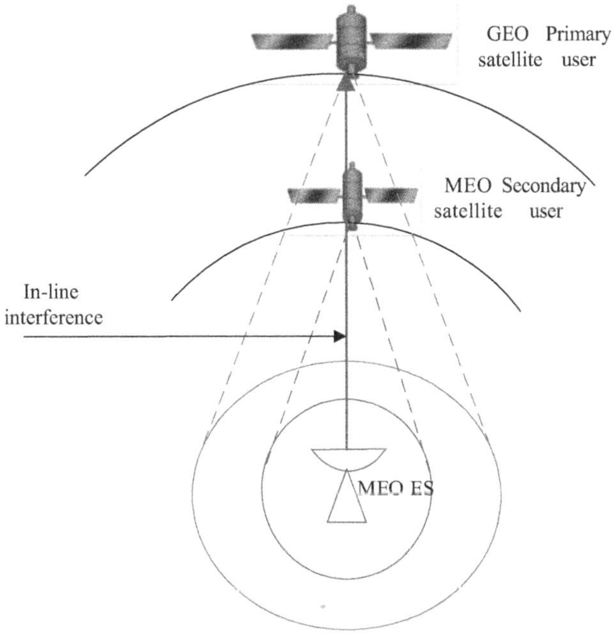

Figure 14.5 In-line interference

- **In-line interference** [16] is produced when an NGEO satellite passes through a line of sight path between an ES and the GEO satellite in the coexistence scenarios of GEO and NGEO networks. This type of interference is shown in Figure 14.5.
- **Terrestrial interference** is produced due to the fact that some frequency bands allocated to SATCOMs are often also allocated to terrestrial communications, particularly at Ka-band [17].
- **Intentional interference** is generated when an interfering signal is designed to degrade the performance of the satellite system. The most known type of intentional interference is the jamming.

From the above analysis, the uplink satellite interference can be also classified in terms of the nature of the interference source into intentional interference (e.g., jamming) and unintentional interference (e.g., cochannel, crosspoll, adjacent channel, adjacent system, in-line and terrestrial interference). The satellite operators have estimated that 90% of all interference events are due to unintentional interference, while intentional interferences correspond to 10% of them [14,15].

Finally, the types of unintentional interference can be further classified according to the service that the interfering signals belong (e.g., broadcasting satellite service, fixed satellite service, VSAT, etc.). According to SES data [18], a VSAT interference is the most critical and with the most important contribution. Each VSAT terminal transmits a low power signal; however, there is a large number of geographically distributed VSAT terminals, and hence, the aggregated interference from many of them has an important impact on SATCOMs.

14.4 Interference detection techniques

We consider a common SATCOM system, where the satellite, the desired terminal and the interferer are equipped with one antenna. The goal is to detect the uplink radio frequency interference. Hence, the detection problem can be formulated as the following binary hypothesis test, which is a baseband symbol sampled model:

$$\mathcal{H}_0 : \mathbf{y} = h\mathbf{s} + \mathbf{w}, \tag{14.1}$$

$$\mathcal{H}_1 : \mathbf{y} = h\mathbf{s} + \mathbf{w} + \mathbf{i}, \tag{14.2}$$

where h denotes the uplink channel, $\mathbf{s} = [s(1) \cdots s(N)]^T$ denotes an $N \times 1$ vector, referred to as the signal transmitted by the desired terminal with power P_s or energy E_s, $\mathbf{i} = [i(1) \cdots i(N)]^T$ denotes an $N \times 1$ vector, referred to as the received signal from the interferer, $\mathbf{w} = [w(1) \cdots w(N)]^T$ denotes an $N \times 1$ vector referred to as the additive noise at the receiving antenna of the satellite, modelled as an independent and identically distributed (i.i.d.) complex Gaussian vector with zero mean and covariance matrix given by $E\{\mathbf{w}\mathbf{w}^H\} = \sigma_w^2 \mathbf{I}_N$, where $\mathbf{I}_N$ denotes an identity matrix of size N, and $\mathbf{y} = [y(1) \cdots y(N)]^T$ denotes an $N \times 1$ vector, referred to as the total received signal at the satellite, at the $1_{st} \cdots N_{th}$ time instant, respectively. The desired transmitted signal $\mathbf{s}$ is a modulated signal consisted of an amount of N_p number of pilot symbols $\mathbf{s}_p$, interleaved with an N_d number of data streams $\mathbf{s}_d$. Therefore, $N = N_p + N_d$, with N denoting the total number of samples. Regarding the adopted model for the distribution of $\mathbf{i}$, note that it can be considered as a general model, where the vector $\mathbf{i}$ can be the aggregated signal of one or more independent interference sources, which are further independent over time. This model can be considered as a valid one for the performance evaluation of the developed detector; however as shall be shown later, the calculation of the detection threshold is independent from the distribution of the interfering signal(s) and can be applied to any scenario.

14.4.1 Conventional energy detector

The interference detection module may be designed in terms of one of the common spectrum sensing techniques discussed in the literature [19], including matched filter detection [20], cyclostationary detection [21] and energy detection [22–25]. Matched filter detection is an optimal detection approach; however, it requires a priori information of the interfering signal, e.g., modulation, coding, etc., which is often not available in practice. Furthermore, cyclostationary detection needs the knowledge of the cyclic frequencies of the interfering signal and increases the complexity, which makes it difficult for practical implementation. On the other hand, the ED does not require a priori knowledge of the interfering signal, and it is the most popular detector due to its simplicity, resulting in low complexity algorithms, which constitutes a crucial factor for on-board processing.

The ED measures the energy of the received signal and compares it with a properly selected threshold in order to decide on the presence or absence of the interference. Therefore, if we apply the ED in the hypothesis test of (14.1) and (14.2) as follows:

$$T(\mathbf{y}) = \|\mathbf{y}\|^2 = \sum_{n=1}^{N} |y(n)|^2 \begin{array}{l} < \gamma_{ced} \rightarrow \mathscr{H}_0 \\ > \gamma_{ced} \rightarrow \mathscr{H}_1 \end{array}, \tag{14.3}$$

where γ_{ced} is the decision threshold under the CED, the distribution of the test statistic $T(\mathbf{y})$ follows a noncentral chi-square distribution with $2N$ degrees of freedom under both hypotheses $\mathscr{H}_0$ and $\mathscr{H}_1$ and the probability of false alarm (P_{FA}) and probability of detection (P_D) can be expressed in closed form as

$$P_{FA} = Q_N\left(\sqrt{\rho_{\mathscr{H}_0}}, \sqrt{\frac{2\gamma_{ced}}{\sigma_w^2}}\right), \tag{14.4}$$

$$P_D = Q_N\left(\sqrt{\rho_{\mathscr{H}_1}}, \sqrt{\frac{2\gamma_{ced}}{\sigma_i^2 + \sigma_w^2}}\right), \tag{14.5}$$

where $Q_m(a, b)$ is the generalized Marcum-Q function and the noncentrality parameter ρ is given by $\rho_{\mathscr{H}_0} = (2\,|h|^2 E_s)/\sigma_w^2$ and $\rho_{\mathscr{H}_1} = (2\,|h|^2 E_s)/(\sigma_w^2 + \sigma_i^2)$, respectively.

However, in practice, the noise and signal power are usually unknown. Then, the P_{FA} and P_D under the condition of noise and signal power uncertainty can be expressed in closed form as

$$P_{FA_u} = Q_N\left(\sqrt{\frac{2\eta_h |h|^2 E_s}{\eta_w \sigma_w^2}}, \sqrt{\frac{2\gamma_{ced_u}}{\eta_w \sigma_w^2}}\right), \tag{14.6}$$

$$P_{D_u} = Q_N\left(\sqrt{\rho_{\mathscr{H}_1}}, \sqrt{\frac{2\gamma_{ced_u}}{\sigma_i^2 + \sigma_w^2}}\right), \tag{14.7}$$

where γ_{ced_u} is the selected threshold under the uncertainty scenario of the conventional ED and the uncertainty factor can be defined as $B = 10\log_{10}\eta$, with B to be in dB. The estimated noise variance is $\hat{\sigma}_w^2 = \eta_w \sigma_w^2$, where η_w reflects on how accurate the estimate is. Similar for η_h. Also, the indexes h and w represent the channel and noise, respectively.

The ED is an efficient technique, especially for strong interference scenarios. However, its main drawback for the detection of interference is its sensitivity to the noise variance and desired signal power uncertainties [4]. However, it is considered as the adopted detection technique because it does not require information about the interfering signal, and its practical implementation is simple and cost effective.

14.4.2 Energy detector with imperfect signal cancellation in the pilot domain

As mentioned earlier, the ED is a very popular detection technique, however, it usually faces difficulties to detect low values of ISNR, because it requires the knowledge of the noise and signal power to correctly set the threshold. However, the accurate knowledge of the noise and signal power in practice is not available; hence, the phenomenon of the ISNR wall [26] appears, above which the accurate detection of interference cannot be carried out. Furthermore, even if this knowledge is accurate, the conventional ED needs a large number of samples, which inhibits the fast detection of interference, and further increases the energy consumption on-board the satellite, which is a critical factor for any in-orbit processing technique.

To overcome these issues, [4] proposes a method which exploits the frame structure knowledge of the SATCOM standards, which employ pilot symbols for the transmission. This detector is well suited to the DTP payloads.

Algorithm 1: ED with imperfect signal cancellation exploiting the pilot symbols.

1. After the time and frame synchronization, the pilot signal is known at the satellite and the hypothesis test of (14.1) and (14.2) is reformulated as follows:

$$\mathcal{H}_{0_p}: \mathbf{y}_p = h\mathbf{s}_p + \mathbf{w}_p, \tag{14.8}$$

$$\mathcal{H}_{1_p}: \mathbf{y}_p = h\mathbf{s}_p + \mathbf{w}_p + \mathbf{i}_p, \tag{14.9}$$

   where $\mathbf{s}_p = \left[s_p(1)\cdots s_p(N_p)\right]^T$ denotes an $N_p \times 1$ vector, referred to as the pilot symbols with power P_p or energy E_p, $\mathbf{i}_p = \left[i_p(1)\cdots i_p(N_p)\right]^T$ denotes an $N_p \times 1$ vector, referred to as the interfering signal related to pilots' position, where $\mathbf{i}_p \sim \mathcal{CN}\left(0, \sigma_{i_p}^2 \mathbf{I}_{N_p}\right)$ with $\sigma_{i_p}^2 = \sigma_i^2$ denoting the variance of $\mathbf{i}_p$, $\mathbf{w}_p = \left[w_p(1)\cdots w_p(N_p)\right]^T$ denotes an $N_p \times 1$ vector referred to as the AWGN related to pilots' position, where $\mathbf{w}_p \sim \mathcal{CN}\left(0, \sigma_{w_p}^2 \mathbf{I}_{N_p}\right)$ with $\sigma_{w_p}^2 = \sigma_w^2$ denoting the variance of $\mathbf{w}_p$ and $\mathbf{y}_p = \left[y_p(1)\cdots y_p(N_p)\right]^T$ denotes an $N_p \times 1$ vector, referred to as the total received signal related to pilots' position.
2. Then, estimate the channel using the pilot symbols.
3. Furthermore, remove the pilot symbols from the total received signal and the new hypothesis test can be written as

$$\mathcal{H}'_{0_p}: \mathbf{y}'_p = \mathbf{w}_p - \varepsilon_{\mathcal{H}_0}\mathbf{s}_p, \tag{14.10}$$

$$\mathcal{H}'_{1_p}: \mathbf{y}'_p = \mathbf{i}_p + \mathbf{w}_p - \varepsilon_{\mathcal{H}_1}\mathbf{s}_p. \tag{14.11}$$

   where $\varepsilon_{\mathcal{H}_0}$ and $\varepsilon_{\mathcal{H}_1}$ denote the channel estimation error under the hypothesis $\mathcal{H}_0$ and $\mathcal{H}_1$, respectively.
4. Finally, apply an ED in the remaining signal as follows:

$$T(\mathbf{y}'_p) = \left\|\mathbf{y}'_p\right\|^2 = \sum_{n=1}^{N_p} \left|y'_p(n)\right|^2 \begin{array}{l} < \gamma_p \rightarrow \mathcal{H}'_{0_p} \\ > \gamma_p \rightarrow \mathcal{H}'_{1_p} \end{array}, \tag{14.12}$$

   where γ_p denotes a properly defined threshold for the algorithm of exploiting the pilot symbols, responsible for the detection of interference.

It is worth mentioning that for the success of this method, time synchronization is required to find the limits of the symbols and also frame synchronization to find where the pilots are in the frame.

To evaluate the detector of (14.12), we need to find the distribution of correlated chi-squared variables [27,28]. Then, based on some manipulations the probabilities of false alarm and detection of the ED with imperfect signal cancellation exploiting the pilot symbols, in this case P_{FA_p} and P_{D_p}, respectively, are given as follows:

$$P_{FA_p} = \frac{\Gamma\left(N_p - 1, (\gamma_p/\sigma_{w_p^2})\right)}{\Gamma(N-1)}, \tag{14.13}$$

$$P_{D_p} = \frac{\Gamma\left(N_p - 1, (\gamma_p/(\sigma_{w_p^2} + \sigma_{i_p^2}))\right)}{\Gamma\left(N_p - 1\right)}, \tag{14.14}$$

which looks like an ED with one less degree of freedom. The corresponding equations for the noise uncertainty case are given by

$$P_{FA_{pu}} = \frac{\Gamma\left(N_p - 1, (\gamma_{u_p}/\eta_{w_p}\sigma_{w_p^2})\right)}{\Gamma\left(N_p - 1\right)}, \tag{14.15}$$

$$P_{D_{pu}} = \frac{\Gamma\left(N_p - 1, (\gamma_{u_p}/(\sigma_{w_p^2} + \sigma_{i_p^2}))\right)}{\Gamma\left(N_p - 1\right)}, \tag{14.16}$$

where γ_{u_p} is the selected threshold under the uncertainty scenario of the ED with imperfect signal cancellation exploiting the pilot symbols.

Therefore, we can notice that the proposed ED with signal cancellation technique is affected only by the noise uncertainty compared to the conventional ED which has to take into account the noise and signal power uncertainty.

14.4.3 Energy detector with imperfect signal cancellation in the data domain

The idea of ED with signal cancellation was introduced earlier by exploiting the frame structure and pilot symbols of the SATCOM standards. This technique provides reliable detection of weak interfering signals under the assumption of adequate number of samples, namely, pilots. However, sometimes, the detection at low values of ISNR may require more samples than the number of pilots supported by the standard. Furthermore, if the interference is intermittent during the frame, the samples related to the position of the pilot symbols may not be affected, and the previous method will not provide a good detection of interference.

To address these concerns, [29] proposes a detection scheme based on the concept of ED with signal cancellation, which does not require pilots symbols. It focuses on the data domain, demodulating the desired data signal, removing it from the total received signal and applying an ED in the remaining signal for the detection of interference. This technique needs a partially regenerative satellite (at least for the SMU), where the received signal can be demodulated.

Algorithm 2: ED with imperfect signal cancellation exploiting the data

1. The hypothesis test of (14.1) and (14.2) is reformulated as follows:

$$\mathcal{H}_{0_d}: \mathbf{y}_d = h\mathbf{s}_d + \mathbf{w}_d, \tag{14.17}$$

$$\mathcal{H}_{1_d}: \mathbf{y}_d = h\mathbf{s}_d + \mathbf{w}_d + \mathbf{i}_d, \tag{14.18}$$

where h denotes the scalar flat fading channel from the desired terminal to the satellite, which is assumed to be known at the satellite receiver (i.e., estimated in advance), and it is assumed to be real after the phase compensation with channel power γ, $\mathbf{s}_d = [s_d(1)\cdots s_d(N_d)]^T$ denotes an $N_d \times 1$ vector, referred to as the transmitted data signal by the desired terminal with power P_d or energy E_d, $\mathbf{i}_d = [i_d(1)\cdots i_d(N_d)]^T$ denotes an $N_d \times 1$ vector, referred to as the interfering signal related to data symbols, where $\mathbf{i}_d \sim \mathcal{CN}\left(0, \sigma_{i_d}^2 \mathbf{I}_{N_d}\right)$ with $\sigma_{i_d}^2 = \sigma_i^2$ denoting the variance of $\mathbf{i}_d$, $\mathbf{w}_d = [w_d(1)\cdots w_d(N_d)]^T$ denotes an $N_d \times 1$ vector referred to as the AWGN related to data symbols, where $\mathbf{w}_d \sim \mathcal{CN}\left(0, \sigma_{w_d}^2 \mathbf{I}_{N_d}\right)$ with $\sigma_{w_d}^2 = \sigma_w^2$ denoting the variance of $\mathbf{w}_d$ and $\mathbf{y}_d = [y_d(1)\cdots y_d(N_d)]^T$ denotes an $N_d \times 1$ vector, referred to as the total received signal related to data symbols.
2. Then, recover the transmitted signal by the desired terminal: $\hat{\mathbf{s}}_d$ denotes the recovered or estimated signal.
3. Furthermore, remove this estimated signal from the total received signal at the satellite: $\mathbf{y}'_d = \mathbf{y}_d - h\hat{\mathbf{s}}_d$
4. Finally, apply an ED in the remaining signal as follows:

$$T\left(\mathbf{y}'_d\right) = \left\|\mathbf{y}'_d\right\|^2 = \sum_{n=1}^{N_d} \left|y'_d(n)\right|^2 \begin{matrix} < \gamma_d \rightarrow \mathcal{H}_{0_d} \\ > \gamma_d \rightarrow \mathcal{H}_{1_d} \end{matrix}, \tag{14.19}$$

where γ_d denotes a properly defined threshold for the algorithm of recovering the data symbols, responsible for the detection of interference.

This algorithm can be applied for any modulation scheme supported by DVB-S2X [30] standard (QPSK, 8PSK, 16APSK, etc.), but in this subsection, we focus on QPSK modulated signals. However, for simplicity, we start our analysis considering a BPSK signal, which as shall be shown later, can be easily extended to QPSK scenario.

Applying the first three steps of the algorithm under the BPSK case, the hypothesis test of (14.17) and (14.18) can be reformulated as follows:

$$\mathcal{H}_{0_B} = \begin{cases} \mathcal{H}_{00_B}: y'_d(n) = w_d(n), \\ \mathcal{H}_{01_B}: y'_d(n) = 2hs_d(n) + w_d(n), \end{cases} \tag{14.20}$$

$$\mathcal{H}_{1_B} = \begin{cases} \mathcal{H}_{10_B}: y'_d(n) = i_d(n) + w_d(n), \\ \mathcal{H}_{11_B}: y'_d(n) = i_d(n) + 2hs_d(n) + w_d(n), \end{cases} \tag{14.21}$$

where $n = 0, 1, \ldots, N-1$, the index B denotes the BPSK scenario, $\mathcal{H}_{00_B}$ and $\mathcal{H}_{10_B}$ represent the hypothesis when the received signal is recovered correctly and the interference is absent and present, respectively, while $\mathcal{H}_{01_B}$ and $\mathcal{H}_{11_B}$ correspond to the wrong recovering case when the interference is absent and present, respectively.

Then, the probability of false alarm of the ED with imperfect signal cancellation recovering the data symbols under the BPSK case; in this case, $P_{FA_{d_B}}$ is given as follows:

$$P_{FA_{d_B}} = \sum_{k=0}^{N_d} \binom{N_d}{k} P_{k_B} P_{e_B}^k \left(1 - P_{e_B}\right)^{N_d - k}, \tag{14.22}$$

where k denotes the number of wrong recovered bits, P_{e_B} is the probability of bit error for BPSK [31] and P_{k_B} is the probability of false alarm for the case that k bits are recovered wrongly which can be approximated as follows:

$$P_{k_B} = Q\left(\frac{\gamma_d - \mu_{\mathcal{H}_{0_B}}}{\sqrt{V_{\mathcal{H}_{0_B}}}}\right), \tag{14.23}$$

where $\mu_{\mathcal{H}_{0_B}}$ and $V_{\mathcal{H}_{0_B}}$ are the mean and variance of the test statistic $T\left(\mathbf{y}'_d \left|\mathcal{H}_{0_B}\right.\right)$, respectively, which are also related to k.

However, the calculation of the detection threshold γ_d, through (14.22), may be complicated, particularly as the number of samples increases. Nevertheless, the probability of false alarm can be approximated by

$$P_{FA_{d_{B_a}}} = Q\left(\frac{\gamma_d - N_d\left(1 - P_{e_B}\right)\mu_{\mathcal{H}_{00_B}} - N_d P_{e_B} \mu_{\mathcal{H}_{01_B}}}{\sqrt{N_d\left(1 - P_{e_B}\right) V_{\mathcal{H}_{00_B}} + N_d P_{e_B} V_{\mathcal{H}_{01_B}}}}\right), \tag{14.24}$$

where $\mu_{\mathcal{H}_{00_B}}, \mu_{\mathcal{H}_{01_B}}, V_{\mathcal{H}_{00_B}}$ and $V_{\mathcal{H}_{01_B}}$ are the mean and variance of the test statistic $T\left(y'_d \left|\mathcal{H}_{00_B}\right.\right)$ and $T\left(y'_d \left|\mathcal{H}_{01_B}\right.\right)$, respectively, where y'_d means only one sample, the index B_a denotes approximation under the BPSK scenario, and hence, this equation approximates and simplifies (14.22), based on the fact that for a large number of samples, the expected number of correct and wrong recovered bits is $N_d(1 - P_{e_B})$ and $N_d P_{e_B}$, respectively. Now, the calculation of the threshold γ_d is straightforward, based on the inverse function of the $P_{FA_{d_{B_a}}}(\cdot)$.

The corresponding probability of detection P_{d_B} is given by

$$P_{D_{d_{B_{apr}}}} = Q\left(\frac{\gamma_d - N_d\left(1 - P'_{e_B}\right)\mu_{\mathcal{H}_{10_B}} - N_d P'_{e_B} \mu_{\mathcal{H}_{11_B}}}{\sqrt{N_d\left(1 - P'_{e_B}\right) V_{\mathcal{H}_{10_B}} + N_d P'_{e_B} V_{\mathcal{H}_{11_B}}}}\right), \tag{14.25}$$

where $P'_{e_B} = Q\left(\sqrt{\frac{\gamma P_d}{\sigma_{w_d}^2 + \sigma_{I_d}^2}}\right)$.

14.4.3.1 Probability of false alarm for QPSK signals

In the previous subsection, we derived the probability of false alarm under the BPSK scenario. Now, the extension of (14.22) to QPSK case is straightforward and it is given as follows:

$$P_{FA_{d_Q}} = \sum_{k=0}^{2N_d} \binom{2N_d}{k} P_{k_Q} P_{e_Q}^k \left(1 - P_{e_Q}\right)^{2N_d - k}, \tag{14.26}$$

where $P_{k_Q} = P_{k_B}$ and $P_{e_Q} = P_{e_B}$. Hence, the only difference with (14.22) is the factor 2, due to the fact that a QPSK signal constitutes of two orthogonal BPSK ones. From the other side, the approximated P_{FA} of (14.24) can be expressed as follows:

$$P_{FA_{dQ_a}} = Q\left(\frac{\gamma_d - a\mu_{\mathscr{H}_{00_Q}} - b\left(\mu_{\mathscr{H}_{01_Q}} + \mu_{\mathscr{H}_{02_Q}}\right) - c\mu_{\mathscr{H}_{03_Q}}}{\sqrt{aV_{\mathscr{H}_{00_Q}} - b\left(V_{\mathscr{H}_{01_Q}} + V_{\mathscr{H}_{02_Q}}\right) - cV_{\mathscr{H}_{03_Q}}}}\right) \tag{14.27}$$

where $a = \left(1 - P_{e_Q}\right)^2$, $b = \left(1 - P_{e_Q}\right)P_{e_Q}$, $c = P_{e_Q}^2$, the index Q denotes the QPSK scenario, P_{e_Q} is the probability of bit error for QPSK and is the same as for BPSK, $\mathscr{H}_{00_Q}$ denotes that both the real and imaginary parts are recovered correctly, $\mathscr{H}_{01_Q}$ means that the real part is recovered wrongly and the imaginary part is recovered correctly, $\mathscr{H}_{02_Q}$ means that the real part is recovered correctly and the imaginary part is recovered wrongly, while $\mathscr{H}_{03_Q}$ denotes that both the real and imaginary parts are recovered wrongly. Furthermore, we can easily see that $\mu_{\mathscr{H}_{00_Q}} = 2\mu_{\mathscr{H}_{00_B}}$ and $V_{\mathscr{H}_{00_Q}} = 2V_{\mathscr{H}_{00_B}}$, $\mu_{\mathscr{H}_{01_Q}} = \mu_{\mathscr{H}_{00_B}} + \mu_{\mathscr{H}_{01_B}}$ and $V_{\mathscr{H}_{01_Q}} = V_{\mathscr{H}_{00_B}} + V_{\mathscr{H}_{01_B}}$, $\mu_{\mathscr{H}_{02_Q}} = \mu_{\mathscr{H}_{00_B}} + \mu_{\mathscr{H}_{01_B}}$ and $V_{\mathscr{H}_{02_Q}} = V_{\mathscr{H}_{00_B}} + V_{\mathscr{H}_{01_B}}$ and finally, $\mu_{\mathscr{H}_{03_Q}} = 2\mu_{\mathscr{H}_{01_B}}$ and $V_{\mathscr{H}_{03_Q}} = 2V_{\mathscr{H}_{01_B}}$.

Regarding the P_D under the scenario that the desired transmitted signal is QPSK modulated is given by (14.25) by substituting again $\sigma_{w_d}^2$ with $\sigma_{w_d}^2 + \sigma_{i_d}^2$ in the related parts. Finally, the probabilities of false alarm and detection for both BPSK and QPSK under the uncertainty scenario can be derived similarly as in the previous sections.

14.5 Current localization techniques

Although there are numerous works in the area of localization, here, we focus on the localization literature in the SATCOMs domain. In the thesis [32], the interferometry technique is used in order to find the AoA of the interfering signal. In the interferometry method, the difference in the phase of an incoming signal at two spatially separated antennas is measured. Based on this measurement, the AoA of the interfering radiation relation to an interferometric baseline is derived. As mentioned in [32], feeds with orthogonal polarization can be used in order to perform interferometry measurements. The signal level on the feed with opposite polarization is 30 dB lower than the signal level on the feed with same polarization as the signal.

The authors of [33] perform time difference of arrival (TDOA) along with phase measurements to localize an unknown interferer using two geostationary (GEO) satellites. In [34], three out of four satellites exposed to interference are used to derive TDOA measurements for localizing an unknown interferer. The performance of the localization in Eutelsat satellites is presented in [35] where TDOA and frequency difference of arrival (FDOA) measurements are used to localize an unknown interferer. To improve the accuracy of the previous works, the altitude constraint in [36,37] is considered to improve the localization accuracy by employing TDOA and/or FDOA technique(s). Two antennas on a spinning satellite in [38] are used to localize an unknown interferer. FDOA measurements done by more than two satellites in [39] are used to localize an interferer. It is shown that in contrast to TDOA, FDOA accuracy is not affected by the bandwidth of the interference signal.

Apart from the scientific papers, there are numerous related patents in the field of satellite localization. In [40], repetitions of the TDOA and FDOA techniques are used to localize a target on the earth. The patent suggest of using two GEO satellites for this purpose. In [41], TDOA and FDOA are used in order to localize a target. A known reference signal is used in order to compensate for the phase noise and the frequency drift in the unknown signal. The reference signal is used to remove sources of error and operational limitations. The patent [41] does not require the velocity and position of the satellites with the accuracy in [40]. Furthermore, it works with more inclined satellites, up to 3°. In the patent [42], repetitions of TDOA and FDOA measurements by two GEO satellites and reference signal are used to localize an emitter on the earth. In this work, an emitter with varying frequency is considered. The reference signal is used to remove sources of error and operational limitations. It gives improved accuracy and extends the range of conditions over which measurements can be made.

The patent [43] uses two TDOA and two FDOA measurements collected by three satellites along with a known reference signal to localized an unknown emitter on the earth. Weights for the errors in the TDOA and FDOA measurements are determined, and the weights are applied in a weighted error function. The weights account for the errors in the measurements and the errors in the satellite positions and velocities, and are dependent on the localization geometry. In [44], a very similar approach to [43] is followed. Three satellites are used to perform two TDOA and two FDOA measurements. The location of an emitter can be determined from minimizing a cost function of the weighted combination of the six solutions derived from the two TDOA measurements and the two FDOA measurements, where the weight of each solution in the combination is determined based on the intersection angle of the two curves that define the possible locations of the emitter based on the TDOA and/or FDOA measurements. Recently, the GLOWLIMK company has registered a patent which uses one satellite to localize an unknown emitter [45]. It is worth mentioning that this a ground-based localization approach.

In addition to the scientific papers and the patents, we introduce satellite localization products. The SIEMENS industry has come up with a technique, SIECAMS® ILS ONE [http://www.siemens.com, accessed on 04/12/2017], in order to localize an emitter on the earth surface using only one satellite. SIECAMS ILS ONE works by analyzing signal distortions that are primarily caused by satellite movement, atmospheric or weather influences and many other environmental factors. By comparing such signal distortions of the interference signal with known signals, SIECAMS ILS ONE is able to identify the precise area of the interference source resulting in a significant increase in resolved interference issues well beyond the limits of traditional satellite interference localization systems. The GLOWLINK company also claims that they can localize an emitting target on the earth using only one satellite having a product with the name "Single Satellite Geolocation" [http://www.glowlink.com, accessed on 04/12/2017]. In addition, there are two satellite Geo-location products built by this company.

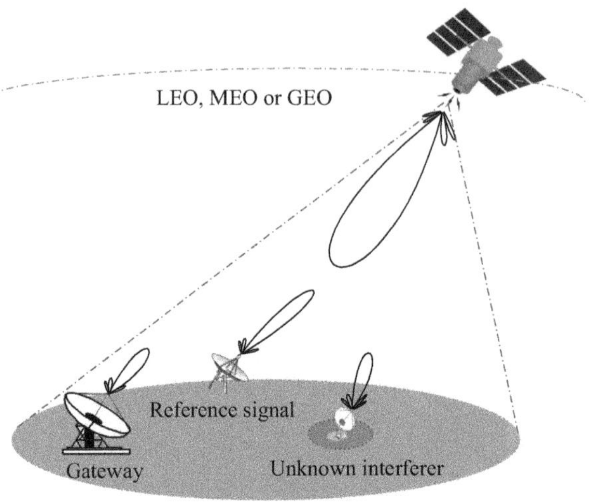

Figure 14.6 An affected or a localization-dedicated satellite receiving interference and reference signals in uplink

14.6 Interference localization using frequency of arrival via a single satellite

In this part, we mention the system model along the algorithm for localization an interferer while using only the affected satellite. We first mention the on-ground method and then proceed to the on-board approach. We consider a transparent satellite which receives uplink signal from a gateway within the Ka band. Concurrently, the satellite receives narrow band uplink interference from an unknown transmitter within the same frequency band as the uplink signal from the gateway. A reference signal is transmitted to the satellite to compensate for the errors. The whole scenario is summarized in Figure 14.6. The central frequency of the interference signal is shown by f_u and since it is interfering with the main uplink signal, we assume that f_u is known. Although f_u may be changed intentionally and/or due to instability of the electronics, for the sake of simplicity, f_u is considered to be fixed through the time. Also, we assume that the derived signal is turned off during sampling the interference signal. All the vectors in this section and Section 14.7 are in Cartesian coordinates. The subscripts u, r, s, gw, ul and dl are used in the equations instead of the terms: unknown interferer, reference transmitter, satellite, gateway, uplink and downlink, respectively.

The frequency of the nth sample in time of the interfering signal by the satellite is

$$f_{n_{u,s}} = f_u \left(1 + \frac{\mathbf{v}_{n_{ul}}^T \mathbf{k}_{n_{u,s}}}{c_n} \right), \tag{14.28}$$

where $f_{n_{u,s}}$ is the frequency of the nth sample of the interfering signal at the satellite, $\mathbf{v}_{n_{ul}}$ is the velocity of the satellite when sampling, c_n is the propagation speed of the signal in the space and $\mathbf{k}_{n_{u,s}}$ is the normalized unit vector pointing from the satellite toward the unknown interferer defined as

$$\mathbf{k}_{n_{u,s}} = \frac{\mathbf{u} - \mathbf{s}_{n_{ul}}}{\left\| \mathbf{u} - \mathbf{s}_{n_{ul}} \right\|}, \tag{14.29}$$

where $\mathbf{s}_{n_{ul}}$ is the position of the satellite during uplink and $\mathbf{u} = [u_1, u_2, u_3]$ is the location of the unknown interferer. Afterwards, the satellite down converts $f_{n_{u,s}}$ into

$$f_{n_{u,s}} - f_T = f_u \left(1 + \frac{\mathbf{v}_{n_{ul}}^T \mathbf{k}_{n_{u,s}}}{c_n} \right) - f_T, \tag{14.30}$$

where f_T is the amount of the frequency down conversion for the nth sample. Subsequently, the satellite forwards the down converted signal to the gateway. Using (14.30), the frequency of the received signal at the gateway is

$$\begin{aligned} f_{n_{u,g}} &= \left(f_u + f_u \frac{\mathbf{v}_{n_{ul}}^T \mathbf{k}_{n_{u,s}}}{c_n} - f_T \right) \left(1 + \frac{\mathbf{v}_{n_{dl}}^T \mathbf{k}_{n_{s,g}}}{c_n} \right) \\ &= f_{n_{dl}} + f_u \frac{\mathbf{v}_{n_{ul}}^T \mathbf{k}_{n_{u,s}}}{c_n} + f_{n_{dl}} \frac{\mathbf{v}_{n_{dl}}^T \mathbf{k}_{n_{s,g}}}{c_n} \\ &\quad + f_u \frac{\mathbf{v}_{n_{ul}}^T \mathbf{k}_{n_{u,s}}}{c_n} \frac{\mathbf{v}_{n_{dl}}^T \mathbf{k}_{n_{s,g}}}{c_n}, \end{aligned} \tag{14.31}$$

where $f_{n_{dl}} = f_u - f_T$ and $\mathbf{k}_{n_{s,g}} = (\mathbf{s}_{gw} - \mathbf{s}_{n_{dl}})/(\left\| \mathbf{s}_{gw} - \mathbf{s}_{n_{dl}} \right\|)$ with $\mathbf{s}_{gw}$ being the position of the gateway and $\mathbf{s}_{n_{dl}}$ being the position of the satellite when forwarding the nth sampled interference to the gateway. The last term in (14.31) is very small compared to the other terms when it comes to GEO satellites with a very slow drift and has been neglected in [45]. However, we keep it since its effect increases as the velocity of the satellite goes higher, especially for low Earth orbit (LEO), medium Earth orbit (MEO) or retro GEO satellites.

The gateway estimates the frequency of the nth sampled interference after receiving it from the satellite. Due to the movement of the satellite, each estimated frequency includes a specific amount of Doppler shift which relates to the position of the unknown interferer. Hence, a location-related equation can be made between each estimated frequency and the location of the unknown interferer. To this end, the gateway requires satellite's positions and velocities during uplink and downlink of the nth sample, the frequency of the satellite's down conversion oscillator, and the frequency of the interference signal while it is being emitted. However, the values related to the oscillator frequency, positions and velocities are different from their real values due to equipment impairments. To compensate for these errors, the gateway needs to calibrate the estimated frequency of the nth sample. For this purpose, a reference signal from a known location on the earth can be transmitted to the satellite and then forwarded to the gateway in one of the following approaches:

1. The reference signal is uplinked in the same frequency as the interference signal after a delay. Due to the delay, the reference and interference signals experience different mismatches.

2. The reference signal is uplinked in a different frequency from the interference signal and the satellite samples the interference and reference signals simultaneously.

Here, the second approach is followed to transmit the reference signal. By following a similar procedure as in (14.28)–(14.31), the frequency of the nth sample of the reference signal at the gateway is obtained by

$$\begin{aligned} f_{n_{r,g}} &= \left(f_r + f_r \frac{\mathbf{v}_{n_{ul}}^T \mathbf{k}_{n_{r,s}}}{c_n} - f_T\right)\left(1 + \frac{\mathbf{v}_{n_{dl}}^T \mathbf{k}_{n_{s,g}}}{c_n}\right) \\ &= f'_{n_{dl}} + f_r \frac{\mathbf{v}_{n_{ul}}^T \mathbf{k}_{n_{r,s}}}{c_n} + f'_{n_{dl}} \frac{\mathbf{v}_{n_{dl}}^T \mathbf{k}_{n_{s,g}}}{c_n} \\ &\quad + f_r \frac{\mathbf{v}_{n_{ul}}^T \mathbf{k}_{n_{r,s}}}{c_n} \frac{\mathbf{v}_{n_{dl}}^T \mathbf{k}_{n_{s,g}}}{c_n}, \end{aligned} \tag{14.32}$$

where $f_{n_{r,g}}$ is the estimated frequency of the reference signal at the gateway and $\mathbf{k}_{n_{r,s}}$ is the normalized unit vector pointing from the satellite toward the reference transmitter defined as $\mathbf{k}_{n_{r,s}} = (\mathbf{r} - \mathbf{s}_{n_{ul}})/(\|\mathbf{r} - \mathbf{s}_{n_{ul}}\|)$ with $\mathbf{r}$ being the location of the reference transmitter. Next, the gateway calculates the expected frequency of the reference signal using the available erroneous data as

$$\begin{aligned} f_{n_{r,g,exp}} &= f_r - f_{n_{T_e}} + f_r \frac{\mathbf{v}_{n_{ule}}^T \mathbf{k}_{n(r,s)_e}}{c_n} + f_{n_{dle}} \frac{\mathbf{v}_{n_{dle}}^T \mathbf{k}_{n(s,g)_e}}{c_n} \\ &\quad + f_r \frac{\mathbf{v}_{n_{ule}}^T \mathbf{k}_{n(r,s)_e}}{c_n} \frac{\mathbf{v}_{n_{dle}}^T \mathbf{k}_{n(s,g)_e}}{c_n}, \end{aligned} \tag{14.33}$$

where $f_{n_{r,g,exp}}$ is the expected frequency of the nth sampled reference signal at the gateway, $\mathbf{k}_{n_{(r,s)_e}} = (\mathbf{r} - \mathbf{s}_{n_{ule}})/(\|\mathbf{r} - \mathbf{s}_{n_{ule}}\|)$ and $\mathbf{k}_{n_{(s,g)_e}} = (\mathbf{s}_{gw} - \mathbf{s}_{n_{dle}})/(\|\mathbf{s}_{gw} - \mathbf{s}_{n_{dle}})\|$. The frequency mismatch for the nth sample is derived using (14.32) and (14.33) as

$$\delta_n = \frac{f_u}{f_r}\left[f_{n_{r,g}} - f_{n_{r,g,exp}}\right], \tag{14.34}$$

where δ_n is the amount of the frequency mismatch, the factor f_u/f_r is used to convert the frequency of the reference signal into the frequency of the unknown emitter since the reference signal has a different frequency and undergoes a different amount of mismatches. Using (14.34), the calibrated frequency of the nth received interference at the gateway is obtained by $\widetilde{f}_{n_{u,g}} = f_{n_{u,g}} - \delta_n$ where $\widetilde{f}_{n_{u,g}}$ is the calibrated frequency. The difference in the location of the unknown and the reference transmitters leads into different values for $\mathbf{k}_{u,s}$ and $\mathbf{k}_{r,s}$. Hence, the satellite velocity in the uplink will have different values and errors in the directions of $\mathbf{k}_{u,s}$ and $\mathbf{k}_{r,s}$, which means that the reference signal does not go through the same amount of mismatches as the unknown interference signal. To improve this, we can perform iterative localization and choose a closer reference transmitter to the unknown interferer after each localization step. After calibration, the known information at the gateway is used to reduce the estimated frequency (14.31) as

$$\hat{f}_{n_{u,g}} = \tilde{f}_{n_{u,g}} - f_u + f_{n_{T_e}} - f_{n_{dl_e}} \frac{\mathbf{v}_{n_{dl_e}}^T \mathbf{k}_{n_{(s,g)_e}}}{c_n} \qquad (14.35)$$

where $\hat{f}_{n_{u,g}}$ is the reduced calibrated frequency of the nth sample at the gateway and $\mathbf{k}_{n_{(u,s)_e}} = (\mathbf{u} - \mathbf{s}_{n_{ul_e}})/(\|\mathbf{u} - \mathbf{s}_{n_{ul_e}}\|)$. The gateway uses (14.35) a long with the available data to build an analytical location-related equation as

$$\hat{f}_{n_{u,g}} = f_u \frac{\mathbf{v}_{n_{ul_e}}^T \mathbf{k}_{n_{(u,s)_e}}}{c_n} + f_u \frac{\mathbf{v}_{n_{ul_e}}^T \mathbf{k}_{n_{(u,s)_e}}}{c_n} \frac{\mathbf{v}_{n_{dl_e}}^T \mathbf{k}_{n_{(s,g)_e}}}{c_n}. \qquad (14.36)$$

Remark 14.1. *The value of f_T changes for each sample due to the instability of satellite's electronics. Due to the difference between f_u and f_r, the error of f_T cannot be accurately derived, which reduces the localization accuracy. As a solution, we can use on-board spectrum monitoring to do on-board localization. Therefore, the sampled interference is not required to be down converted and thus its frequency is not influenced by the drift in the oscillator. Hence, the localization accuracy can be improved by on-board localization.*

In the following part, we describe the procedure to calculate the location of the interferer using the estimated and calibrated frequencies at the gateway.

14.7 Localization algorithm and solution

Since it is already known that the unknown interferer is located on the earth, at least two equations as in (14.37) plus the equation of the earth surface are required to get an estimation for the location of the unknown interferer.

$$f_n(\mathbf{u}) = \frac{f_u}{c_n}\left[\mathbf{v}_{n_{ul_e}}^T \mathbf{k}_{n_{(u,s)_e}} \left(1 + \frac{\mathbf{v}_{n_{dl_e}}^T \mathbf{k}_{n_{(s,g)_e}}}{c_n}\right)\right] - \hat{f}_{n_{u,g}}. \qquad (14.37)$$

To make a system of location-related equations, N of the estimated frequencies at the gateway, with $N \geq 2$, are randomly selected. This system of nonlinear equations is solved using an iterative algorithm with the initial guess $\mathbf{u}_0$. To this end, the first-order Taylor series approximation around $\mathbf{u}_0$ is applied on each location-related equation to obtain

$$\mathbf{f}(\mathbf{u}) \approx \mathbf{f}(\mathbf{u}_0) + \mathbf{F}'(\mathbf{u}_0)(\mathbf{u} - \mathbf{u}_0), \qquad (14.38)$$

where $\mathbf{f}(\mathbf{u}) = \left[f_1(\mathbf{u}), \ldots, f_N(\mathbf{u}), \|\mathbf{u}\|^2 - r^2\right]$, $\|\mathbf{u}\|^2 = r^2$ is the surface of the earth equation, r is the earth radius, $\mathbf{f}(\mathbf{u}_0) = \left[f_1(\mathbf{u}_0), \ldots, f_N(\mathbf{u}_0), \|\mathbf{u}_0\|^2 - r^2\right]$, and $\mathbf{F}'(\mathbf{u}_0)$ is the partial derivative matrix calculated at the initial guess as

$$\mathbf{F}'(\mathbf{u}_0) = \begin{bmatrix} \frac{\partial f_1(\mathbf{u}_0)}{\partial u_1} & \frac{\partial f_1(\mathbf{u}_0)}{\partial u_2} & \frac{\partial f_1(\mathbf{u}_0)}{\partial u_3} \\ \cdot & \cdot & \cdot \\ \cdot & \cdot & \cdot \\ \cdot & \cdot & \cdot \\ \frac{\partial f_N(\mathbf{u}_0)}{\partial u_1} & \frac{\partial f_N(\mathbf{u}_0)}{\partial u_2} & \frac{\partial f_N(\mathbf{u}_0)}{\partial u_3} \\ 2u_1 & 2u_2 & 2u_3 \end{bmatrix}. \tag{14.39}$$

The partial derivatives of f_n with respect to u_m for $m = 1,\ 2,\ 3$ are derived as $\partial f_n(\mathbf{u})/\partial u_m = (f_u/c_n)\eta_n\left[\mathbf{v}_{n_{ule}}^T\,\mathbf{a}_m\right]$ where

$$\mathbf{a}_1 = \begin{bmatrix} -\frac{g_n - \left(u_1 - s_{n1_e}\right)^2 g_n^{-1}}{g_n^2} \\ \frac{\left(u_1 - s_{n1_e}\right)\left(u_2 - s_{n2_e}\right)}{g_n^3} \\ \frac{\left(u_1 - s_{n1_e}\right)\left(u_3 - s_{n3_e}\right)}{g_n^3} \end{bmatrix}, \tag{14.40}$$

$$\mathbf{a}_2 = \begin{bmatrix} \frac{\left(u_2 - s_{n2_e}\right)\left(u_1 - s_{n1_e}\right)}{g_n^3} \\ -\frac{g_n - \left(u_2 - s_{n2_e}\right)^2 g_n^{-1}}{g_n^2} \\ \frac{\left(u_2 - s_{n2_e}\right)\left(u_3 - s_{n3_e}\right)}{g_n^3} \end{bmatrix}, \tag{14.41}$$

$$\mathbf{a}_3 = \begin{bmatrix} \frac{\left(u_3 - s_{n3_e}\right)\left(u_1 - s_{n1_e}\right)}{g_n^3} \\ \frac{\left(u_3 - s_{n3_e}\right)\left(u_2 - s_{n2_e}\right)}{g_n^3} \\ -\frac{g_n - \left(u_3 - s_{n3_e}\right)^2 g_n^{-1}}{g_n^2} \end{bmatrix}, \tag{14.42}$$

$\eta_n = \left(1 + ((\mathbf{v}_{n_{dle}}^T \mathbf{k}_{n(s,g)_e})/c_n)\right)$, $g_n = \left\|\mathbf{u} - \mathbf{s}_{n_{ule}}\right\|$ and $\mathbf{s}_{n_{ule}} = \left(s_{n1_e}, s_{n2_e}, s_{n3_e}\right)$. We need to find the point $\mathbf{u} = \mathbf{u}_1$ to have $\mathbf{f}(\mathbf{u}_0) + \mathbf{F}'(\mathbf{u}_0)(\mathbf{u}_1 - \mathbf{u}_0) = 0$ so that $\mathbf{F}'(\mathbf{u}_0)\Delta\mathbf{u} = -\mathbf{f}(\mathbf{u}_0)$, which is a system of linear equations with $\Delta\mathbf{u} = \mathbf{u}_1 - \mathbf{u}_0$. The system of linear equations can be solved via LU and QR factorization techniques. In case of using LU factorization, the complexity is $2n^3/3$ flops where n is the number of location-related equations and the earth equation. After deriving $\Delta\mathbf{u}$, the initial guess is updated as

$$\mathbf{u}_{i+1} = \mathbf{u}_i + \Delta\mathbf{u}, \tag{14.43}$$

and continues till $\|\Delta\mathbf{u}\| < \varepsilon$ where ε depends on the required localization accuracy.

Finally, it is worth mentioning that the maximum likelihood approach recommended here can be expanded to a search grid approach based on maximum likelihood to avoid errors in convergence to different local maximum.

14.8 Numerical results

In this section, we present some results to evaluate the performance of the proposed interference detection and localization techniques.

14.8.1 Performance analysis of interference detection techniques

In the simulations, the channel is considered to be a scalar complex channel of unit power, stable for a long period (i.e., at least for the whole frame), the ISNR varies from -25 to 5 dB, while the probability of false alarm is set to $P_{FA} = 0.1$. Furthermore, the desired transmitted signal is QPSK modulated, while the noise is generated by independent identically distributed (i.i.d.) complex Gaussian random variables with distribution $\mathcal{CN}(0, \sigma_w^2)$, where $\sigma_w^2 = \sigma_{w_p}^2 = \sigma_{w_d}^2$. The reliability of the proposed detectors is based on the ability to correctly set the threshold. Therefore, for simplicity reasons, we assume that the interfering signal is generated by i.i.d. complex Gaussian random variables with distribution $\mathcal{CN}(0, \sigma_i^2)$, where $\sigma_i^2 = \sigma_{i_p}^2 = \sigma_{i_d}^2$. Finally, we mention that the interference can affect both the forward and return link, but here we present simulation results based on the considered return link budget of Table 14.1.

Figure 14.7 presents the probability of detection as a function of the received ISNR comparing the following detection schemes: (i) ED with imperfect signal cancellation with data recovery (EDISC with data), (ii) ED with imperfect signal cancellation exploiting the pilot symbols (EDISC with pilots) and (iii) CED taking also into account the noise variance and signal energy uncertainties. In practice, the uncertainty factor in receiver is typically 1–2 dB [46]. Here, we consider that $B_p = B_{w_d} = B_{E_d} = 1$ dB, while the number of modulated symbols and pilots are set to $N_d = 460$, $N_p = 56$ and $N = 516$ representing a more realistic waveform according to DVB-RCS2 standard. It is observed that in both figures, the interference detection

Table 14.1 Return link budget parameters for the uplink

Parameter	Value
Orbit	GEO circular
Satellite height	35,786 km
G/T	2.5 dB K^{-1}
Uplink carrier frequency	14.25 GHz
VSAT EIRP	39.3 dBW
Uplink free space loss	206.59 dB
Total atmospheric attenuation (clear sky)	0.8 dB
Symbol rate	1 Msps

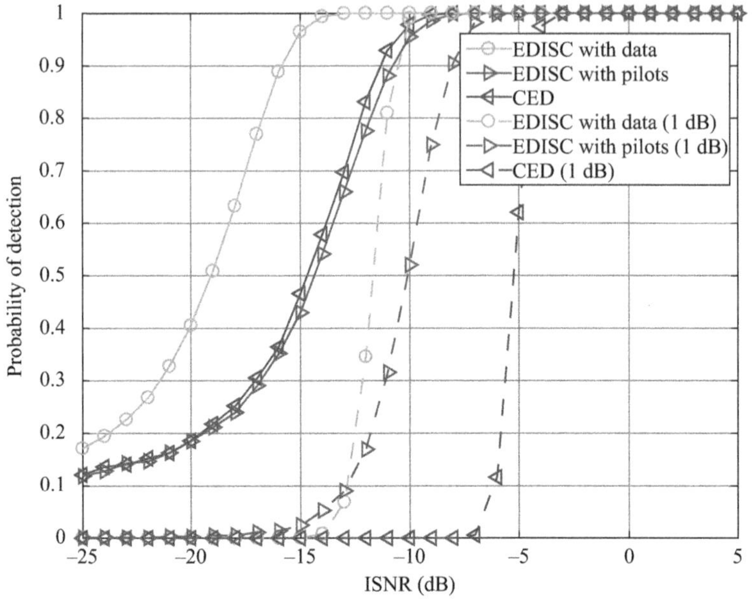

Figure 14.7 Probability of interference detection versus the ISNR for the QPSK scenario comparing the EDISC with two stage, the EDISC with pilots, the EDISC with data and the CED taking into account 1 dB of noise variance and signal energy uncertainties, where $E_p/\sigma^2_{w_p} = E_d/\sigma^2_{w_d} = 6$ *dB*

performance decreases due to the uncertainty. The latter may lead to the ISNR wall phenomenon [24], where beyond a certain ISNR value the detectors cannot robustly detect the interference. Furthermore, we see that the EDISC with data recovery or pilots under the more practical scenario of uncertainty still perform considerably better than the CED with uncertainty, improving the ISNR wall by more than 5 dB.

14.8.2 Performance analysis of interference localization techniques

In this part, it is assumed that the processing at the satellite is quick enough so that the satellite positions and velocities can be considered to be the same during sampling and forwarding the interference and reference signals. The locations of the system elements are shown by the Geographic coordinate system as (*longitude, latitude, altitude*). The errors in the position and velocity of the satellite are shown by vectors $\mathbf{e}_p$ and $\mathbf{e}_v$, which their elements are uniform random variables within the distance $[-e_p, e_p]$ and $[-e_v, e_v]$, respectively. The acronym *OB* is used instead of the term on-board in the legend of the figures to save space.

For LEO, MEO and retro GEO satellites, it is assumed that the satellite moves from (0,0, *altitude*) to (20,0, *altitude*) and samples the interference in every 0.5°

Table 14.2 System parameters

Parameter	Value
Satellite orbit type	LEO, MEO, retro GEO, GEO
Operating band	K_a band
Uplink frequency, GHz	30
Satellite oscillator frequency, GHz	18
Error bound for oscillator frequency, Hz	50
Reference signal frequency, GHz	29
Location of the gateway	(5,14,0)
Location of the unknown interferer	(30,20,0)

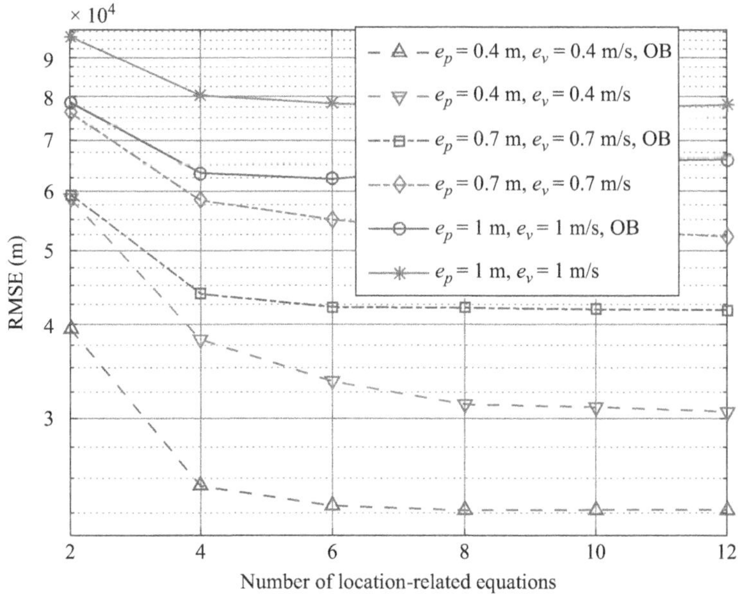

Figure 14.8 Localization RMSE versus the number of location-related equations for the GEO satellite when $\|\mathbf{v}\| = 3.63$ m/s, satellite altitude is 35,786 km, and the position of the reference transmitter is (20,20,0)

which results in 40 samples. Regarding the GEO satellite, it is assumed that the satellite collects 40 samples along a circular path with radius of 50 km which takes one day to complete. The GEO satellite is located right above the intersection of 0° latitude and 0° longitude with the altitude 35,786 km. The rest of the parameters which are common for all the satellites are summarized in Table 14.2.

The localization RMSE with respect to the number of location-related equations for the GEO satellite is presented in Figure 14.8. It can be seen in Figure 14.8 that the localization accuracy improves by both increasing the number of location-related equations and using on-board localization. Since a GEO satellite moves relatively slow, the Doppler shift caused by its movement is small and can be easier influenced by the oscillator error. Hence, using on-board localization can considerably enhance the localization accuracy when a GEO satellite is sampling and forwarding the interference.

14.9 Conclusion

In this chapter, we discussed the benefits of detecting and localizing the interference on-board the satellite. An on-board SMU should be able to implement and calibrate a number of detection algorithms to identify any interference on carriers. Three interference detection algorithms based on the energy detection were proposed, starting with the conventional ED and moving to more advanced algorithms of the ED with imperfect signal cancellation exploiting the pilot symbols or the data decoding. Simulation results showed that the CED is a good detection scheme for strong interference scenarios but not so reliable as both EDISC algorithms for the detection of low values of ISNR. Furthermore, we proposed an FoA technique to localize an unknown interferer while only relying on either the affected satellite, or the satellite dedicated to interference localization. We used a reference signal to calibrate the estimated frequency of the interferer at the gateway, and built location-related equations using the values of satellite's oscillator frequency, velocities and positions. It was shown that increasing the number of location-related equations, i.e., measurements, can improve the localization accuracy. Finally, the simulations showed that using the proposed on-board localization approach can further enhance the localization accuracy since the oscillator error is avoided, particularly for on-board GEO localization.

As for the future work, most of the techniques have been developed for GEO satellites, and hence the research on other types of satellites (LEO, MEO) is considered as a valuable idea. Furthermore, another idea is the study of the benefits and constraints of using multiple antennas.

References

[1] D. Petrolati, *Technology Roadmap: Interference Monitoring, Detection/Isolation, Classification, Localisation and Mitigation*, ESA document, May 2015.

[2] ETSI TS 103 129 V1.1.1 (2013-05), "Digital video broadcasting (DVB); Framing structure, channel coding and modulation of a carrier identification system (DVB-CID) for satellite transmission," May 2013.

[3] C. Politis, S. Maleki, S. Chatzinotas, B. Ottersten, "Harmful interference threshold and energy detector for on-board interference detection," *22nd Ka band and Broadband Communications Conference*, Cleveland, USA, Oct. 2016.

[4] C. Politis, S. Maleki, C. Tsinos, S. Chatzinotas, B. Ottersten, "On-board the satellite interference detection with imperfect signal cancellation," *IEEE*

Intern. Workshop on Sig. Proc. Advanc. in Wirel. Comm., Edinburgh, Scotland, Jul. 2016.

[5] A. Kalantari, S. Maleki, S. Chatzinotas, B. Ottersten, "Frequency of arrival based interference localization using a single satellite," *Advanced Satellite Multimedia Systems Conference and Signal Processing for Space Communications Workshop (ASMS/SPSC)*, Palma de Mallorca, Spain, Sep. 2016.

[6] G. Maral, M. Bousquet, *Satellite Communication Systems*, 5th Ed., Wiley, 2009.

[7] SES, "Satellite Evolution Sparks a Service Revolution," white paper, Jun. 2016.

[8] P. Angeletti, R. De Gaudenzi, M. Lisi, "From bent pipes to software defined payloads: Evolution and trends of satellite communications systems," *Proceedings of the 26th AIAA International Communications Satellite Systems Conference*, 2008.

[9] Thales Alenia Space, "Digital Transparent Processor," data sheet, Mar. 2012.

[10] A. Le Pera, F. Forni, M. Grossi, *et al.*, "Digital transparent processor for satellite telecommunication services," *IEEE Aerospace Conference*, Manhattan, USA, Mar. 2007.

[11] G. Maral, *VSAT Networks*, 2nd Ed., Wiley, 2003.

[12] A. D. Panagopoulos, P. -D. M. Arapoglou, P. G. Cottis, "Satellite communications at Ku, Ka and V bands, propagation impairments and mitigation techniques," *IEEE Communication Surveys and Tutorials*, vol. 6, no. 3rd Quarter, pp. 1–13, Oct. 2004.

[13] T. T. Ha, *Digital Satellite Communications*, McGraw-Hill Communications Series, 1990.

[14] E. Lavan, *Satellite Interference: An Operator's Perspective*, Eutelsat Communications, Jun. 2013.

[15] R. Rideout, "Technologies to identify and/or mitigate harmful interference," *International Satellite Communication Workshop on the ITU-Challenges in the 21st Century: Preventing Harmful Interference to Satellite Systems*, Jun. 2013.

[16] S. K. Sharma, S. Chatzinotas, B. Ottersten, "In-line interference mitigation techniques for spectral coexistence of GEO and NGEO satellites," *International Journal of Satellite Communications and Networking*, vol. 34, no. 1, pp. 11–39, Feb. 2016.

[17] C. Kourogiorgas, A. D. Panagopoulos, K. Liolis, "Cognitive uplink FSS and FS links coexistence in Ka-band: Propagation based interference analysis," *Proc. IEEE ICC 2015 1st International Workshop on Cognitive Radios and Networks for Spectrum Coexistence of Satellite and Terrestrial Systems (CogRaN-Sat)*, London, UK, Jun. 2015.

[18] S. Smith, SES, "Satellite interference commercial industry views," *Presented at Satellite Interference Reduction Group (sIRG)*, Jun. 2013.

[19] S. K. Sharma, T. E. Bogale, S. Chatzinotas, B. Ottersten, L. B. Le, X. Wang, "Cognitive radio techniques under practical imperfections: A survey," *IEEE Communications Surveys & Tutorials*, vol. 17, no. 4, pp. 1858–1884, Nov. 2015.

[20] D. Cabric, S. M. Mishra, R. W. Brodersen, "Implementation issues in spectrum sensing for cognitive radios," *Proc. Asilomar Comf. Signals, Syst., Comput.*, Pacific Grove, CA, USA, pp. 772–776, Nov. 2004.
[21] P. D. Sutton, K. E. Nolan, L. E. Doyle, "Cyclostationary signatures in practical cognitive radio applications," *Journal on Selected Areas in Communications*, vol. 26, no. 1, pp. 13–24, Jan. 2008.
[22] H. Urkowitz, "Energy detection of unknown deterministic signals," *Proceedings of the IEEE*, vol. 55, no. 4, pp. 523–531, Apr. 1967.
[23] F. Digham, M. -S. Alouini, M. K. Simon, "On the energy detection of unknown signals over fading channels," *IEEE Transactions on Communications*, vol. 55, no. 1, pp. 21–24, Jan. 2007.
[24] R. Tandra, A. Sahai, "Fundamental limits on detection in low SNR under noise uncertainty," *Proc. IEEE Int. Conf. Wireless Netw., Commun. Mobile Comput.*, Maui, HI, USA, vol. 1, pp. 464–469, Jun. 2005.
[25] S. Atapattu, C. Tellambura, H. Jiang, "Performance of an energy detector over channels with both multipath fading and shadowing," *IEEE Transactions on Wireless Communications*, vol. 9, no. 12, pp. 3662–3670, Dec. 2010.
[26] R. Tandra, A. Sahai, "SNR walls for signal detection," *IEEE Journal of Selected Topics in Signal Processing*, vol. 2, no. 1, pp. 4–17, Feb. 2008.
[27] C. D. Hou, "A simple approximation for the distribution of the weighted combination of non-independent or independent probabilities," *Statistics and Probability Letters*, vol. 73, no. 2, pp. 179–187, Jun. 2005.
[28] M. S. Alouini, A. Abdi, M. Kaveh, "Sum of gamma variates and performance of wireless communication systems over Nakagami-Fading channels," *IEEE Transactions on Vehicular Technology*, vol. 50, no. 6, pp. 1471–1480, Nov. 2001.
[29] C. Politis, S. Maleki, C. Tsinos, S. Chatzinotas, B. Ottersten, "Weak interference detection with signal cancellation in satellite communications," *IEEE Intern. Conference on Acoustics, Speech and Signal Processing (ICASSP)*, New Orleans, USA, Mar. 2017.
[30] ETSI EN 302 307-2 v1.1.1, "Digital video broadcasting: Second generation framing structure, channel coding and modulation systems for Broadcasting, Interactive Services, News Gathering and other broadband satellite applications; Part 2: DVB-S2 Extensions (DVB-S2X)," Oct. 2014.
[31] D. Tse, P. Viswanath, *Fundamentals of Wireless Communication*, Cambridge University Press, 2005.
[32] J. W. W. Smith. *A Satellite Interference Location System*, Georgia Institute of Technology; 1990.
[33] J. W. W. Smith, P. G. Steffes, "A satellite interference location system using differential time and phase measurement techniques," *IEEE Aerospace and Electronic Systems Magazine*, vol. 6, no. 3, pp. 3–7, Mar. 1991.
[34] K. C. Ho, Y. T. Chan, "Solution and performance analysis of geolocation by TDOA," *IEEE Transactions on Aerospace and Electronic Systems*, vol. 29, no. 4, pp. 1311–1322, Oct. 1993.

[35] D. P. Haworth, N. G. Smith, R. Bardelli, *et al.* "Interference localization for EUTELSAT satellites—the first European transmitter location system," *International Journal of Satellite Communication*, vol. 15, no. 4, pp. 155–183, Jul. 1997.

[36] K. C. Ho, Y. T. Chan, "Geolocation of a known altitude object from TDOA and FDOA measurements," *IEEE Transactions on Aerospace and Electronic Systems*, vol. 33, no. 3, pp. 770–783, Jul. 1997.

[37] T. Pattison, S. I. Chou, "Sensitivity analysis of dual-satellite geolocation," *IEEE Transactions on Aerospace and Electronic Systems*, vol. 36, no. 1, pp. 56–71, Jan. 2000.

[38] T. Li, F. Guo, W. Jiang, "A novel emitter localization method using an interferometer on a spin-stabilized satellite," *Wireless Communications and Signal Processing (WCSP)*. Suzhou, China; 2010.

[39] J. Li, F. Guo, W. Jiang, "A linear-correction least-squares approach for geolocation using FDOA measurements only," *Chinese Journal of Aeronautics*, vol. 25, no. 5, pp. 709–714, May 2012.

[40] J. E. Effland, J. M. Gipson, D. B. Shaffer, *et al.* Method and system for locating an unknown transmitter. Google Patents; 1991. US Patent 5,008,679. Available from: http://www.google.com/patents/US5008679.

[41] D. P. Haworth, Locating the source of an unknown signal. Google Patents; 2000. US Patent 6,018,312. Available from: http://www.google.com/patents/US6018312.

[42] R. M. Rideout, P. R. Edmonds, S. R. Duck, *et al.* Method and apparatus for locating the source of unknown signal. Google Patents; 2004. US Patent 6,677,893. Available from: http://www.google.com/patents/US6677893.

[43] D. K. C. Ho, J. C. Chu, M. L. Downey, Determining a geolocation solution of an emitter on earth based on weighted least-squares estimation. Google Patents; 2010. US Patent 7,663,547; Apr. 2007. Available from: http://www.google.com.br/patents/US7663547.

[44] D. K. C. Ho, J. C. Chu, M. L. Downey, Determining a geolocation solution of an emitter on earth using satellite signals. Google Patents; 2010. US Patent 7,667,640. Available from: http://www.google.com.na/patents/US7667640.

[45] D. K. C. Ho, J. C. Chu, M. L. Downey. Determining transmit location of an emitter using a single geostationary satellite. Google Patents; 2013. US Patent 8,462,044. Available from: http://www.google.com/patents/US8462044

[46] A. Taherpour, M. Nasiri-Kenari, S. Gazor, "Multiple antenna spectrum sensing in cognitive radios," *IEEE Transactions on Wireless Communications*, vol. 9, no. 2, pp. 814–823, Oct. 2010.

Chapter 15

Random access in satellite communications: a background on legacy and advanced schemes

Karine Zidane[1], Jérôme Lacan[2], Mathieu Gineste[3], Marie-Laure Boucheret[4], and Jean-Baptiste Dupé[5]

During the last decade, the population of satellite terminals have been rapidly growing and new networks like the Internet of Things (IoT) have been emerging. That is to say, the access techniques on the return link shall be redesigned to handle denser networks and to resolve massive multiple access problems. Of course, dedicated access is still very useful in areas such as file streaming and uploading of big data. However, in IoT and machine-to-machine (M2M) communications, the traffic profiles are characterised by being sporadic, with short data packets and very low-duty cycles. In such scenarios, the use of random access (RA) techniques on the return link is of interest, as they fit with the unpredictable nature of the traffic and result in more flexible communications. However, the main drawback of using RA is the high risk for packet collisions, therefore, new enhanced RA techniques have been proposed in the literature to resolve this issue.

This chapter presents a background on the various legacy and advanced RA techniques proposed for satellite communications. First, we describe the main motivations for enhancing RA performance on the return link. Then, we present a list of legacy RA techniques used mostly for login purposes. Furthermore, we provide another list describing the recent RA techniques with enhanced performance due to data replication and additional signal processing at the receiver side. These recent RA schemes can be mutually used for login as well as data transmissions over the return link. Finally, we give a global comparison of the performance of enhanced RA and we discuss the application of each scheme with respect to system constraints such as power limitations, lower data rates and synchronisation overhead reduction.

[1]TéSA Laboratory, France
[2]University of Toulouse, ISAE-SUPAERO/DISC and TéSA, France
[3]Thales Alenia Space (TAS), France
[4]University of Toulouse, ENSEEIHT and TéSA, France
[5]Centre National d'Etudes Spatiales (CNES), France

15.1 Introduction

Satellite communications are expected to play an important role in future 5G networks [1]. They will be a fundamental component in many key areas such as complemented and extended coverage for terrestrial cells, traffic routing and backhauling integrated with terrestrial networks, areas of increased security and availability, as well as recently growing IoT and M2M networks.

As a consequence of the emergence of 5G systems, the resources on the return link have to be shared among a significantly growing number of connected terminals. Therefore, preserving bandwidth and time resources on the return link constitute is one of the major challenges. Moreover, the access on the return link should be adapted to the traffic profiles of the data transmitted by terminals in IoT and M2M networks, especially when the terminals used are low-cost and power or energy limited (e.g. battery-powered).

The multiple access scheme widely used nowadays on the return link in satellite communications is multi-frequency-time division multiple access (MF-TDMA). As mentioned in the satellite standards DVB-RCS and DVB-RCS2 [2,3], MF-TDMA combines the advantages of both frequency and time division by allowing the users to transmit their packets on different frequency bands and/or different time slots (TSs). Thus, MF-TDMA can permit the use of lower-cost terminals by requiring lower power emissions compared to time division schemes and fewer modems compared to frequency division schemes. For data transmission, each terminal uses a specific time/frequency slot defined with a carrier frequency, a bandwidth, a start time and a duration. A frame is described as a set of time/frequency slots shared among a certain number of users and occupying a certain portion of the total bandwidth on the return link.

Based on the MF-TDMA scheme, two main access techniques are used on the return link: demand assignment multiple access (DAMA) [4] and RA. In DAMA, resource allocation requests are required prior to data transmission, and each user is assigned one or several time/frequency slots on which it can transmit its data. On the contrary, in RA techniques, the users can access the shared media at randomly chosen time/frequency slots, thus reducing signalling overhead but increasing the risk of packet collisions. For this reason, the use of RA on the return link in existing satellite standards is limited to particular use cases such as the transmission of signalling packets, logins and capacity requests. Despite this, the use of RA techniques can be also interesting for data transmissions in specific communication scenarios. Such scenarios are characterised with short packet lengths, low-duty cycles and random packet arrivals (e.g. following the Poisson process). Therefore, RA techniques can be well suited for the HTTP traffic [5] as well as the traffic in IoT and M2M networks [6].

Moreover, it has been shown in the literature [7] that DAMA techniques can be inefficient and under-utilising for the satellite resources in such types of scenarios. Nevertheless, the use of DAMA for data streaming and files uploading is still required. In addition, the authors in [8] demonstrated the benefits of the integration of the two access strategies DAMA and RA with significant gains in delay and throughput in moderate to high load operating regions. Therefore, using RA combined with DAMA on the satellite return link presents a promising solution for massive access problems and motivates for further research in this field.

With this in mind, enhancements to RA schemes for satellite communications have been gaining more attention not only from academic researchers but also from industrials. In the recent satellite standard DVB-RCS2 [9], a new improved RA technique called contention resolution diversity slotted ALOHA (CRDSA) [10] has been made an optional component of the return link. Another enhanced RA scheme has been deployed recently for commercial use by EutelSat for its Smart LNB system [11,12], which is a low-cost connectivity solution for IoT and M2M applications. The terminals connected through this technology can access the Internet through satellite communications by using a RA technique called enhanced spread spectrum ALOHA (E-SSA) [13]. The recent satellite standards S-MIM [14] and F-SIM [15] designed for S-Band mobile and Ku/Ka-band fixed interactive multimedia, respectively, are based on the E-SSA RA. In fact, the Smart LNB system implements the F-SIM standard.

In this chapter, we provide a background on both legacy and recent RA techniques proposed for satellite communications. In Section 15.2, we present some legacy RA schemes mainly used for transmission of signalling packets and login information. As numerous RA schemes have been proposed recently to enhance the return link performance, we detail in Section 15.3 most of the advanced RA schemes and we distinguish between two main groups: the synchronous and the asynchronous. In Section 15.4, we compare several RA schemes based on a defined set of metrics such as synchronisation, transmit power, throughput and other metrics important for IoT and M2M applications and communications with low cost terminals. Finally in Section 15.5, we draw a general conclusion about the most promising RA schemes to be used in satellite communications integrated in 5G systems, and we discuss some future challenges and perspectives.

15.2 Legacy RA techniques for satellite communications

Legacy RA protocols such as ALOHA and slotted ALOHA (SA) are known to have a high probability of packet collisions. Because packets retransmission delays can be very long in satellite communications especially for geostationary satellites, the use of these protocols can induce latencies and is not well suited for very dense networks. In order to understand their performance and the proposed enhanced solutions, this section provides a description of the main legacy RA protocols used in satellite communications: ALOHA, SA and diversity SA (DSA).

15.2.1 ALOHA

Among the most famous non-slotted RA protocols used for both terrestrial and satellite communications, we cite the ALOHA protocol [16], which was proposed by Norman Abramson in the 1960s at the University of Hawaii. The basic principle of ALOHA as shown in Figure 15.1, is the following:

- If a user has a backlogged packet, it sends it at a random time instant.
- At the receiver side, if the packet has encountered collisions from other users, it is considered destructed and non-decoded.

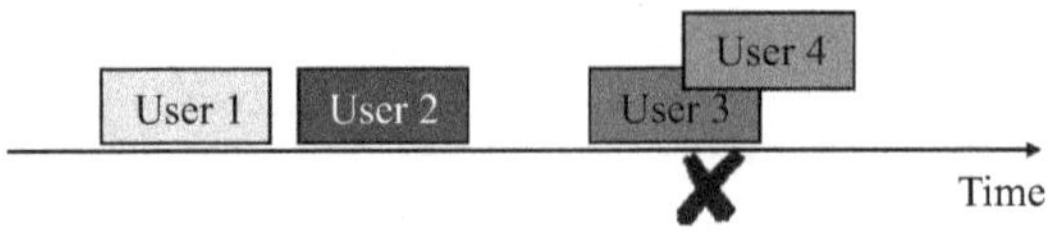

Figure 15.1 Example of the ALOHA RA scheme

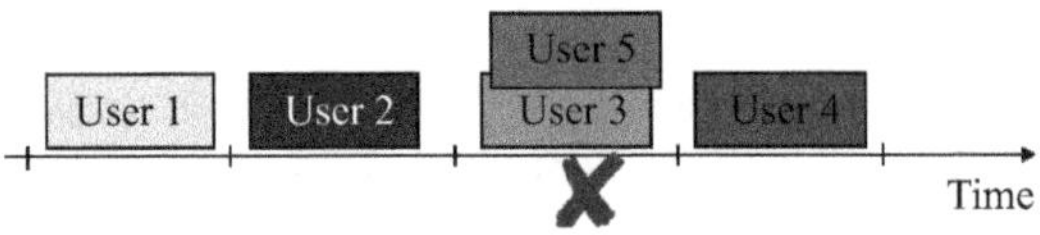

Figure 15.2 Example of the slotted ALOHA RA scheme

Due to the high rate of packet collisions, ALOHA is not well suited for data transmissions on the return link in satellite communications with high signal propagation delays, particularly for systems with very dense populations. Therefore, in satellite communications, ALOHA is mainly used for signalling transmissions, logins and resource allocation requests.

15.2.2 Slotted versions ALOHA

As seen previously in ALOHA, each user can transmit a packet at any random instant on the frame. The slotted versions of ALOHA permit to avoid partial interference among packets by allowing each user to transmit only at the beginning of a TS. A TS is defined as a limited slot of time sufficient to transmit one physical-layer packet taking into account predefined guard intervals. The information for the TSs planning among the terminals should be transmitted on the forward-link by the network control centre (NCC) when initialising a communication. Two main slotted versions of ALOHA can be cited: SA [17] and DSA [18].

15.2.2.1 Slotted ALOHA

As shown in Figure 15.2, the main difference between ALOHA and SA is the transmission of packets only at the beginning of predefined TSs. Even if each packet is received within one TS, timing offsets can still occur among received packets due to the transmitter and receiver not being perfectly synchronised. Considering the collision channel model, the maximum throughput achieved with SA is doubled compared to the throughput of ALOHA. However, the performance of SA is still poor for satellite communications scenarios due to the packet loss ratio (PLR) which decreases to acceptable values only at very low loads.

15.2.2.2 Diversity slotted ALOHA

Diversity is added to SA by replicating each packet by a certain number of replicas N_{rep}. Thus, in low load regimes, DSA can increase the probability of receiving at

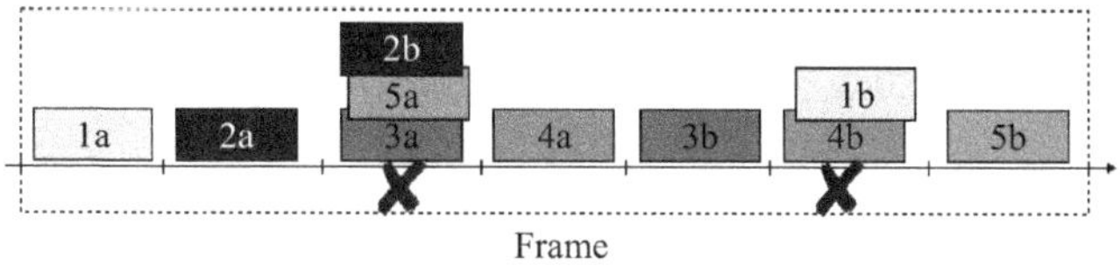

Figure 15.3 A received slotted frame with the diversity slotted ALOHA RA scheme and $N_{rep} = 2$ replicas per packet

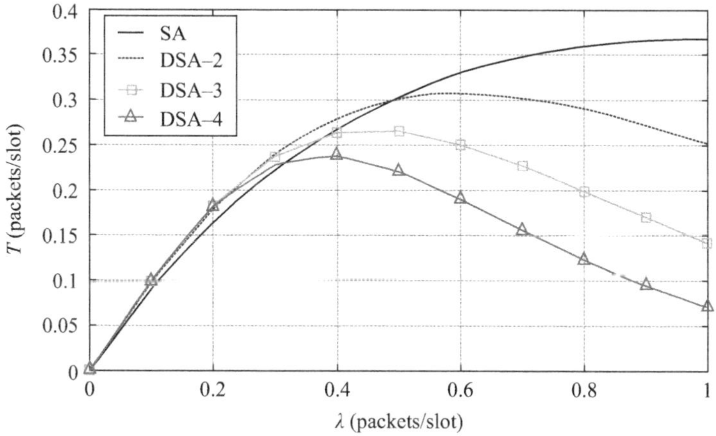

Figure 15.4 Analytical throughput vs. normalised channel load for SA and DSA (source: [19])

least one packet replica without collisions. As shown in Figure 15.3, each user can transmit several replicas of the same packet on randomly chosen TSs of the frame. At the receiver side, each packet has a slightly higher probability of having at least one replica received without collisions, due to the increased diversity in the TS selection, especially in low load regimes.

15.2.3 Conclusion on legacy RA techniques for the return link

To conclude, Figure 15.4 shows a comparison of the MAC layer analytical throughput achieved with SA and DSA with the assumption of a collision channel model. The notation *DSA-N_{rep}* is used to denote DSA using N_{rep} replicas per packet. Given the results observed in Figure 15.4, the maximum throughput obtained with DSA-2 is higher than SA, DSA-3 and DSA-4, for a MAC channel load $\lambda < 0.5$. However, the performance degrades rapidly for higher channel loads. Figure 15.4 also shows that the maximum throughput achieved with all the RA techniques presented previously is relatively low and achieved with high levels of packet losses, which is not practical for satellite scenarios. It is worth noting that the performance of SA and DSA was also evaluated analytically and via simulations for an Additive White Gaussian Noise (AWGN) channel model in [20]. Both analytical and simulation results are matching. The maximum throughput obtained with a coding rate 1/2 at $E_s/N_0 = 7$ dB using equi-powered packets is around 0.37 packet/slot for SA and 0.6 packet/slot for DSA-2. Then again, a PLR of 10^{-2} is achieved only with a load lower than 0.1 packet/slot. Therefore, legacy RA methods such as ALOHA, SA and DSA are not good candidates for satellite communications, especially in applications that do not tolerate large packet retransmissions delays. For this reason, researchers have been studying new RA protocols for satellite communications, which can cope with packet collisions and increase the MAC layer throughput. These protocols are presented in the following section.

15.3 Advanced RA techniques for satellite communications

As previously discussed, there is a clear interest in using enhanced RA techniques on the return link. However, advanced approaches should be considered in order to improve the performance of these techniques and mitigate the impact of packet collisions. Recently, several RA techniques for satellite communications have been proposed in the literature, and some of them have been already implemented for commercial purposes. In general, these recent techniques propose to cope with packet collisions at the receiver side by using information redundancy (in frequency, time or spreading codes) as well as the principle of successive interference cancellation (SIC). In this section, we will present several advanced RA schemes, while dividing them in two main categories:

- Synchronous RA: the frame is composed of several TSs, and each user can transmit its packet only at the beginning of a TS. Thus, only full packet collisions can occur.
- Asynchronous RA: the division of the frame into timeslots is dismissed, and the packets can be received at any time instant. Therefore, any given packet can experience both partial and full collisions.

First, let us present the main system metrics used for the performance evaluation of advanced RA techniques.

15.3.1 Main metrics for the evaluation of advanced RA schemes via simulations

The main simulation metrics used in the literature to evaluate the performance of advanced RA protocols designed for satellite communications are:

- The normalised MAC-layer load (λ in packets per slot and G in bits per symbol);
- The normalised MAC-layer throughput (T in in bits/symbol or packets/slot depending on whether we would like to compare several modulations and coding schemes or not);
- The MAC-layer PLR.

The normalised MAC-layer load expressed in unique packets/slot is denoted by λ and computed as follows:

$$\lambda = \frac{N_u}{N_s} \quad \text{(packets/slot)}, \tag{15.1}$$

with N_u being the average total number of users on a frame and N_s being the total number of TSs on a frame. λ is normalised to the number of packet replicas, so that different RA schemes using different values of N_{rep} can be fairly compared. In order to compare different systems using different types of modulation and coding schemes, the normalised MAC-layer load G (in bits/symbol) is used. G is computed as shown below:

$$G = \lambda R \log_2(M) \quad \text{(bits/symbol)}, \tag{15.2}$$

with M being the modulation order (e.g. $M = 4$ for quadrature phase shift keying), and R being the coding rate. The normalised MAC-throughput T obtained at a certain load G and computed using simulations, can be expressed as

$$T = \lambda\,(1 - \text{PLR}(\lambda)) \quad \text{(packets/slot)}, \tag{15.3}$$

or

$$T = G\,(1 - \text{PLR}(G)) \quad \text{(bits/symbol)}, \tag{15.4}$$

where PLR is the packet loss ratio, i.e. the percentage of non-decoded packets on a frame for a given load G in bits/symbol or λ in packets/slot, and a given signal-to-noise ratio (SNR) per packet.

15.3.2 Advanced synchronous RA techniques

The main motivation behind using synchronous RA schemes relies on the fact that they are more practical for the detection of the start of the received packets on a frame. In the following, we describe a list of several advanced synchronous RA techniques and we discuss their performance on the return link.

15.3.2.1 Contention resolution diversity slotted ALOHA

CRDSA [10] is an enhanced version of DSA included as optional in the DVB-RCS2 standard. The main concept of CRDSA is based on packets replication at the transmitter side combined with SIC at the receiver side. The number of replicas per packet N_{rep} is the same for all users. As shown in Figure 15.5, each user can transmit two or more replicas of the same packet on randomly chosen TSs of the frame. Each packet contains a signalling field with pointers to the locations of its replicas. At the receiver side, the frame is stored and scanned iteratively. Then, the SIC process is applied for each successfully decoded packet. In other words, when one packet is decoded successfully, it is removed from the frame, then the decoded pointers are used to localise its replicas. The decoded bits of the first recovered replica are used to reconstruct its remaining copies on the localised TSs. Hence, the interference contribution on the

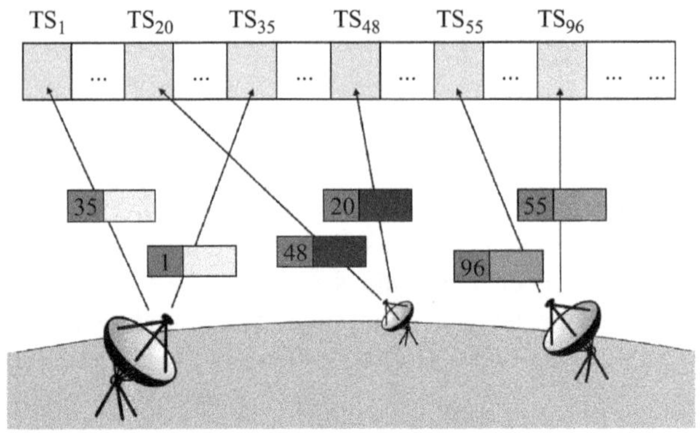

Figure 15.5 CRDSA transmission scheme with $N_{rep} = 2$ replicas per user

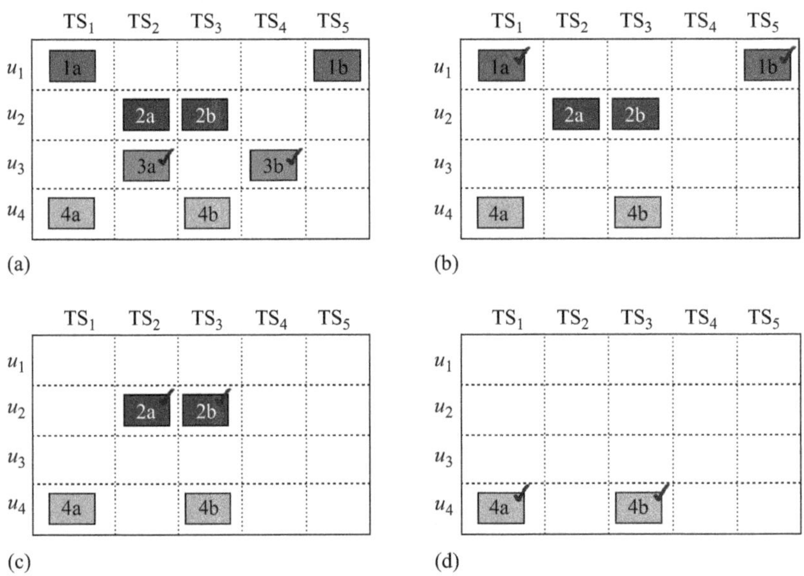

Figure 15.6 CRDSA example: 4 users (u) sharing a frame of 5 time slots (TS). (a)–(d) represent the successive steps of interference cancellation

remaining non-decoded packets on the frame is reduced. This process is repeated over the entire frame until a maximum number of iterations are reached.

Example 15.1 (CRDSA example). *Figure 15.6 illustrates an example of a frame treated with CRDSA, where all the packets are considered received with the same power level. Figure 15.6(a) and (b) shows how the packets of user 3 and 1 are successfully decoded in the first CRDSA iteration: first, the packet replica (3b) is* *decoded successfully on the fourth TS (TS_4) because it is received without collisions. Its replica (3a) is localised with the decoded pointer then reconstructed and removed from TS_2. In a similar way, the packet replica of user 1 on TS_5 is decoded successfully and removed from the frame as well as its copy received on TS_1. In a second CRDSA iteration (see Figure 15.6(c) and (d)), the packets corresponding to users 2 and 4 are successively decoded and removed from the frame. Thus, the SIC process permits to retrieve all the packets successfully. It is worth noting that without the replicas pointers and the SIC, a DSA scheme is obtained and the decoding process stops only after decoding the replicas (1b) and (3b). Depending on the channel code rate, the remaining packets on the frame can be lost.*

The authors of CRDSA showed that with a QPSK modulation, a forward error correction (FEC) code of rate 1/3, and $E_s/N_0 = 10$ dB, replicas experiencing one packet collision can be resolved successfully. In that case, with $N_{rep} = 3$ replicas per packet, the maximal throughput of CRDSA can reach 1.2 packets/slot which is equivalent to an efficiency of 0.8 bits/symbol.

First evaluations of the CRDSA scheme [10] have considered that the packets are all received at the same power level (i.e. equi-powered). Later studies showed that diversifying the packets power can lead to major performance improvements for RA schemes using interference cancellation (IC) [7,20,21]. In fact, packets power unbalance allows the receiver to detect the strongest packets first, and to decode them with a higher success probability. This phenomenon is called the capture effect [17,22] because the strongest packets are 'in capture' and can be decoded successfully even when they are undergoing collisions from other packets. Exploiting the capture effect together with the SIC process allows to resolve more packet collisions on the frame and increase the MAC-layer throughput. In particular, for CRDSA, the impact of packets power unbalance has been evaluated in [20,21].

In fact, in realistic channel conditions, power unbalance among different transmitters is unavoidable. The terminal equivalent isotropic radiated power (EIRP) may randomly vary around a certain value and the path losses experienced by each user may be different depending on the area of coverage. It has been shown in [23] that in mobile communications channels, packets power approximately follows a truncated lognormal distribution of parameters $\mu = 0$ dB and σ varying between 2 and 3 dB, depending on the channel characteristics. Nevertheless, the replicas corresponding to a same packet can still be considered equi-powered over the duration of one frame. The authors showed that the performance of CRDSA with lognormally distributed packets power is significantly enhanced compared to the equi-powered packets case.

Figure 15.7 depicts the performance of CRDSA with $N_{rep} = 3$ replicas per packet, in terms of normalised MAC throughput in bits/symbol and PLR, with several values of the standard deviation σ for the truncated lognormal packets power distribution ($\sigma = 0$ dB refers to the case of equi-powered packets). The results are shown using a QPSK modulation with a 3GPP/UMTS turbo code [24] of rate 1/3 and $E_s/N_0 = 10$ dB. An error floor appears in Figure 15.7(b) with $\sigma = 3$ dB, because with packet power distribution with a higher variance, there is a higher probability of receiving packets at lower values of E_s/N_0.

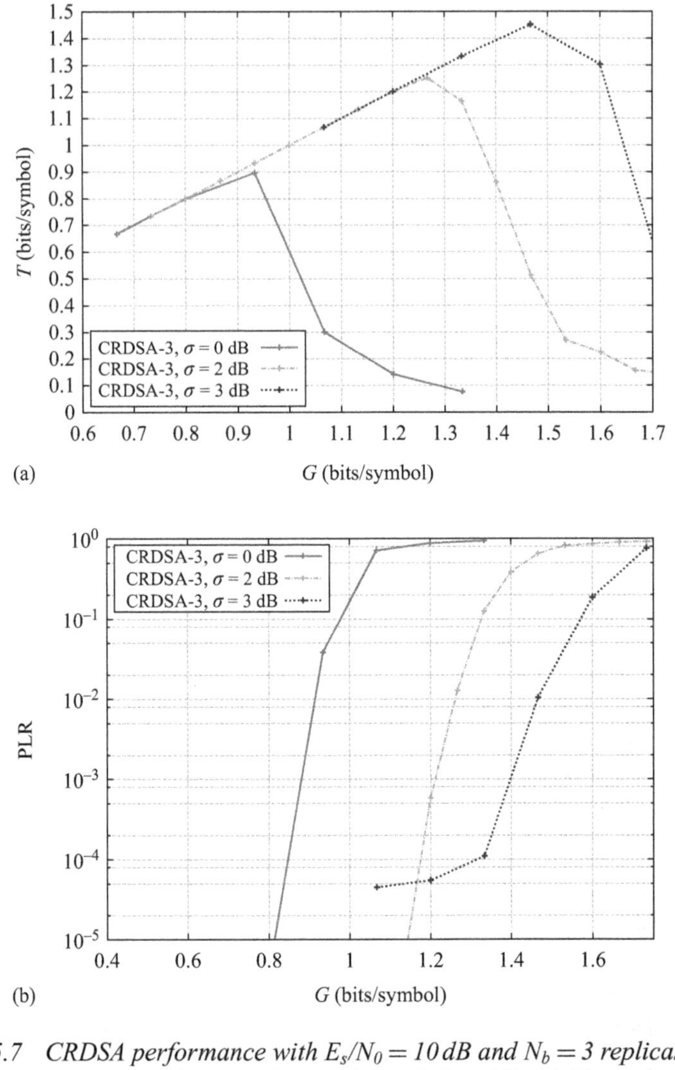

Figure 15.7 CRDSA performance with $E_s/N_0 = 10\,dB$ and $N_b = 3$ replicas. QPSK modulation, 3GPP turbo code $R = 1/3$. (a) Throughput and (b) PLR

15.3.2.2 Irregular repetition slotted ALOHA

Irregular repetition SA (IRSA) [25,26] is a variant of CRDSA, in which the users transmit an irregular number of replicas per packet. The evaluation approach for IRSA exploited the analogy between the SIC process and iterative erasure decoding of graph-based codes [27,28]. The SIC process of CRDSA was described with a bipartite graph. In [29,30], the analysis of [25] has been extended in order to optimise the burst repetition rate distributions in order to maximise the throughput. Under

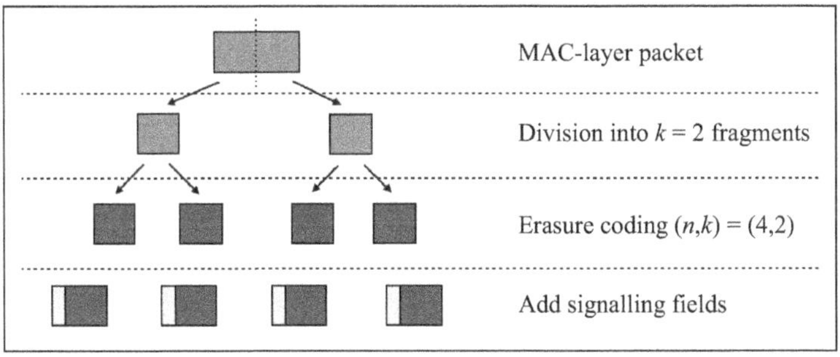

Figure 15.8 CSA transmission scheme

the same channel model, IRSA enhances the performance of CRDSA in terms of throughput but its performance for a PLR around 10^{-3} is lower when compared to CRDSA with 3 or 4 replicas.

15.3.2.3 Coded slotted ALOHA

Coded SA (CSA) [31–34] is a generalisation of IRSA, which encodes and divides the packets prior to their transmission rather than simply replicating them. Of course, the SIC process is also applied at the receiver side. The analytical throughput of this scheme was evaluated in [35] considering the collision channel model. At the transmitter side, each packet is constructed as shown in Figure 15.8. The main steps for packet construction in CSA are listed as follows:

1. The packet is divided into k information fragments.
2. The k fragments are encoded via a local packet-oriented code which generates n encoded fragments ($n > k$). The code rate in this case is $R = k/n$.
3. A control header is added to the beginning of each encoded fragment. This header shall contain signalling information about the other fragments locations on the frame.

At the receiver side, if the packet fragments of a given user are received with collisions from other fragments, they are considered lost (hence, the analogy with erasure decoding). However, the receiver may recover the information received from other fragments of the same packet that have not experienced collisions. Thus, the packet can be decoded and its interference contribution is subtracted from the frame (i.e. all the corresponding fragments are removed). In [35], it is shown that the asymptotic throughput (i.e. the throughput obtained when the number of users $N_u \to \infty$ and the number of TSs $N_s \to \infty$) can reach up to 0.9 packets per slot with an erasure code of rate $R = 2/7$. Furthermore, an analytical study in [36] demonstrated that for asymptotically large frames and asymptotically large maximum number of replicas, CSA can reach 1 packet/slot under the collision channel, which is equivalent to the performance of an orthogonal scheme.

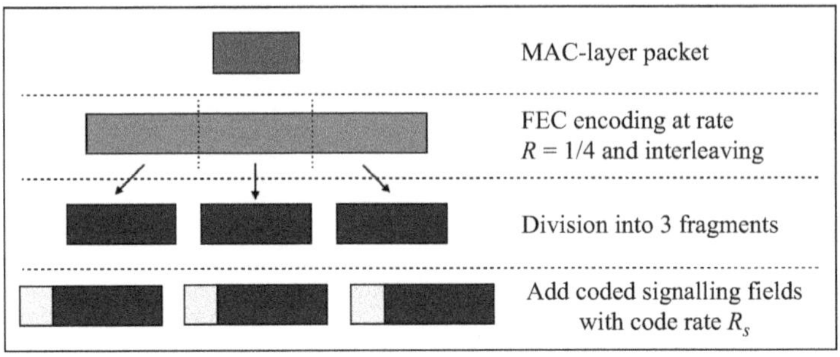

Figure 15.9 MuSCA transmission scheme

15.3.2.4 Multi-slot coded ALOHA

Multi-slot coded ALOHA (MuSCA) [37] is another advanced synchronous RA method proposed in 2012 by Bui *et al.* Figure 15.9 illustrates the main operations performed at the transmitter side in MuSCA. First, the transmitter encodes the packet with a robust FEC code of rate R. Then, the codeword is bit-interleaved and modulated. The resulting codeword is divided into multiple fragments and a signalling field is added to the beginning of each fragment. This signalling field serves to localise packet fragments on the frame and it is encoded separately from the data field. In order to be able to localise the fragments of a given packet even when collisions occur, the signalling field should be encoded with a robust error correcting code. In [37], a Reed–Muller code of rate R_s has been used for the signalling field.

In MuSCA, the operations performed at the receiver side can be divided into two main phases:

1. **Decoding the signalling fields:** At first, the decoder scans the frame and attempts to decode the signalling fields on each TS. The SIC process is applied in order to remove each successfully decoded signalling field. In other words, whenever a signalling field for a given packet is successfully decoded, the signalling fields for the same packet on other TSs are reconstructed and subtracted subsequently from the frame. The frame is scanned iteratively until no additional signalling fields can be retrieved. At the end of this phase, if all the headers are successfully decoded, the receiver knows the locations of all the fragments as well as the level of collisions on each TS. Thus, the receiver can proceed to the next phase to decode the data fields.
2. **Decoding the data fields:** At this phase, the fragments corresponding to each packet are re-assembled and the physical layer packet is reconstructed, demodulated, de-interleaved and decoded. Then, SIC is performed using the successfully decoded packets in order to remove their corresponding data fragments from all the corresponding TSs. Of course, the frame is scanned iteratively until no additional packets can be recovered.

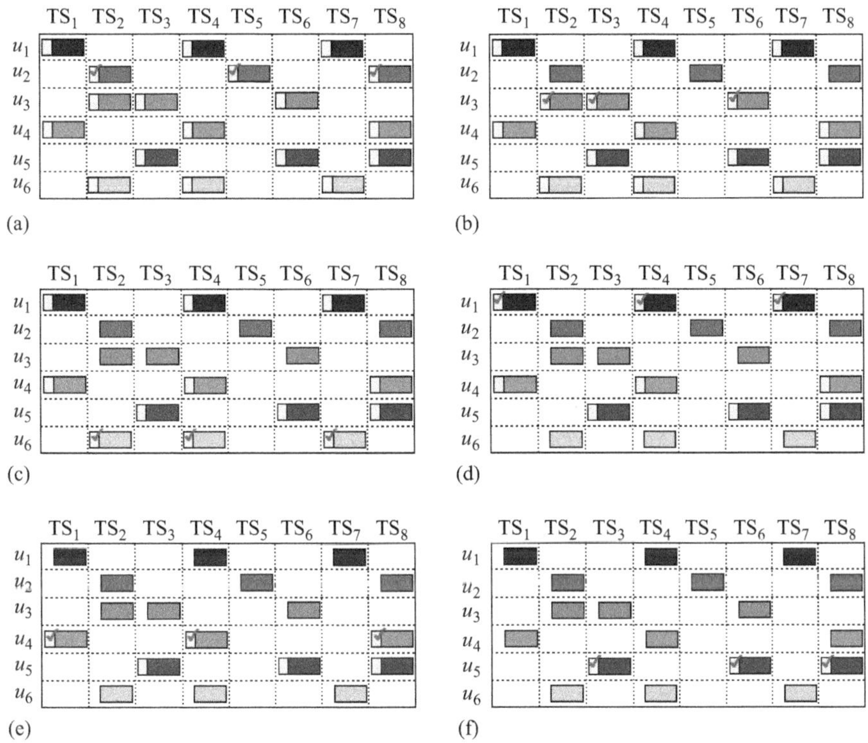

Figure 15.10 *MuSCA example: signalling fields decoding phase. 6 users (u) sharing a frame of 8 time slots (TS). (a)–(f) represent the frame after successive interference cancellation of the decoded signalling fields*

Example 15.2 (MuSCA example). *Figures 15.10 and 15.11 show an example of the two-phases decoding process at the receiver side in MuSCA. In Figure 15.10(a), the decoder finds the packet of user 2 on slot 5 free of collisions, so it decodes its corresponding signalling field successfully and removes the signalling parts of its fragments in slots 2 and 8. The packet of user 3 in slot 2 becomes in collision with only one other packet. Therefore, the receiver can decode its signalling field successfully given the robust Reed–Muller code used. Once decoded, the other signalling fields in slot 3 and 6 can be removed. The decoder continues this process iteratively until all the signalling fields are decoded.*

Figure 15.11 depicts the useful information decoding phase. The decoder starts by choosing the packet which is less interfered on all its fragments. In the case of Figure 15.11(a), it starts by decoding the packet of user 2. The fragments of the packet are collected from slots 2, 5 and 8, then the codeword is reconstructed, demodulated and decoded. Given the robust FEC code used on the payload part of each packet, the successful decoding probability can be considered relatively high. Once decoded

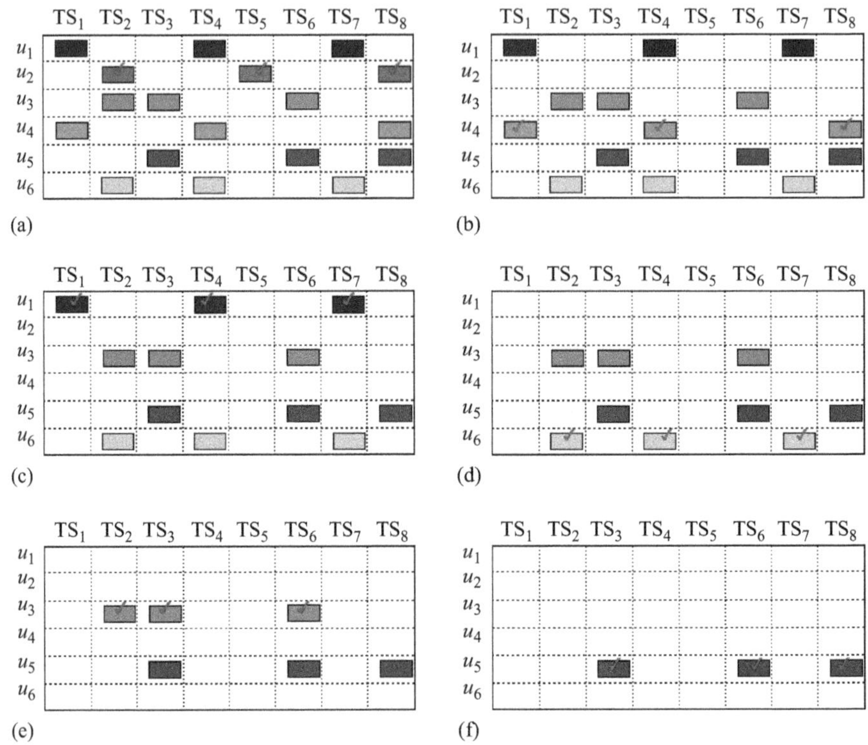

Figure 15.11 MuSCA example: useful information decoding phase. 6 users (u) sharing a frame of 8 time slots (TS). (a)–(f) represent the frame after successive interference cancellation of the decoded payload fields

successfully, the packet and all its fragments are removed from slots 2, 5 and 8. Then, the decoder tries to resolve the packet fragments of user 5. If the decoding attempt fails, then the decoder passes to user 4 and so on until all the packets on the frame are successfully retrieved.

Figure 15.12 shows the performance of MuSCA in terms of throughput T (in packets/slot) obtained with a QPSK modulated payload and a Consultative Committee for Space Data Systems (CCSDS) turbocode [38] of rate $R = 1/6$ and a number of fragments per packet N_f equal to 3. Several levels of E_s/N_0 are compared. The results are shown for a system with perfect channel state information (CSI). It is observed that the performance is significantly enhanced if we compare it to the RA techniques presented previously. The maximum throughput obtained is higher than 1 packet/slot starting from $E_s/N_0 = 1$ dB. The authors of MuSCA have also proposed an irregular version of the scheme called irregular MuSCA [39], where each user sends a random number of packet fragments on the frame. The maximum throughput achieved with Irregular MuSCA could reach 1.4 packets/slot for an optimal probability distribution of the number of fragments sent by each user. However, in both regular and irregular

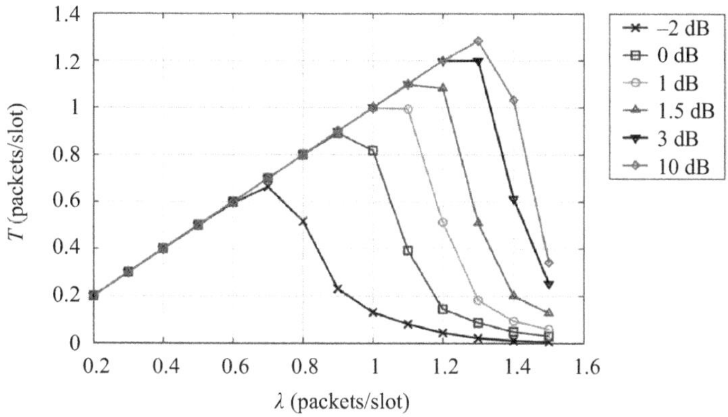

Figure 15.12 Throughput T vs. channel load λ for MuSCA in perfect channel conditions with several values of E_s/N_0. QPSK modulation, FEC code $R = 1/6$, $N_f = 3$ fragments per packet, packet length 456 bits and $N_s = 100$ slots (Source: [19])

versions of MuSCA, the throughput results do not account for the header overhead added for fragments localisation. Thus, the cost for the better performance is the increased signalling overhead required to localise the packets fragments on the frame.

15.3.2.5 Multi-replica decoding using correlation based localisation

Multi-replicA decoding using coRrelation baSed locALisAtion (MARSALA) [40] is an advanced synchronous RA method which proposes to enhance the throughput compared to previous RA schemes such as CRDSA, IRSA and others without using additional signalling overhead on the packet. More precisely, MARSALA proposes a new replicas localisation and decoding technique to be combined with CRDSA at the receiver side. In the following, we will explain the different steps of MARSALA and we will present some performance evaluation results.

The transmitter side in MARSALA is the same as the one described previously for CRDSA. The system modifications are only made at the receiver side and are explained as follows: after the entire frame is stored, CRDSA is applied in order to retrieve non-decoded packets on the frame. Whenever a packet is successfully decoded, the SIC process is applied to remove the decoded packet and its corresponding replicas from the frame. After scanning the entire frame, the same packets detection and SIC process is repeated iteratively in order to recover more packets after the previous IC. However, in high load regimes, some packets might not be recovered with CRDSA due to strong collisions. At this point, MARSALA is applied. The procedure is illustrated in Figure 15.13 and goes as follows:

1. A reference TS TS_{ref} is selected.
2. MARSALA attempts to localise the replicas of the packets present on TS_{ref} by using a cross-correlation technique. Replicas localisation is done under the

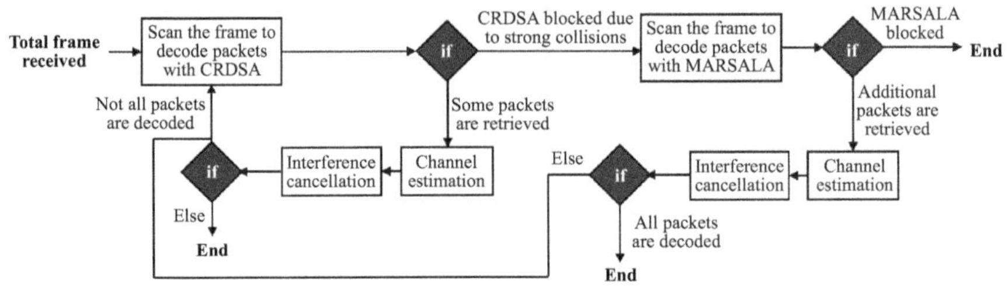

Figure 15.13 Frame processing scheme combining CRDSA and MARSALA at the receiver side

assumption of randomly varying phase shifts and timing offsets for different replicas of a same packet. However, the amplitude and frequency offsets are supposed to remain constant over the duration of one frame.

3. Once the replicas of a given packet on TS_{ref} are localised, the following procedures are applied:
   i. Replicas synchronisation in time and phase.
   ii. Replicas combining with or without maximum ratio combining (MRC) [41,42].
   iii. Channel estimation on the combined replicas.
   iv. Demodulation and decoding.
   v. Cancellation of the successfully decoded replicas (SIC).

MARSALA can be applied until at least one packet is successfully recovered, then the receiver can switch back to CRDSA. Thus, MARSALA can play a major role in releasing CRDSA non-decoded packets from collisions and triggering additional SIC iterations.

The simulations of MARSALA combined with CRDSA showed significant performance gains compared to CRDSA alone [43,44]. Figure 15.14 shows the results obtained considering the following simulation parameters: three replicas per packet, the DVB-RCS 2 turbo code for linear modulation [3] with a code rate $R = 1/3$, a burst length of 456 symbols modulated with QPSK (DVB-RCS2 waveform id 3) and an equal power level for all received packets. The results shown in Figure 15.14 are obtained for several values of E_s/N_0 without MRC. Both scenarios considering perfect CSI or a real channel model are considered.

Several enhancement schemes have also been applied to MARSALA [45], such as using packets power unbalance, MRC or the exploitation of different coding schemes other than DVB-RCS2. Table 15.1 summarises some performance results obtained in [45].

Remark 15.1 (No phase noise assumption). *It is worth clarifying that although frequency, timing and phase shifts among replicas were taken into account in [43–45], the fluctuating phase noise was not. Phase noise can be represented as a stochastic process of short-term frequency variations. In fact, phase noise fluctuations depend*

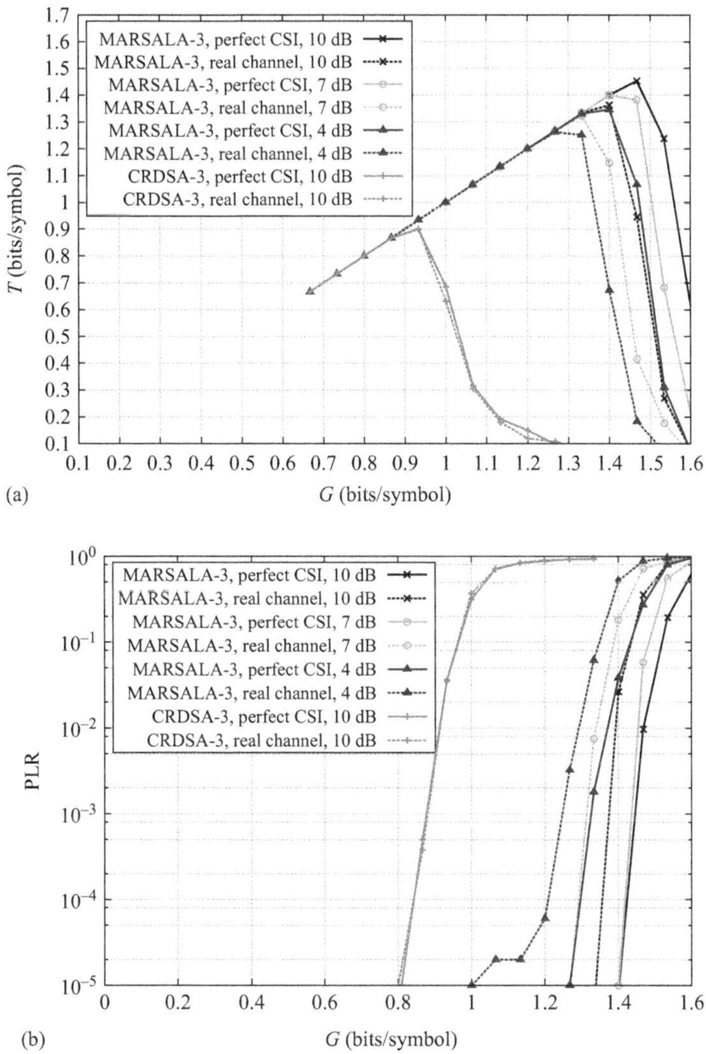

Figure 15.14 MARSALA-3 in real channel conditions compared to perfect CSI. QPSK modulation with DVB-RCS2 Turbocode $R = 1/3$. (a) Throughput and (b) PLR

Table 15.1 Throughput of MARSALA-3 in bits/symbol with $E_s/N_0 = 10\,dB$, a target PLR of 10^{-4} and QPSK modulation. The coding rate is 1/3

	DVB-RCS2 (waveform id 3)	**3GPP/UMTS**
No MRC	1.3	–
With MRC	1.5	1.7
With MRC and login $\sigma = 3$ dB	2.35	2.75

on the symbol rate, i.e., lower symbol rates induce higher phase noise. Therefore, phase noise can have an impact on the accuracy of replicas localisation using correlation as well as replicas combining and demodulation. It is important to study this impact on MARSALA with low data rates in future work.

15.3.2.6 Multi-frequency contention resolution diversity slotted ALOHA

The users in CRDSA share the same frequency bandwidth B, but transmit their packet replicas on distinct TSs. In this case, if the number of TSs N_s is below 100 slots, a visible degradation shall be expected due to higher probability of occurrence of the loop phenomenon.[1] As all the packets occupy the same bandwidth B in CRDSA, then the symbol rate R_s (in Baud, i.e. symbols per second) for each packet is equal to

$$R_s = \frac{B}{1+\alpha} = N_{symb} \times \frac{N_s}{T_f} \tag{15.5}$$

with α being the roll-off of the shaping filter, N_{symb} being the number of symbols per packet and T_f being the total frame duration in seconds. The level of E_s/N_0 at the receiver side is proportional to the received signal power (C) and inversely proportional to the symbol rate R_s, as shown in the following equation:

$$\left[\frac{E_s}{N_0}\right]_{dB} = \left[\frac{C}{N_0}\right]_{dBHz} \times \left[\frac{1}{R_s}\right]_{Baud}, \tag{15.6}$$

with C/N_0 expressed in decibel-Hertz (dBHz) and referring to the ratio of the carrier power and the noise power per unit bandwidth. Consequently, to achieve a given level of E_s/N_0 at the receiver side while reducing the transmitted carrier power, the symbol rate shall be reduced. Thus, in order to target low cost terminals and reduce the EIRP, multi-frequency CRDSA (MF-CRDSA) was proposed in [46]. Figure 15.15 illustrates the three comparative schemes of CRDSA, E-SSA and MF-CRDSA. In MF-CRDSA, the frame is not only divided in a number of TSs N_s^{MF} but also in sub-frequency bands denoted by B^{MF} such as $B^{MF} = B/N^{MF}$ with N^{MF} being the total number of sub-frequency bands. Therefore, the transmitted symbol rate per packet R_s^{MF} is reduced to

$$R_s^{MF} = \frac{B^{MF}}{1+\alpha} = \frac{R_s}{N^{MF}} = N_{symb} \times \frac{N_s^{MF}}{T_f}. \tag{15.7}$$

Obviously, for a number of fixed symbols per packet and a fixed frame duration, we can conclude that the total number of TSs in CRDSA should be equal to the total number of TSs in MF-CRDSA multiplied by the number of sub-frequency bands N^{MF}, as shown in the following equations:

$$\frac{R_s^{MF}}{N_s^{MF}} = \frac{R_s}{N_s} \rightarrow \frac{R_s}{N^{MF} \times N_s^{MF}} = \frac{R_s}{N_s} \rightarrow N_s = N^{MF} \times N_s^{MF}. \tag{15.8}$$

[1]A loop phenomenon occurs when all the replicas of a given packet could not be recovered and the collisions on these packets could not be resolved.

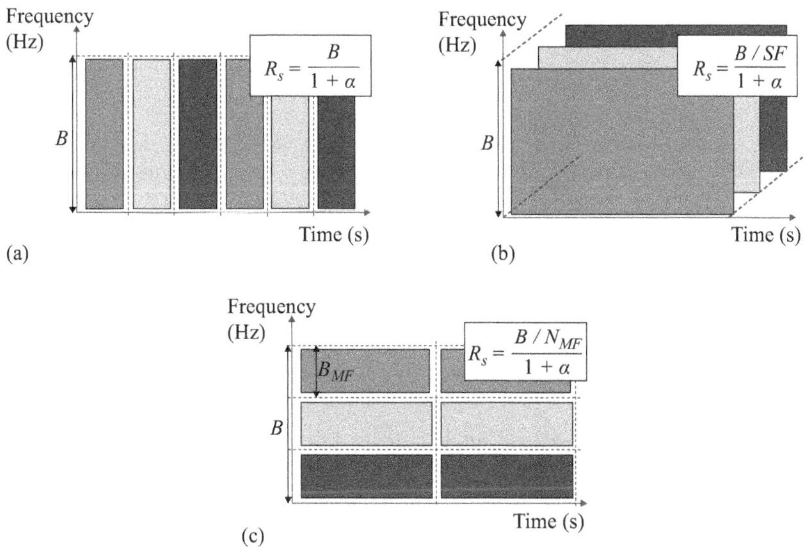

Figure 15.15 Comparison of time and frequency diversity as well as symbol rates per packet for (a) CRDSA, (b) E-SSA and (c) MF-CRDSA

Thus, the required C/N_0 in MF-CRDSA becomes

$$\left(\frac{C}{N_0}\right)_{MF} = \frac{E_s}{N_0} \times R_s^{MF} = \frac{E_s}{N_0} \times \frac{N_s^{MF}}{N_s} = \frac{C}{N_0} \times \frac{N_s^{MF}}{N_s}, \tag{15.9}$$

with $N_s^{MF}/N_s < 1$. Clearly, the lowest EIRP level can be obtained by taking $N_s^{MF} = N_{rep}$. The authors in [46] showed that MF-CRDSA can achieve a similar throughput performance compared to CRDSA but with a slightly higher PLR. This result is explained by the increased probability of the loop phenomenon given that different replicas of a same packet should be sent on different TSs in order to limit the level of signal distortion due to multi-carrier amplification.

15.3.2.7 Summary and brief discussion for advanced synchronous RA

In this subsection, we presented several advanced synchronous RA techniques for satellite communications. The first technique presented was CRDSA which performs packets replication and SIC at the receiver side. CRDSA showed significant PLR reduction compared to SA and DSA even at system loads around 0.8 bits/symbol. IRSA proposed a diversification of the number of replicas per user and CSA uses packets encoding and fragmenting combined with SIC. But for both IRSA and CSA, the performance has been only assessed for the collision channel model in which the throughput limit is 1 packet/slot. In order to further enhance the PLR in very high load regimes, MuSCA was proposed. Indeed, the performance in terms of throughput was significantly improved; however, the cost is the important overhead added to the packets. Then, MARSALA presented a solution to enhance CRDSA throughput and PLR without requiring additional overhead. Instead, MARSALA proposed packet localisation using correlation and replicas combining at the receiver side. Nevertheless, its performance may be sensitive to phase noise, especially in low data rates communications (in the order of kbps). Moreover, MF-CRDSA was presented as a solution which achieves a performance similar to CRDSA but with the advantage of requiring much less EIRP at the terminal side.

It is clear that adding advanced signal processing at the receiver side permits to significantly enhance the performance of legacy slotted RA techniques. As seen previously, very low PLRs have been achieved, which make RA schemes much more suitable to be used on the return link in satellite communications. However, in order to achieve the promised performance, synchronous RA techniques can present some limitations as explained in the following:

- They usually require higher energy per symbol.
- Given that TSs and frames are time-limited, slotted RA is well-suited for transmissions at high symbol rates. Otherwise, the frames would be too long.
- For each terminal, the EIRP in time-slotted RA shall be oversized with respect to non-slotted access. Although MF-CRDSA addresses this point by reducing the transmitted symbol rate at the terminal side when using the multi-frequency diversity.
- Terminal synchronisation at the frame and TSs level is a burden on the communication especially in networks with very large number of terminals and transmissions with low-duty cycles.

15.3.3 Advanced asynchronous RA techniques

As shown previously, legacy asynchronous RA techniques like ALOHA suffer from both partial and full packet collisions at the receiver side. However, recent studies showed that this problem can be mitigated by using appropriate modulation and coding at the transmitter as well as applying advanced signal processing and SIC at the receiver. Thus, packet collisions can be resolved and the performance in terms of PLR can be significantly enhanced.

In particular, asynchronous RA techniques are known to require minimal synchronisation among terminals and the receiver. This characteristic makes them very appealing to be used in communications with large propagation delays, with a large population of users and with very low duty cycles. In the following, we will present the main advanced asynchronous RA techniques proposed for satellite communications as well as the main results of their performance evaluation.

15.3.3.1 Enhanced spread spectrum ALOHA

E-SSA was proposed in 2008 [13,23] for mobile and M2M services using the L and S bands on the satellite return link. It is an advanced variant of spread spectrum ALOHA (SSA) [47] which is inspired from code spreading in code division multiple access (CDMA). For ease of understanding, let us provide a brief background on CDMA and SSA in the following.

Background on CDMA and SSA

CDMA is a widely known multiple access protocol based on spread spectrum. It allows the users to transmit their packets simultaneously over the same carrier frequency, but using different spreading codes [48]. A spreading code can be defined as a unique code given to each user and has a higher rate (R_c) than the actual rate of the transmitted bit streams (R_b), hence, the term 'spreading'. In CDMA, spreading codes are multiplied with the data streams of each user and permit to distinguish the packets corresponding to different users at the receiver. Each user is supposed to choose a unique binary code among a set of orthogonal codes (for synchronous CDMA) or pseudo-random codes (for asynchronous CDMA). The spreading factor is denoted by SF, with $SF = R_c/R_b$.

In SSA, it was shown that using different codes in a CDMA system is not necessary to distinguish the different users. Instead, a large value of the spreading factor SF can be used because the probability of collision with a spreading sequence aligned at the chip level is reduced when the spreading factor increases. SSA uses the same waveform configuration as defined in the terrestrial standard '3GPP Wideband CDMA' [49,50].

A subtraction algorithm is also applied in SSA at the receiver side. In other words, upon the reception of several signals at the same time, the receiver determines the strongest or earliest signal, then demodulates and decodes it and subtracts it from the frame. The same process is repeated on the remaining signals of the other users. The fact that the signals of different users are not synchronised enables to distinguish between them by correlation with the known spreading code.

E-SSA is a recent version of SSA that introduces a sliding window-based approach at the receiver side, combined with iterative IC on each window. At the transmitter side, the performed operations are listed as follows:

1. Define a preamble sequence to be added at the beginning of the packet. The preamble shall be a pre-defined sequence known at the receiver side, and its main purposes are the detention of the start time of a packet and channel estimation. Following E-SSA S-MIM consideration [14], a preamble of at least 96 symbols is required for packets detection at very low SNRs. Then, the preamble sequence is spread with the spreading factor SF.
2. Spread the BPSK-modulated data sequence using the selected spreading code.
3. Transmit the packet only if the downlink signal quality is good. This decision is made based on an open loop called SNR-plus-interference ratio (SNIR)-driven uplink transmit packet control (SDUTPC). The role of SDUPTC is to control the received packets power level in order to avoid receiving packets with a very low SNIR which could not be decoded successfully even without interference. SDUPTC procedure at the terminal side can be described as follows:
   i. The terminal performs a data-aided (DA) SNIR estimation on the received signal on the forward link, in order to decide if it can transmit or not on the return link.

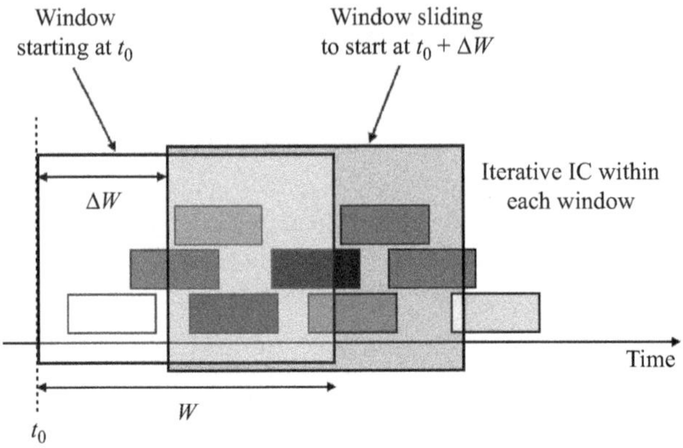

Figure 15.16 E-SSA sliding window and iterative IC algorithm

ii. If the estimated SNIR at a certain instant is within a certain window representative of line of sight (LOS) conditions, the packet is transmitted on the channel. More details on the numerical procedure are provided in [13].

At the receiver side (see Figure 15.16), the signal received on one window duration W is stored (usually W is equal to the duration of 3 packets), then when the E-SSA process on one window is finished, the receiver slides the actual window by a predefined step ΔW. On each window duration, the detector in E-SSA performs the following operations iteratively until N_{max} iterations are reached:

1. Detection of the packet with the strongest SNIR;
2. Channel estimation then demodulation and decoding of the strongest packet;
3. If decoding is successful after cyclic redundancy check:
   i. Re-encoding and modulation of the decoded packet;
   ii. IC.

The E-SSA performance was evaluated in [13]. At this point, it is important to note that the normalised load G and the throughput T in E-SSA are computed in bits/chip in order to take into account the spreading factor value. Thus, the load G is derived as shown in the equation below, and the throughput is computed using G in bits/chip.

$$G = \frac{\lambda R \log_2(M)}{SF}(\text{bits/chip}), \tag{15.10}$$

The authors showed that the throughput with E-SSA can reach up to 1.7 bits/symbol for a target PLR of 10^{-4} with a lognormal packets power distribution ($\sigma = 2$ dB), BPSK modulation, 3GPP turbo code of rate 1/3 and a spreading factor $SF = 256$. This result presents a significant gain compared to SSA which achieves a maximal throughput of only 0.5 bits/symbol.

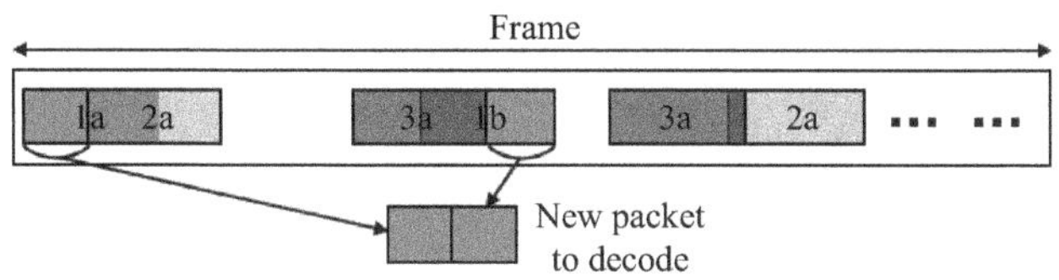

Figure 15.17 ECRA scheme

Remark 15.2 (E-SSA in several satellite standards). *E-SSA is the RA technique proposed for two satellite standards: S-band mobile interactive multimedia (S-MIM) [14,51] and its variant for fixed terminals Fixed Satellite Interactive Multimedia (F-SIM) [15]. In S-MIM, the data structure in a packet is constituted of two separate channels:*

- *A channel for useful data transmission, called physical data channel (PDCH).*
- *A channel for control data transmission, called physical control channel (PCCH). This channel is used to transmit the signalling information and the pilot symbols used for coherent demodulation of the data sequence.*

The PDCH information is encoded with a 3GPP/UMTS turbo code, and both the PDCH and PCCH contents are modulated and spread (using orthogonal codes). The PDCH and PCCH channels are multiplied by unbalanced power coefficients, i.e. the control channel has normally a lower power level than the data channel. Then, the two channels are superposed by transmitting one in-phase and the other in quadrature-phase. The resulting signal is scrambled using a Gold code and a preamble is added to the beginning of the message.

The F-SIM standard has been proposed to be used by Eutelsat Broadcast Interactive System [11,12]. F-SIM is also based on E-SSA RA and it uses the same PDCH and PCCH channels superposition as in S-MIM. However, F-SIM is designed for fixed terminals and higher frequency bands (Ku–Ka) as well as different data rates than S-MIM.

15.3.3.2 Enhanced contention resolution ALOHA

Enhanced contention resolution ALOHA (ECRA) is another asynchronous RA protocol proposed in 2013 [52]. As its name reveals, it is an enhanced version of a former RA scheme called contention resolution ALOHA (CRA) [53]. In CRA, the authors proposed the removal of the slots boundaries in CRDSA by allowing asynchronous packets transmissions at any time instant on one frame. However, the concept of frame-level synchronisation is still present. ECRA adds the idea of replicas 'best parts' combination to CRA and the frame is only defined at the terminal side; hence, frame synchronisation among users is not needed. As a matter of fact, given the asynchronous nature of the communication, replicas corresponding to a same packet may experience partial interference on different parts of the packet as shown in Figure 15.17. The example in the figure shows that the first replica of user 1 experiences collisions only on the right-most part of the packet; however, the second replica is interfered only on the left part. In the case of CRA, both replicas can be lost if the interference power is too high. For this reason, ECRA proposes a solution to this problem by combining the non-interfered symbols of the replicas into a new packet. Thus, it is obvious that the new packet would have a higher successful decoding probability. The decoding process of ECRA can be detailed as follows:

1. The frame is stored at the receiver side and the SIC process is applied. The frame is scanned in an iterative way in order to detect and decode packets. Whenever a packet is decoded successfully, it is removed from the frame and the decoded pointers are used to localise its replicas and remove them as well.
2. When no further packets could be decoded on the frame, ECRA intervenes in order to attempt to decode the remaining packets using the following procedure:
   i. If some parts of a given packet encounter interference in all the replicas, then the parts (or the symbols) encountering the lowest interference power are used to construct a new packet. Therefore, ECRA shall perform symbol-by-symbol SNIR estimation in order to correctly select the parts of the replicas to combine.
   ii. If the new constructed packet is successfully decoded, the packet and its replicas are removed from the frame.

The authors of ECRA showed that it can achieve a maximum normalised throughput of 1.2 bits/symbol[2] with $N_b = 2$ replicas, a QPSK 1/2 modulation coding scheme and $E_s/N_0 = 10$ dB. In [54], the authors proposed a localisation technique for asynchronous packet replicas using on a two-step threshold-based approach: first, cross-correlation with a known sequence to detect the packets start time, then non-coherent cross-correlation to detect the locations of the replicas corresponding to a same packet. They evaluated the packets detection probability according to a pre-defined threshold and studied the performance in terms of throughput when MRC is used for replicas combination. The results showed that the throughput is slightly degraded when the two-phase detection and combining technique is applied. However, when the load is higher than 1 bit/symbol, the PLR seems to be more affected as the throughput starts to decrease compared to the ideal detection case. Recent improvements to ECRA were presented in [55] and an analytical approximation of the PLR performance for asynchronous RA schemes was provided. A significant gain can be observed in the throughput of ECRA which can reach up to 2.5 bits/symbol.

15.3.3.3 Asynchronous contention resolution diversity ALOHA

Another asynchronous RA method recently proposed in the literature is asynchronous contention resolution diversity ALOHA (ACRDA) [56]. ACRDA is a modified asynchronous version of CRDSA. The operations at the transmitter and the receiver sides have similarities with both CRDSA and E-SSA in order to cope with asynchronous transmissions. In ACRDA, TSs and frame boundaries are not defined in

[2]The results in [52] are obtained by using the Shannon Bound, i.e. a decoding threshold based on Shannon capacity. This assumption can degrade the results because all the packets received below this threshold are discarded.

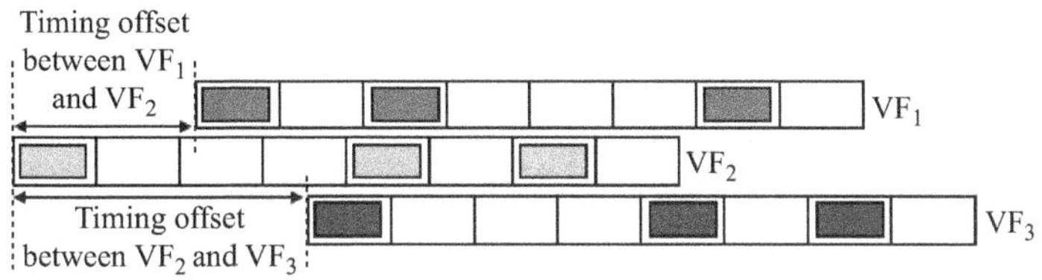

Figure 15.18 ACRDA virtual frames scheme

reference to the global timeline at the centralised gateway (i.e. NCC) demodulator. Instead, the delimitation of TSs and frame are local to each transmitter and completely asynchronous among different transmitters. Thus, unlike CRDSA, frame-level synchronisation among users is not needed. The term 'virtual frame' (VF) is used to refer to the local frame at each transmitter. Each VF contains N_{slots} and each slot has a duration T_{slot}, so that the duration of a VF is $T_{VF} = N_{slots}T_{slot}$. Figure 15.18 illustrates the reception of 3 asynchronous VFs corresponding to different transmitters with completely independent timing offsets at the receiver. If all the transmitters have the same timing offset, then the classical CRDSA scheme is obtained. The ACRDA scheme at the transmitter side is detailed as follows:

1. Before transmitting a packet on the RA channel, N_b replicas are generated and N_b TSs are randomly selected within the duration of one VF.
2. Similarly to CRDSA, the information concerning the location of the other replicas is added to each packet. In the case of ACRDA, the location information is the TS offset relative to the start of the current packet.
3. The start time of a VF is chosen randomly at the transmitter side, and no wide centralised synchronisation is needed.
4. A preamble containing a known sequence common to all transmitters is added to the beginning of each packet replica. This common preamble is used for packets detection and channel estimation at the receiver side.
5. Each packet replica is transmitted on the selected TSs of the local VF.

At the receiver side, the same window-based memory processing as done in E-SSA is applied as it was shown in Figure 15.16. The operations of ACRDA at the receiver side are detailed below.

1. The received signal is down-converted, filtered and sampled.
2. For each sliding window,
    i. The signal covering a duration of W VFs is stored in the receiver memory (in general $W = 3$ is assumed).
    ii. The ACRDA process is repeated iteratively on each window, as explained below:
        a. First, the common packet preamble is searched using a cross-correlation matched to the preamble sequence.
        b. Each time a preamble sequence is detected, the packet demodulation and decoding is attempted.

c. If a packet is successfully decoded, channel estimation is performed using the full packet content. Then, the packet is removed from the frame.
d. The successfully decoded packet is also used to locate its other replicas, reconstruct their corresponding signals and remove them from the frame.
e. If the currently decoded packet points to a replica that is not in the current window, the packet information is stored until the sliding window finds the corresponding replica.

iii. Once the ACRDA process on one window is completed, the window is shifted towards the next ΔWT_{VF}.

The authors in [56] have concluded that ACRDA performs slightly better than CRDSA in terms of throughput and PLR, particularly with $N_b = 2$ replicas per packet. However, in the asynchronous mode, the implementation complexity at the receiver side is higher. The performance simulations done in [56] have considered a QPSK modulation and a 3GPP FEC code of rate $R = 1/3$ and $E_s/N_0 = 10$ dB. It has been shown that the maximum normalised throughput is achieved with $N_b = 2$ replicas, and it can reach 0.9 bit/symbol for a $PLR < 10^{-4}$. With a lognormal packets power distribution of $\sigma = 3$ dB, the throughput can increase up to 1.5 bits/symbol for a $PLR < 10^{-4}$. At the same time, the authors have shown that significant gains are achieved with ACRDA in terms of packets transmission delays.

15.3.3.4 Summary and brief discussion for advanced asynchronous RA

This subsection presented the advanced asynchronous RA techniques proposed for satellite communications. The first technique presented was E-SSA, in which spectrum spreading is used to resolve packet collisions at the receiver side. With the adequate choice of the spreading factor, E-SSA permits to significantly enhance the performance even compared to CRDSA, and it targets low symbol rates communications. Another unslotted RA presented was ECRA, which uses packets replication at the transmitter and packets 'best parts' combining at the receiver side. The throughput obtained in ECRA is enhanced compared to CRDSA, however the PLR evaluation while taking into account the packet replicas misdetection probability was not analysed in [55]. ACRDA is another asynchronous RA presented, and it refers to an asynchronous version of CRDSA, where each user defines a VF and a sliding window approach is applied at the receiver side. ACRDA achieves almost the same throughput as CRDSA but at the same time, it permits to mitigate the PLR floor seen in CRDSA thanks to the significant reduction of the probability of loop phenomenon.

Advanced asynchronous RA techniques present many interesting use cases for satellite communications. They are mainly characterised with

- Decreased signalling overhead for terminals synchronisation with the satellite or the gateway.
- Better PLR performance thanks to the resolution of partial packets collisions.
- Reduced access delays.
- Suitable for low symbol rate communications.

An amount of complexity for packets detection shall be added at the receiver side, however this complexity is affordable as long as it is done at the network infrastructure side (gateway). A general discussion and conclusion for recent RA schemes will be given in the following.

15.4 General comparison metrics for different advanced RA techniques

Following this detailed review on advanced RA techniques, a general comparison can be done in order to distinguish the suitable RA solutions to be used in future 5G systems. This comparison will be based on four main metrics: the power limitations at the connected terminal side, the communications with very low data rates, the high throughput requirement and the signalling overhead reduction. In the same context, a detailed review has been also given in [57].

15.4.1 Power limitations at the terminal side

As 5G networks are targeting more low cost terminals, power limitations appear on the terminal side. For this reason, advanced RA should be able to resolve packets collisions and ensure a low PLR even with low energy per symbol. RA schemes such as E-SSA, CRDSA combined with MARSALA and ACRDA showed relatively good performance with lower levels of E_s/N_0 compared to other RA schemes.

15.4.2 Communications at very low data rates

In IoT and M2M communications, the terminals transmit signals at very low data rates in order to consume lower power and also because those types of services do not require higher data rates. Therefore, in such environments, the use of low rate RA like E-SSA and MF-CRDSA is of interest. Otherwise, a frame containing several TSs on one frequency band would cause large delays for packets transmission.

15.4.3 High throughput performance at MAC-layer level

Obviously, in order to handle very dense networks, the RA scheme used shall be able to provide low PLR in high load regimes. RA schemes such as CRDSA combined with MARSALA, ACRDA and E-SSA presented an average throughput performance of more than 1 bit/symbol for a PLR as low as 10^{-3}.

15.4.4 Signalling overhead

The traffic on the return link is shifting towards more sporadic profiles with very low duty cycles per terminal. Therefore, the transmission of signalling packets for the purpose of synchronisation should be maintained relatively low. For this reason, the use of asynchronous RA instead of synchronous RA is of interest. Moreover, as previously pointed out in this chapter, asynchronous packets reception permits

Table 15.2 Comparison of some metrics for advanced RA schemes: the required EIRP, the targeted symbol rates, the throughput (in bits/symbol) at low PLR and the possibility for a return link only communication

EIRP requirement (from lower to higher)	**Throughput (from lower to higher)**	**Possibility for return only channel**
ESSA SMIM (mainly due to low symbol rates – 5 kbps)	ECRA (2.5) with equi-powered packets	Only for asynchronous RA
ESSA FSIM (higher symbol rates than SMIM – 160 kbps)	MARSALA (2.3)	
MF-CRDSA	ESSA SMIM (1.9)	
ECRA	ESSA FSIM (1.9)	
MARSALA	ACRDA (1.5)	
ACRDA	MF-CRDSA (1.4)	
CRDSA	CRDSA (1.3)	

to achieve a lower PLR on the frame because partial packet collisions induce less interference on the useful packet to decode.

15.4.5 Comparative table

In Table 15.2, we present a brief general qualitative comparison of different advanced RA schemes with respect to: the required EIRP, the targeted symbol rates, the throughput at low PLR and the possibility for a return link only communication. The throughput shown in the table is expressed in bits/symbol and it is obtained for a PLR around 10^{-4} with lognormal power distribution of variance $\sigma = 3$ dB.

In the end, it is worth mentioning that there are variants of E-SSA like ME-SSA [58] which may provide higher throughput then conventional E-SSA (around 50% of throughput gain).

15.5 General summary and final remarks

In this chapter, we provided an overview of several advanced RA techniques proposed for satellite communications and envisioned to provide access solutions for terminals in 5G networks. Many of these techniques target networks with very large populations integrating IoT and M2M communications, either for fixed or mobile scenarios.

For future deployments, one of the main challenges is to remain as much as possible coherent with existing terrestrial communications standards such as 3GPP [24] and terrestrial IoT technology leaders like Sigfox and LoRA. Other challenges arise particularly related to the targeted low data rates, low power consumption and energy harvesting. As a matter of fact, in such environments, phase noise has a higher impact on the received signal and can cause packet losses especially at low SNIR levels.

Moreover, many open questions remain unanswered on whether to use packet redundancy or spectrum spreading, or whether to target low data rates with low power requirements on the terminal side and high phase noise or high data rates requiring higher power but less sensitive to phase noise. Furthermore, future cross-layer system evaluation should be considered in order to ensure enhancing the global system performance in future 5G networks.

References

[1] Evans BG. The role of satellites in 5G. In: 2014 7th Advanced Satellite Multimedia Systems Conference and the 13th Signal Processing for Space Communications Workshop (ASMS/SPSC); 2014. p. 197–202.

[2] DVB Document ETSI EN 301 790, Digital Video Broadcasting (DVB); Interaction Channel for Satellite Distribution Systems; 2003.

[3] DVB Document ETSI A155-1, Digital Video Broadcasting (DVB); Second Generation DVB Interactive Satellite System (RCS2); Part 2; 2011.

[4] P M Feldman. An Overview and Comparison of Demand Assignment Multiple Access (DAMA) Concepts for Satellite Communications Networks. Santa Monica, CA: RAND Corporation: RAND; 1996.

[5] Dainotti A, Pescape A, Ventre G. A packet-level characterization of network traffic. In: 2006 11th International Workshop on Computer-Aided Modeling, Analysis and Design of Communication Links and Networks; 2006. p. 38–45.

[6] Nikaein N, Laner M, Zhou K, *et al.* Simple traffic modeling framework for machine type communication. In: ISWCS 2013; The Tenth International Symposium on Wireless Communication Systems; 2013. p. 1–5.

[7] Gaudenzi RD, del Rio Herrero O. Advances in random access protocols for satellite networks. In: International Workshop on Satellite and Space Communications, 2009; 2009. p. 331–336.

[8] Kissling C, Munari A. On the integration of random access and DAMA channels for the return link of satellite networks. In: 2013 IEEE International Conference on Communications (ICC); 2013. p. 4282–4287.

[9] Digital Video Broadcasting (DVB); Second Generation DVB Interactive Satellite System (DVB-RCS2); Guidelines for Implementation and Use of LLS. EN 301 545-2; 2012.

[10] Casini E, Gaudenzi RD, Herrero ODR. Contention resolution diversity slotted ALOHA (CRDSA): an enhanced random access scheme for satellite access packet networks. IEEE Transactions on Wireless Communications. 2007 April;6(4):1408–1419.

[11] SmartLNB Connected TV services via satellite [Online PDF]. Paris: Eutelsat. Available from: http://www.eutelsat.com/files/contributed/news/media_library/brochures/EUTELSAT_SmartLNB.pdf.

[12] SmartLNB M2M & IoT via satellite [Online PDF]. Paris: Eutelsat;. Available from: http://www.eutelsat.com/files/contributed/news/media_library/brochures/EUTELSAT_SmartLNB_M2M.pdf.
[13] O del Rio Herrero and R De Gaudenzi. A high efficiency scheme for quasi-real-time satellite mobile messaging systems. In: 2008 10th International Workshop on Signal Processing for Space Communications; 2008. p. 1–9.
[14] Satellite Earth Stations and Systems; Air Interface for S-band Mobile Interactive Multimedia (S-MIM); V 1.1.1; 2011.
[15] Arcidiacono A, Finocchiaro D, Collard F, *et al.* From S-band mobile interactive multimedia to fixed satellite interactive multimedia: making satellite interactivity affordable at Ku-band and Ka-band. International Journal of Satellite Communications and Networking. 2016 July;34(4):575–601. Available from: https://doi.org/10.1002/sat.1158.
[16] Abramson N. The ALOHA system: another alternative for computer communications. In: Proceedings of the November 17–19, 1970, Fall Joint Computer Conference. AFIPS '70 (Fall). New York, NY: ACM; 1970. p. 281–285. Available from: http://doi.acm.org/10.1145/1478462.1478502.
[17] Roberts LG. ALOHA packet system with and without slots and capture. ACM, SIGCOMM Computer Communication Review. 1975;5(Feb.): 28–42.
[18] Choudhury Gagan L, Rappaport Stephen S. Diversity ALOHA–A Random Access Scheme for Satellite Communications. IEEE Transactions on Communications. 1983 March;31(3):450–457.
[19] Bui HC. Méthodes d'accès basées sur le codage réseau couche physique [Phd]. Institut Supérieur de l'Aéronautique et de l'Espace (ISAE). University of Toulouse; 2012. Available from: http://www.theses.fr/2012ESAE0031/document.
[20] O del Rio Herrero and R De Gaudenzi. Generalized analytical framework for the performance assessment of slotted random access protocols. IEEE Transactions on Wireless Communications. 2014 February;13(2): 809–821.
[21] O del Rio Herrero and R De Gaudenzi. A high-performance MAC protocol for consumer broadband satellite systems. IET Conference Proceedings; 2009. p. 512–512(1).
[22] K Whitehouse, A Woo, F Jiang, J Polastre and D Culler. Exploiting the capture effect for collision detection and recovery. In: 45–52, editor. The Second IEEE Workshop on Embedded Networked Sensors, 2005. EmNetS-II.; 2005.
[23] O Del Rio Herrero and R De Gaudenzi. High efficiency satellite multiple access scheme for machine-to-machine communications. IEEE Transactions on Aerospace and Electronic Systems. 2012 October;48(4):2961–2989.
[24] Evolved Universal Terrestrial Radio Access (E-UTRA); Physical Channels and Modulation, 3GPP TS 36.211 v8.9.0; 2009.
[25] Liva G. Contention resolution diversity slotted ALOHA with variable rate burst repetitions. In: 2010 IEEE Global Telecommunications Conference GLOBECOM 2010; 2010. p. 1–6.

[26] Berioli M, Cocco G, Liva G, *et al.* Modern random access protocols. Foundations and TrendsÂ® in Networking. 2016;10(4):317–446. Available from: http://dx.doi.org/10.1561/1300000047.

[27] Richardson TJ, Shokrollahi MA, Urbanke RL. Design of capacity-approaching irregular low-density parity-check codes. IEEE Transactions on Information Theory. 2006 September;47(2):619–637. Available from: http://dx.doi.org/10.1109/18.910578.

[28] Gallager RG. Low-Density Parity-Check Codes; 1963. Cambridge, MA: M.I.T. Press.

[29] Liva G. A slotted ALOHA scheme based on bipartite graph optimization. In: 2010 International ITG Conference on Source and Channel Coding (SCC); 2010. p. 1–6.

[30] Liva G. Graph-based analysis and optimization of contention resolution diversity slotted ALOHA. IEEE Transactions on Communications. 2011 February;59(2):477–487.

[31] Paolini E, Liva G, Chiani M. High throughput random access via codes on graphs: coded slotted ALOHA. In: IEEE International Conference on Communications (ICC) 2011; 2011. p. 1–6.

[32] Paolini E, Liva G, Chiani M. Graph-based random access for the collision channel without feedback: capacity bound. In: IEEE Global Telecommunications Conference (GLOBECOM 2011); 2011. p. 1–5.

[33] Paolini E, Liva G, Chiani M. Coded slotted ALOHA: a graph-based method for uncoordinated multiple access. IEEE Transactions on Information Theory. 2015 December;61(12):6815–6832.

[34] Paolini E, Stefanovic C, Liva G, *et al.* Coded random access: applying codes on graphs to design random access protocols. IEEE Communications Magazine. 2015 June;53(6):144–150.

[35] Chiani M, Liva G, Paolini E. The marriage between random access and codes on graphs: coded slotted ALOHA. In: Satellite Telecommunications (ESTEL), 2012 IEEE First AESS European Conference on; 2012. p. 1–6.

[36] Narayanan KR, Pfister HD. Iterative collision resolution for slotted ALOHA: an optimal uncoordinated transmission policy. In: 2012 7th International Symposium on Turbo Codes and Iterative Information Processing (ISTC); 2012. p. 136–139.

[37] Bui HC, Lacan J, Boucheret ML. An enhanced multiple random access scheme for satellite communications. In: Wireless Telecommunications Symposium (WTS), 2012; 2012. p. 1–6.

[38] CCSDS. TM Synchronization and channel coding: recommended standard. CCSDS 1310-B-2; Blue Book. 2011.

[39] Bui HC, Lacan J, Boucheret ML. Multi-slot coded ALOHA with irregular degree distribution. In: 2012 IEEE First AESS European Conference on Satellite Telecommunications (ESTEL); 2012. p. 1–6.

[40] Lacan J, Bui HC, Boucheret M-L. Reception de donnees par paquets a travers un canal de transmission a acces multiple; 2013. Brevet d'invention. BFF130051GDE.

[41] Lo TKY. Maximum ratio transmission. In: Communications, 1999. ICC '99. 1999 IEEE International Conference on. vol. 2; 1999. p. 1310–1314.

[42] Jasper SC, Birchler MA, Oros NC. Diversity reception communication system with maximum ratio combining method. Google Patents; 1996. US Patent 5,553,102. Available from: http://www.google.com/patents/US5553102.

[43] Bui HC, Zidane K, Lacan J, Boucheret ML. A multi-replica decoding technique for contention resolution diversity slotted ALOHA. In: Vehicular Technology Conference (VTC Fall), 2015 IEEE 82nd; 2015. p. 1–5.

[44] Zidane K, Lacan J, Gineste M, Bes C, Deramecourt A, Dervin M. Estimation of timing offsets and phase shifts between packet replicas in MARSALA random access. In: 2016 IEEE Global Communications Conference; 2016.

[45] Zidane K, Lacan J, Gineste M, Bes C, Bui C. Enhancement of MARSALA random access with coding schemes, power distributions and maximum ratio combining. In: 2016 8th Advanced Satellite Multimedia Systems Conference and the 14th Signal Processing for Space Communications Workshop (ASMS/SPSC); 2016.

[46] Mengali A, Gaudenzi RD, Arapoglou PD. Enhancing the physical layer of contention resolution diversity slotted ALOHA. IEEE Transactions on Communications. 2017;65(10):4295–4308.

[47] Abramson N. Spread ALOHA CDMA data communications. Google Patents; 1996. US Patent 5,537,397. Available from: https://www.google.com/patents/US5537397.

[48] Ozluturk F, Jacques A, Lomp G, *et al.* Code division multiple access communication system. Google Patents; 1998. WO Patent App. PCT/US1998/004,716. Available from: http://www.google.com/patents/WO1998040972A2?cl=en.

[49] Universal Mobile Telecommunications System (UMTS); Physical channels and mapping of transport channels onto physical channels (FDD); 1999.

[50] 3GPP Spreading and modulation (FDD); 1999.

[51] Scalise S, Niebla CP, Gaudenzi RD, *et al.* S-MIM: a novel radio interface for efficient messaging services over satellite. IEEE Communications Magazine. 2013 March;51(3):119–125.

[52] Clazzer F, Kissling C. Enhanced contention resolution ALOHA – ECRA. In: Proceedings of 2013 9th International ITG Conference on Systems, Communication and Coding (SCC); 2013. p. 1–6.

[53] C Kissling. Performance enhancements for asynchronous random access protocols over satellite. In: 2011 IEEE International Conference on Communications (ICC); 2011. p. 1–6.

[54] Clazzer F, Lazaro F, Liva G, *et al.* Detection and combining techniques for asynchronous random access with time diversity. In: SCC 2017; 11th International ITG Conference on Systems, Communications and Coding; 2017. p. 1–6.

[55] Clazzer F, Kissling C, Marchese M. Enhancing contention resolution ALOHA using combining techniques. IEEE Transactions on Communications. 2017;PP(99):1–1.

[56] De Gaudenzi R, del Río Herrero O, Acar G, Garrido Barrabés E. Asynchronous contention resolution diversity ALOHA: making CRDSA truly asynchronous. IEEE Transactions on Wireless Communications. 2014 November;13(11):6193–6206.

[57] R. De Gaudenzi, O. Del Rio Herrero, G. Gallinaro, S. Cioni, P.-D. Arapoglou. Random access schemes for satellite networks, from VSAT to M2M: a survey. International Journal of Satellite Communications and Networking. 2018;36(1),66–107.

[58] Gallinaro G, Alagha N, Gaudenzi RD, *et al.* ME-SSA: an advanced random access for the satellite return channel. In: 2015 IEEE International Conference on Communications (ICC); 2015. p. 856–861.

Chapter 16

Interference avoidance and mitigation techniques for hybrid satellite-terrestrial networks

Konstantinos Ntougias[1], Dimitrios K. Ntaikos[1], George K. Papageorgiou[1], and Constantinos B. Papadias[1]

16.1 Introduction

16.1.1 5G radio access technologies

An exponential growth in the volume of the mobile data traffic has been noticed over the past few years [1]. This trend, which is mainly attributed to video data transfers, is expected to continue at an increasing pace in the foreseeable future, as it is indicated by the attributes and the corresponding requirements of the envisioned fifth Generation (5G) services [2]. More specifically, a 1,000× increase of the provided capacity is anticipated to take place by 2020 [3], in comparison to the capacity of today's long-term evolution (LTE) systems. This number refers to the cellular downlink, where the base stations (BSs) act as transmitters (TXs), whereas the user terminals (UTs) play the role of the receivers (RXs).

The apparent scarcity of the sub-6-GHz radio spectrum [4], which is typically utilized for cellular access and other long-range wireless communication applications, calls for the use of a number of complementary radio access technologies, in order to reach the aforementioned capacity target of the next-generation cellular mobile radio communication networks [5].

One approach is the exploitation of the abundant available bandwidth at the centimeter-wave (3–30 GHz) and millimeter-wave (30–300 GHz) segments of the spectrum [6]. Another direction is the use of advanced multiple-input–multiple-output (MIMO) technologies, which leverage the spatial dimension that is provided by the use of multiple antennas at the UTs or/and the BSs in order to increase the spectral efficiency of the system. Examples include coordinated multipoint (CoMP) [7] and massive MIMO [8].

Spectrum sharing paradigms also target the improvement of the spectrum utilization's efficiency by allowing the reuse of the spectrum at different systems on a noninterfering basis. For example, licensed shared access (LSA) [9] enables a mobile

[1]Broadband Wireless and Sensor Networks (B-WiSE) Lab, Athens Information Technology (AIT), Greece

network operator, which is called an LSA licensee under this context, to access the band operated by an incumbent (or part of it) when and where the latter does not make use of it, according to a commonly agreed set of spectrum usage rules. Hence, LSA unlocks spectrum that was reserved for exclusive use, despite its severe underutilization in the time, space, and frequency dimensions, as it has been reported manifold [10]. Examples of frequency bands where spectrum sharing, in one form or another, may take place include the 2.3–2.4-GHz band in Europe and the 3.5 GHz band in the United States [11], which are currently utilized by media and entertainment services and military communication systems, as well as frequency bands that are exploited by satellite communication systems and microwave point-to-point or point-to-multipoint systems worldwide (e.g., 19, 28 GHz, etc.).

Yet another trend is the densification of the radio access network, i.e., the deployment of dense small-cell networks. These so-called ultra-dense networks enable the aggressive reuse of the available frequencies across the service area [12].

Finally, the off-loading of the traffic to 4G and Wi-Fi networks constitutes another capacity-enhancing strategy that is considered in the 5G framework [4].

16.1.2 MIMO communication technologies

The single-user MIMO (SU-MIMO) and single-cell multiuser MIMO (MU-MIMO) technologies have been extensively used in LTE networks. SU-MIMO exploits the multipath propagation in order to increase the capacity of a point-to-point link that is established in the downlink between a multiantenna BS (TX) and a multiantenna UT (RX), at no extra cost in terms of link bandwidth or transmission power [13]. To this end, this communication paradigm enables the concurrent transmission of multiple data streams toward a scheduled user over a single frequency band. Moreover, it relies on appropriate precoding and postcoding signal-processing operations, in order to enable the mitigation of the resulting interstream interference. The achieved throughput depends on the employed precoding, postcoding, and power allocation schemes.

Besides its capacity-enhancement capability, the SU-MIMO technology presents also a number of performance limitations and implementation difficulties: (a) it requires a scattering environment for the spatial multiplexing of the data streams to take place; (b) it requires the use of multiantenna UTs, for the spatial demultiplexing and consequent detection of the received signals to become possible; (c) it requires the orthogonalization of the users in the time or/and frequency domain, since it lacks a mechanism that enables the handling of the multiuser interference (MUI); and (d) its performance deteriorates when the channel matrix is not full rank or it is ill-conditioned, since in this case, the number of the effective spatial subchannels that are created between the transmit and receive antennas due to the multipath is reduced.

All of the aforementioned issues are addressed by single-cell MU-MIMO [13]. This technology takes advantage of the physical separation between the users, in order to make the spatial sharing of the channel among them possible. That is, it enables the transmission of multiple data signals toward a group of active users on a single time-frequency resource and the reduction or elimination of the resulting intracell co-channel interference (CCI) with the help of precoding and user selection schemes. The aggregated or sum-rate (SR) throughput of the corresponding point-to-multipoint link in the downlink, which is also known as MIMO broadcast channel (BC), depends on the employed precoding, user selection and power allocation schemes. Single-cell MU-MIMO can be applied to both line-of-sight (LOS) and non-LOS (NLOS) scenarios, as well as in situations where single-antenna UTs are utilized. This is because its operation is based on the spatial separation of the users and the use of multiuser spatial multiplexing. However, it cannot control the intercell interference (ICI), which degrades the system-wide spectral efficiency.

Multicell MU-MIMO radio access technologies are extensions of the single-cell MU-MIMO paradigm. They have the advantage of managing both the intracell and the intercell MUI, instead of treating the ICI as noise. This feature allows for more aggressive frequency reuse. Regarding ICI management, CoMP, which has been introduced in the LTE-Advanced standard, relies on coordinated or joint transmissions (JTs) between neighboring BSs [7], while massive MIMO is based on the use of an excessive number of transmit antennas [8]. Both of these technologies are expected to be utilized at the radio access segment of 5G networks.

16.1.3 Flexible hybrid satellite-terrestrial backhaul

The enormous capacity requirements of the 5G networks place a heavy burden on the mobile backhaul (MBH) as well. Therefore, the use of technological enhancements and novel network architectures is required, in order to avoid any possible discrepancy between the capacity demands and the provided backhaul capacity.

A recent solution that utilizes several of the aforementioned radio access technologies is the use of hybrid terrestrial-satellite MBH networks, as depicted in Figure 16.1. In this approach, a part of the terrestrial backhaul traffic is off-loaded to the satellite segment. This solution takes advantage of the high data rates that are provided by the contemporary high throughput satellite communication systems. These systems operate at the Ka-band (28–40 GHz) and utilize multibeam technology, which enables aggressive frequency reuse across multiple relatively narrow spot beams [14].

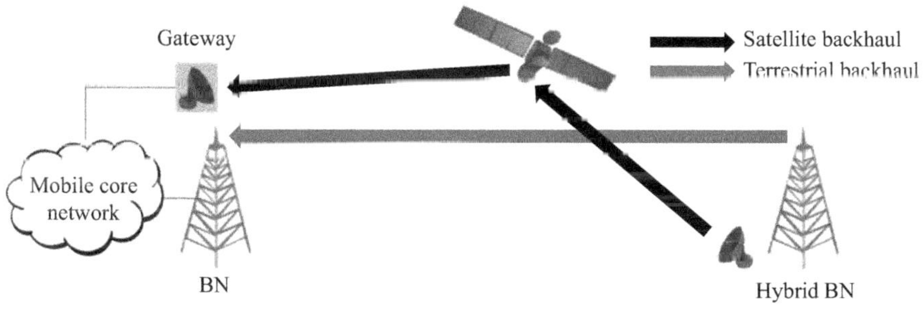

Figure 16.1 A high-level overview of a hybrid satellite-terrestrial mobile backhaul (MBH) system. The satellite segment off-loads part of the traffic that is transported over the terrestrial segment

To maximize the spectral efficiency, the two MBH segments may be allowed to share the same spectrum. In this case, MIMO technology and CCI mitigation techniques play a very important role [14].

From a network architecture point of view, such a hybrid MBH network requires, besides the use of conventional terrestrial backhaul nodes (BN) (connected to the BSs), the utilization of some intelligent BNs (iBN), which are equipped with both antennas for the terrestrial links and satellite dishes. Additionally, it includes a hybrid network manager (HNM) that determines the routing of the traffic and the load-balancing between the satellite and the terrestrial MBH segments [14].

More specifically, the HNM calculates the topology instances (routes) based on information collected from the iBNs and the adopted quality of service (QoS) policies, in order to optimize the transport of traffic over the MBH (in terms of QoS metrics such as throughput, latency, and packet error rate) and enhance the system's resiliency. The HNM is equipped with various interfaces to integrate different modules, such as external radio resource managers and interference analyzers, which provide real-time analysis and management capabilities and allow the performance of direct actions in an autonomous manner. Examples include the avoidance of interference or the handling of link failure and congestion issues. The decisions of the HNM are directed to the iBNs, which then take the corresponding actions.

The use of antenna arrays instead of drum antennas and the adoption of MIMO communication technologies are integral features of the aforementioned hybrid MBH system that present several advantages. First, this approach is usually a better fit for dense small-cell networks than the use of highly directive point-to-point links. This is attributed to the fact that the MIMO technology can be applied in NLOS propagation environments, which are typically found in dense small-cell setups where the antennas are placed at street-level (e.g., on lamp posts), while the utilization of point-to-point links requires a LOS environment. We should also mention here that such setups favor in general the application of SU-MIMO, in contrast to radio access setups. This is because it is relatively easy to equip the BNs with, at least, a few antennas, while it is difficult in general to pack multiple antennas in a device with such a small form factor as a UT. Moreover, in such ultradense network setups, the employment of CoMP allows for the coordination of the severe ICI that is caused by the small distances between the deployed BNs. Furthermore, the use of antenna arrays allows the adaptive shaping and steering of the radiation patterns (beams). This flexibility, in turn, enables the dynamic establishment/reconfiguration of the links, thus facilitating the implementation of the routing decisions taken by the HNM.

The application of multiantenna communication techniques in the wireless terrestrial MBH segment should take into consideration the characteristics of the involved technologies. More specifically, the spatial multiplexing, directivity, and interference management capabilities of the iBNs depend on the provided number of array degrees-of-freedom (DoF). This quantity, which determines essentially the achieved throughput of the employed MIMO transmission schemes, is equal to the number of BN antennas when digital antenna arrays (DAA) are utilized [15]. For instance, the capacity of a MIMO link grows linearly with the minimum number of antennas at both ends of the link [16,17]. As another example, the capacity of a MIMO-BC channel scales with the number of the BS antennas, provided that there are at least as many users as transmit antennas, irrespective of the number of antennas that are installed at the UTs [18–20]. The same statement holds for the capacity of a "super MIMO-BC" that is formed in a CoMP setup where the JT scheme is utilized [21]. Finally, it should be noted that the massive MIMO concept is based on the dependence of the SR capacity on the array's DoF—that is, on the number of service antennas.

In practice, the number of antennas that can be installed at the iBNs is limited by size constraints as well as by hardware complexity, cost, and energy consumption limitations, thus reducing the capacity enhancement potential of the applied MIMO schemes. The size constraint is associated with the requirement of having a sufficient interelement spacing, in order to avoid the occurrence of mutual coupling that can degrade the radiation efficiency. The latter limitations are related with the requirement of connecting each antenna element to a radio frequency (RF) unit. These constraints are more strict for systems that operate at high frequencies, such as the wireless terrestrial MBH systems, as well as in scenarios where compact, low-cost, low-power BNs, such as small-cell BSs, are considered.

In order to overcome these problems, various hybrid analog-DAA solutions have been proposed. The core of this technology is the use of a limited number of active antenna elements (and, therefore, RF modules)—thus leading to cost, complexity, and energy consumption savings—and the addition of passive antenna elements. Load-controlled parasitic antenna arrays (LC-PAAs) constitute a representative example of these antenna systems [21]. The main feature of this technology that differentiates it from other hybrid antenna array paradigms is that it actually exploits the mutual coupling among the antennas in order to provide radiation pattern reconfiguration capabilities with a small number of RF units, instead of trying to mitigate its effects.

16.1.4 Chapter objectives and structure

Motivated by the discussion in the previous section, in this chapter, we present various techniques that aim at avoiding or mitigating the CCI in the hybrid satellite-terrestrial MBH network. We consider SU-MIMO and CoMP setups comprised by BNs that are equipped with single-RF or multi-RF LC-PAAs. The utilization of user-level and symbol-level precoding schemes is also presented.

The structure of the chapter is the following: First, the LC-PAA technology is described. Then, a method that provides robust arbitrary channel-dependent precoding with such antenna arrays is presented. In the following section, we propose a low-complexity procedure for efficient spectrum reuse in the considered CoMP setup. After introducing the relevant mathematical signal and interference models, the chapter continues with the description of the proposed linear precoding techniques. The subsequent section deals with symbol-level precoding. In cases where joint processing of data or/and control information is not possible, SU-MIMO may be utilized instead. Under this context, a precoding and power allocation technique for SU-MIMO terrestrial links that coexist with satellite links is presented. This technique takes into account the interference threshold at the satellite RX. Next, a proposed LC-PAA prototype design is described. The chapter concludes with a comparative performance evaluation of the studied techniques via numerical simulations that use realistic node topologies, operating parameters, antenna radiation patterns, and channel models. The simulations focus on the (bandwidth-normalized) capacity for various signal-to-noise-ratio (SNR) values, as the main performance metric of interest.

16.2 Load-controlled parasitic antenna arrays

LC-PAAs are hybrid analog-digital antenna systems that comprise a few active antenna elements (i.e., antennas that are fed by an RF chain) and a number of passive antenna elements. A single-fed antenna array of this type is also known as a load-controlled single-active multiple-passive (LC-SAMP) array, while for such antenna systems with multiple active elements we use the term load-controlled multiple-active multiple-passive (LC-MAMP).

In LC-PAAs, the parasitic antenna elements are placed deliberately in the vicinity of the active ones and are terminated to tunable capacitive or inductive loads. Due to the strong mutual coupling among the antenna elements, which is caused by the small interelement distance, the feeding voltages induce currents to the so-called parasitic elements. Thus, by adjusting the impedance of the parasitic loads with the help of a

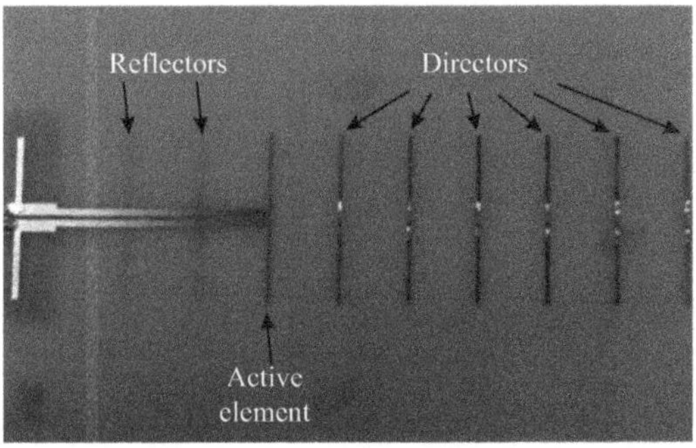

Figure 16.2 An example of a printed single-RF load-controlled parasitic antenna array with eight passive elements (planar Yagi–Uda with six directors on the front side and two reflectors on the back side)

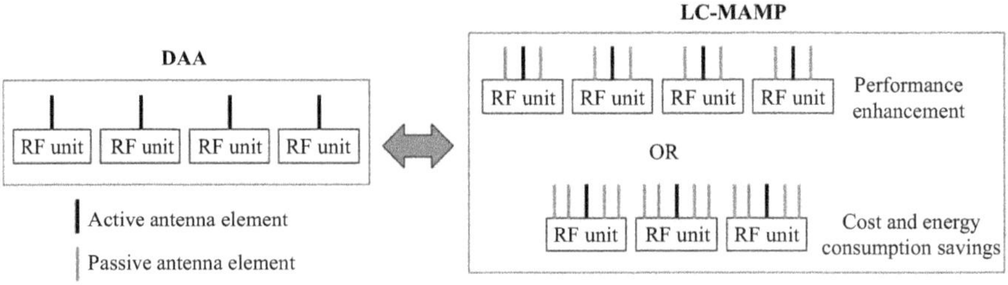

Figure 16.3 Benefits of LC-MAMP technology over DAA technology

low-cost digital control circuit, the amplitude and the phase of the antenna currents are controlled. Hence, we can dynamically shape and steer the far-field radiation pattern as desired, within the array's capability [21]. Figure 16.2 presents a single-RF LC-PAA (planar Yagi–Uda antenna).

LC-PAAs provide a higher number of array DoF than DAAs for a given number of RF modules. This is due to the additional effective DoF of the parasitic elements, which essentially give better control on the beamforming operation. Equivalently, such antenna arrays achieve a specific target number of array DoF with fewer RF units than DAAs. Hence, this technology leads to performance enhancement with minimal additional cost as well as to cost and energy consumption savings for a given performance level. Figure 16.3 presents schematically the benefits of a LC-MAMP array versus a DAA.

16.3 Robust arbitrary channel-dependent precoding method

In order to use the LC-PAA technology in single-cell and multicell MU-MIMO setups, we should be able to perform channel-dependent precoding at the TXs, which are equipped with such arrays, in order to manage the MUI.

Let us consider a LC-MAMP array with L active antenna elements and M antennas in total (active and passive elements). The relation between the currents and the voltages associated with the antenna elements of this array is given by the generalized Ohm's law [22]:

$$\mathbf{i} = (\mathbf{Z} + \mathbf{Z}_L)^{-1}\,\mathbf{v}, \tag{16.1}$$

where $\mathbf{i}$ is the $(M \times 1)$ vector of the currents that run on the antenna elements; $\mathbf{Z}$ is the $(M \times M)$ mutual coupling matrix whose diagonal entry Z_{mm} represents the self-impedance of the mth antenna element, while the off-diagonal entry Z_{mk} denotes the mutual impedance between the mth and the kth antenna element; $\mathbf{Z}_L$ is the $(M \times M)$ diagonal load matrix whose diagonal elements are the source resistances $R_1, \ldots, R_L$ and the impedances of the parasitic loads jX_m $(m = M - L + 1, \ldots, M)$, with $j = \sqrt{-1}$ denoting the imaginary unit; and $\mathbf{v}$ is the $(M \times 1)$ voltage vector that holds the L feeding voltages $v_1, \ldots, v_L$. Figure 16.4 illustrates the equivalent diagram of a LC-SAMP. In this special case $L = 1$, $V_1 = V_s$, $V_2 = \cdots = V_M = 0$ and $m = 2, \ldots, M$.

A $(M, (K, 1))$ MU MIMO setup formed between a TX with M transmit antennas and K single-antenna RXs is described, from an antenna point of view, by [22]

$$\mathbf{y} = \mathbf{H}\mathbf{i} + \mathbf{n}, \tag{16.2}$$

where $\mathbf{y}$ is the $(K \times 1)$ vector of open-circuit voltages at the receive antennas, $\mathbf{i}$ represents the $(M \times 1)$ vector of currents that run on the transmit antennas, $\mathbf{H}$ denotes the $(K \times M)$ composite channel matrix whose entry h_{km} relates the mth input current with the kth open-circuit output voltage, and $\mathbf{n}$ constitutes a $(K \times 1)$ additive white Gaussian noise vector with covariance matrix $\mathbf{R}_n = \sigma_n^2\mathbf{I}_K$, where σ_n^2 is the noise variance and $\mathbf{I}_K$ is the $(K \times K)$ identity matrix.

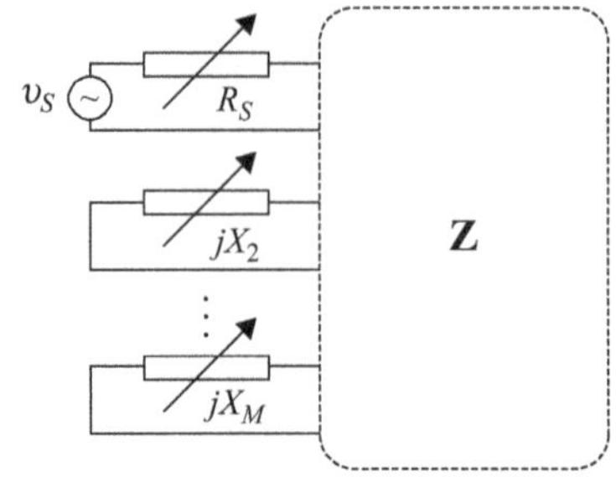

Figure 16.4 Equivalent circuit diagram of a single-RF load controlled parasitic antenna array

Assuming the application of channel-aware linear precoding, (16.2) is written as

$$\mathbf{y} = \mathbf{HWs} + \mathbf{n}, \tag{16.3}$$

where $\mathbf{W}$ is the $(M \times M)$ precoding matrix and $\mathbf{s}$ is the $(M \times 1)$ input signal vector. Hence, in order to apply channel-aware precoding to a LC-MAMP, we have to map the precoded symbols to the antenna currents [23],

$$\mathbf{i} = \mathbf{Ws}, \tag{16.4}$$

and then calculate the corresponding impedances (loading values) that can generate these currents. Additionally, we should ensure that the input resistance, which depends on the loads, is positive, in order to guarantee that the antenna system will not reflect power back [24] to the RF units. Since the loading values depend on the precoded signals, it becomes apparent that we cannot guarantee that this design condition will be met for arbitrary input signal constellation/precoding scheme combinations.

A workaround to this problem was proposed in [25]. In this work, the precoded signal is approximated by another one which leads to a feasible set of loading values, according to the previously mentioned design guideline, under the condition that the mean square error of the approximation is minimum. Nevertheless, this method is neither computationally light nor robust.

On the other hand, it is well known that the LC-PAAs can admit any input signal in transmit beamforming applications, since the desired array manifold does not depend on the given input. In this case, the loads play essentially the role of the beamforming weights, and the only restriction is that their impedance should lie within a range of feasible values. Based on this remark, an alternative approach that enables the performance of robust, low-complexity, arbitrary channel-aware precoding with such arrays can be followed [26]:

1. First, we apply transmit beamforming using any valid method.
2. Then, we perform channel-aware precoding over the employed beam(s).

In other words, by taking advantage of the radiation pattern's reconfiguration capabilities of the LC-PAAs through the decoupling of the problem to a beamforming and a precoding part, we overcome the circuit stability and implementation complexity issues associated with arbitrary precoding.

We should mention that this precoding method can be applied also to single-RF LC-PAAs when CoMP is utilized, since then the data signals can be centrally processed at the baseband units pool.

16.4 Low-complexity communication protocol for single-cell MU-MIMO/CoMP setups

The problem of dynamic load computation discourages the application of the LC-PAA technology. In terrestrial wireless MBH setups, though, where the nodes are fixed, we can overcome this issue by using a number of fixed loading sets that correspond to predetermined radiation patterns (i.e., beams) and switching through these sets instead of utilizing tunable loads.

In this section, we describe a communication protocol that takes into account the aforementioned implementation as well as the dynamic link establishment capability/radiation pattern reconfigurability of the BNs. This protocol can be applied in both single-cell MU-MIMO and coordinated multicell MU-MIMO setups. The system operation is divided in three phases [26]:

1. **Learning phase:** For each beam combination, the TX(s) sends a pilot signal. Then, the RXs measure their signal-to-interference-plus-noise-ratio (SINR) or estimate the gain of the direct and cross channels and report back this channel quality metric.
2. **Beam-selection phase:** After switching through all possible beam combinations, the TX(s) selects the optimum one, in terms of the achieved SR throughput, based on the information reported by the RXs.
3. **Transmission phase:** The TX(s) transmits precoded signals over the selected beams.

In Figure 16.5, we present an example of a CoMP setup, where each transmitting node selects one out of four possible beams.

Note that the use of SINR feedback has been included as a low-feedback alternative to the conventional channel state information (CSI) feedback procedure. Of course, after the SINR-feedback-based beam selection, a CSI-feedback phase for the selected composite "beam-channel" should follow, in order to enable the use of channel-dependent precoding.

16.5 Signal and interference modeling

16.5.1 SU-MIMO setup

Consider a SU-MIMO setup, where two different links are colocated and operate at the same frequency band. For example, these links may correspond to (a) one

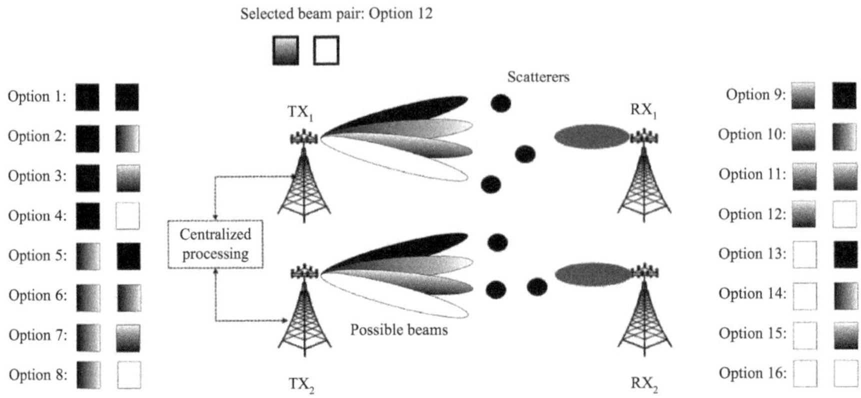

Figure 16.5 A CoMP setups comprised by two transmit and two receive BNs equipped with LC-PAAs. Each transmit LC-PAA can generate at each timeslot one out of four possible beam patterns. The best beam combination is selected jointly by the TX nodes based on SINR or CSI feedback from the RX nodes. Then, precoded transmission over these beams takes place

terrestrial and one satellite MBH link or (b) two terrestrial MBH links in a hybrid satellite-terrestrial MBH network. The first link is denoted by A-link, $TX_A - RX_A$ and the second one is the B-link, $TX_B - RX_B$. Next, we model the signals from each link, by taking into account the possible interference that is caused from the cross channels. As in standard MIMO setups, we assume that each TX/RX is equipped with an antenna array of multiple antennas. Let the A-link consist of k antenna (active) elements for the TX and ℓ for the RX, while the B-link consists of m antenna elements for the TX and n for its RX. The received signals at the A-link's RX are modeled as

$$\mathbf{y}_A = \mathbf{H}_A\mathbf{s} + \mathbf{H}_{BA}\mathbf{x} + \mathbf{n}, \tag{16.5}$$

and at the B-link's RX as

$$\mathbf{y}_B = \mathbf{H}_B\mathbf{x} + \mathbf{H}_{AB}\mathbf{s} + \mathbf{v}. \tag{16.6}$$

The transmitted signals for the A-link and B-link are denoted as $\mathbf{s} \in \mathbb{C}^k$ and $\mathbf{x} \in \mathbb{C}^m$, respectively, with zero mean complex Gaussian distribution; the channel gain from the jth TX to the ith RX's element is denoted as h_{ij}. Thus, for the channels of each link in Figure 16.6 we have $\mathbf{H}_A \in \mathbb{C}^{\ell\times k}$, $\mathbf{H}_B \in \mathbb{C}^{n\times m}$, $\mathbf{H}_{BA} \in \mathbb{C}^{\ell\times m}$, and $\mathbf{H}_{AB} \in \mathbb{C}^{n\times k}$ and are assumed fixed and frequency flat. We have also considered that $\mathbf{n} \sim \mathcal{CN}(0, \sigma_n^2\mathbf{I}_n)$ and $\mathbf{v} \sim \mathcal{CN}(\mathbf{0}, \sigma_v^2\mathbf{I}_\ell)$ are additive white circularly complex Gaussian noise processes. Both the signals and the noise are assumed uncorrelated with each other.

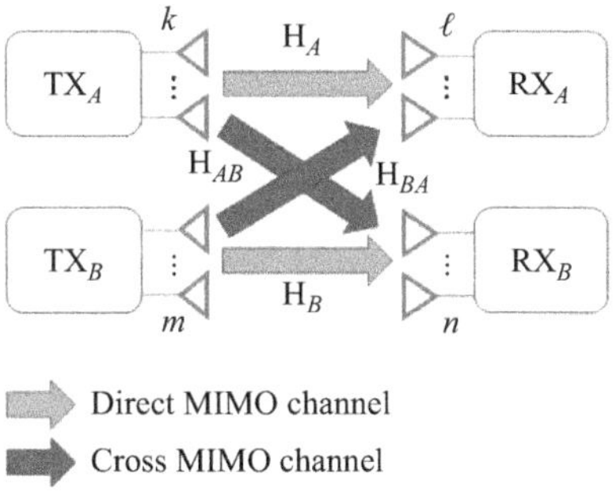

Figure 16.6 An example of two collocated MIMO terrestrial links and the corresponding direct-channels and cross-channels

16.5.2 Single-cell MU-MIMO/JT CoMP setup

The input–output relationship of a $(K, (K, 1))$ MIMO BC formed between a TX with $M = K$ transmit antennas (which might be a composite TX formed in a JT CoMP scenario) and K single-antenna RXs, assuming that linear precoding is utilized and that the channel is modeled as narrowband and quasistatic, is given by

$$y_k = \mathbf{h}_k^\dagger \left(\sum_{m=1}^{K} \mathbf{w}_m \sqrt{p_m} s_m \right) + n_k, \quad k = 1, 2, \ldots, K \tag{16.7a}$$

$$\mathbf{y} = \mathbf{HWP}^{1/2}\mathbf{s} + \mathbf{n}, \tag{16.7b}$$

where $\mathbf{y}$ is the $(K \times 1)$ vector of received signals y_k; $\mathbf{H}$ denotes the $(K \times K)$ channel matrix, whose rows $\mathbf{h}_k$ are $(1 \times K)$ vectors that hold the channels h_{km} between the kth RX and each one of the K transmit antennas; $\mathbf{W}$ represents the $(K \times K)$ precoding matrix, whose column $\mathbf{w}_k$ is the $(K \times 1)$ BF vector for the kth RX; $\mathbf{P}$ is the $(K \times K)$ power allocation matrix, whose element p_k is the power allocated to the kth RX; $\mathbf{s}$ refers to the $(K \times 1)$ symbol vector, with s_k being the data symbol intended for the kth RX; and $\mathbf{n}$ is the additive noise vector, whose elements n_k represent the noise at the kth RX.

The SINR at the kth RX is expressed as

$$\mathrm{SINR}_k = \frac{\left| \mathbf{h}_k^\dagger \mathbf{w}_k \right|^2 p_k}{\sum_{m \neq k} \left| \mathbf{h}_k^\dagger \mathbf{w}_m \right|^2 p_m + \sigma_n^2}, \quad k = 1, 2, \ldots, K. \tag{16.8}$$

The data rate of the kth RX is given by

$$R_k = \log_2(1 + \mathrm{SINR}_k), \tag{16.9}$$

and the SR throughput is

$$R = \sum_{i=1}^{K} R_k = \sum_{i=1}^{K} \log_2(1 + \mathrm{SINR}_k). \tag{16.10}$$

16.6 Joint precoding schemes

The application of precoding requires CSI at the TX. The capacity-achieving strategy for the MIMO-BC is dirty paper coding (DPC), a nonlinear multiuser precoding scheme that exploits the noncausal knowledge of all the data to subtract the interference prior to transmission [13]. Despite its optimality, though, this precoding method is rarely used in practice due to its high computation complexity, especially for high number of users.

Linear precoding methods are suboptimal alternatives that present a reasonable trade-off between performance and complexity. They constitute multiuser beamforming techniques that strike a balance between focusing the signal power at the intended users (and, thus, maximizing the received SNR) and reducing the interference toward nonintended users [27]. In the ideal case, this statement is translated as using beamforming vectors that match the channel vectors of the intended users and are orthogonal to the channel vectors of the non-intended users. However, it is very difficult to meet both objectives in practice, due to the amount of required DoF. Therefore, typically simple heuristics are utilized. We should note that for a given linear precoding scheme, the optimization of the system's performance, in terms of the achieved average SR throughput, depends on the applied user scheduling and power-allocation methods.

The precoding schemes that are employed in CoMP are typically generalizations/extensions of the linear precoding methods that are utilized in single-cell MU-MIMO.

So far, we have implicitly assumed the utilization of user-level linear precoding methods which aim at reducing or even eliminating the CCI. However, as is shown in [28], at symbol level, the CCI may be constructive in some cases instead of destructive, in the sense that it may increase the received SNR at no expense of transmit power. Several symbol-level extensions of the user-level linear precoding schemes have been studied in the literature (e.g., [28]).

16.6.1 Linear precoding schemes

Zero-forcing beamforming (ZFBF) is a characteristic example of user-level linear precoding schemes. This precoding technique makes use of beamforming vectors that are orthogonal to the subspace of other users' channel vectors (i.e., their inner product with these channel vectors is null), in order to eliminate the MUI:

$$\left\| \mathbf{h}_k^{\dagger} \mathbf{w}_m^{(\mathrm{ZF})} \right\|^2 = 0, \quad k, m = 1, 2, \ldots, K, \; m \neq k. \tag{16.11}$$

The zero-forcing condition implies the use of the Moore–Penrose pseudoinverse of the composite channel matrix as the precoding matrix:

$$\mathbf{F}^{(\mathrm{ZF})} = \mathbf{H}^{+} = \mathbf{H}^{\dagger}\left(\mathbf{H}\mathbf{H}^{\dagger}\right)^{-1}. \tag{16.12a}$$

$$\mathbf{W}^{(\mathrm{ZF})} = \frac{\mathbf{F}^{(\mathrm{ZF})}(\,:, k)}{\|\mathbf{F}^{(\mathrm{ZF})}(\,:, k)\|}, \quad k = 1, 2, \ldots, K. \tag{16.12b}$$

This precoding scheme attains a significant portion of the DPC capacity in the high SNR regime, especially when single-antenna RXs are utilized [13]. Also, it approaches the capacity as the number of users grows toward infinity, since in this case, user selection benefits from the abundance of spatial directions and the multiuser diversity effect (i.e., users that have both sufficient spatial separation and high-gain channels are scheduled). The main drawback of ZFBF is that it is power-inefficient, since the beamforming vectors do not match to the users' channels. Thus, the performance of ZFBF deteriorates at low SNR values.

Regularized ZFBF (R-ZFBF) is an extension of ZFBF that introduces a controllable amount of MUI. The value of the coefficient that controls the level of the residual MUI is typically set such that the SINR at the users is maximized. More specifically, in R-ZFBF, we have

$$\mathbf{v}_k^{(\mathrm{RZF})} = \mathbf{H}^{\dagger}\left(\frac{1}{p_k}\mathbf{I}_K + \mathbf{HH}^{\dagger}\right)^{-1}, \quad k = 1, 2, \ldots, K. \tag{16.13a}$$

$$\mathbf{w}_k^{(\mathrm{RZF})} = \frac{\mathbf{v}_k^{(\mathrm{RZF})}}{\|\mathbf{v}_k^{(\mathrm{RZF})}\|}, \quad k = 1, 2, \ldots, K. \tag{16.13b}$$

R-ZFBF is asymptotically optimal at both low and high SNR and performs reasonably well at intermediate SNR values. Moreover, due to the regularization of the channel, it is more robust against pathological situations such as the existence of ill-conditioned channel matrices, where channel inversion without regularization might be problematic. However, the introduction of residual MUI complicates the power allocation procedure.

16.6.2 Symbol-level precoding

In [28], a symbol-level variant of the ZFBF scheme is presented, which is called constructive-interference ZFBF (CI-ZFBF). Similar to DPC, CIZF takes advantage of the availability of all data symbols at the BS prior to downlink transmission in order to predict the interference and "zero-force" only the destructive one, while leaving the CI unaffected [28].

Consider a K-user multiple-input single-output system. Let us define the $(K \times K)$ channel cross-correlation matrix $\mathbf{R}$ as [28,29]

$$\mathbf{R} = \mathbf{HH}^{\dagger}. \tag{16.14}$$

The symbol-to-symbol CCI from s_k to s_m is then expressed as

$$\mathrm{CCI}_{km} = s_k \rho_{km}, \quad k, m = 1, 2, \ldots, K, \; m \neq k \tag{16.15}$$

while the cumulative CCI on s_k from all symbols is given by

$$\mathrm{CCI}_k = \sum_{k=1}^{K} s_k \rho_{km}, \quad m = 1, 2, \ldots, K, \; m \neq k \tag{16.16}$$

where

$$\rho_{km} = \frac{\mathbf{h}_k \mathbf{h}_m^\dagger}{\|\mathbf{h}_k\| \, \|\mathbf{h}_m\|} \tag{16.17}$$

is the (k, m)th element of $\mathbf{R}$ that represents the cross-correlation factor between the kth user's channel and the mth transmitted data stream.

In CI-ZFBF, the precoding matrix has the following form [28,29]:

$$\mathbf{W}^{(\mathrm{CIZF})} = \mathbf{W}^{(\mathrm{ZF})}\mathbf{T} = \mathbf{H}^\dagger \mathbf{R}^{-1}\mathbf{T}. \tag{16.18}$$

The received signal at the kth user is given by

$$y_k = \tau_{kk}\sqrt{p_k}s_k + \sum_{m \neq k} \mathrm{CI}_{km} + n_k, \quad k, m = 1, 2, \ldots, K \tag{16.19}$$

where $\mathrm{CI}_{km} = \tau_{km}\sqrt{p_m}s_m$ denotes the constructive CCI from the mth user to the kth user, and τ_{km} is the (k, m) element of the $K \times K$ matrix $\mathbf{T}$. Then, the kth user's SINR is given by

$$\mathrm{SINR}_k^{(\mathrm{CIZF})} = \sum_{m=1}^{K} |\tau_{km}|^2 p_m, \quad k = 1, 2, \ldots, K. \tag{16.20}$$

$\mathbf{T}$ is calculated on a symbol-by-symbol basis as follows [28,29]: First, $\mathbf{R}$ is calculated according to (16.14) and next, assuming for simplicity, the use of binary phase shift keying (BPSK) modulation (i.e., $s_k = \pm 1, k = 1, 2, \ldots, K$), the $(K \times K)$ matrix $\mathbf{G}$ is computed as

$$\mathbf{G} = \mathrm{diag}(\mathbf{s})\mathbf{R}\mathrm{diag}(\mathbf{s}). \tag{16.21}$$

Then, $\tau_{kk} = \rho_{kk}$ and $\tau_{km} = 0$ if $g_{km} < 0$ or $\tau_{km} = \rho_{km}$ otherwise.

Since we do not have a Gaussian input but a finite-alphabet one, we do not calculate the SR capacity through the Shannon formula, but we use instead the following relationship [28,29]:

$$R = (1 - \mathrm{BLER})m, \tag{16.22}$$

where $m = 1$ symbol for BPSK and the block error rate (BLER) is given by $\mathrm{BLER} = 1 - (1 - P_e)^{N_f}$, with P_e being the symbol error rate (SER) of BPSK and N_f being the frame size.

The generalization of the previous analysis to higher order modulation schemes is straightforward. For instance, if we assume quadrature PSK (QPSK), we should replace (16.21) with

$$\mathrm{Re}\{\mathbf{G}\} = \mathrm{Re}\{\mathrm{diag}(\mathbf{s})\mathbf{R}\}\mathrm{Re}\{\mathrm{diag}(\mathbf{s})\}. \tag{16.23}$$

$$\mathrm{Im}\{\mathbf{G}\} = \mathrm{Im}\{\mathrm{diag}(\mathbf{s})\mathbf{R}\}\mathrm{Im}\{\mathrm{diag}(\mathbf{s})\}. \tag{16.24}$$

Also, in (16.22), $m = 2$ and P_e refers to the SER of QPSK in this case.

We should mention that symbol-level precoding is typically utilized in conjunction with low-order modulation schemes, such as BPSK and QPSK. This is because these modulation schemes are commonly used when the channel quality is not very good, and this is exactly the case where the effect of increasing the receive SNR without wasting transmit power becomes more prominent.

16.7 Optimal transmission technique under an interfered receiver constraint

In this section, we provide an optimal transmission technique that can be used in cases where cooperative transmission/interference coordination schemes do not apply. Our goal is to *maximize the capacity of the A-link under the interference constraint imposed by coexistence with the B-link*, according to the architecture of Figure 16.6. The approach comprises a power allocation strategy and its precoding technique for a MIMO link, under the presence of an interfered RX constraint, which is typical of spectrum sharing setups operating at the same frequency band. The problem is formulated and solved via convex optimization techniques. The algorithm maximizes the mutual information of one link, while satisfying an interference constraint. While the derivation presented below was derived independently for the problem at hand, we realized recently that the solution is equivalent to one presented much earlier in [30] in the context of cognitive radio.[1] Due to its large potential in various applications, we believe that the method plays an important role in the adoption of spectrum sharing/coexisting techniques in next generation communications networks.

16.7.1 Problem formulation

Let us denote the second part of (16.5) as $\mathbf{z} = \mathbf{H}_{BA}\mathbf{x} + \mathbf{n}$. The covariance matrix of the signal received by the A-link's RX can be written as

$$\mathbf{R}_{\mathbf{y}_s} := \mathrm{E}\{\mathbf{y}_s\mathbf{y}_s^\dagger\} = \mathbf{H}_A\mathbf{R}_s\mathbf{H}_A^\dagger + \mathbf{R}_z, \tag{16.25}$$

where $\mathbf{R}_s$ is the covariance matrix of the A-link's transmitted signal; $\mathbf{R}_z$ is the covariance matrix of the vector $\mathbf{z}$, i.e., $\mathbf{R}_z = \mathbf{H}_{BA}\mathbf{R}_x\mathbf{H}_{BA}^\dagger + \sigma_n^2\mathbf{I}_n$, and $\mathbf{R}_x$ is the covariance matrix of the B-link's transmitted signal. Our goal is to maximize the mutual information of the A-link. In our analysis, and with respect to Figure 16.6, we have considered the following:

- The covariance matrix, $\mathbf{R}_x$, of the B-link's transmitted signal is known to the system.
- The channels $\mathbf{H}_A$, $\mathbf{H}_{AB}$, and $\mathbf{H}_{BA}$, are known to the system.
- In order to simplify the analysis, we have also considered that $\sigma_n = 1$.

[1]The main difference between the two derivations is that ours models explicit the spatial color of the interference, while in [30], the authors assume a prewhitened interference vector.

The maximum *mutual information* of the A-link (disregarding any constraint on the interference caused to RX_A), see [31], is given by

$$\begin{aligned} I(\mathbf{y}_s;\mathbf{s}) &= \log_2 \det\left(\pi e \mathbf{R}_{\mathbf{y}_s}\right) - \log_2 \det(\pi e \mathbf{R}_{\mathbf{z}}) \\ &= \log_2 \det(\mathbf{I}_m + \mathbf{H}_A^{\dagger}\mathbf{R}_z^{-1}\mathbf{H}_A\mathbf{R}_s), \end{aligned} \tag{16.26}$$

for $n \geq m$. At this point, we consider the eigendecomposition of matrix $\mathbf{H}_A^{\dagger}\mathbf{R}_z^{-1}\mathbf{H}_A$, i.e.,

$$\mathbf{H}_A^{\dagger}\mathbf{R}_z^{-1}\mathbf{H}_A = \mathbf{U}\boldsymbol{\Lambda}\mathbf{U}^{\dagger}, \tag{16.27}$$

where $\mathbf{U}$ is a unitary matrix and $\boldsymbol{\Lambda}$ is the diagonal matrix with the positive eigenvalues, i.e., $diag(\lambda_1, \ldots, \lambda_r)$, where r is the rank of the decomposed matrix $\boldsymbol{\Lambda}$. Therefore, by imposing the A-link's transmitted signal to be of the form $\mathbf{s} = \mathbf{U}\mathbf{s}_w$, where $\mathbf{s}_w$ is spatially white, leads to $\mathbf{R}_s = \mathrm{E}\left(\mathbf{s}\mathbf{s}^{\dagger}\right) = \mathbf{U}\mathbf{D}\mathbf{U}^{\dagger}$, where $\mathbf{D} = \mathrm{E}(\mathbf{s}_w\mathbf{s}_w^{\dagger})$ is diagonal. Thus, (16.26) is simplified to

$$I(\mathbf{y}_s;\mathbf{s}) = \log_2 \det(\mathbf{I}_r + \boldsymbol{\Lambda}\mathbf{D}). \tag{16.28}$$

Hence, the standard mutual information maximization task for the A-link's transmitted signal is given by

$$\underset{\mathbf{D}}{\text{maximize}} \quad \log_2 \det(\mathbf{I}_r + \boldsymbol{\Lambda}\mathbf{D}) \tag{16.29a}$$

$$\text{subject to} \quad \mathbf{D} \succeq 0, \tag{16.29b}$$

$$tr(\mathbf{D}) \leq 1, \tag{16.29c}$$

where without loss of generality (avoiding an equivalent normalization), we have considered that the maximum transmission power of the A-link's MIMO antenna array is 1. The optimization task in (16.29a)–(16.29c) admits the standard water-filling solution,[2] which is given by

$$d_i = (\rho - \lambda_i^{-1})^{+}, \quad i = 1, \ldots, r, \tag{16.30}$$

where ρ is the water-level chosen to satisfy the power constraint with equality, i.e., $\sum_{i=1}^{r} d_i = 1$.

However, in the presence[3] of the B-link, i.e., $TX_B - RX_B$, transmission of TX_A with a power level imposed by (16.30) may cause harmful interference to the B-link's RX. In order to avoid causing excessive interference to RX_B, an additional constraint should be satisfied, which can be expressed in view of (16.6) as

$$\mathrm{tr}\left(\mathbf{H}_{AB}\mathbf{R}_s\mathbf{H}_{AB}^{\dagger}\right) = \mathrm{tr}\left(\widetilde{\mathbf{H}}_{AB}\mathbf{D}\widetilde{\mathbf{H}}_{AB}^{\dagger}\right) \leq P_I, \tag{16.31}$$

where $\widetilde{\mathbf{H}}_{AB} = \mathbf{H}_{AB}\mathbf{U}$ and $P_I > 0$ is the maximum value of interference that is tolerable to the B-link's RX, due to TX_A. Thus, our goal now is to find a solution for (16.29a)–(16.29c) under the additional constraint in (16.31).

[2]The task can be equivalently transformed to a convex optimization one (since the cost function is concave); thus, a unique solution exists.

[3]We assume that the two links interfere with each other, due to coexistence of their RXs as in.

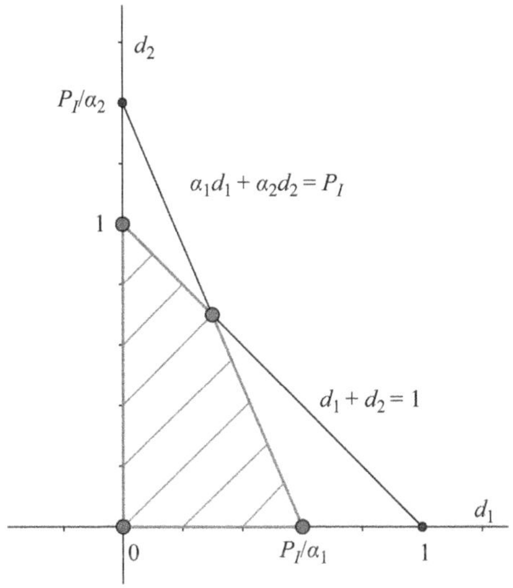

Figure 16.7 Representation of the optimization task's feasible region for $r = 2$

16.7.2 Derivation of the solution

The new optimization power allocation with the additional interference constraint task is now formulated as

$$\underset{d_i}{\text{minimize}} \quad -\sum_{i=1}^{r} \log_2(1 + \lambda_i d_i) \tag{16.32a}$$

$$\text{subject to} \quad d_i \geq 0,\ i = 1, \ldots, r, \tag{16.32b}$$

$$\sum_{i=1}^{r} d_i \leq 1, \tag{16.32c}$$

$$\sum_{i=1}^{r} \alpha_i d_i \leq P_I, \tag{16.32d}$$

where $\alpha_i = \left\| \tilde{\mathbf{h}}_i \right\|_2^2$, $i = 1, \ldots, r$ is the squared norm of the column vectors of matrix $\widetilde{\mathbf{H}}_{AB}$. The objective function in (16.32a) is convex, and the constraints in (16.32b)–(16.32d) define a polyhedron, as demonstrated in Figure 16.7 for $r = 2$. Thus, the optimization task is convex; hence, it attains a unique minimum.

16.7.2.1 Optimality conditions

For the solution of this convex optimization task, we use the Karush–Kuhn–Tucker (KKT) conditions (also known as optimality conditions), see [32,33]. In order to maximize the capacity, (16.32c) should be satisfied with equality. Let ν denote the

Lagrange multiplier corresponding to the constraint of (16.32c), μ the Lagrange multiplier corresponding to the constraint of (16.32d), and $\xi_1, \ldots, \xi_r$ denote the Lagrange multipliers corresponding to the constraints that force the powers to be positive. The Lagrangian form for the solution of the optimization task (16.32a)–(16.32d) is

$$L(d_i; \nu, \mu, \xi_i) = -\sum_{i=1}^{r} \log_2(1 + \lambda_i d_i) + \nu\left(\sum_{i=1}^{r} d_i - 1\right) + \mu\left(\sum_{i=1}^{r} \alpha_i d_i - P_I\right) - \sum_{i=1}^{r} \xi_i d_i, \tag{16.33}$$

where $\nu,\ \mu,\ \xi_i,\ i = 1, \ldots, r$ are the Lagrange multipliers associated with the constraints. Hence, at the minimum, we imply that (16.32b) and (16.32d) hold, as well as

$$\xi_i \geq 0, \text{ for all } i = 1, \ldots, r \tag{16.34a}$$

$$\xi_i d_i = 0,\ i = 1, \ldots, r, \tag{16.34b}$$

$$\sum_{i=1}^{r} d_i = 1, \tag{16.34c}$$

$$\mu \geq 0 \tag{16.34d}$$

$$\mu\left(\sum_{i=1}^{r} \alpha_i d_i - P_I\right) = 0, \tag{16.34e}$$

$$\frac{\lambda_i \log_2 e}{(1 + \lambda_i d_i)} + \xi_i = \nu + \mu\alpha_i,\ i = 1, \ldots, r \tag{16.34f}$$

By observing (16.34f), we first notice that $\nu + \mu\alpha_i > 0$, since $\lambda_i > 0$.

16.7.2.2 Solution

Next, we provide the solution to the power allocation task under the interference constraint.

Restriction 1: From (16.34b), it is observed that if $d_i > 0$, then $\xi_i = 0$. Thus, according to (16.34d) and (16.34f), we have the restriction that $\log_2 e/(\lambda_i^{-1} + d_i) - \nu \geq 0$, which leads to $\lambda_i^{-1} < \log_2 e/\nu$.

Restriction 2: If $\lambda_i^{-1} \geq \log_2 e/\nu$, then from the derived inequality (Restriction 1), we obtain that $d_i = 0$.

Thus, the derived solution of the first stage is given by (16.30) for $\rho = \log_2 e/\nu$. Next, we should differentiate between the two following cases in the power allocation:

- Case 1: If $\sum_{i=1}^{r} \alpha_i \left(\frac{\log_2 e}{\nu} - \frac{1}{\lambda_i}\right)^+ \leq P_I$, then the power allocation $d_i = \left(\left((\log_2(e))/\nu\right) - (1/\lambda_i)\right)^+,\ i = 1, \ldots, r$ is a valid solution that satisfies all the KKT conditions. It should also be noted that in this case, $\mu = 0$.

- Case 2: If $\sum_{i=1}^{r}\alpha_i\left((\log_2 e/\nu)-(1/\lambda_i)\right)^+ > P_I$, then $\mu > 0$ and thus according to (16.34e), we have $\sum_{i=1}^{r}\alpha_i d_i = P_I$. Thus, two options exist:
  - If $\lambda_i^{-1} \geq \log_2 e/(\nu+\mu\alpha_i)$, $d_i = 0$. This is proved by contradiction, since if we assume $d_i > 0$, it would lead to $\xi_i = 0$ and thus $\lambda_i^{-1} < \log_2 e/(\nu+\mu\alpha_i)$.
  - If $\lambda_i^{-1} < \log_2 e/(\nu+\mu\alpha_i)$, $\xi_i = 0$. This can also be proved by contradiction. Let us instead assume that $\xi_i > 0$. Thus, from (16.34b) $d_i = 0$ and according to (16.34f) leads to $\lambda_i^{-1} > \log_2 e/(\nu+\mu\alpha_i)$, which is a contradiction.

  Summarizing Case 2, the solution of the second equality is given by

$$d_i = \left(\frac{\log_2 e}{\nu+\mu\alpha_i}-\frac{1}{\lambda_i}\right)^+, \; i=1,\ldots,r, \tag{16.35}$$

where μ is obtained from

$$\sum_{i=1}^{r}\alpha_i\left(\frac{\log_2 e}{\nu+\mu\alpha_i}-\frac{1}{\lambda_i}\right)^+ = P_I. \tag{16.36}$$

It should be noted that the solution to (16.36) cannot be obtained in closed form; however, it can be solved iteratively. Existence and uniqueness of the its solution is derived in Section 16.7.2.3.

According to the aforementioned analysis, we provide the following result for the solution of our the constrained optimization task.

Theorem 16.1. *The solution to the optimization task* (16.32a)–(16.32d) *is*

$$d_i = \begin{cases} \left(\frac{\log_2 e}{\nu}-\frac{1}{\lambda_i}\right)^+, & \text{if } \sum_{i=1}^{r}\alpha_i\left(\frac{\log_2 e}{\nu}-\frac{1}{\lambda_i}\right)^+ \leq P_I, \\ \left(\frac{\log_2 e}{\nu+\mu\alpha_i}-\frac{1}{\lambda_i}\right)^+, & \text{otherwise} \end{cases}, \tag{16.37}$$

where the Lagrange multipliers are obtained from the two-stage procedure. First, ν is obtained by solving (16.34c) *and, if required, μ is obtained by solving* (16.36) *with the value of ν obtained from the previous stage.*

It should be noted that for Case 1, the value $\rho = \log_2 e/\nu$ can be interpreted as the standard water level of the water-filling power allocation method. However, for Case 2, the initial water level violates the second condition, i.e., (16.32d) and the initial water level is penalized by the term $\mu\alpha_i$, which is different for each channel, since it depends on α_i's. Moreover, it can be readily seen that, for the new power level and the ν obtained at the first stage, $\sum_{i=1}^{r} d_i < 1$, for any $\mu > 0$.

16.7.2.3 The algorithm

The established iterative scheme for the power allocation task under the B-link's RX interference constraint is presented in Algorithm 1. It should be noted that this is a generic method, whose standard water-filling algorithmic part is only a special case. For coexisting links, which is the case of our interest, the interference constraint is not necessarily satisfied by the standard water-filling solution.

Algorithm 1: Interference constrained water-filling algorithm

1: **procedure** ICWF(λ_i, α_i, P_I)
2: $\quad d_i = \left(\frac{\log_2 e}{\nu} - \frac{1}{\lambda_i}\right)^+$, where ν is obtained from $\sum_{i=1}^{r} d_i = 1$
3: $\quad$ **if** $\sum_{i=1}^{r} \alpha_i d_i > P_I$ **then** $p \leftarrow 1$
4: $\quad\quad$ **while** $p \leq r$ **do**
5: $\quad\quad\quad \gamma_p = \left(P_I + \sum_{i=1}^{r-p+1} \left(\frac{\alpha_i}{\lambda_i}\right)\right) / \log_2 e$
6: $\quad\quad\quad$ Find μ as the solution of $g_p(\mu) = 0$ in (16.38)
7: $\quad\quad\quad$ Compute: $d_i = \left(\frac{\log_2 e}{\nu + \mu\alpha_i} - \frac{1}{\lambda_i}\right)^+$
8: $\quad\quad\quad p \leftarrow p + 1$
9: $\quad$ **Output:** d_i, for $i = 1, \ldots, r$

At the first stage, the algorithm computes a ν, which is related to a specific water level, according to the standard water-filling solution. At the second stage, a decision is taken; the derived solution can either satisfy the interference power constraint or not. In the latter case, given the ν that is already computed, the algorithm computes a $\mu > 0$ from (16.36), which is equivalent to obtaining the root of the following function:

$$g_p(\mu) := \sum_{i=1}^{r-p+1} \frac{\alpha_i}{\nu + \mu\alpha_i} - \gamma_p, \tag{16.38}$$

for $p = 1, \ldots, r$, where γ_p is given in the fifth row of Algorithm 1. At this point, one should notice that the function g_p is strictly decreasing for $\mu \geq 0$. Moreover, $g_p(0) > 0$ (Case 2 of Section 16.7.2.2) and $\lim_{\mu\to\infty} g_p(\mu) = -\gamma_p < 0$. Thus, $g_p(\mu) = 0$ has a unique solution for every ν obtained from the first stage of the algorithm, which can be derived via an iterative method, such as the bisection or the Newton's method.

16.8 Proposed LC-MAMP design

In Figure 16.8, we present a novel LC-MAMP design for the considered hybrid terrestrial-satellite MBH. It is based on a bowtie patch antenna operating at 19.25 GHz. Each antenna element has a small gap between its two branches. In this gap, either a port or a load (capacitor or inductor) is placed. For the case of the port (SMA connector), the antenna element is considered to be active (denoted with a dark grey circle), while for the case of the load, it is considered to be passive (denoted with a light grey circle). There are 4 active and 40 parasitic elements in total, in the specific example.

The LC-MAMP array was designed and simulated with an electromagnetic analysis simulation software that implements the finite element method. In Figure 16.9, we present the three main view planes of the 3D far field radiation pattern of the MAMP antenna, provided that only one antenna is active at any given time.

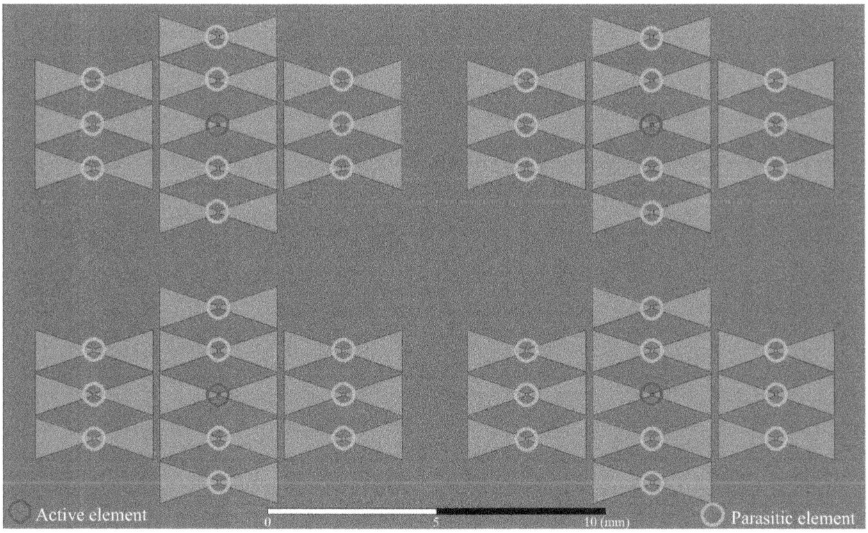

Figure 16.8 LC-MAMP with 4 active and 40 parasitic elements. The array lays on the X–Y plane. The Z axis is pointing toward the reader

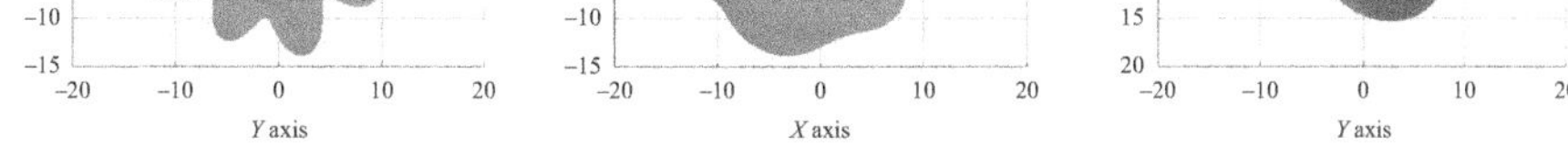

Figure 16.9 3D far field radiation patterns. Left to right: Y–Z, X–Z, X–Y cut plane

16.9 Numerical simulations

16.9.1 SU-MIMO setup

For the numerical evaluation of the SU-MIMO setup, we have performed two sets of experiments. First, we consider two 3 × 3 MIMO backhaul terrestrial interfering links; next, we consider a hybrid setup, consisting of a 4 × 4 MIMO backhaul terrestrial link that coexists with a satellite SISO one. All MBH links (terrestrial and hybrid) operate at 19 GHz, and transmitting at a narrow band. For both setups, we compare the capacity obtained from Algorithm 1 with the standard water-filling power allocation and evaluate the capacity loss (percentage) for various values of the interference constraint. Moreover, we evaluate the cumulative distribution function

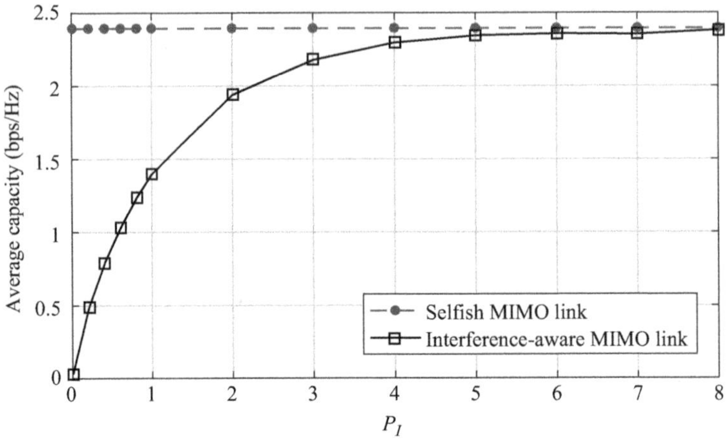

Figure 16.10 Capacity of a 3 × 3 MIMO link for various values of the interference constraint P_I. The dashed line corresponds to the capacity of a selfish (unconstrained) link (standard water-filling power allocation), while the solid one corresponds to the capacity of the interference-aware method, which guarantees that the interference constraint is satisfied for the second terrestrial link (B-link) of the MBH

(CDF) for various interference threshold values. For the entire set of simulations, we have considered omnidirectional antennas (for each active element of the array) at the TX/RX, whereas for the satellite link, we have considered a horn antenna of 10 dB gain for the satellite TX and a dish of 30 dB gain for the terrestrial RX [14]. For the latter link, apart from the antenna gains, we have also considered the free-space path loss. Moreover, we have assumed Rayleigh fading for the channels and for each value of interference constraint, P_I, we have performed 10,000 Monte Carlo runs and averaged the results.

16.9.1.1 Intrasystem interference scenario

For the evaluation of the derived power-allocation technique, we perform the following experiment. We consider two 3 × 3 MIMO backhaul terrestrial links equipped with omnidirectional antennas, the A-link and the B-link ($n, m, k, \ell = 3$) and attempt to maximize the A-link's capacity. In Figure 16.10, we have evaluated the capacity for various values of the interference constraint, P_I.[4] The dashed (red) line corresponds to the capacity according to the selfish power allocation (standard water-filling solution), i.e., without the consideration of the interference constraint ($P_I = \infty$). The solid (black) line corresponds to the achieved capacity, according to the interference-aware

[4]Due to the chosen normalization, we have considered that the power sums up to $P = 1$; however, if the sumpower constraint was chosen equal to $P \neq 1$, one should measure the capacity for different values of the ratio P_I/P.

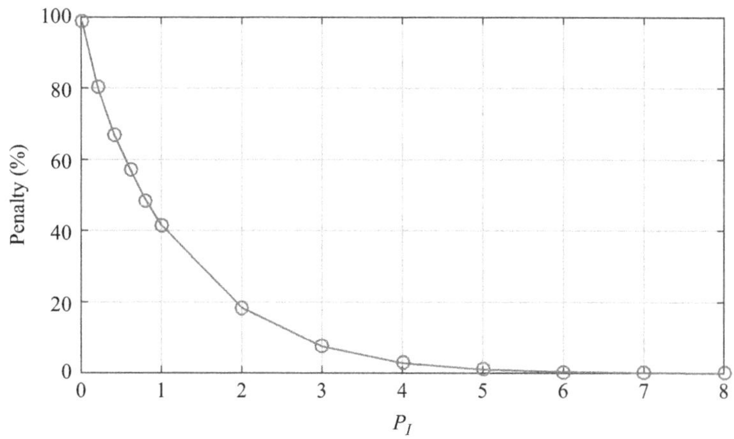

Figure 16.11 The penalty corresponds to the percentage of the capacity loss of a 3 × 3 terrestrial MIMO link for various values of the interference threshold P_I

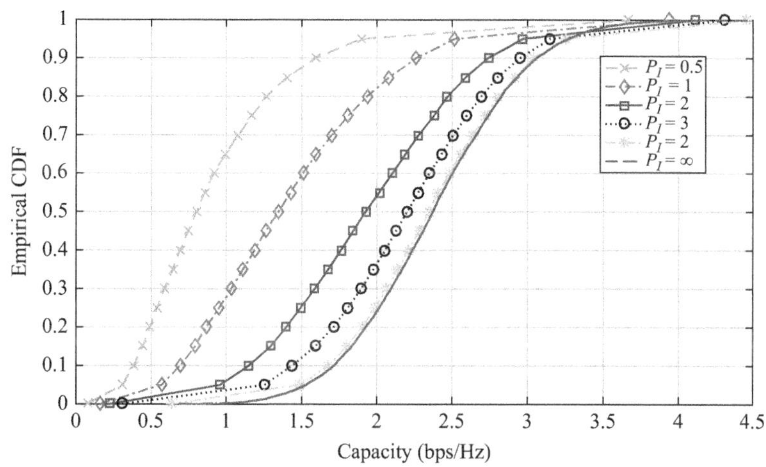

Figure 16.12 Empirical CDF of a 3 × 3 MIMO link for various values of the interference threshold P_I. The dashed line corresponds to the interference constraint-free case ($P_I = \infty$)

power allocation scheme in Algorithm 1, under the B-link's interference constraint. Moreover, in Figure 16.11, we have computed the percentage of capacity loss for the A-link that is caused due to the interference constraint of the B-link.

Finally, in Figure 16.12, we present the empirical CDF for each interference value P_I. The dashed line corresponds to the selfish (unconstrained) power allocation. It is observed that for small values of P_I, the CDFs are far from the ideal case of the unconstrained task.

From our analysis for the case of terrestrial links, it is clear that a tighter constraint translates to a greater penalty. However, the A-link is guaranteed to coexist, under agreed spectrum sharing rules, with the B-link, without causing excessive interference to the latter. It should be noted that, we do not claim that no interference is caused to the B-link, but that it is kept under a certain threshold, which in turn guarantees the A-link's QoS.

16.9.1.2 Intersystem interference scenario

In the current section, the interference-aware power allocation technique is evaluated for the case where the terrestrial MBH link coexists with a satellite one. This can be viewed as links with collocated RXs.

We consider the desired A-link to be a 4×4 MIMO one ($n, m = 4$), equipped with omnidirectional antennas. The distance between the TX_A and RX_A is 500 m. For the interfering B-link, which we aim to protect, we have considered a satellite SISO link ($k, l = 1$) with antennas of 40 dB gain, operating on the same frequency band. The link corresponds to a GEO satellite, which is in orbit at the height of 35,786 km. The satellite ground terminal, RX_B, is collocated (in the x, y plane) with the RX_A at a height of 10 m above the latter RX.

In Figure 16.13, we have evaluated the achieved capacity for various values of interference constraint P_I. The dashed line corresponds to the achieved capacity according to the selfish power allocation, i.e., without an interference constraint ($P_I = \infty$), and the solid line corresponds to the capacity achieved by the interference-aware power allocation of Algorithm 1. Moreover, in Figure 16.14, we have computed the percentage of capacity loss of the A-link that is caused by the interference constraint imposed by the satellite link. Finally, in Figure, 16.15, we present the empirical CDF for various values of interference, P_I. The dashed (red) line corresponds to the selfish power allocation. It is observed that almost all CDFs are closer to the ideal case of the unconstrained standard water-filling power allocation for this hybrid setup.

By comparing Figure 16.13 with Figure 16.10, we observe that the terrestrial link (A-link) benefits more (in terms of capacity) for the case of coexistence with the satellite one rather than with a terrestrial one (B-link). Thus, the interference-aware method is rendered more suitable for this type of hybrid scenario, under a spectrum sharing/reuse setup.

16.9.2 CoMP setup

In this section, we evaluate the performance of the proposed arbitrary precoding framework and communication protocol for the CoMP setup that is shown in Figure 16.5 through numerical simulations that take into account the radiation pattern of the considered LC-PAAs, the scattering environment and the propagation mechanisms in the 19.25 GHz band. The SANSA channel model simulator is utilized to this end.

First, we assume that we activate only one out of the four patches of the LC-PAA described in Section 16.8 in order to shape a beam. We see in Figure 16.16 that the R-ZFBF scheme outperforms the ZFBF method, but their performance converges at

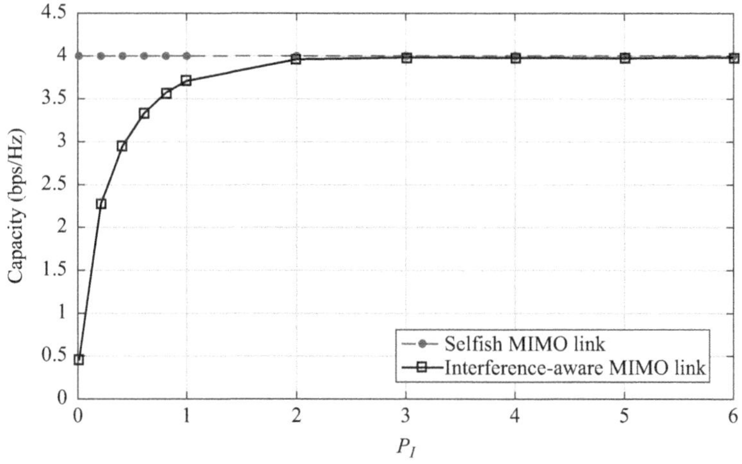

Figure 16.13 *Capacity of a 4 × 4 MIMO link for various values of the interference constraint, P_I. The dashed line corresponds to the capacity of a selfish (unconstrained) link (standard water filling power allocation), while the solid one corresponds to the capacity of the interference-aware method, which guarantees that the interference constraint is satisfied for the satellite link (B-link)*

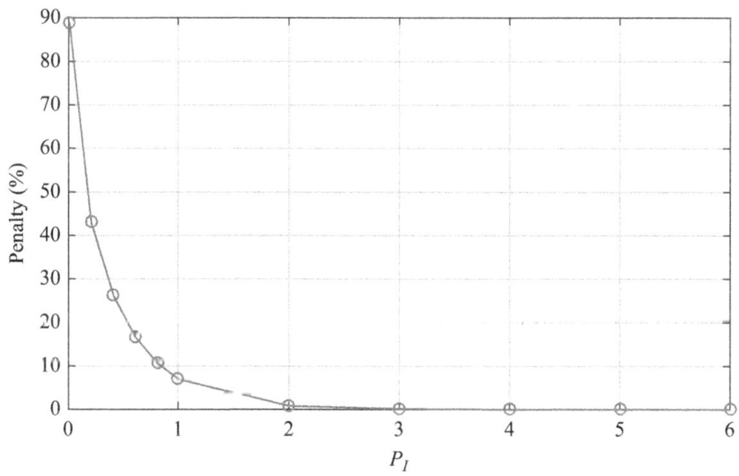

Figure 16.14 *The penalty corresponds to the percentage of the capacity loss of a 4 × 4 terrestrial MIMO link for various values of the interference threshold P_I*

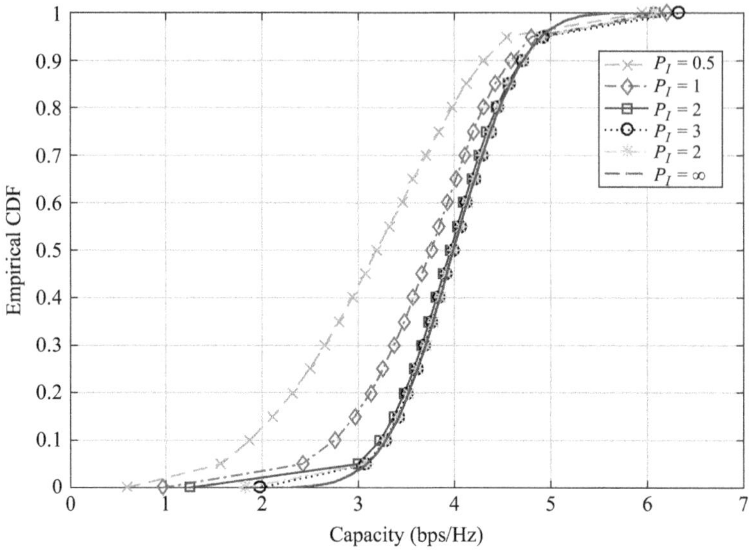

Figure 16.15 The empirical CDFs of a 4 × 4 MIMO link (terrestrial A-link) for different capacity values, which corresponds to interference levels P_I. The dashed line corresponds to the case where no interference constraint exists. ($P_I = \infty$)

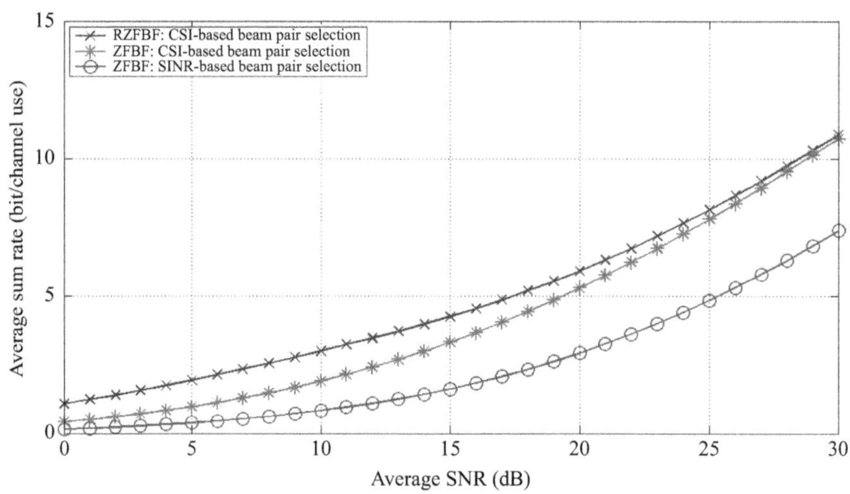

Figure 16.16 Performance of the communication protocol described in Section 16.4 assuming the use of equivalent patch bow-tie LC-SAMP arrays with 10 parasitic elements operating at 19.25 GHz

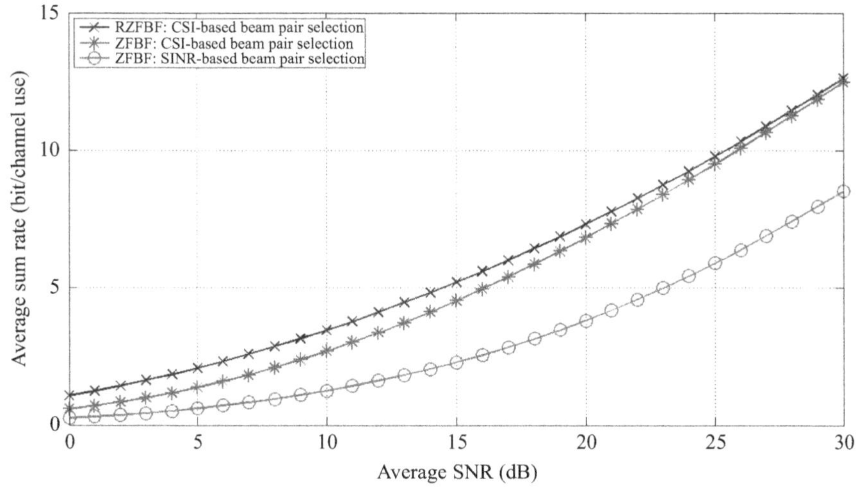

Figure 16.17 Performance of the communication protocol described in Section 16.4 assuming the use of the LC-MAMP described in Section 16.8

high SNRs, as it was expected. Also, we notice that CSI-feedback beam pair selection improves the performance of the ZFBF scheme, in comparison with the case where the selection of the optimum beam pair, in terms of the achieved SR throughput, is based on SINR-feedback.

Then, we assume that all four patches are activated in order to shape a beam. We observe a similar behavior as before; however, in this case, we notice in Figure 16.17 that the achieved SR throughput is higher, due to the higher gain of this antenna configuration.

16.9.3 Symbol-level ZFBF

Finally, in Figure 16.18, the performance of CI-ZFBF against ZFBF in the aforementioned CoMP setup is presented. The use of BPSK modulation is assumed. We note that CI-ZFBF performs much better than its user-level counterpart.

16.10 Summary

In this chapter, we described interference avoidance and mitigation techniques for hybrid satellite-terrestrial MBH systems. More specifically, we considered a hybrid backhaul setup where the satellite segment off-loads the terrestrial one and enhances

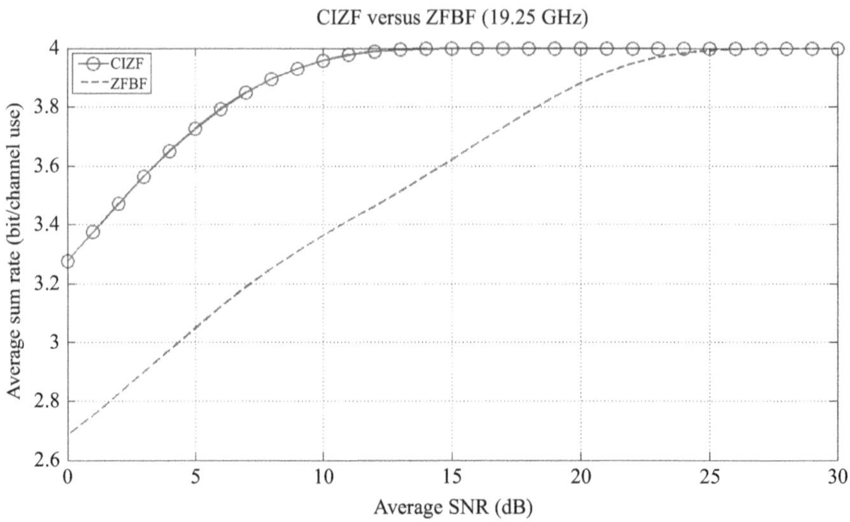

Figure 16.18 Performance of CI-ZFBF versus ZFBF for the considered CoMP setup

the overall capacity of the system. Moreover, we assumed that these two segments share the same spectrum, in order to utilize more efficiently this scarce and expensive resource. In addition, in the proposed system, the backhaul nodes are equipped with antenna arrays instead of drum antennas and make use of multiantenna communication techniques. This technology enables the transmission of multiple data signals in a point-to-point or point-to-multi-point setup on a single time-frequency resource, thus increasing the spectral efficiency, and the management of the intrasystem (terrestrial) and the intersystem (satellite-terrestrial) CCI that is attributed to the concurrent transmission of signals over the same frequency band by colocated nodes. These antenna systems might be conventional DAAs or LC-PAAs. The latter provide a larger number of effective DoF for a given number of RF modules than the former one, which essentially allows for better control of the beam-shaping and beam-steering operations. Equivalently, LC-PAAs provide the same effective DoF with equivalent DAAs by using fewer RF units, thus leading to hardware cost and energy-consumption savings. Also, in the proposed setup, there are deployed some intelligent backhaul nodes which are equipped with both satellite dish antennas and antenna arrays for the terrestrial links. A HNM determines at each timeslot the topology instances based on information such as congestion, link failures, interference levels, etc., in order to accomplish efficient transport of the data over the hybrid backhaul system and enhance the resiliency. Such dynamic routing decisions are facilitated by the ability of the antenna arrays to reconfigure their radiation pattern and establish on-the-fly links; the ability of the terrestrial backhaul nodes to operate in a SU-MIMO, single-cell MU-MIMO, or coordinated/cooperative multi-cell MU-MIMO (CoMP) mode; and the ability of the intelligent backhaul nodes to transmit signals over both the terrestrial and the satellite backhaul segments.

Clearly, the use of this setup suits well to scenarios where the backhaul nodes are small-cell BSs, since in this case, we have often to deal with an NLOS radio environment.

Regarding the interference management techniques, we considered two use cases of practical interest. In the first one, we assumed a scenario where cooperation/interference coordination cannot take place in the terrestrial segment. Under this context, we considered a setup where the backhaul nodes are equipped with conventional DAAs and operate in a SU-MIMO mode. We focused on two sub-scenarios, one where two terrestrial SU-MIMO links utilize the same time-frequency resources and another where a terrestrial SU-MIMO link coexists with a satellite link. In both cases, we proposed a power allocation scheme that can be used in conjunction with the SVD-based precoding that is utilized on the SU-MIMO link of interest, in order to maximize the throughput on this link while at the same time respecting the interference threshold of the colocating terrestrial or satellite RX (depending on the sub-scenario). Through numerical simulations and under realistic assumptions about the operating parameters, channel model, etc., we observed that the proposed interference-aware power allocation method mitigates significantly the capacity loss associated with the use of selfish transmission in such a setup and enables spectrum sharing within the terrestrial segment as well as between the terrestrial and satellite segments in the aforementioned hybrid backhaul system.

The second use case corresponds to a setup where LC-PAAs are installed on the backhaul nodes and the system operates in CoMP mode. First, we described a method that enables us to perform arbitrary channel-dependent precoding with such antenna systems. Then, we presented a communication protocol that relaxes the need to dynamically compute the loading values and yet provides the best possible performance, in terms of the achieved SR throughput, within the given spatial resolution (which corresponds to a performance-complexity trade-off). We also described a LC-PAA design for the operating frequency of interest (19.25 GHz). Based on the above, we ran numerical simulations considering the aforementioned setup, the radiation pattern of the proposed LC-PAAs, and the propagation characteristics. We assumed the use of various linear precoding and symbol-level precoding schemes. We also considered low feedback overhead alternatives in the communication protocol as well as a scenario where only one RF of the proposed LC-MAMP is utilized in beamforming, as suboptimal communication strategies that might be used in practice due to limitations associated with the feedback channel or the antenna arrays. The simulation results showcase the feasibility of the proposed precoding method and communication protocol and illustrate the ability of this setup to manage the CCI and increase the spectral efficiency of the terrestrial backhaul segment.

The considered network architecture, use cases, antenna array, and multiantenna communication technologies; the proposed precoding method, communication protocol, and LC-PAA design; and the studied interference mitigation/avoidance techniques constitute a framework which demonstrates that satellite communication technology can be used efficiently (in terms of spectrum occupancy) together with wireless terrestrial systems to meet the enormous capacity demands of future backhaul networks.

References

[1] Ericsson. Mobility Report; 2017.
[2] NGMN. 5G White Paper; 2015.
[3] 5G-PPP. 5G Vision; 2015.
[4] Mueck M, Jiang W, Sun G, Cao H, Dutkiewicz E, Choi S. White Paper: Novel Spectrum Usage Paradigms for 5G. IEEE SIG CR in 5G; 2014.
[5] DOCOMO. White Paper: 5G Radio Access: Requirements, Concept and Technologies. NTT DOCOMO; 2014. Available from: https://www.slideshare.net/allabout4g/docomo-5g-whitepaper.
[6] Rappaport TS, Heath Jr. RW, Daniels RC, Murdock JN. Millimeter Wave Wireless Communications. Prentice Hall; 2015.
[7] Marsch P, Fettweis GP. Coordinated Multi-Point in Mobile Communications: From Theory to Practice. Cambridge: Cambridge University Press; 2011.
[8] Larsson EG, Edfors O, Tufvesson F, *et al.* Massive MIMO for Next Generation Wireless Systems. IEEE Communications Magazine. 2014 February; 52(2):186–195.
[9] Nokia, Qualcomm, editors. Authorised Shared Access (ASA) – An Evolutionary Spectrum Authorisation Scheme For Sustainable Economic Growth And Consumer Benefit; 2011. Input Document FM(11)116, 72nd Meeting of the WG FM; 2011.
[10] FCC. Spectrum Policy Task Force Report No. 02-155; 2002.
[11] Morgado A, Gomes V, Frascolla K, *et al.* Dynamic LSA for 5G networks: The ADEL perspective. In: 24th European Conference on Networks and Communication (EuCNC 2015). Paris, France; 2015. p. 190–194.
[12] Kamel M, Hamouda W, Youssef A. Ultra-Dense Networks: A Survey. IEEE Communications Surveys Tutorials. 2016 Fourthquarter;18(4):2522–2545.
[13] Huang H, Papadias CB, Venkatesan S. MIMO Communication for Cellular Networks. New York: Springer-Verlag; 2012.
[14] Horizon2020 Project. Shared Access Terrestrial-Satellite Backhaul Network Enabled by Smart Antennas. Available from: http://www.sansa-h2020.eu/10-sansa/8-home.
[15] Winters JH. Smart Antennas for Wireless Systems. IEEE Personal Communications Magazine. 1998;5(1):23–27.
[16] Foschini GJ, Gans MJ. On Limits of Wireless Communications in a Fading Environment When Using Multiple Antennas. Wireless Personal Communications. 1998;6:311–335.
[17] Telatar E. Capacity of Multi-Antenna Gaussian Channels. European Transactions on Telecommunications. 1999 November;10(6):585–596.
[18] Caire G, Shamai S. On achievable rates in a multi-antenna broadcast downlink. In: Proceedings 38th Annual Allerton Conference on Communications, Control and Computing; 2000. p. 1188–1193.
[19] Yu W, Cioffi JM. Sum capacity of a Gaussian vector broadcast channel. In: International Symposium on Information Theory; 2002. p. 498.

[20] Sharif M, Hassibi B. A Comparison of Time-Sharing, DPC, and Beamforming for MIMO Broadcast Channels with Many Users. IEEE Transactions on Communications. 2007 January;55(1):11–15.

[21] Kalis A, Kanatas AG, Papadias CB, editors. Parasitic Antenna Arrays for Wireless MIMO Systems. New York: Springer-Verlag; 2014.

[22] Barousis VI, Papadias CB, Müller RR. A new signal model for MIMO communication with compact parasitic arrays. In: 2014 6th International Symposium on Communications, Control and Signal Processing (ISCCSP); 2014. p. 109–113.

[23] Alexandropoulos GC, Barousis VI, Papadias CB. Precoding for multiuser MIMO systems with single-fed parasitic antenna arrays. In: 2014 IEEE Global Communications Conference; 2014. p. 3897–3902.

[24] Barousis VI, Papadias CB. Arbitrary Precoding with Single-Fed Parasitic Arrays: Closed-Form Expressions and Design Guidelines. IEEE Wireless Communications Letters. 2014 April;3(2):229–232.

[25] Zhou L, Khan FA, Ratnarajah T, *et al.* Achieving Arbitrary Signals Transmission Using a Single Radio Frequency Chain. IEEE Transactions on Communications. 2015 December;63(12):4865–4878.

[26] Ntougias K, Ntaikos D, Papadias CB. Coordinated MIMO with single-fed load-controlled parasitic antenna arrays. In: 2016 IEEE 17th International Workshop on Signal Processing Advances in Wireless Communications (SPAWC); 2016. p. 1–5.

[27] Bjornson E, Jorswieck E. Optimal Resource Allocation in Coordinated Multi-Cell Systems. Foundations and Trends in Communications and Information Theory. 2013 January;9(2–3):113–381.

[28] Masouros C, Alsusa E. Dynamic Linear Precoding for the Exploitation of Known Interference in MIMO Broadcast Systems. IEEE Transactions on Wireless Communications. 2009 March;8(3):1396–1404.

[29] Ntougias K, Ntaikos D, Papadias CB. Robust low-complexity arbitrary user- and symbol-level multi-cell precoding with single-fed load-controlled parasitic antenna arrays. In: 2016 23rd International Conference on Telecommunications (ICT); 2016. p. 1–5.

[30] Zhang R, Liang YC. Exploiting multi-antennas for opportunistic spectrum sharing in cognitive radio networks. In: 2007 IEEE 18th International Symposium on Personal, Indoor and Mobile Radio Communications; 2007. p. 1–5.

[31] Paulraj A, Nabar R, Gore D. Introduction to Space-Time Wireless Communications. Cambridge: Cambridge University Press; 2003.

[32] Theodoridis S. Machine Learning: A Bayesian and Optimization Perspective. Orlando, FL: Academic Press; 2015.

[33] Boyd S, Vandenberghe L. Convex optimization. Cambridge: Cambridge University Press; 2004.

Chapter 17

Dynamic spectrum sharing in hybrid satellite–terrestrial systems

Marko Höyhtyä[1] and Sandrine Boumard[1]

The focus of this chapter is on dynamic spectrum sharing in hybrid satellite–terrestrial systems. We start by classifying the scenarios for these systems. The most important dynamic spectrum-sharing techniques such as spectrum sensing, databases, beamforming, beam hopping, and adaptive frequency and power allocation are discussed and their applicability in different scenarios is analysed. Interference analysis shows how Ka band sharing between satellite and terrestrial systems can be enabled. Autonomous ships are defined as an interesting emerging application area for hybrid satellite–terrestrial systems. In order to make them operate reliably and safely both close to shoreline and in deep sea, multiple communication technologies are needed. Interference management and spectrum-sharing techniques could be used, e.g. to prevent blocking or hijacking of the control signalling of a ship. In addition, we discuss shortly the citizens broadband radio service (CBRS) concept in the 3.5-GHz band. Ideas to use CBRS and other database techniques in millimetre wave bands to enable spectrum sharing between satellite and terrestrial components of a future 5G system are given.

17.1 Introduction

The fifth generation of mobile communications technology (5G) will enable a fully mobile and connected society and aim to address the business and technology demands of 2020 and beyond. To materialize this vision, 5G needs to support and exploit the integration of heterogeneous networks such as terrestrial and satellite. Satellites will play an important role by providing resiliency and coverage to sparsely populated areas, e.g., for maritime users. In recent years, there have been significant technological advances such as the spot beam technology of satellites and software-defined networking (SDN) to facilitate efficient implementation and operation of hybrid satellite–terrestrial systems. One important aspect that needs to be considered is spectrum management and spectrum sharing for those systems due to demand for broadband access and more bandwidth.

[1]VTT Technical Research Centre of Finland Ltd, Finland

Regulatory decisions and spectrum-sharing studies conducted in satellite bands during the last years have clearly shown the importance of developing sharing techniques according to satellite system specific characteristics such as long links and transmission latencies and wide coverage areas, e.g., see [1,2]. A lot of spectrum-sharing analyses have been conducted in the terrestrial domain, but due to many differences, some of the techniques are not applicable in satellite bands, and other techniques need modifications. Some of the most promising spectrum-sharing techniques include power control, beamforming, beam hopping, and spectrum databases [1–15]. Recently, spectrum-sharing considerations have evolved from the licence-exempt approach with uncontrolled interference environment towards more controlled setups with better operational conditions. Database techniques have been favoured both in terrestrial and satellite domains over spectrum sensing in obtaining awareness on the current spectrum use since they provide better protection to incumbent users. One of the examples is a three-tier CBRS model developed for the 3.5-GHz band [16,17].

Hybrid satellite–terrestrial systems are needed to fulfil the demands of many emerging applications such as autonomous driving both on the roads and in marine environment, massive machine-type communications, and also to provide high quality Internet anywhere in the world [18–21]. Satellites are currently the only option in marine environment far away from the shoreline, especially if broadband connectivity is needed. Due to high delays, the most important roles for satellites regarding the automated cars is positioning, timing, and providing backbone. For low-latency operation such as remote driving of a car, the connectivity has to be provided by future terrestrial 5G systems. High throughput satellite (HTS) networks in Ka band and the use of higher frequency bands such as Q/V/W are important use cases for dynamic spectrum sharing techniques in the future. Also, recent surge of announcements about planned mega-constellation satellite networks composed of hundreds of LEO satellites, like SpaceX, OneWeb, and LeoSat, will increase the need for dynamic spectrum management [22].

This chapter classifies hybrid satellite–terrestrial spectrum sharing scenarios and discusses applicability of different techniques in those scenarios. We will also take a look at the core network functionality that is needed for seamless cooperation between different radio interfaces as well as to QoS and priority control of the hybrid network. We will describe the method for interference analysis between satellite and terrestrial systems and provide some results regarding the situation in Poland. We will describe the needs for hybrid systems in autonomous ship case, review some connectivity challenges, and describe how dynamic spectrum sharing techniques can improve the reliability of remote controlled and autonomous ships. Another interesting use case to look at is the CBRS system and its counterpart licensed shared access (LSA) system in Europe. Unlike LSA, the CBRS system uses also spectrum sensing in addition to the database to enable spectrum sharing between different radio systems. This chapter gives a descriptive overview of what the main components and features of the system are, gives some numerical results based on the work conducted with live LTE networks in Finland, and specifically addresses spectrum sharing between cellular and fixed satellite services (FSS) systems.

17.2 Classification of hybrid satellite–terrestrial spectrum sharing scenarios

Many studies and practical implementations of systems have combined satellite and terrestrial components together to provide both high throughput and wide coverage. The same trend seems to continue in 5G that is a multi-radio system built upon both new high-capacity and low-latency interfaces and convergence of existing radio technologies such as LTE and Wi-Fi to a ubiquitous radio access network. 3GPP has started a study item on non-terrestrial networks [23] aiming to define satellite deployment scenarios and related-system parameters such as architecture, altitude, orbit, etc. Both integrated satellite–terrestrial systems and stand-alone satellite networks are considered in the study. Satellites are seen to: (1) provide coverage to the areas that cannot be covered by terrestrial 5G systems, (2) reinforce 5G service reliability by providing service continuity for M2M/IoT devices and passengers on mobile platforms and ensuring service availability anywhere, e.g., for critical communication users and to railway/maritime/aeronautical communications, and (3) provide multicast/broadcast resources to network edges and terminals. Many possible use cases are depicted in the document such as broadband connectivity between the core network and the cells on board a moving platform (e.g. aircraft or vessels).

From the spectrum sharing point of view, the hybrid satellite–terrestrial systems can be divided into two main scenarios:

1. Uncoordinated systems: Coexistence of terrestrial and satellite systems in the same frequency band.
2. Coordinated satellite–terrestrial systems with CR techniques.

The classification includes different concepts from the literature. An uncoordinated scenario includes, e.g., two totally independent systems operating in the same frequency band [7,24]. In a coordinated system, satellites can be used to improve the performance of terrestrial networks [25,26]. Cognitive radio (CR) techniques are also used to improve the operation of a coordinated system where both satellite and terrestrial components are providing services to the end users [9,27]. Each category is discussed in detail in the following.

17.2.1 Uncoordinated systems: coexistence of terrestrial and satellite

The majority of CR research has focused on investigating how two independent systems could coexist in the same frequency band. Considering the coexistence of terrestrial and satellite systems, there are two major application areas to consider:

1. Satellite system is a primary user (PU) of the spectrum, and the terrestrial system is the secondary user (SU) that can dynamically use the temporally or spatially available spectrum resources without interfering with the PU [5,28,29].

2. Satellite is an SU of the spectrum, using highly directional antennas to access the terrestrial spectrum [4,8,30]. Typically, the terrestrial system in the latter is a microwave link, and the sharing is possible due to spatial separation of the signals. Dynamic carrier and power allocation can be used to enhance the operation of the secondary system.

Satellite system as a PU: An example of this application area is given in [28] where a secondary terrestrial mobile communication system accesses the C-band spectrum primarily allocated to the FSS system using lowpower small cells or device-to-device transmission mode. The system model is shown in Figure 17.1.

The FSS system sends data in the downlink (DL) direction to the end users either directly or via the gateway. The secondary system is not allowed to interfere with the FSS receivers. Locations of the victim FSS earth stations need to be known in order to facilitate database-assisted spectrum sharing in the band. The location awareness requirement leads to the use of licenced C-band stations. Depending on the national conditions and the existing use on the band, the database implementation may range from the use of simple protection areas to the application of frequency-specific and location-specific restrictions on the maximum permitted radiated power of individual mobile-base station sectors.

In [28], the frequency band is given as 3,400–3,600 MHz, but it can be extended to cover band up to 3,800 MHz to support studies in the main 5G pioneer bands 3,400–3,800 MHz defined by European Commission in [31]. The band is seen suitable for broadband connectivity in urban areas providing bandwidths of 100 MHz and peak data rates up to Gbps for mobile cellular users. However, there are existing FSS users in the band and thus, there should be mechanisms to share the band in a way that both coexisting systems are served.

The obtained results suggest that using measurement-based path loss model for urban environment there is a possibility to achieve protection distances smaller than 500 m even in the aggregate interference case where multiple simultaneous terrestrial users are interfering the satellite reception. In open areas the protection zone is drastically larger. Another example is the coexistence between a cellular network and FSSs in an mmWave scenario in Ka band [29]. As in [28], both single and aggregate interference scenarios were analysed. Authors included a beamforming scheme at the transmitters with the assumption of no interaction among FSS and BSs. It is shown that the coexistence between cellular and satellite services is possible, while interference at the FSS antenna can be kept below recommended levels in both studied bands.

Satellite system as an SU: Perhaps, the most actively studied frequency band on coexistence of satellite and terrestrial systems has been the Ka band, i.e. satellite DL in the 17.7–19.7 GHz band and uplink in the 27.5–29.5 GHz band, see e.g. [1,5–8,29,30,32,33]. Spectrum sharing in the satellite forward link is already allowed by the European Conference on Postal and Telecommunications administrations. The uplink case is more challenging due to interference caused at the terrestrial system. That band is shared between uncoordinated FSS Earth stations and fixed service (FS). The locations of terrestrial nodes are fixed and can be found in national registries. The main incentive for sharing in this band is to increase the satellite system capacity,

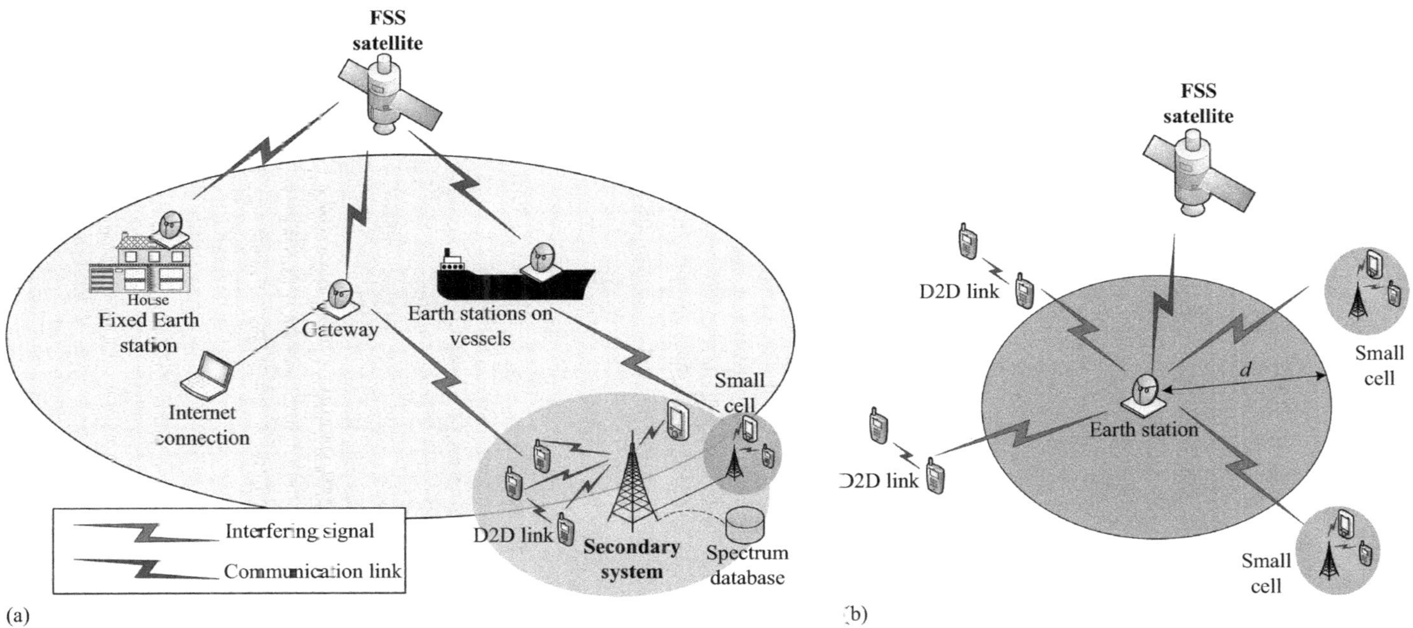

Figure 17.1 (a) Secondary spectrum use of the FSS satellite band 3.4–3.8 GHz, (b) protection zone when multiple interferers exist

and according to [8], FSS earth stations deployed in very high FS density zone are able to use more than 65% of the 17.7–19.7 GHz band at the worst location. In rural areas, 95% of the spectrum is available at the worst location for the FSS earth stations.

Database-assisted spectrum access provides means to control the coexistence in Ka band [7,8]. The database gives permission to the FSS earth station to use the band at a certain location. Exclusion zones or protection zones around FS stations are defined based on the antenna pointing and transmission power of FSS stations. Outside these zones, FSS earth stations could operate with given maximum transmission power values. The zones are frequency dependent, i.e., defined by the transmission frequency of FS stations. A suitable exclusion zone calculation method needs to be incorporated in the controller software giving access to the spectrum to requesting end users. The overall interference modelling concept is presented in Section 17.4.

17.2.2 Coordinated systems: coexistence of terrestrial and satellite

There are a couple of hybrid concepts where the idea has been to use the satellite to assist the operation of a terrestrial cognitive network. It is proposed in [25,34] that the satellite would be a central controller, i.e., in charge of the spectrum allocation and management. Satellites having a wide footprint depending on their orbital height have a wide knowledge about the users and the network in its service region. Therefore, the satellites can serve the terrestrial network in two different ways [34]. First, it can enable the policy and software updates more easily by a policy update message broadcast. Second, being aware of the environment and network state in its footprint, the satellite can manage the spectrum use and allocation in its coverage zone. Spectrum awareness is obtained by collecting environment status reports regularly from the base stations in the footprint.

Satellites are used to connect terrestrial cells together in [26] where two main applications are considered: (1) the IEEE802.22 CR based wireless regional area networks for long-range communication, and (2) the CR-based ultra-wide band communication systems for short-range personal area networks. The proposed hybrid satellite–terrestrial systems apply CR techniques in the satellite uplink and in the terrestrial segment, sharing the same frequency resources. The satellite uplink is not causing excessive interference to the terrestrial systems because Earth stations use directional transmissions. The satellite DL is assumed not to adopt any dynamic spectrum sharing capabilities due to the large footprint of the satellite.

5G is expected to bring more flexibility to network access. Not only several different kinds of technologies could be used to provide the radio access but also mechanism for service integration and security solutions would be more dynamic and local in 5G. While IoT is one major driving force for the convoluted radio access, satellite is also likely to play part in providing heterogeneous access, especially in sparsely populated or hard-to-reach locations. Satellites will extend the 5G cellular networks to sea, air, and remote land areas [35]. Satellite connection together with the localized 5G multi-access edge computing services will provide high potential to enable mission critical services in disaster areas, and it might be the most valuable enabler for introducing basic internet services also in developing third world countries.

Satellite networks will provide resiliency to improve the decreased level of service due to disruptive environmental conditions such as earthquakes, or capacity overload such as temporary surge of number of users.

The convergence of satellite and terrestrial network components can be seen as the inclusion of satellite infrastructure into the 5G system as a network capability and technical enabler for new service creation [36,37]. The satellite segment can provide a backup solution to the terrestrial backhaul links in the case of failure, in the case of emergency, or for load balancing in dense areas with high-peak traffic demands, e.g., in a sport stadium where big events take place [32]. Creation of a functional hybrid system will need smart CR-based technologies i.e. using the available information in selecting the best radio interface to use in any environment and use case. Load balancing in satellite/terrestrial networks based on the network type, signal strength, data rate, and network load was studied in [38]. A step towards integrated 5G systems has been architecture descriptions and interference studies conducted in hybrid LTE networks in [9,27]. Dynamic resource management functions are used to handle the coexistence especially in a single frequency network (SFN) case where both terrestrial and satellite components use the same frequency [9].

17.3 Satellite band sharing techniques

There are numerous techniques that can be used to enable the depicted hybrid satellite–terrestrial scenarios. The following spectrum-sharing techniques are not options that exclude others out but can also be used jointly to achieve the required goals. For example, spectrum awareness can be obtained using spectrum sensing, or through the spectrum database or as a combination of these techniques. Then, resources can be allocated using the spectrum awareness, e.g. using jointly frequency allocations and beamforming. In each category, we will provide also an example on how the considered technique is applicable in the depicted scenarios.

17.3.1 Spectrum sensing

Spectrum sensing can be defined as a task of obtaining awareness about the spectrum utilization in a given geographical area. The main goal of sensing is to decide between the two hypotheses, namely

$$x(t) = \begin{cases} n(t) & H_0 \\ hs(t) + n(t) & H_1 \end{cases}$$

where $x(t)$ is the complex signal received by the spectrum sensor, $s(t)$ is the transmitted signal of the incumbent user, $n(t)$ is the additive white Gaussian noise (AWGN), and h is the complex gain of the ideal wireless channel between the transmitter and the sensor, i.e., there is no multipath fading. If the channel is not ideal, h and $s(t)$ are convolved instead of multiplied. The null hypothesis H_0 states that no incumbent users are present in the observed frequency band, and the alternative hypothesis H_1 indicates that a signal exists.

A clear advantage of sensing compared to spectrum databases is the ability to provide spectrum information autonomously. The cognitive system senses the spectrum and can use this information directly without the need to cooperate with other systems. This is also a disadvantage due to the possible hidden node problem which can be caused for example by an obstacle between the transmitter and the sensor. The secondary system may interfere with the primary receiver due to better channel between them while not being able to detect the primary transmitter.

Energy detection is the most used technique due to its simplicity. Advantages and disadvantages of the sensing techniques such as energy detection, feature detection, and matched filter detection have been discussed intensively, e.g. in [39]. If the incumbent signal is not known a priori, the optimal detector is an energy detector that measures the energy of the received waveform over an observation time window. First, the input signal is filtered with a bandpass filter to select the bandwidth of interest. The filtered signal is squared and integrated over the observation interval. Finally, the output of the integrator is compared to the threshold to decide whether the incumbent signal is present or not.

Spectrum sensing is a complex task that often requires sampling over multiple dimensions such as time, frequency, and even space. There are challenges related to measurement equipment, setting the decision threshold, removal of shadowing, and multipath fading by averaging. For a given algorithm, the processing time will increase along with the volume of data being analysed. Increasing the resolution in any dimension will increase the accuracy of the results but also reduce the efficiency of the calculations. Thus, the accuracy-efficiency trade-off needs to be dynamically balanced based on the computational resources available while achieving the target detection performance including missed detection probability and false alarm probabilities.

Sensing has been studied in satellite communication, e.g., in [40] where satellites are performing sensing and in [41] where satellite signals are sensed. Devices using energy detection methods might be required to use highly directed antennas towards the satellites to reliably detect satellite signals [41] and also to detect any change in the interference situation such as new interferers. That might require separate sensing stations with parabolic antennas to be used for detection purposes. Feature detection and matched filter detection methods are able to detect signals under the noise floor, but they require a priori information on the signal to be detected. Feature detection may exploit features such as sine-wave carrier, symbol rate, and modulation type of the signal. Matched filter detection is the optimal method if a priori information at both physical and medium access control (MAC) layers is available, such as the pulse shape, modulation type, and the packet format.

17.3.2 Spectrum databases

Spectrum sensing cannot guarantee certain QoS nor interference free operation for coexisting systems. This has been the main driver for development of spectrum databases that are currently favoured both in terrestrial and satellite spectrum sharing systems for spectrum awareness. The basic principle of a spectrum database approach in any frequency band is that the SU is not allowed to access the spectrum until it has successfully received information from the database that the channel it intends to operate is free at the location of the user.

Spectrum database models that have been implemented and demonstrated in practice include TV white space database [42], LSA [43] and spectrum access system (SAS) [16,43]. The latter two are so-called licenced sharing approaches where both coexisting users will operate based on the licence and thus will obtain guaranteed QoS. Under the LSA approach, the incumbent operators are required to provide a priori information about their spectrum use over the area of interest to the database. They tell explicitly where, when, and which parts of the frequency bands are available for the secondary use. This requires most probably a third party to operate the LSA system since operators are often not willing to share the information about their spectrum use to other spectrum operators.

Spectrum databases have been actively developed for hybrid satellite–terrestrial scenarios [1,6–8]. Even though databases have been implemented and demonstrated successfully with terrestrial systems, there are still challenges in tailoring the proposed database systems to the satellite bands, e.g. due to large coverage areas and the latencies associated with long transmission links. Sensing can be used to enhance the operation of the database, and this kind of hybrid approach is, e.g. the SAS described in [16,17].

Recently, ECC report describing the operational guidelines for implementing the LSA system in the part of the 5G pioneer band (3.6–3.8 GHz) was published [44]. Figure 17.2 shows the overall procedure described in the report. The aim in

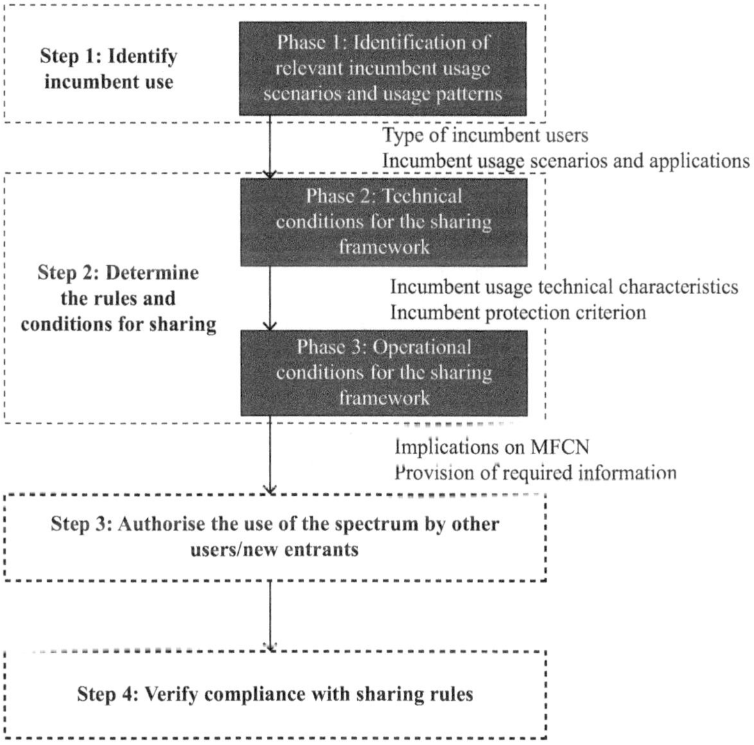

Figure 17.2 Step-by-step approach for implementing LSA [44]

the adoption of the LSA is to have interference-free spectrum for both satellite and terrestrial systems in the band. The implementation of LSA framework implies the agreement of both the incumbent and of the mobile operator on the conditions of use of the spectrum. This kind of controlled sharing is an attractive option since in some cases, it can save both the use of spectrum to current services as well as the position of incumbent operator for current operators. In the worst case, the political pressure might lead to losing the spectrum assets to other wireless services if they are considered more valuable to the society. By allowing sharing, incumbents could continue their operations in the bands to fulfil their obligations defined by the society with minimum additional investment.

17.3.3 Beamforming and smart antennas

One of the main resources is the space since smart antennas and beamforming techniques enable multiple users to exploit same frequency resources at the same time and in the same geographical area. Beamforming has been studied, e.g., in [10–12] that addresses the problem of beamforming and combining based amplify-and-forward relaying in a hybrid satellite–terrestrial cooperative system. In this set-up, a multiple antenna-based relay node forwards the received satellite signals to the destination, by using a beamforming vector, and multiple antenna-based destination node uses maximal ratio combining. Advantage of beamforming and use of smart antennas is that this technique enables denser networks and produces less interference to unwanted directions. A disadvantage is the need for more complex and expensive equipment and may also require location information from satellite terminals. Transmitter-based interference mitigation is called also precoding that can be seen as generalization of beamforming to support multi-stream transmission in multi-antenna wireless communications [45].

Traditionally, outdoor-base stations have been equipped with directional sector antennas. Horizontal beamforming has been enabled with active antenna technology providing the possibility to steer the antenna beam to the desired direction. Recently, the possibility to dynamically control the beam in both azimuth and elevation dimension has been studied and demonstrated [46–48]. Several possible applications of vertical beamforming are depicted in Figure 17.3. With the most advanced active antennas, multiple antenna beams can be generated and controlled independently improving the spectral efficiency and the system capacity within a given area.

Vertical beamforming (or 3D beamforming [47]) enables dynamic adaptation of the vertical beam pattern for several different possibilities whether it is separate carrier tilting, cell splitting, or radio access technology (RAT) tilting, offering promising performance improvements. It can be used to increase signal strength by pointing the vertical main lobe directly at the receiver at any location. In addition, it reduces inter-cell interference when serving users closer to the base stations. Improved spectrum use is achieved either by improving signal quality or by increasing the number of simultaneously served users over a certain geographic area. Differences of pure horizontal and vertical splitting are presented in Figure 17.4. Colours represent different resources such as frequencies to be allocated among users in a cell.

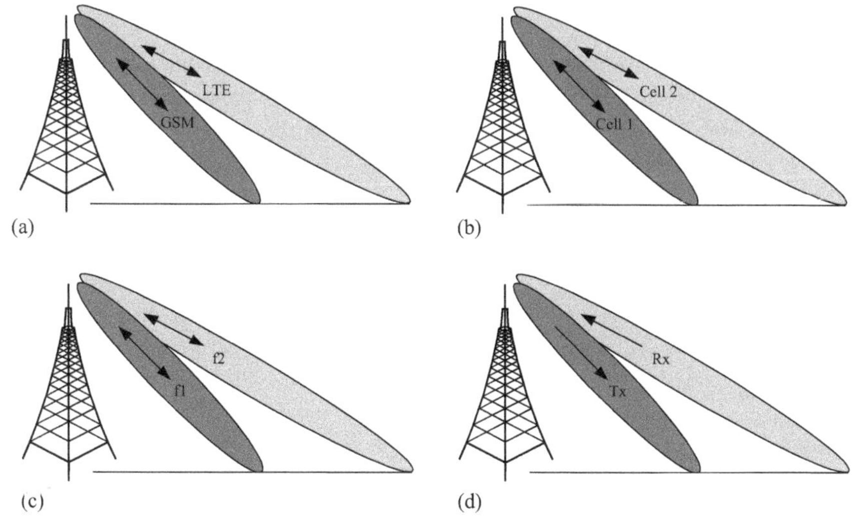

Figure 17.3 Vertical beamforming applications: (a) separate service/RAT tilting, (b) vertical/horizontal cell split, (c) separate carrier tilting, (d) separate Tx–Rx tilting

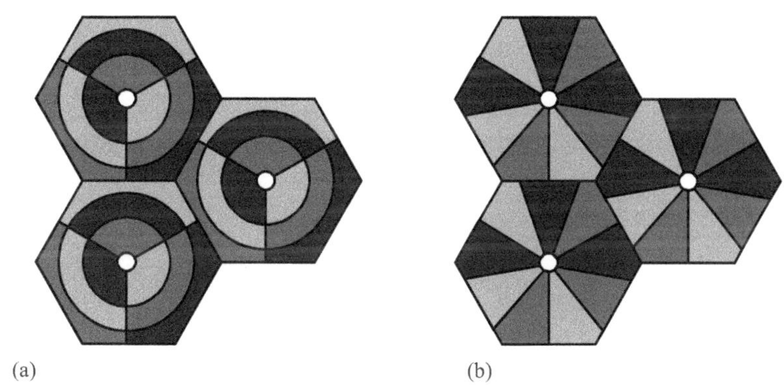

Figure 17.4 Vertical and horizontal splitting, from [47]: (a) vertical sorting, (b) horizontal sorting

Another approach based on the antenna techniques is to use massive multiple-input–multiple-output (MIMO) systems where the number of antennas in a base station is much larger than the number of users [49]. This is a new version of the older concept called Space Division Multiple Access or multiuser MIMO. Massive MIMO becomes also possible [50], but there will be large challenges with energy efficiency and equipment cost.

17.3.4 Beam hopping

Latest satellites have adopted spot beam technology that allows much narrower spots compared to the conventional satellites. Beam hopping is an emerging technology that provides an ability to switch the transmitting power from beam to beam as a function of time. This will improve the flexibility, agility, and throughput of satellite systems. With beam hopping, each beam is adaptively activated and deactivated according to the actual traffic demands. Illumination typically consists of only a subset of the satellite beams through an appropriately designed beam illumination pattern.

A cognitive beam hopping satellite can enhance the spectrum use assuming that the secondary gateway is aware of the primary's beam hopping pattern [15]. In this case, a primary satellite with its own gateway and the secondary satellite with a different gateway are collocated in the same geostationary (GEO) orbit. The gateways are connected with the help of a high-speed terrestrial link (e.g. optical fibre and microwave). This is called a cognition link to emphasize the information sharing between gateways. The system model from [15] is depicted in Figure 17.5. The main difference between conventional and cognitive multi-beam systems is that in the latter multiple beams within a cluster share available spectrum in the time domain instead of the frequency domain. Since the primary satellite only illuminates a small fraction of beams out of a large number of beams deployed under beam hopping systems, the rest of the beams remain idle at that time waiting for their transmission slots. Then, another system with smaller beams can operate in the same area and use the free resources. This might be only possible if both satellites are operated by the same operator since the primary operator may not share the beam hopping sequence with others.

Another possibility would be to use the non-illuminated beams by a secondary terrestrial system/terrestrial component of a coming 5G system again assuming that the primary operator would provide the information about the available and used

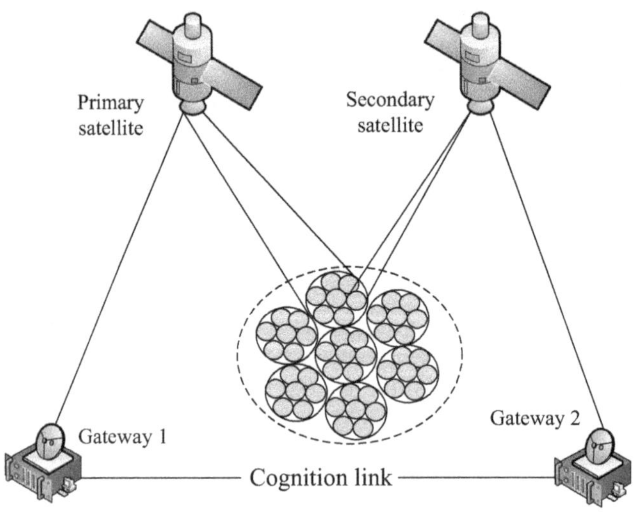

Figure 17.5 Coexistence of two satellite systems, adopted from [15]

areas. This might be especially suitable for an integrated 5G system where both the terrestrial and satellite component could be under the same operator.

17.3.5 Frequency and power allocations

Spectrum sharing can be enabled by fixed transmission powers with carefully calculated protection zones around the interference receivers. However, power adaptation provides more opportunities for sharing. Frequency and power should be allocated in a way that optimizes the use of available resources while keeping the interference at the acceptable level. In this context, SUs can adaptively optimize their spectrum use among available channels. A possible disadvantage of adaptive allocation strategies is aggregated interference towards sensitive satellite receivers. Also chaotic situations are possible if the secondary systems are autonomously learning their allocations. Thus, care has to be taken in developing methods. Transmission power of interfering transmitters has to be limited based on the requirements set, e.g. in ITU-R recommendations in order to satisfy interference criteria [7]. In [4], power control of cognitive satellite terrestrial systems is considered and capacity is maximized using the water-filling principle. In [30], a joint beamforming and carrier allocation scheme to support secondary satellite link is developed. Compared to multi antenna techniques that are complex and expensive with high performance gains, the frequency and power allocations are cheap but with limited gain.

17.3.6 Core network functionality

The core network will undergo fundamental changes in the future, with increased levels of abstraction allowing for further reconfiguration of the network. It is crucial to harmonize the core network structures in order to enable seamless cooperation between the terrestrial and satellite segments [9]. Core network enables QoS management of data transmission, e.g., by dedicating part of the resources to applications with higher priority. The current trend of softwarization and all-IP networking are advancing the hybrid network concept. IP-based operation provides well-standardized interfaces between radio access and the services layer already now in many satellite systems such as Inmarsat Fleet Xpress system designed especially to support maritime users. In addition, SDN technologies will enable a separate control plane as a logically centralized software controller that manages and alters the routing of data through the 5G network. Integrated satellite–terrestrial networks have been studied actively in EU and ESA programmes and core network functionality development is an on-going activity, e.g., see [37,51].

Network sharing is a way for operators to share the heavy deployment costs for wireless networks, especially in the roll-out phase. One interesting concept for sharing in 3GPP specifications is multi-operator core networks where each network operator has its own core network, but the radio access network is shared [52]. Another option defined in the same document is the gateway core network approach. The network operators also share core network nodes such as service gateway or mobility management entity, which is responsible for bearer management and connection management between the mobile terminal and the network.

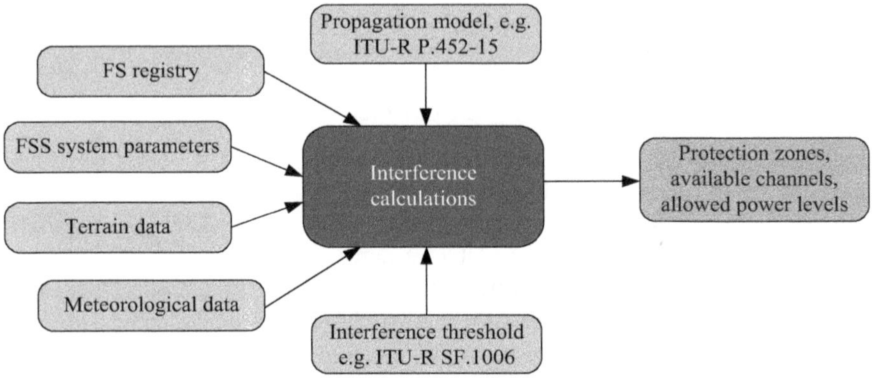

Figure 17.6 Interference modelling on FS–FSS sharing database [7]

17.4 Interference analysis

In the coexistence scenario, as introduced in Section 17.2.1, interference analysis determines the potential level of interference from and to the legitimate users. Detailed interference characteristics can be used to study possible interference mitigation techniques at the SU. The overall interference modelling concept is given in Figure 17.6. In full modelling case, also terrain data and meteorological data are included to obtain the most reliable figure about the sharing possibilities. Modelling without terrain data is a worst case from the FSS point of view, and inclusion of the terrain data improves the sharing possibilities significantly [7].

In the following, the focus is on the forward link of a FSS system in the 17.7–19.7 GHz band in which the PUs are FS links.[1] The terrestrial services deployment databases can be obtained from countries administrations. Ideally, the database include information on the positions of the FS stations, their antenna heights, diameter, and peak gain, their pointing direction, transmission power, carrier frequency and bandwidth, as well as the type of service, point-to-point (P2P) or point-to-multipoint (P2MP).

Link power budgets are used to assess the gains and losses for a link between a transmitter and a receiver. As the interest here is on the interference level, a simple link budget can be used, discarding the transmitter and receiver circuits' losses, and focusing on the antenna gains and path losses. The trans-horizon link phenomena such as surface ducting and tropospheric scattering can be included. The polarization can also be taken into account in the calculation. The precipitation effects can be discarded if the focus is on the worst case situation [53].

[1]Part of this work has been done during the ESA-funded activity 'Antennas and Signal Processing Techniques for Interference Mitigation in Next Generation Ka Band High Throughput Satellites', contract/grant number: AO/1-7821/14/NL/FE. The view expressed herein can in no way be taken to reflect the official opinion of the European Space Agency.

The total power of the interference at the receiver is given by [54,55]

$$P_{\text{interf}} = \sum_{n=0}^{N_{\text{interf}}} \text{eirp}_{TX}(n) - G_{TX,\max}(n) + G_{TX}(n) + L_{\text{path}}(n) + G_{RX}(n) + B_{\text{corr}}(n)$$

where N_{interf} is the number of interferers, $\text{eirp}_{TX}(n)$ is the transmitter eirp in dB W/MHz, $G_{TX}(n)$ is the interferer's antenna gain in the direction of the receiver, and $G_{TX,\max}(n)$ is the maximum antenna gain, $L_{\text{path}}(n)$ is the path loss between the interferer and the receiver, $G_{RX}(n)$ is the antenna gain of the receiver in the direction of the interferer. An additional correcting factor $B_{\text{corr}}(n)$ has been added to the link budget to take into account the fact that the bandwidths occupied by the interferer and the receiver might not fully overlap. Empirical approaches to calculate $B_{\text{corr}}(n)$ have been defined in [55] for two general situations, when the interfering bandwidth is wider than the victim bandwidth and vice versa. The worst case situation is assumed here, and the simplest equation used in which $B_{\text{corr}}(n)$ is equal to the bandwidth of the interferer carrier that overlaps with the bandwidth of the interfered. The transmitter and receivers antenna gain will change according to the location of the interferers with respect to the receiver as well as the antenna direction, etc. P_{interf} is often expressed in dB W/MHz. Only the interference from FS links are accounted for at the FSS, the possible interference from other FSS stations are not taken into account since it is a system design problem. The time variation of the interference from the FS links is not taken into account since there is no single standard for FS links, and this prevents making any assumptions about the protocol in use.

Using the path loss model defined in ITU-R P. 452 [56] but simplifying it to consider only short link distances and worst case interference conditions, the path loss can be written as [8]

$$L_{\text{path}} = 92.5 + 20\log f + 20\log d + L_d(p) + A_g + A_h + E_{\text{sp}}(p)$$

where f is the frequency, d is the path length, $L_d(p)$ is the diffraction loss where p refers to the time percentage for which the calculated basic transmission loss is not exceeded, and A_g is the total gaseous absorption. A_h is the height gain correction, which accounts for the additional diffraction losses at antennas which are embedded in local ground clutter, and $E_{\text{sp}}(p)$ is the correction for multipath and focusing effects at p percentage times. A_g is defined as $A_g = [\gamma_0 + \gamma_\omega(\rho)]$ with γ_0 and $\gamma_\omega(\rho)$ specific attenuation due to dry air and water vapour, respectively, and are found in ITU-R P. 676 [57], ρ is the water vapour density in g/m^3, and ω is the fraction of the total path over water. The diffraction loss can be taken into account by using a spherical earth model described in ITU-R P.526 [58]. Eleven different clutter categories are defined in ITU-R P.452 [56]. Focusing on the worst case scenario, the clutter that leads to the least losses is chosen [53]. It is to be noted that the path loss increases at a higher pace for distances above 25 km [59], since the trans-horizon link phenomena are not taken into account. This means that in the simulation, interferers at distances longer than 25 km will not be included in the interference level calculations.

In the full ITU-R P452-15 propagation model [56], the path loss L_{path} includes the terrain information as in the case of the CoRaSat project [55, Section 4.2].

The diffraction mechanism in ITU-R P452 is the Delta-Bullington model. The path profile is computed using topographic data. The modified Shuttle Radar Topography Mission '90m' data available from the CGIAR-CSI GeoPortal [60], the US National Geospatial-Intelligence Agency, or the Advanced Spaceborne Thermal Emission and Reflection Radiometer Global Digital Elevation Model Version 2 can be used. The interpolation method used between points in the topography grid must be carefully chosen and applied.

The antenna patterns of ITU-R F.699 [61] and ITU-R F.1336 [62] are applied to the P2P FS stations and the P2MP FS stations, respectively. Other implemented antenna radiation patterns for FS can be taken from ITU-R F.1245 [63], which fits P2P FS stations. The antenna pattern of ITU-R S.465 [64] is applied for the FSS stations.

The study of the level of interference created at the FSS station by the FS stations requires the 17.7–19.7 GHz frequency band to be split into 32 channels of 62.5 MHz bandwidth, corresponding to relevant values of forward link channel bandwidth used in digital video broadcasting (DVB)-S2. The FSS stations must be pointing to a

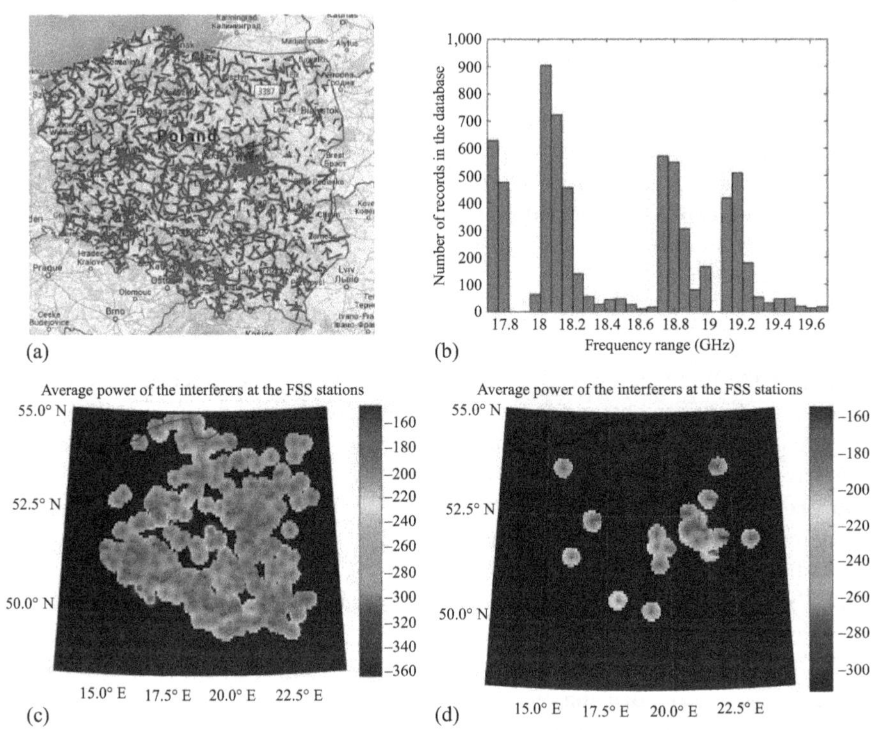

Figure 17.7 Example results for FS–FSS interference study: (a) distribution of FS links in Poland in the 17.7–19.7 GHz band per 62.5 MHz channels; (b) FS links carrier distribution; (c) average interference power in dB W/MHz in channel 19; (d) average interference power in dB W/MHz in channel 14

fixed direction, which depends on the country and the satellite. In order to create interference maps for a country, the country is divided into grids. In each grid, link power budgets are calculated at randomly placed FSS stations. An example is shown in Figure 17.7. The FS links as provided by the Polish administration are shown, together with the distribution of the links versus the 62.5 MHz wide FSS bands. Average power of the interferers at the FSS stations is calculated for each grid over 100 randomly placed FSS stations within this grid.

We can see from the pictures that some channels are not used by many FS links and hence the interference at FSS stations is very small in most areas. It is clear from the interference maps that certain zones and channels are free of interference and can easily be used by FSS. In other areas, different mitigation techniques can be utilized to remove the interference from the FS links if no channel is free. However, the sheer size of countries, the number of FS and FSS stations per grid, and the number of FSS channels lead to very long simulation times. For time limited countrywide studies, a map representing the FS links and the distribution of the carrier frequency of the links can provide enough information to define interesting interference scenarios. Then on, the interference levels can be calculated in a more localized problematic area, using a finer grid.

17.5 Practical application scenarios

Hybrid satellite–terrestrial systems and dynamic spectrum sharing are needed in many practical scenarios. We will review two potential ones, namely autonomous ships and the CBRS system in the following.

17.5.1 Autonomous ships

Autonomous and remote-controlled ships are becoming reality, and e.g. in Finland, an open test area has been opened for them recently [65]. There is a need for a Remote Operation Centre (ROC) and remote operators that could be called 'virtual captains', able to steer multiple ships, simultaneously. An autonomous ship should be able to monitor its own health and environment, communicate obtained information, and make decisions based on that without human supervision. However, the 'virtual captain' from the ROC will perform critical or difficult operations [66].

A critical enabling component of the autonomous ship concept is the connectivity. Such communication will need to be bidirectional, accurate, scalable, and supported by multiple systems – creating redundancy and minimizing risk. The connectivity solution has to guarantee sufficient communication link capacity for sensor monitoring and remote control. It is natural to design a hybrid satellite–terrestrial architecture for such a ship, and testing of different communication technologies will start soon in the testing area.

In-ship communications, e.g. communicating information from the sensors that are monitoring the health of the ship (status of engines, propulsion systems, ballast tank, cargo) as well as from the environment detection and collision avoidance sensors [light detection and ranging (LiDARs), radars, optical cameras] can be handled partly

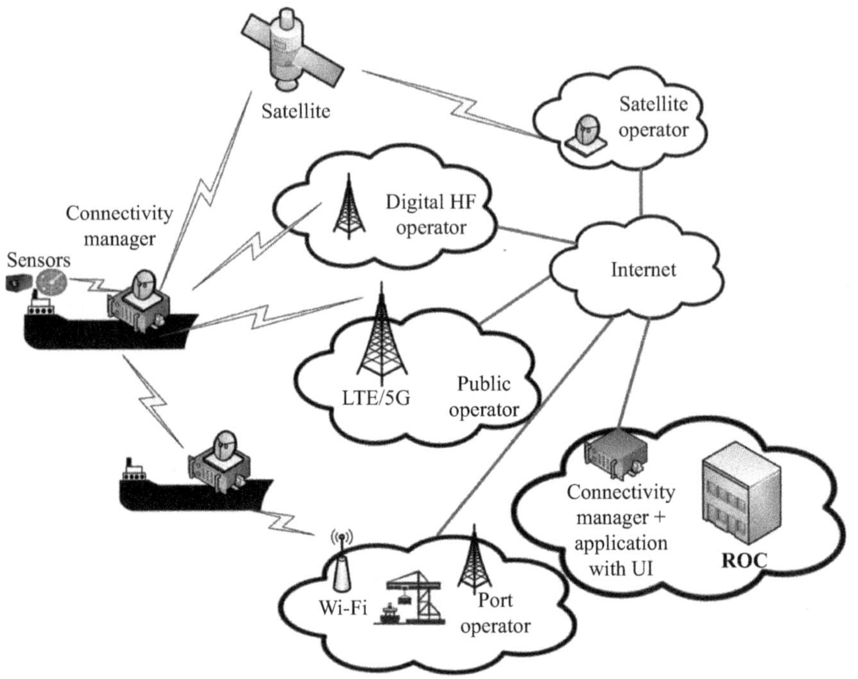

Figure 17.8 High-level communications architecture for an autonomous/remote controlled ship

wirelessly and partly with cables. Short-range solutions such as small cells and Wi-Fi will be used for this purpose. On the other hand, communication from the open sea requires long-range technologies such as high frequency (HF) communications or satellites. A unique challenge in autonomous shipping is that the uplink data rate requirements for sending the sensor data to ROC can be several Mbit/s, whereas DL data requirement for the control data is only some kilobits/s [67]. Traditionally, the communication systems have been designed for mirrored needs.

The high-level architecture for ship-to-ship and ship-to-shore communications is presented in Figure 17.8. The system is a hybrid architecture, comprising of satellite and terrestrial components. An essential part of the architecture is the connectivity manager that decides which data is sent over which route depending on the QoS requirements and availability of the links. Connection should be very robust to ensure safety at sea and efficient transport of goods anytime, anywhere in the world. Therefore, interference management and dynamic spectrum sharing techniques should be used to guarantee continuous connectivity. Some identified topics include

1. Security. Traditionally, the ship control system has been closed. Due to remote operation centre, it has to be opened to outside world. Thus, cybersecurity and redundancy using, e.g., either HF link or satellite link are needed to avoid hijackers taking control or jamming the link to be unusable. The system should be able

to detect the jamming signal and change frequency automatically using cognitive principles.

2. Geolocation. The ship should not only rely on satellite geolocation since it can be jammed easily. Simple mechanisms such as shielding the GPS receiver from the horizontal interference help in increasing the resistance to the interference.
3. Spectrum sharing. It is possible that the hybrid 5G system will be a SFN system i.e. the satellite and terrestrial component use the same frequency. Whether it is a SFN or a multi-frequency network used in a ship one has to develop clever frequency allocation strategies to fulfil the QoS requirements of sensor and control data transmissions.
4. Enhanced collision detection. Large ships are required to use automatic identification system (AIS) and send data such as unique identification, position, course, and speed so that other ships will obtain the information to avoid collisions. Radars, optical cameras, and other sensors can be used to find smaller boats that are not using the AIS system. An interesting idea could be to use also spectrum sensing to find small boats around the ship, e.g. by detecting the signals coming from mobile phones.

17.5.2 Citizens broadband radio service

In the United States, the prevailing approach for spectrum sharing is the CBRS governed by SAS in the 3,550–3,700 MHz band [16,17,43]. SAS is a three-tier model. The first tier is the incumbent system. The second tier includes priority access users, such as mobile network operators, that are allocated to exclusive channels protected from general authorized access (GAA) users. The GAA tier facilitates opportunistic spectrum use, multiple users can use a given channel, and thus there is no interference protection. The interference protection between the tiers is reduced top down.

At the core of the SAS concept is a database system. The incumbent user may provide its spectrum usage information, such as duration of operation, and/or the operational parameters such as transmitter identity, location, antenna height, transmission power, interference tolerance capability, and protection contour to be included in the database [16,17]. The SAS can use either database, or a database-plus-sensing approach to identify the available spectrum opportunities. CBRS devices (CBSDs), that can be, e.g. LTE base stations, are required to be authorized and coordinated by one or more authorized SASs in order to access the spectrum. SAS enforces exclusion and protection zones around incumbent users that are Department of Defense shipborne radars operating in coastal areas and non-federal FSS earth stations.

In order to protect FSS earth stations, Federal Communications Commission has adopted a rule that requires satellite operators to register their stations annually [68]. In the case of shipborne radars, SAS uses information from environment sensing capability devices to ensure that CBSDs operate in a manner that does not interfere with incumbents as well as facilitates information exchange between multiple SASs.

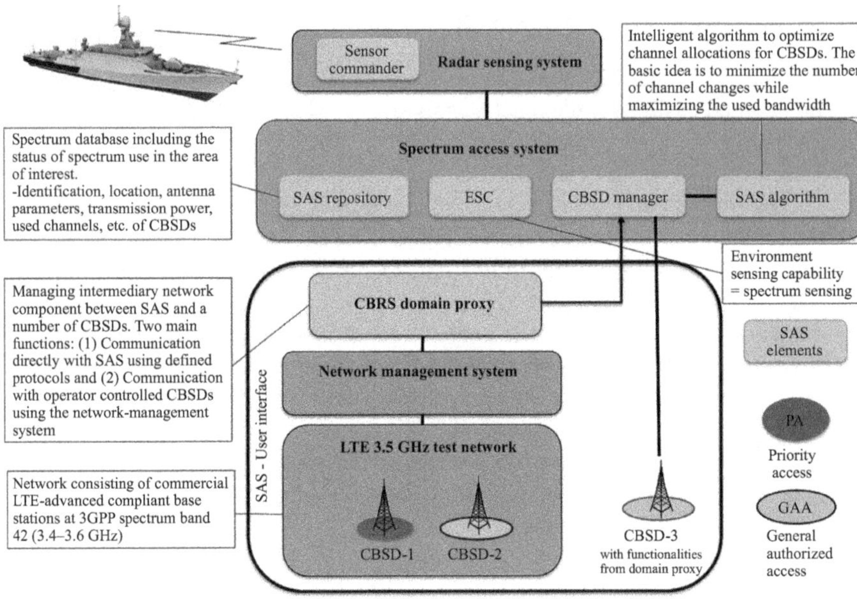

Figure 17.9 Architecture of the implemented CBRS system in Finland [17]

The CBRS/SAS trial environment in Finland and its key building blocks are depicted in Figure 17.9. The building blocks are developed and governed by multiple organizations and are located in different places in Finland. The trial environment can be managed remotely to allow live demonstrations at, e.g. conferences and events. During the trials, it was noticed that a frequency change from a band to another one including evacuation of the current one may take around 3 min due to the current commercial hardware that has not been designed for fast frequency changes. Detailed analysis of steps in the evacuation and frequency change process can be found in [17]. However, this CBRS system provides means for sharing the spectrum in the 5G pioneer band 3.4–3.8 GHz between cellular and satellite users as well as between cellular and radar users.

A recent research paper proposed a precoding design to address a MIMO ship-borne radar and a MIMO commercial coordinate multipoint (CoMP) communication system coexistence scenario, which is applicable for the LTE-advanced system [69]. This work may provide some insights to the future spectrum sharing in the 5G pioneer band 3.4–3.8 GHz. Simulation results in the 3.5 GHz band show promising results but assume also modifications to the incumbent signal. MIMO radar is an emerging area of research and a possible upgrade option of legacy radar systems. Unlike the standard phased-array radar that transmits scaled versions of a single waveform, MIMO radars transmit multiple probing signals that can be chosen freely [69]. This gives MIMO radars significant additional degrees of freedom compared to phased-array radars, allowing them to track more targets with better performance, while simultaneously better eliminating clutter and interference.

17.6 Future recommendations

17.6.1 Spectrum sensing

One of the defining features of 5G and beyond networks is local small range networks. Local and indoor networks like pico- and smaller femtocells with intelligent devices and sensors call for simple and non-complex channel search methods that are easy to implement [70]. Therefore, spectrum sensing is clearly a method that needs to be developed to support that kind of operation, also in the bands that are shared with the satellites. Also, hybrid methods that combine databases and sensing are foreseen in the future, e.g., in military applications. One example of such systems is the CBRS concept described in this chapter, but more work is needed to develop the most suitable solutions to other bands.

17.6.2 Spectrum databases

In addition to the discussed 5G pioneer band below 4 GHz, the future 5G systems will operate in mm W bands to achieve high capacity demands. One example is the 24.25–27.5 GHz pioneer band [31] that is expected to include both cellular and satellite users. LSA approaches such as CBRS and LSA are very promising database-enabled approaches in mm W bands, having also industry support behind them. It is important to conduct exclusion zone studies with targeted systems such as small cells with 5G New Radio to understand the aggregate interference effect and be able to define the protection distances around FSS earth stations.

17.6.3 Beamforming

Beamforming is critical for coverage at higher frequencies both for terrestrial and satellite systems. According to [71], hybrid multiple-antenna transceivers, which combine large-dimensional analogue pre/post-processing with lower dimensional digital processing, are the most promising approaches for reducing the hardware cost and training overhead in massive MIMO systems. Due to high promises of hybrid beamforming approach, it should be studied also in the context of hybrid satellite–terrestrial systems. A natural place, e.g., for the 26 GHz sharing would be to have a massive MIMO system with hybrid beamforming at the terrestrial 5G system.

17.6.4 Beam hopping

Classical multi-beam satellite system relies on a semi-permanent allocation of frequencies in beams. A chunk of spectrum is allocated to each beam and spectrum allocated to one beam may be reused in another beam according to the beam allocation table but not in adjacent beams in order to avoid interference.

In a beam-hopping system, the complete spectrum is allocated to all the beams, enabling a full reuse of all the frequencies. Multiplexing of different beams is made in time and space domains by allocating transmission time slots to each beam. The beam-hopping pattern is based on frames and may change dynamically several times per hour. The benefits of this approach are a better spectral efficiency of the system and more flexibility to adjust the system capacity to the user demand. Spectrum sharing between a beam-hopping satellite system and a terrestrial system is an interesting study item, assuming the satellite to be either primary or SU of the spectrum.

17.6.5 Frequency and power allocations

The main challenge in CR operation is to take into account all the available information – such as locations of devices, sensing information, regulations, database information, etc. – and make decisions about where in the spectrum to operate at any given moment and how much power to use in that band. Frequency and power allocation method is a simpler and cheaper method than use of sophisticated multi-antenna systems and should always be considered in any spectrum sharing scenario. Since mobility and automated driving is increasing all the time, there is a need to include mobility in the sharing equation. Joint mobility and spectrum prediction could be used, e.g., in assisting the base station selection, minimize the number of handovers, and improve the quality of vehicle-to-everything (V2X) communications.

17.6.6 Core network functionality and network slicing

5G enablers such as SDN will separate control and data plane of the network and thus the same control plane can be used to manage intelligently the use of multiple RATs. Network slicing is another interesting concept to support different applications' needs in the same physical network [72]. Slicing means partitioning of existent physical network resources in an efficient manner, e.g., to support applications of different QoS requirements in different slices. The main aspects of slicing are resource allocation and isolation. Network slicing can be implemented in different levels such as spectrum, infrastructure, and network. In the spectrum case, different applications operate in different spectrum slices. Infrastructure slicing means that part of the physical network such as antenna is sliced, whereas in end-to-end network slicing, there can be isolated bit pipes throughout whole network for different services. Therefore, it is recommended to study how slicing will affect the way hybrid satellite–terrestrial systems will be implemented and used in the near future.

17.6.7 Implementation challenges

In order to achieve the best gains, many sharing scenarios assume collaboration between operators instead of competition. Even though this is slowly happening, there are still challenges, e.g., in collaboration between satellite and terrestrial operators. Moreover, costs of more complex solutions such as smart antennas in order to achieve performance gains are clearly a trade-off that affects how the networks are built. There should be a gain of doing investments with advanced technologies, e.g., by attracting and being able to support more users and their applications.

17.7 Conclusions

Terrestrial and satellite systems can share the spectrum when they apply techniques that minimize the interference between the systems. We have reviewed the application scenarios and techniques that could be used in those scenarios and discussed practical use cases. It is envisioned that the coming 5G systems will be hybrid systems, combining both satellite and terrestrial parts seamlessly together. Satellite component will support several applications including in-flight services, asset tracking, rapidly deployed public safety communication networks, autonomous driving, and high data rate broadcast services. Even though there are some existing systems including both satellite and terrestrial components, such as DVB-next generation handheld (DVB-NGH), there are still many challenges before seamlessly integrated satellite–terrestrial networks will become reality. The main obstacle in the past has been the cost of satellite services to the end users, and in the spectrum sharing case, the high possibility for interference. The adoption of new technology such as SDN and NFV as well as licenced sharing approaches is driving the development, and the use of hybrid networks in the near future and hybrid systems will be adopted both below 6 GHz and in the millimetre wave bands.

References

[1] European Commission, 'Promoting the shared use of radio spectrum resources in the internal market,' Brussels, Belgium, September 2012.

[2] Höyhtyä M., Mämmelä A., Chen X., *et al.* 'Database-assisted spectrum sharing in satellite communications: A survey,' *IEEE Access*, 2017;5: 25322–25341.

[3] Maleki S., Chatzinotas S., Evans B., *et al.* 'Cognitive spectrum utilization in Ka band multibeam satellite communications,' *IEEE Communications Magazine*, 2015;**53**(3): 24–29.

[4] Vassaki S., Poulakis M. I., Panagopoulos A. D. 'Optimal iSINR-based power control for cognitive satellite terrestrial networks,' *Transactions of Emerging Telecommunications Technologies*, 2017;**28**(2): 1–10.

[5] Höyhtyä M., Kyröläinen J., Hulkkonen A., Ylitalo J., Roivainen A. 'Application of cognitive radio techniques to satellite communication,' in *Proceedings of Dynamic Spectrum Access Networks Conference*; Bellevue, WA, USA, October 2012, pp. 540–551.

[6] Sharma S. K., Chatzinotas S., Ottersten B. 'Satellite cognitive communications: Interference modelling and techniques selection,' in *Proceedings of Advanced Satellite Multimedia Systems Conference*; Baiona, Spain, September 2012, pp. 111–118.

[7] Höyhtyä M., Mämmelä A. 'Spectrum database for coexistence of terrestrial FS and FSS satellite system in the 17.7–19.7 GHz band,' in *Proceedings of 21st Ka and Broadband Communications Conference*; Bologna, Italy, October 2015.

[8] ECC Report 232, 'Compatibility between fixed satellite service uncoordinated receive earth stations and the fixed service in the band 17.7–19.7 GHz,' May 2015.

[9] Ylitalo J., Hulkkonen A., Höyhtyä M., *et al.* 'Hybrid satellite systems: Extending terrestrial networks using satellites,' in Chatzinotas S., Ottersten B., De Gaudenzi R. (eds.). *Cooperative and Cognitive Satellite Systems* (London, Elsevier Ltd, 2015), pp. 337–371.

[10] Yuan C., Lin M., Ouyang J., Bu Y. 'Beamforming schemes for hybrid satellite-terrestrial cooperative networks,' *International Journal of Electronics and Communications*, 2015;**69**(8): 1118–1125.

[11] ESA ASPIM project: Antennas and signal processing techniques for interference mitigation in next generation Ka band high throughput satellites. Contract/grant number: AO/1-7821/14/NL/FE; 2017.

[12] Arti M. K., Bhatnagar M. R. 'Beamforming and combining in hybrid satellite-terrestrial networks,' *IEEE Communication Letters*, 2014;**18**(3): 483–486.

[13] Chatzinotas S., Evans B., Guidotti A., *et al.* 'Cognitive approaches to enhance spectrum availability for satellite systems,' *International Journal of Satellite Communications and Networking*, 2017;**35**(5): 407–442.

[14] Icolari V., Guidotti A., Tarchi D., Vanelli-Coralli A. 'An interference estimation technique for satellite cognitive radio systems,' in *Proceedings of International Conference on Communications*; London, UK, June 2015.

[15] Sharma S. K., Chatzinotas S., Ottersten B. 'Cognitive beamhopping for spectral coexistence of multibeam satellites,' *International Journal of Satellite Communications and Networking*, 2015;**33**(1): 69–91.

[16] Sohul M. M., Miao Y., Yang T., Reed J. H. 'Spectrum access system for the citizen broadband radio service,' *IEEE Communications Magazine*, 2015;**53**(7): 18–25.

[17] Palola M., Höyhtyä M., Aho P., *et al.* 'Field trial of the 3.5 GHz citizens broadband radio service governed by a spectrum access system (SAS),' in *Proceedings of Dynamic Spectrum Access Networks Conference*; Baltimore, MD, USA, March 2017.

[18] Kong L., Khan M. K., Wu F., Chen G., Zeng P. 'Millimeter-wave wireless communications for IoT-cloud supported autonomous vehicles: Overview, Design, and Challenges,' *IEEE Communications Magazine*, 2017;**55**(1): 62–68.

[19] Choi J., Va V., Gonzalez-Prelcic N., Daniels R., Bhat C. R., Heath R. W. 'Millimeter-wave vehicular communication to support massive automotive sensing,' *IEEE Communications Magazine*, 2016;**54**(12): 160–167.

[20] Rødseth Ø. J., Kvamstad B., Porathe T., Burmeister H.-C. 'Communication architecture for an unmanned merchant ship,' in *Proceedings of IEEE OCEANS Conference*; Bergen, Norway, June 2013.

[21] Charou E., Bratsolis E., Gyftakis S., Giannakopoulos T., Perantonis S. 'Use of Sentinel-1 data for maritime domain awareness: Preliminary results,' in *Proceedings of International Conference on Information, Intelligence, Systems and Applications*; Corfu, Greece, July 2017.

[22] De Selding P. B. 'European governments boost satcom spending,' *SpaceNews*. January 2016.

[23] 3GPP TR38.811 v0.2.1, 'Study on New Radio (NR) to support non terrestrial networks (Release 15),' November 2017.

[24] Suffriti R., Corazza G. E., Guidotti A., *et al.* 'Cognitive hybrid satellite-terrestrial systems,' in *Proceedings of the 4th International Conference on Cognitive Radio and Advanced Spectrum Management*; Barcelona, Spain, October 2011.

[25] Gozupek D., Bayhan S., Alagöz F. 'A novel handover protocol to prevent hidden node problem in satellite assisted cognitive radio networks,' in *Proceedings of the 3rd International Symposium on Wireless Pervasive Computing*; Santorini, Greece, May 2008.

[26] Kandeepan S., De Nardis L., Di Benedetto M.-G., Guidotti A., Corazza G. E. 'Cognitive satellite terrestrial radios,' in *Proceedings of the Global Communications Conference*; Miami, FL, USA, December 2010.

[27] Roivainen A., Ylitalo J., Kyröläinen J., Juntti M. 'Performance of terrestrial network with the presence of overlay satellite network,' in *Proceedings of the International Conference on Communications*; Budapest, Hungary, June 2013.

[28] Höyhtyä M. 'Sharing FSS satellite C band with secondary small cells and D2D communications,' in *Proceedings of the International Conference on Communications, CogRAN-Sat workshop*; London, UK, June 2015.

[29] Guidolin F., Nekovee M., Badia L., Zorzi M. 'A study on the coexistence of fixed satellite service and cellular networks in a mmWave scenario,' in *Proceedings of International Conference of Communications*; London, UK, June 2015.

[30] Lagunas E., Sharma S. K., Maleki S., Chatzinotas S., Ottersten B. 'Resource allocation for cognitive satellite communications with incumbent terrestrial networks,' *IEEE Transactions on Cognitive Communications and Networking*, 2015;**1**(3): 305–317.

[31] European Commission, Radio Spectrum Policy Group, 'Strategic roadmap towards 5G in Europe: Opinion on spectrum related aspects for next-generation wireless systems (5G),' November 2016.

[32] Artiga X., Vazquez M. A., Perez-Neira A., *et al.* 'Spectrum sharing in hybrid terrestrial-satellite backhaul networks in the Ka band,' in *Proceedings of the European Conference on Networks and Communications*; Oulu, Finland, June 2017.

[33] EU FP7, CoRaSAT project: Cognitive radio for satellite communications (CoRaSAT),' 2015. Available from: https://sites.google.com/a/ict-corasat.eu/corasat/home.

[34] Bayhan S., Gur G., Alagoz F. 'Satellite assisted spectrum agility concept,' in *Proceedings of Military Communications Conference*; Orlando, FL, USA, October 2007.

[35] NetWorld2020, 'The role of satellites in 5G,' white paper, version 5, July 2014. [Online]. Available: http://networld2020.eu/wp-content/uploads/2014/02/SatCom-in-5G_v5.pdf.

[36] EU H2020, 'Shared access terrestrial-satellite backhaul network enabled by smart antennas (SANSA)'. http://www.sansa-h2020.eu/.

[37] EU H2020, 'Virtualized hybrid satellite-terrestrial systems for resilient and flexible future networks (VITAL),' 2015–2017. http://www.ict-vital.eu.

[38] Ali M. *Load balancing in heterogeneous wireless communications networks: Optimized load aware vertical handovers in satellite-terrestrial hybrid networks incorporating IEEE 802.21 media independent handover and cognitive algorithms*, Ph. D. thesis, University of Bradford, 2014.

[39] Yücek T., Arslan H. 'A survey of spectrum sensing algorithms for cognitive radio applications,' *IEEE Communications Surveys & Tutorials*, 2009;**11**(1): 116–130.

[40] Jia M., Liu X., Yin Z., Guo Q., GU X. 'Joint cooperative spectrum sensing and spectrum opportunity for satellite cluster communication networks,' *Ad Hoc Networks*, 2017;**58**(4): 231–238.

[41] Höyhtyä M. 'Secondary terrestrial use of broadcasting satellite services below 3 GHz,' *International Journal of Wireless and Mobile Networking*, 2013;**5**(1): 1–14.

[42] Murty R., Chandra R., Moscibroda T., Bahl P. 'Senseless: A database-driven white spaces network,' *IEEE Transactions on Mobile Computing*, 2012;**11**(2): 189–203.

[43] Tehrani R. H., Vahid S., Triantafyllopolou D., Lee H., Moessner K. 'Licensed spectrum sharing schemes for mobile operators: A survey and outlook,' *IEEE Communications Surveys & Tutorials*, 2016;**18**(4): 2591–2623.

[44] ECC Report 254, 'Operational guidelines for spectrum sharing to support the implementation of the current ECC framework in the 3600–3800 MHz range,' November 2016.

[45] Vazquez M. A., Perez-Neira A., Christopoulos D., *et al.* 'Precoding in multibeam satellite communications: Present and future challenges,' *IEEE Wireless Communications*, 2016;**23**(6): 88–95.

[46] Caretti M, Crozzoli M., Dell'Aera G. M., Orlando A. 'Cell splitting based on active antennas: Performance assessment for LTE system,' *Proceedings of the IEEE 13th Annual Wireless and Microwave Technology Conference*; Cocoa Beach, FL, USA, April 2012.

[47] Koppenborg J., Halbauer H., Saur S., Hoek C. '3D beamforming: Performance improvements in cellular networks,' *Bell Labs Technical Journal*, 2013;**18**(2): 37–56.

[48] Heikkilä M., Kippola T., Jämsä J., Nykänen A., Matinmikko M., Keskimaula J. 'Active antenna system for cognitive network enhancement,' in *Proceedings of 5th IEEE Conference on Cognitive Infocommunications*; Vietri sul Mare, Italy, November 2014.

[49] Marzetta T. L. 'Noncooperative cellular wireless with unlimited numbers of base station antennas,' *IEEE Transactions on Wireless Communications*, 2010;**9**(11): 3590–3600.

[50] Jungnickel V., Manolakis K., Zirwas W., *et al.* 'The role of small cells, coordinated multipoint, and massive MIMO in 5G,' *IEEE Communications Magazine*, 2014;**52**(5): 44–51.

[51] ESA ARTES 1 'Scenarios for integration of satellite components in future networks (INSTINCT),' 2014–2016. https://artes.esa.int/projects/instinct.

[52] 3GPP TS 23.251 v14.0.0, 'Network sharing; architecture and functional description, (Release 14),' March 2017.

[53] Mohamed A., Lopez-Benitez M., Evans B. 'Ka band satellite terrestrial co-existence: A statistical modelling approach,' in *20th Ka and Broadband Communications, Navigation and Earth Observation Conference*; October 2014, 8 p.

[54] ITU-R F.758-5, 'System parameters and considerations in the development of criteria for sharing or compatibility between digital fixed wireless systems in the fixed service and systems in other services and other sources of interference,' 2012.
[55] CoRaSat project, Deliverable D3.2, 'Performance evaluation of existing cognitive techniques in satellite context,' 2015.
[56] ITU-R P.452-15, 'Prediction procedure for the evaluation of interference between stations on the surface of the Earth at frequencies above about 0.1 GHz,' September 2013.
[57] ITU-R P.676-10, 'Attenuation by atmospheric gases,' 2013.
[58] ITU-R P.526-13, 'Propagation by diffraction,' 2013.
[59] VTT and Elektrobit, 'Note on FS/FSS interference – Ka-band,' 2012. [Online]. Available http://cept.org/Documents/fm-44/5764/INFO005_Note-on-FSFSS-interference-Ka-band.
[60] SRTM 90m Digital Elevation Data. Available from http://srtm.csi.cgiar.org/ as of February 2014; Jarvis A., Reuter H. I., Nelson A., Guevara E., 'Hole-filled seamless SRTM data V4,' International Centre for Tropical Agriculture (CIAT), 2008, available from http://srtm.csi.cgiar.org.
[61] ITU-R F.699-7, 'Reference radiation patterns for fixed wireless system antennas for use in coordination studies and interference assessment in the frequency band from 100 MHz to about 70 GHz,' 2006.
[62] ITU-R F.1336-4, 'Reference radiation patterns of omnidirectional, sectoral and other antennas for the fixed and mobile services for use in sharing studies in the frequency range from 400 MHz to about 70 GHz,' 2014.
[63] ITU-R F.1245-2, 'Mathematical model of average and related radiation patterns for line-of-sight point-to-point fixed wireless system antennas for use in certain coordination studies and interference assessment in the frequency range from 1 GHz to about 70 GHz,' 2012.
[64] ITU-R S.465-6, 'Reference radiation pattern for earth station antennas in the fixed-satellite service for use in coordination and interference assessment in the frequency band from 2 to 31 GHz,' 2010.
[65] World Maritime News, 'First test area for autonomous ships opened in Finland,' 15th August 2017. Available: http://worldmaritimenews.com/archives/227275/first-test-area-for-autonomous-ships-opened-in-finland/.
[66] Ahvenjarvi S. 'The human element and autonomous ships,' *International Journal on Marine Navigation and Safety of Sea Transportation*, 2016;**10**(3), 517–521.
[67] Höyhtyä M., Huusko J., Kiviranta M., Sohlberg K., Rokka J. 'Connectivity for autonomous ships: Architecture, use cases, and research challenges,' in *Proceedings of the 8th International Conference on ICT Convergence*, Jeju Island, Korea, October 2017.
[68] FCC GN Docket No. 12-354, 'Amendment of the Commission's rules with regard to commercial operations in the 3550–3650 MHz band,' May 2016.
[69] Mahal J. A., Khawar A., Abdelhadi A., Clancy T. C. 'Spectral coexistence of MIMO radar and MIMO cellular system,' *IEEE Transactions on Aerospace and Electronic systems*, 2017;**53**(2): 655–668.

[70] Vartiainen J., Höyhtyä M., Vuohtoniemi R., Ramani V. L. 'The future of spectrum sensing,' in *Proceedings of the 8th International Conference on Ubiquitous and Future Networks*; Vienna, Austria, July 2016.

[71] Molisch A. F., Ratnam V. V., Han S., Li Z., Nguyen S. L. H., Li L., Haneda K. 'Hybrid beamforming for massive MIMO: A survey,' *IEEE Communications Magazine*, 2017;**55**(9): 134–141.

[72] Richart M., Baliosian J., Serrat J., Gorricho J.-L. 'Resource slicing in virtual wireless network: A survey,' *IEEE Transactions on Network and Service Management*, 2016;**13**(3): 462–476.

Chapter 18
Two-way satellite relaying

Arti M.K.[1]

Satellite communications provide various benefits like wideband transmission capability, large coverage area, and navigation assistance. Because of these benefits, these communication systems have always received a great attention. Since satellites provide ubiquitous broadband coverage over a large area of thousands of square kilometres, these are very useful for disaster recovery, which requires the establishment of broadband access from a disaster area to the rest of the world. Signal latency is the delay between requesting data and the receipt of a response, or in the case of one-way communication, between the actual moment of a signal's broadcast and the time it is received at its destination. The amount of latency depends upon the distance travelled and the speed of light. In terrestrial networks, signal latency is negligible; however, there is a big problem of signal latency associated with satellite communications. Satellite communications experiences very high latency because the signal needs to travel a very long distance [e.g. 35,786 km for geostationary earth orbit (GEO) satellite] to a satellite orbit and back to Earth again. For example, the round-trip latency of a GEO satellite communications network is almost 20 times that of a terrestrial link-based network.

This chapter discusses three major issues of two-way satellite relaying (TWSR): (1) a differential modulation-based TWSR for satellite communication systems, (2) beamforming and combining-based TWSR systems, and (3) TWSR with imperfect channel state information (CSI). Two-way amplify-and-forward (AF) relaying is a well-known concept for the relay-assisted communication systems. In the two-way AF relaying, two users can exchange their signals in two orthogonal time-slots/phases through a relaying node. This mode of communication is divided into two phases. In the first phase/time-interval, the relay simultaneously accesses by both users; this phase is also called the multiple access phase. In the next phase/time-interval, the relay broadcasts whatever it has received during the multiple access phase to both users. On the other hand, if two users want to exchange their messages by using one-way relaying, then it would require four phases/time-intervals. In this way, two-way relaying spends half number of time slots than those required in one-way relaying based two-way communication between two users. This property of the two-way relaying is very suitable for the satellite communication which

[1]Department of Electronics and Communication, Ambedkar Institute of Advanced Communication Technologies and Research, India

involves significant latency in bidirectional communication between two earth stations (ESs). However, the two ESs require knowledge of channel coefficients of all links involved with the TWSR, which is almost infeasible. In satellite communication, the uplink and downlink are non-reciprocal so it is not easy to generate the information of all links in the ground receivers in TWSR. A differential modulation-based TWSR protocol for two-way AF satellite communication may be a possible solution, which avoids the need for having CSI in the destination receiver. Further, in order to avoid the performance loss due to the differential modulation, a channel estimation-based TWSR technique will also be explained. This chapter will also discuss different beamforming-and-combining-based schemes for a TWSR system, where two multiple antenna-based ESs exchange their data through a single antenna-based transparent satellite relay.

18.1 Background

Subscribers prefer to use satellite communications because of the fact that it offers communication services using satellite as a relay. It is becoming an important and hot point in wireless communication field, and the research of channel transmission properties, latency, masking effect, and channel model are the significant aspects for engineering exploitation and design of satellite mobile communication system from practical perspective. The properties both in satellite and mobile channels are fading, multipath effect, Doppler shift, etc. [1]. There are many channel models to characterize the satellite link like Loo's model, Lutz model, Corazza model. Recently another important channel model, i.e., Shadowed Rician (SR) model is proposed by Abdi *et al.* According to the SR model proposed in [1], the entries of the line-of-sight component $\bar{\mathbf{h}}$ can be modelled as independent and identically distributed (i.i.d.) Nakagami-m random variables (RVs); and scattered component $\bar{\mathbf{h}}$ are assumed to be i.i.d. complex Gaussian RVs with zero-mean and unit variance.

Satellite systems can be broadly divided into two types: one is bent-pipe satellites, which are similar to AF-based cooperative communication systems, and the others are on board processing satellites, which are similar to decode-and-forward (DF)-based satellite systems. Bent-pipe satellites receive signals over the uplink channel from a source ES, downconvert this and forward them to the destination ES. The bent-pipe satellites are very commonly used because of their small size, low weight, and low cost. The on-board processing satellites need complex circuitry and therefore are heavier in weight as compared to the transparent satellites. Simple circuitry and computational complexity are two major factors in satellite system designing. These factors decide the power requirement and consequently the weight and cost of the satellite systems. The on-board processing satellites are useful where the de-noising property of the DF protocol is essential, but in most of the applications like broadcasting, global telephone, etc., non-regenerative/transparent satellite systems are used, The on-board processing satellites [2] are also used but represent a very limited but important class of satellites, e.g., their usage in military, defence, and emergency services. We will concentrate over AF relaying in this chapter.

Carrier-in-carrier technology is one of the latest techniques in the satellite industry to enable customers to save bandwidth costs. It allows a full duplex satellite link to be allocated the same transponder space as a single carrier. In this way, space segment saving can be increased as compared to conventional methods of duplexing [3–6].

In the terrestrial relay-assisted communication systems [7], two-way AF relaying is a well-known concept. In this relaying protocol, two terrestrial users can exchange their signals in two orthogonal time-slots/phases through a terrestrial relaying node. There are two transmission phases – in the first phase, both users transmit their data to the relay, simultaneously. In the second phase, the relay broadcasts the received data. It can be easily observed that if two users exchange their data by using one-way relaying, then four time slots are needed, whereas in two-way relaying only two time slots are required. Therefore, two-way relaying spends half number of time slots than those required in one-way relaying-based two-way communication between two users. Since latency is a big problem in satellite communication, therefore, the property of the two-way relaying is very suitable for the satellite communication which involves significant latency in bidirectional communication between two ESs. The concept of two-way relaying is explored in satellite communication in [8–11]. Similar to terrestrial two-way relaying, two ESs communicate via satellite by TWSR. However, in most of existing works [8–10], common assumption is that each ES has perfect knowledge of the CSI, which is required by the destination to cancel self-interference and decoding of transmitted data.

Since the two-way relaying utilizes smaller (half) number of time slots than those required in one-way relaying, two-way relaying is useful for reducing the delay involved in the transmission of the data of both users. The delay reduction property of two-way relaying is suitable for reducing the delay in satellite communications. Moreover, the concept of two-way relaying can be extended to multi-way relaying, where multiple ESs can communicate to each other via satellite node for further reduction in delay. It is well known that to detect the transmitted symbol, CSI is required at the destination ES. Other techniques like beamforming and combining also requires CSI at transmitting and receiving ES. In practical set-ups, perfect CSI is not available at different nodes; therefore, CSI needs to be estimated.

18.2 Two-way satellite relaying

Channel estimation, differential modulation, and beamforming and combining are major challenges in TWSR. We focus on these challenges in this chapter. First, we discuss the problems associated with channel estimation. Non-reciprocity and high value of latency are two main challenges in CSI estimation at ESs in satellite communication systems. In satellite communication, different frequency bands are assigned for uplink and downlink transmissions. The downlink denotes for the transmission from the satellite to the ES, and uplink stands for the transmission from the ES to satellite. Therefore, the property of reciprocity does not exist in between the uplink and downlink channels in the satellite communication. Consequently, the uplink and downlink channels need to be estimated separately. This scenario is different from the terrestrial communication systems, where the estimation of downlink channel is sufficient due to reciprocity between uplink and downlink channels. However, reverse training-based estimation of uplink channel in ES is not possible, as the satellite *cannot* transmit over the uplink frequencies due to the large link budget requirement and other practical limitations [12]. Channel estimation can be avoided by using differential modulation. In differential modulation, the symbol is detected with the help of previous symbol. It is assumed in many works of detector design that the CSI can be reliably estimated at different nodes, either by training or blind estimation techniques. However, channel estimation is not straightforward in many cases, especially in the case of satellite communication systems, where channels are fast fading. The need for differential or non-coherent modulation techniques to bypass channel estimation in wireless communication systems has been explored in [13–15].

Another big problem in case of satellite communication is high value of latency and fast variation of the channel due to its low elevation angle and atmospheric fluctuations. These problems can be solved by using TWSR. As can be seen from the previous discussion that it is very difficult to generate the information of the uplink channel in the ES because of the large round trip delays involved with the satellite communications. Therefore, the practical implementation of the two-way relaying protocol in the satellite links is not an easy task. In [11], the CSI estimation is bypassed by using differential modulation-based TWSR; however, it leads to heavy penalty in terms of the error performance. If channel estimation is performed, then with the help of estimated channel gains beamforming and combining can be used to improve the error performance in satellite communication systems.

A two-way cooperative system with two ESs (with single antenna in case of channel estimation and differential relaying and multiple antennas for beamforming and combining) and a satellite with single antenna is considered as shown in Figure 18.1. Both ESs are involved in a two-way communication via a satellite, i.e. they wish to exchange their signals via a satellite. For realizing the TWSR scheme between the two ESs, we assume that both ESs are lying in a common beam of the satellite. We assume that all links have the same block fading duration, but they can fade differently. Moreover, it is assumed that both ESs are geometrically separated by a very large distance, and hence, a direct communication between them is not feasible. We assume perfect synchronization between both ESs and the satellite.

In two-way relaying, transmission of data occurs in two phases; during the first phase, both ESs transmit their data to the satellite. In the second phase, the bent-pipe type satellite amplifies the received signals with a fixed transponder gain, and then broadcasts to both ESs.

The received signal at the satellite in the first phase is given by

$$y_s = \sum_{i=1}^{2} h_i s_i + w_s, \tag{18.1}$$

where w_s represents the additive white Gaussian noise (AWGN) at the satellite, containing zero mean complex Gaussian noise elements with variance σ_s^2; s_i is the symbol

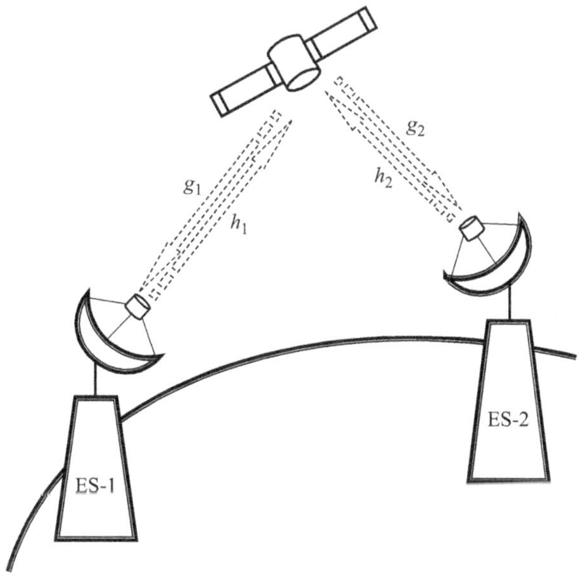

Figure 18.1 Two-way satellite relaying between two Earth stations

of ES-i from M-ary phase shift keying (M-PSK) constellation with E_s energy; h_i denotes the uplink channel coefficient of ES-i.

The satellite receives these signals, scales the received amplitude by its transponder gain a, downconverts the carrier frequency, and broadcasts the signal to both ESs. Consequently, the data received at the ES-i is given by

$$y_i = ag_i y_s + w_i, \tag{18.2}$$

where w_i is the AWGN noise with variance σ_i^2. The downlink channel coefficient of ES-i is represented by g_i.

From (18.1) and (18.2), we can write the signal received at ES-i in a simplified form as

$$y_i = ag_i h_i s_i + ag_i h_j s_j + ag_i w_s + w_i. \tag{18.3}$$

It can be seen from (18.3) that the received signal contains a self-interference term $ag_i h_i s_i$. In this term, ES-i has perfect knowledge of a and s_i (its own symbol). But it does not have knowledge of its uplink and downlink channel coefficients h_i and g_i, respectively. Therefore, the self-interference term is unknown at the ES. In order to decode the symbol s_j, this interference term need to be removed. For this, ES-i requires perfect information of g_i and h_j. Further, after removal of the self-interference term, s_j can be decoded if the uplink channel of ES-j, i.e., h_j is also perfectly known along with g_i. In the succeeding section, we discuss the challenges involved in the estimation of these channel gains in ESs.

All channel gains are modelled as SR fading channels. The probability distribution function (PDF) of $|g_i|^2$ is given by [1]

$$f_{|g_i|^2}(x) = \alpha_i e^{-\beta_i x}_1 F_1(m_i; 1; \delta_i x), \quad x > 0, \tag{18.4}$$

where $i = 0, 1$, $\alpha_i = 0.5(2b_i m_i/(2b_i m_i + \Omega_i))^{m_i}/b_i$, $\beta_i = (0.5/b_i)$, $\delta_i = 0.5\Omega_i/(2b_i^2 m_i + b_i \Omega_i)$, the parameter Ω_i is the average power of LOS component, $2b_i$ is the average power of the multipath component, and $0 \le m_i \le \infty$ is the Nakagami parameter, for $m_i = 0$ and $m_i = \infty$, the envelope of h_i follows the Rayleigh and Rician distribution, respectively; and ${}_1F_1(a; b; z)$ is the confluent hypergeometric function [16, Eq. (9.210)]. The PDF of h_i can be obtained from (18.4) by replacing $m_i, b_i, \alpha_i, \beta_i, \delta_i, \Omega_i$ with $\tilde{m}_i, \tilde{b}_i, \tilde{\alpha}_i, \tilde{\beta}_i, \tilde{\delta}_i, \tilde{\Omega}_i$, respectively.

18.3 Training-based two-way satellite relaying system

The optimal channel estimation and training design for reciprocal terrestrial two-way relay networks is studied in [17]. The channel estimators proposed in [17] either require brute force search of channel estimates or perfect information of the second-order statistics of uplink and downlink channels and noise in the receiving nodes. In satellite communication, the uplink and downlink are non-reciprocal, so, it is not easy to generate the information of statistics of all links in the ground receiver in TWSR. Some other techniques for terrestrial two-way communication systems are given in [18,19]. But, these are not suitable for TWSR because of latency and non-reciprocal nature of links. Further, existing orthogonal training for terrestrial two-way systems is not PAPR efficient; therefore, it is not suitable for TWSR.

In this section, a training design is discussed for TWSR systems. We would first discuss the detection of symbol in a two-way satellite system and derive a maximum likelihood (ML) for the transmitted symbol of ES-j at ES-i. Based on this optimal detector, we present a training design for these systems.

Let us first rewrite (18.3) as

$$y_i = aG_i s_i + aG_{i,j} s_j + a g_i w_s + w_i, \tag{18.5}$$

where $G_i = g_i h_i$ and $G_{i,j} = g_i h_j$ denote the cascaded channel gains of the self and cooperative link of ES-i. From (18.5), the decision metric for the symbol s_j can be written by maximizing conditional PDF, as

$$\hat{s}_j = \arg\min_{s_j} \left| y_i - aG_i s_i - aG_{i,j} s_j \right|^2. \tag{18.6}$$

In (18.6), it is an ML detector under the assumption with perfect CSI. However, channel knowledge is not available at ES-i, ES-i contains perfect information of s_i (its own symbol) and a (fixed transponder gain of satellite), but it does not have information of G_i and $G_{i,j}$. Therefore, G_i and $G_{i,j}$ need to be estimated.

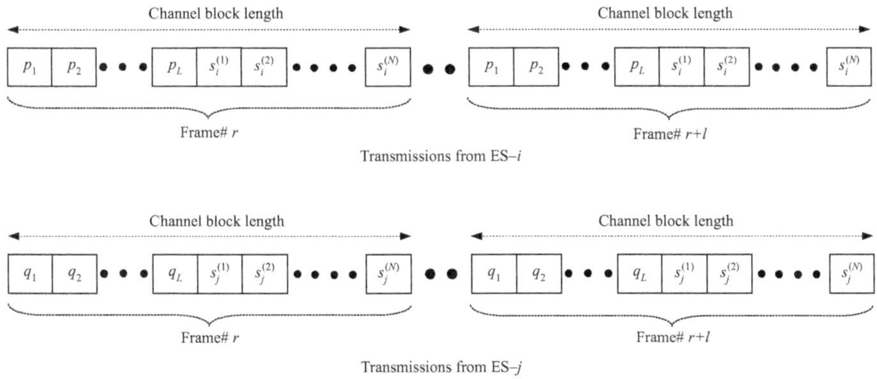

Figure 18.2 Transmission of training and data symbols in two-way satellite relaying from both ESs

These channel estimates are used in the place of exact CSI in (18.6), and the decision metric with estimated channel gains can be written as

$$\hat{s}_j = \arg\min_{s_j} \left| y_i - a\hat{G}_i s_i - a\hat{G}_{i,j} s_j \right|^2, \tag{18.7}$$

where $\hat{G}_i$ and $\hat{G}_{i,j}$ are the estimates of G_i and $G_{i,j}$, respectively.

It is apparent from the discussion in the previous section that the satellite link can be assumed block fading over sufficiently large number of symbol transmissions because of the very large bandwidth of the satellite links, though this block fading period is smaller than the round trip propagation delay of the signal. This assumption is approximately satisfied because both ESs are in the common beam of the satellite and hence fade almost simultaneously.

Let us assume that both ESs have frames of symbols to be exchanged via the satellite by using the two-way relaying. The length of each frame is equal to the block fading length of the satellite links. The data frames of both ESs are shown in Figure 18.2. It can be seen from Figure 18.2 that in the beginning of each frame, we embed L, $L \in \mathcal{Z}$, training symbols p_k and q_k in the data frames of the ES-i and ES-j, respectively, where $k = 1, 2, \ldots, L$. In Figure 18.7, $s_i^{(n)}$ and $s_j^{(n)}$ represent the symbols transmitted by ES-i and ES-j, respectively, in the n-th, $n = 1, 2, \ldots, N$, time interval in a frame of duration $L + N$ symbol transmission time intervals.

The signals received in the ES-i during the training period ($k = 1, \ldots, L$) can be written by using (18.5) as

$$z_i^{(k)} = aG_i p_k + aG_{i,j} q_k + ag_i w_s + w_i. \tag{18.8}$$

Further, from (18.5), the signals received during the data transmission phase ($n = 1, 2, \ldots, N$) will be

$$y_i^{(n)} = aG_i s_i^{(n)} + aG_{i,j} s_j^{(n)} + ag_i w_s^{(n)} + w_i^{(n)}, \tag{18.9}$$

where $w_s^{(n)}$ and $w_i^{(n)}$ are the AWGN with σ_s^2 and σ_i^2 variances, respectively. The ML detector of $s_j^{(n)}$ in ES-i can be obtained by using (18.7) as

$$\hat{s}_j^{(n)} = \arg\min_{s_j^{(n)}} \left| y_i^{(n)} - a\hat{G}_i s_i^{(n)} - a\hat{G}_{i,j} s_j^{(n)} \right|^2. \tag{18.10}$$

We can write a matrix relation by putting all received signals at ES-i during the training period [given in (18.8)] together in the form of a column vector:

$$\mathbf{z}_i = aG_i\mathbf{p} + aG_{i,j}\mathbf{q} + ag_i\mathbf{w}_s + \mathbf{w}_i. \tag{18.11}$$

In (18.11), $\mathbf{z}_i = [z_i^{(1)}, z_i^{(2)}, \ldots, z_i^{(L)}]^T$, $\mathbf{p} = [p_1, p_2, \ldots, p_L]^T$, $\mathbf{q} = [q_1, q_2, \ldots, q_L]^T$ are $L \times 1$ column vectors, (here $(\cdot)^T$ denotes the transpose) and $\mathbf{w}_s$ and $\mathbf{w}_i$ contain AWGN elements. Let $\mathbf{v}_i \in \mathbb{C}^{L\times 1}$ be the combining vector which is used for processing the received signal $\mathbf{z}_i$ in the receiver. After left multiplying $\mathbf{z}_i$ by $\mathbf{v}_i^H$, where $(\cdot)^H$ stands for hermitian, we get

$$\mathbf{v}_i^H\mathbf{z}_i = aG_i\mathbf{v}_i^H\mathbf{p} + aG_{i,j}\mathbf{v}_i^H\mathbf{q} + ag_i\mathbf{v}_i^H\mathbf{w}_s + \mathbf{v}_i^H\mathbf{w}_i. \tag{18.12}$$

In order to remove the contribution of $G_{i,j}$ from (18.12), we should choose $\mathbf{v}_i$ such that $\mathbf{v}_i^H\mathbf{q} = 0$ and $\mathbf{v}_i^H\mathbf{p} \neq 0$. Similarly, for removing the contribution of G_i from (18.12), we should have $\mathbf{v}_i^H\mathbf{q} \neq 0$ and $\mathbf{v}_i^H\mathbf{p} = 0$. If $\mathbf{p}$ and $\mathbf{q}$ are orthogonal vectors, then $\mathbf{p}^H\mathbf{q} = \mathbf{q}^H\mathbf{p} = 0$. This property allows for using $\mathbf{v}_i = \mathbf{p}$ and $\mathbf{v}_i = \mathbf{q}$ for removing the contributions of $G_{i,j}$ and G_i, respectively, from (18.12). Under the assumption that $\mathbf{p}$ and $\mathbf{q}$ are orthogonal to each other, let us put $\mathbf{v}_i = \mathbf{p}$ in (18.12) and get

$$\mathbf{p}^H z_i = aG_i\mathbf{p}^H\mathbf{p} + ag_i\mathbf{p}^H\mathbf{w}_s + \mathbf{p}^H\mathbf{w}_i. \tag{18.13}$$

We can rewrite (18.13) as

$$z_i' = aPG_i + ag_iw_s' + w_i', \tag{18.14}$$

where $z_i' = \mathbf{p}^H\mathbf{z}_i$, $w_s' = \mathbf{p}^H\mathbf{w}_s$, $w_i' = \mathbf{p}^H\mathbf{w}_i$, and $P = \mathbf{p}^H\mathbf{p}$. From (18.14), we get the ML estimate of G_i by

$$\hat{G}_i = \frac{z_i'}{aP}. \tag{18.15}$$

Now after putting $\mathbf{v}_i = \mathbf{q}$ in (18.12), we get

$$z_i'' = aQG_{i,j} + ag_iw_s'' + w_i'', \tag{18.16}$$

where $z_i'' = \mathbf{q}^H\mathbf{z}_i$, $w_s'' = \mathbf{q}^H\mathbf{w}_s$, $w_i'' = \mathbf{q}^H\mathbf{w}_i$, and $Q = \mathbf{q}^H\mathbf{q}$. From (18.14), the ML estimate of $G_{i,j}$ will be

$$\hat{G}_{i,j} = \frac{z_i''}{aQ}. \tag{18.17}$$

The mean square errors in the channel estimates are given by

$$\begin{aligned} \Sigma_{G_i} &= E\left\{\left(\hat{G}_i - G_i\right)\left(\hat{G}_i - G_i\right)^*\right\}, \\ \Sigma_{G_{i,j}} &= E\left\{\left(\hat{G}_{i,j} - G_{i,j}\right)\left(\hat{G}_{i,j} - G_{i,j}\right)^*\right\}, \end{aligned} \tag{18.18}$$

where $E\{\cdot\}$ denotes the expectation over the AWGN. It can be easily shown after some algebra and from (18.13)–(18.18) that

$$\Sigma_{G_i} = \frac{\left(a^2|g_i|^2\sigma_s^2 + \sigma_i^2\right)}{a^2\mathbf{p}^H\mathbf{p}},$$

$$\Sigma_{G_{i,j}} = \frac{\left(a^2|g_i|^2\sigma_s^2 + \sigma_i^2\right)}{a^2 q^H q}. \tag{18.19}$$

Let us put a constraint over the training power by $\mathbf{p}^H\mathbf{p} \leq S$ and $q^H\mathbf{q} \leq S$. The efficiency of the power amplifier is an important factor in the satellite communications. The power amplifier performs efficiently if the peak-to-average power ratio (PAPR) is small. For example, the PAPR of the transmitted signals can be reduced by avoiding zero transmissions at different time instances. Similarly, the PAPR would be minimum if the power of the training symbols is constant. After these observations, we state the following optimization problem:

$$\begin{aligned} &\text{minimize } \Sigma_{G_i}, \Sigma_{G_{i,j}}, \text{PAPR} \\ &\text{such that } \mathbf{p}^H\mathbf{p} \leq S,\ q^H\mathbf{q} \leq S \\ &\mathbf{p}^H\mathbf{q} = q^H\mathbf{p} = 0. \end{aligned} \tag{18.20}$$

Note that Σ_{G_i} and $\Sigma_{G_{i,j}}$ are minimized for $\mathbf{p}^H\mathbf{p} = \mathbf{q}^H\mathbf{q} = S$, as can be seen from (18.19). Therefore, let us only keep equality in the constraints and then we can rewrite the optimization problem of (18.20) as

$$\begin{aligned} &\text{minimize PAPR} \\ &\text{such that } \mathbf{p}^H\mathbf{p} = S, q^H\mathbf{q} = S \\ &\mathbf{p}^H\mathbf{q} = q^H\mathbf{p} = 0. \end{aligned} \tag{18.21}$$

There are (possibly infinitely) many solutions of the optimization problem of (18.21). Few possible solutions of the optimization problem in (18.21) are

$$\mathbf{p} = \sqrt{\frac{S}{L}}[1, -1, 1, -1, \ldots]^T, \quad \mathbf{q} = \sqrt{\frac{S}{L}}[1, 1, 1, 1, \ldots]^T$$

or

$$\mathbf{p} = \sqrt{\frac{S}{L}}[1, 1, 1, 1, \ldots]^T, \quad \mathbf{q} = \sqrt{\frac{S}{L}}[1, -1, 1, -1, \ldots]^T$$

or

$$\mathbf{p} = \sqrt{\frac{S}{2L}}[1+j, -1-j, 1+j, -1-j, \ldots]^T,$$

$$\mathbf{q} = \sqrt{\frac{S}{2L}}[1+j, 1+j, 1+j, 1+j, \ldots]^T$$

or

$$\mathbf{p} = \sqrt{\frac{S}{2L}}[1+j, 1+j, 1+j, 1+j, \ldots]^T,$$

$$\mathbf{q} = \sqrt{\frac{S}{2L}}[1+j, -1-j, 1+j, -1-j, \ldots]^T \tag{18.22}$$

It can be seen from (18.22) that the PAPR of all the training sequences is one, which is the minimum value of the PAPR. In general, the MSE and PAPR optimal training sequences for the considered TWSR system are given by

$$\mathbf{p} = \sqrt{\frac{S}{(u^2+v^2)L}}[u+jv, -u-jv, u+jv, -u-jv, \ldots]^T,$$

$$\mathbf{q} = \sqrt{\frac{S}{(u^2+v^2)L}}[u+jv, u+jv, u+jv, +jv, \ldots]^T, \tag{18.23}$$

where u and v are arbitrary real values.

18.3.1 Average BER

Let us assume that the ESs use Gray coding for encoding $\log_2 M$ bits in a symbol belonging to the M-PSK constellation. It is established in [20], by using signal-space concepts, that decoding of each bit of the bit-mapping Gray code of an M-PSK symbol can be performed by using independent binary hard decisions. Therefore, the instantaneous bit error rate (BER) at the ES-i for the M-PSK constellation is given by

$$\mathrm{Pe}_i(\gamma_i) = \xi_M \sum_{k=1}^{\eta_M} Q(\sqrt{g_k \gamma_i}), \tag{18.24}$$

where $Q(\cdot)$ denotes the q-function; $\xi_M = 2/\max(\log_2 M, 2)$, $\eta_M = \max(M/4, 1)$, and $g_k = 2\sin^2((2k-1)\pi/M)$ are the modulation specific parameters.

The average BER can be obtained after many algebra is given by

$$\begin{aligned}
\mathrm{Pe}_i(\bar{\gamma}) \cong\ & \xi_M \sum_{k=1}^{\eta_M} \sum_{l=1}^{3} \frac{B_l \alpha_i \tilde{\alpha}_j}{(\tilde{\beta}_j - \tilde{\delta}_j)^{\tilde{d}_j}} \sum_{l_j=0}^{\tilde{c}_j} \binom{\tilde{c}_j}{l_j} \tilde{\beta}_j^{\tilde{c}_j - l_j} \sum_{l_i=0}^{c_i} \binom{c_i}{l_i} \\
& \times \beta_i^{c_i - l_i} \left(\frac{g_k A_l \bar{\gamma}}{(1 + (2E_s/a^2 S))} \right)^{l_j} \Big(\mathcal{D}\big(l_i, d_i, l_j, \tilde{d}_j, \kappa_{j,k}\big) + \epsilon_i \delta_i \\
& \times \mathcal{D}\big(l_i, d_i + 1, l_j, \tilde{d}_j, \kappa_{j,k}\big) + \frac{\tilde{\epsilon}_j \tilde{\delta}_j}{\tilde{\beta}_j - \tilde{\delta}_j} \mathcal{D}\big(l_i, d_i, l_j, \tilde{d}_j + 1, \kappa_{j,k}\big) \\
& + \frac{\epsilon_i \delta_i \tilde{\epsilon}_j \tilde{\delta}_j}{\tilde{\beta}_j - \tilde{\delta}_j} \mathcal{D}\big(l_i, d_i + 1, l_j, \tilde{d}_j + 1, \kappa_{j,k}\big) \Big),
\end{aligned} \tag{18.25}$$

where

$$\mathcal{D}\left(l_i, d_i, l_j, \tilde{d}_j, \kappa_{j,k}\right) = \sum_{m=0}^{\tilde{d}_j - l_j} \binom{\tilde{d}_j - l_j}{m} \frac{\kappa_{j,k}^{-(l_j+m+d_i-l_i)}}{\Gamma(d_i)\,\Gamma\left(\tilde{d}_j\right) c^{l_j+m}}$$

$$\times G_{23}^{22}\left(\frac{\beta_i - \delta_i}{\kappa_{j,k}} \Big| \begin{matrix} 1-d_i, 1-\left(l_j+m+d_i-l_i\right) \\ 0, \tilde{d}_j - l_j - m - d_i + l_i, 1 - d_i + l_i \end{matrix}\right), \tag{18.26}$$

$$\kappa_{j,k} = \frac{g_k A_l \bar{\gamma} + \left(1 + \left(2E_s/a^2 S\right)\right)\left(\tilde{\beta}_j - \tilde{\delta}_j\right)}{\left(1 + \left(2E_s/a^2 S\right)\right)\left(\tilde{\beta}_j - \tilde{\delta}_j\right)c}, \tag{18.27}$$

and $G_{p,q}^{m,n}\left(\cdot \,|\, \begin{matrix}\cdots\\ \cdots\end{matrix}\right)$ is the Meijer-G function [16, Eq. (9.301)]. The details of derivation are given in [21].

18.3.2 Ergodic capacity

Ergodic capacity, the average capacity of TWSR scheme for ES-i, can be written in terms of MGF by putting $p=2$ and $q=1$ in [22, Eq. (10)] as

$$C_i = \frac{B}{\ln 2} \sum_{n=1}^{N} v_n U_1(s_n) \left\{ \frac{\delta}{\delta s} M_{\gamma_i}(s)|_{s \to s_n} \right\}, \tag{18.28}$$

where B stands for the bandwidth, $M_{\gamma_i}(s) = E_{\gamma_i}\{e^{-s\gamma_i}\}$ denotes the MGF of the received signal-to-noise-ratio (SNR) at ES-i, i.e. γ_i,

$$s_n = \tan\left(\frac{\pi}{4}\cos\left(\frac{2n-1}{2N}\pi\right) + \frac{\pi}{4}\right), \tag{18.29}$$

$$v_n = \frac{\pi^2 \sin(((2n-1)/2N)\pi)}{4N\cos^2((\pi/4)\cos(((2n-1)/2N)\pi) + (\pi/4))}, \tag{18.30}$$

$$U_1(s_n) = -H_{3,2}^{1,2}\left[\frac{1}{s_n} \Big| \begin{matrix} (1,1), (1,1), (1,1) \\ (1,1), (0,1) \end{matrix}\right], \tag{18.31}$$

where $H_{3,2}^{1,2}[\cdot]$ is the Fox's H function [23] and N is a positive integer. We have an alternative representation for $U_1(s_n)$ in the form of the Meijer G function:

$$U_1(s_n) = -G_{2,1}^{0,2}\left(\frac{1}{s_n} \,\Big|\, \begin{matrix} 1,1 \\ 0 \end{matrix}\right). \tag{18.32}$$

We get the MGF of γ_i after many algebra, as

$$M_{\gamma_i}(s) \cong \frac{\alpha_i \tilde{\alpha}_j}{(\tilde{\beta}_j - \tilde{\delta}_j)^{\tilde{d}_j}} \sum_{l_j=0}^{\tilde{c}_j} \binom{\tilde{c}_j}{l_j} \tilde{\beta}_j^{\tilde{c}_j - l_j} \sum_{l_i=0}^{c_i} \binom{c_i}{l_i} \beta_i^{c_i - l_i}$$

$$\times \left(\frac{s\bar{\gamma}}{1 + \left(2E_s/a^2 S\right)}\right)^{l_j} \left(\mathcal{J}\left(l_i, d_i, l_j, \tilde{d}_j, \vartheta_{j,k}\right) + \epsilon_i \delta_i\right.$$

$$\times \mathcal{J}\left(l_i, d_i + 1, l_j, \tilde{d}_j, \vartheta_{j,k}\right) + \frac{\tilde{\epsilon}_j\tilde{\delta}_j}{\tilde{\beta}_j - \tilde{\delta}_j}\mathcal{J}\left(l_i, d_i, l_j, \tilde{d}_j + 1, \vartheta_{j,k}\right)$$

$$+ \frac{\epsilon_i\delta_i\tilde{\epsilon}_j\tilde{\delta}_j}{\tilde{\beta}_j - \tilde{\delta}_j}\mathcal{J}\left(l_i, d_i + 1, l_j, \tilde{d}_j + 1, \vartheta_{j,k}\right)\Bigg), \tag{18.33}$$

where

$$\mathcal{J}\left(l_i, d_i, l_j, \tilde{d}_j, \vartheta_{j,k}\right) = \sum_{m=0}^{\tilde{d}_j - l_j} \binom{\tilde{d}_j - l_j}{m} \frac{\vartheta_{j,k}^{-(l_j+m+d_i-l_i)}}{\Gamma(d_i)\,\Gamma\left(\tilde{d}_j\right)c^{l_j+m}}$$

$$\times G_{23}^{22}\left(\frac{\beta_i - \delta_i}{\vartheta_{j,k}}\middle| \begin{matrix} 1 - d_i, 1 - (l_j + m + d_i - l_i) \\ 0, \tilde{d}_j - l_j - m - d_i + l_i, 1 - d_i + l_i \end{matrix}\right), \tag{18.34}$$

$$\vartheta_{j,k} = \frac{s\bar{\gamma} + \left(1 + \frac{2E_s}{a^2S}\right)\left(\tilde{\beta}_j - \tilde{\delta}_j\right)}{\left(1 + \frac{2E_s}{a^2S}\right)\left(\tilde{\beta}_j - \tilde{\delta}_j\right)c}. \tag{18.35}$$

The capacity of the training-based TWSR system can be calculated by using (18.28) and first order derivative of MGF.

18.3.3 Numerical results and discussion

A TWSR system with two ESs with single antenna at each of them and a satellite node with a single antenna is considered for simulation and analysis. The bent-pipe transponder with unity gain is considered. All links are assumed to be the SR fading LMS links. Three SR fading scenarios are considered for all numerical results: (1) frequent heavy shadowing (FHS) ($b_i = 0.063$, $m_i = 0.739$, $\Omega_i = 8.97 \times 10^{-4}$), (2) average shadowing (AS) ($b_i = 0.126$, $m_i = 10.1$, $\Omega_i = 0.835$), and (3) infrequent light shadowing (ILS) ($b_i = 0.158$, $m_i = 19.4$, $\Omega_i = 1.29$). All these fading scenarios are listed in [1]. The FHS SR fading is the most severe form of shadowing which is due to the heavy snow, rain, or storm, and it almost blocks the satellite transmissions. The ILS SR fading has a light shadowing and satellite link performs the best under this scenario. The combined fading scenario for TWSR system is named like AS/FHS if g_i and h_j undergo the AS and FHS, respectively.

In Figure 18.3, the simulated and analytical BERs versus SNR performance of the TWSR scheme for quadrature phase-shift keying (QPSK) constellation and FHS/FHS, FHS/AS, FHS/ILS, AS/AS, AS/ILS, and ILS/ILS fading scenarios is shown. The length of training sequences is kept as $L = 2$ for simulation and analysis. We use a unit-norm QPSK constellation such that $E_s = 1$. It is assumed that the total power devoted to training is $S = E_sL = L$. Further, we assume that $\sigma_s^2 = \sigma_i^2 = \sigma^2$ and $\sigma^2 = 1/\text{SNR}$. The SNR is shown on the x-axis of all figures. In simulations, we transmit the following training sequences: $\mathbf{p} = [1, 1]^T$ (from ES-1) and $\mathbf{q} = [1, -1]^T$ (from ES-2), in the beginning of each frame. The SR fading channels of h_i, g_i, and h_j

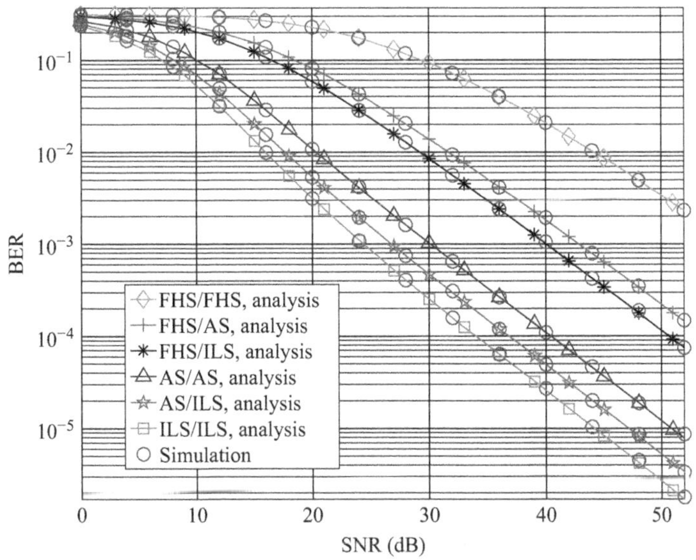

Figure 18.3 BER versus SNR performance of TWSR scheme with $L=2$ and under different fading scenarios

are assumed to fade together over a block of 20 symbol transmission periods. The BER performance of ES-1 is shown in the figure. The analytical BER values are obtained in closed form by using (18.25). A close match of the simulated and analytical BERs is evident from the figure for all considered fading scenarios and SNR values. Hence, the presented BER analysis very accurately predicts the error performance of the training-based TWSR system at all SNR values considered in the figure. Therefore, for exploring the characteristics of TWSR system in detail, we can use the presented analytical BER results.

In Figure 18.4, the analytical BER at ES-1 of the considered TWSR scheme is plotted for QPSK constellation, with different training lengths $L=2,4,6,8,10$, and under ILS/FHS, AS/AS, and ILS/ILS fading scenarios. The BER performance of ES-1 with perfect CSI ($L=\infty$) is also shown in the figure. However, this is just a theoretical scenario and completely bared for practical situations. In practice, we try to minimize the training sequence to save the bandwidth for useful data transmission. It is shown in the figure that $L=10$ provides a channel estimation extremely close to perfect CSI. It can be seen from the figure that the training-based TWSR scheme performs very close to the ideal TWSR scheme with perfect CSI in all considered fading scenarios and at all SNR values. Even for the smallest training length, i.e. $L=2$, it loses only 3 dB SNR gain as compared to the perfect CSI-based TWSR scheme as shown in the figure. It can be seen from the figure that the SNR loss reduces to 1.75 dB for $L=4$ as compared to 3 dB for $L=2$. The SNR loss can be further reduced to 1.2 dB by using only six training symbols, as shown in the figure. Further, the scheme works closer to the ideal TWSR scheme with $L=8$ and $L=10$ training symbols. Looking

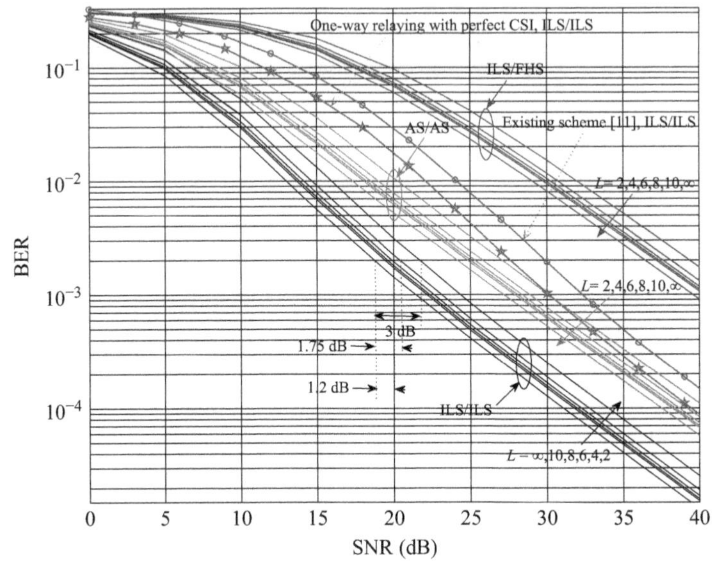

Figure 18.4 BER versus SNR performance of the existing differential TWSR scheme [11] with QPSK constellation, one-way satellite relaying scheme with 16-PSK constellation, and TWSR system with QPSK constellation, $\mathbf{L} = \mathbf{2, 4, 6, 8, 10}, \infty$*, and under different fading scenarios;* $\mathbf{L} = \infty$ *denotes the perfect CSI in ES-1*

at the large bandwidth of the satellite links, a training sequence of length $L = 10$ is very affordable. However, even $L = 6$ is also a very good compromise in all fading scenarios as indicated by the figure. In addition, the simulated BER performance of one-way satellite relaying scheme using 16-PSK constellation and perfect CSI is also shown in Figure 18.4 under ILS/ILS shadowing environment. It can be seen from the figure that the training-based TWSR scheme significantly outperforms the trivial one-way relaying scheme which employs perfect CSI.

The average capacity versus SNR plots for FHS/FHS fading scenario are shown in Figure 18.5. The satellite transmission bandwidth is assumed to be 36 MHz which lies within the specified bandwidth for L, C, Ku, and Ka band. The simulated values of the average capacity of the TWSR are also shown in the figure. Further, the capacity of the TWSR scheme is severely affected due to the poor training, i.e. small value of L. For example, it is shown in the figure that at 11 dB SNR, the considered scheme loses about 46% capacity with $L = 2$ as compared to the casewhen perfect CSI is available at the ESs. Moreover, it can be seen from the figure that the training-based TWSR scheme is able to achieve very large value of capacity, i.e. 7.4 Mbps with training length of $L = 10$ at 12 dB SNR and under FHS/FHS fading scenario (which is the worst fading scenario).

Useful closed-form analytical expressions for the BER and average capacity of the training-based TWSR scheme have been derived. These expressions have been

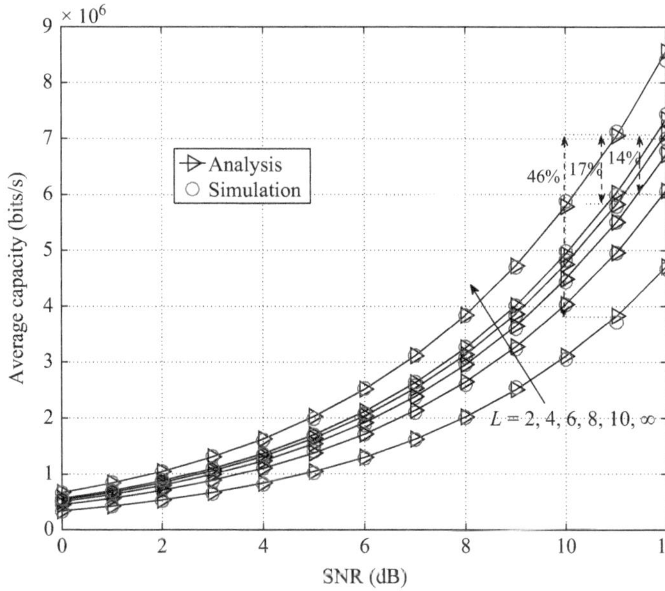

Figure 18.5 Average capacity (bits/s) versus SNR performance of the TWSR system with $L = 2,4,6,8,10,\infty$ under FHS/FHS fading scenarios; $L = \infty$ denotes the perfect CSI in ESs

verified to be very accurate by matching the simulated and analytical values. The derived BER and capacity expressions have been used to explore some useful findings about the scheme. The considered scheme has been found to perform very close (from BER and capacity point-of-view) to the perfect CSI-based TWSR system for training lengths of $L = 8$ and $L = 10$.

18.4 Differential modulation-based TWSR

Differential modulation is very useful because of the fact that it does not require CSI [24,25]. The transmitter introduces correlation in the stream of the transmitted symbols by using some special operations like multiplication or modulo operation such that this correlation can be utilized by the receiver to skip the channel estimation for decoding of the currently transmitted symbol in differential modulation. As seen from previous section, the TWSR-based communication has a big problem with the estimation of CSI of different links, differential modulation may be a solution for this system.

Let $x_i[n]$ be a differentially modulated symbol transmitted by the ES-i, then we can write

$$x_i[n] = x_i[n-1]s_i[n], \tag{18.36}$$

where $s_i[n]$ denotes the information containing symbol to be transmitted in the n-th time interval; $|x_i[n]|^2 = 1$ which denotes that $s_i[n]$ belongs to a unit-norm M-PSK constellation. The signal received at ES-i in three consecutive time intervals $n-2$, $n-1$, and n can be written as

$$\begin{aligned} y_i[n-2] &= ag_ih_ix_i[n-2] + ag_ih_jx_j[n-2] \\ &\quad + ag_ie_s[n-2] + e_i[n-2], \\ y_i[n-1] &= ag_ih_ix_i[n-1] + ag_ih_jx_j[n-1] \\ &\quad + ag_ie_s[n-1] + e_i[n-1], \\ y_i[n] &= ag_ih_ix_i[n] + ag_ih_jx_j[n] + ag_ie_s[n] + e_i[n]. \end{aligned} \tag{18.37}$$

Since $x_i[n]$ is perfectly known in ES-i, we can obtain the following relations from (18.36) and (18.37):

$$\begin{aligned} y_i'[n-1] &= y_i[n-1]x_i^*[n-1] - y_i[n-2]x_i^*[n-2] \\ &= ag_ih_jx_i^*[n-2]\,x_j[n-2]\left(s_i^*[n-1]s_j[n-1]-1\right) \\ &\quad + ag_ie_s[n-1]\,x_i^*[n-1] - ag_ie_s[n-2]\,x_i^*[n-2] \\ &\quad + e_i[n-1]\,x_i^*[n-1] - e_i[n-2]x_i^*[n-2], \end{aligned}$$

$$\begin{aligned} y_i'[n] &= y_i[n]x_i^*[n] - y_i[n-1]x_i^*[n-1] \\ &= ag_ih_jx_i^*[n-1]\,x_j[n-1]\left(s_i^*[n]s_j[n]-1\right) \\ &\quad + ag_ie_s[n]\,x_i^*[n] - ag_ie_s[n-1]\,x_i^*[n-1] \\ &\quad + e_i[n]\,x_i^*[n] - e_i[n-1]\,x_i^*[n-1]. \end{aligned} \tag{18.38}$$

First we define the following intermediate variables:

$$\begin{aligned} h_{i,j}[n-2] &\triangleq ag_ih_jx_i^*[n-2]\,x_j[n-2] \\ z_i[n-1] &\triangleq s_i^*[n-1]s_j[n-1]-1. \end{aligned} \tag{18.39}$$

From (18.36), (18.38), and (18.39), we get

$$\begin{aligned} y_i'[n-1] &= h_{i,j}[n-2]z_i[n-1] + w_i[n-1] \\ y_i'[n] &= h_{i,j}[n-2]s_i^*[n-1]s_j[n-1]z_i[n] + w_i[n], \end{aligned} \tag{18.40}$$

where

$$\begin{aligned} w_i[n-1] &= ag_ie_s[n-1]x_i^*[n-1] - ag_ie_s[n-2]x_i^*[n-2] \\ &\quad + e_i[n-1]x_i^*[n-1] - e_i[n-2]x_i^*[n-2] \\ w_i[n] &= ag_ie_s[n]x_i^*[n] - ag_ie_s[n-1]x_i^*[n-1] \\ &\quad + e_i[n]x_i^*[n] - e_i[n-1]x_i^*[n-1] \end{aligned} \tag{18.41}$$

denote the additive noises. Note that $h_{i,j}[n-2]$ is present in both relations in (18.40). For simplicity, we denote $h_{i,j}[n-2]$ by $h_{i,j}$ in rest of the chapter. It can be seen from (18.41) that noise present in these relations is heavily correlated. After calculating noise correlation matrix $\left(\Lambda_i = E\left\{\begin{bmatrix} w_i[n-1] \\ w_i[n] \end{bmatrix}\left[w_i^*[n-1]w_i^*[n]\right]\right\}\right)$, we can obtain the PDF of $\mathrm{z}_i[n] = [y_i'[n-1], y_i'[n]]^T$. By maximizing this PDF, a ML decoder of the symbols of ES-j can be obtained as

$$\hat{\mathrm{s}}_j = \mathrm{s}_j \in \mathcal{A}^2 \min\left((\mathrm{z}_i[n] - \mathrm{m}_i[n])\Lambda^{-1}(\mathrm{z}_i[n] - \mathrm{m}_i[n])^H\right), \tag{18.42}$$

where the vector s_j contains the symbols $s_j[n-1]$ and $s_j[n]$ transmitted by the ES-j and $(\cdot)^H$ denotes the hermitian. By substituting the value of Λ into (18.42), and after some algebra, we get the following ML decoder of s_j

$$\begin{aligned} \hat{\mathrm{s}}_j &= \mathrm{s}_j \in \mathcal{A}^2 \min(2|y_i'[n-1] - m_i[n-1]|^2 + 2|y_i'[n] - m_i[n]|^2 \\ &\quad + (y_i'[n-1] - m_i[n-1])(y_i'[n] - m_i[n])^* \\ &\quad + (y_i'[n] - m_i[n])(y_i'[n-1] - m_i[n-1])^*). \end{aligned} \tag{18.43}$$

It can be observed that $m_i[n-1]$ and $m_i[n]$ depend upon the effective channel gain $h_{i,j}$; hence, the decoder in (18.43) depends upon $h_{i,j}$. For deriving a differential detector which does not depend upon any channel information, we need to eliminate the dependence upon $h_{i,j}$ in (18.43). One can find an estimate of $h_{i,j}$ by minimizing (18.43) with respect to $h_{i,j}$ under the assumption that $\mathbf{s}_j$ is perfectly known, and then substitute this estimate back in (18.43) to obtain a decoder which is independent of $h_{i,j}$. Then the differential detector of s_j can be obtained by substituting the value of $\hat{h}_{i,j}$ in (18.43).

18.4.1 Constellation rotation angle calculation

In order to reduce the error in the channel estimate $\hat{h}_{i,j}$, we need to minimize the MSE in the channel estimate over all possible relative rotation angles between the two constellations. This observation leads to the following optimization problem:

$$\begin{aligned} &\text{minimize } i \neq j \sum_{i=1}^{2}\sum_{j=1}^{2} E\left\{\left|h_{i,j} - \hat{h}_{i,j}\right|^2\right\} \\ &\text{such that } s_i[k] \in \left\{e^{j\phi}, e^{j(\phi+2\pi/M)}, \ldots, e^{j(\phi+2(M-1)\pi/M)}\right\}, \\ &s_j[k] \in \left\{e^{j(\phi+\theta)}, e^{j(\phi+\theta+2\pi/M)}, \ldots, e^{j(\phi+\theta+2(M-1)\pi/M)}\right\}, \\ &\phi = 0,\ 0 < \theta < \pi, \end{aligned}$$

where expectation is performed upon the RV $h_{i,j} - \hat{h}_{i,j}$ and the unit-norm M-PSK constellations of $s_i[k]$ and $s_j[k]$. It can be seen from (18.17) that the optimization depends upon the PDF of the RV $h_{i,j} - \hat{h}_{i,j}$, which in turn depends upon the PDF of the channel $h_{i,j}$ and its estimate $\hat{h}_{i,j}$. The optimized rotation angle can be numerically calculated from (18.17). The LMS channel parameters for different fading scenarios of satellite links are listed in [1, Table III]. The optimized rotation angles for different constellations and fading scenario are shown in [11, Table II]. If the downlink of ES-i is FHS and the uplink of ES-j is AS, then this fading scenario is named as FHS/AS in Table 2. Similarly, other fading scenarios are named.

The average PEP and diversity order calculation can be derived by using standard procedure. The detailed analysis of these calculations is given in [11].

18.5 Multiple antennas-based TWSR system

In this section, we discuss beamforming and combining in TWSR communication systems, where ESs contain multiple antennas. Beamforming and combining techniques can be broadly divided into two parts: (1) beamforming and combining technique based on local channel information and (2) optimal beamforming and combining technique. Here, these schemes are discussed with the assumption that perfect CSI is available at different nodes; however, perfect CSI is not available, in practice. Therefore, channel estimates are obtained by using the method discussed in Section 18.3, and perfect CSI is replaced by estimated CSI.

We consider a two-way cooperative system with two ESs and a satellite with single antenna, in which ES-i, $i = 1, 2$ has N_i antennas. In this two-way relaying scheme, the transmission of data takes place in two phases. In the first phase, both ESs transmit their beamformed data simultaneously over the satellite uplink. In the second phase, the satellite broadcasts the received signal with fixed gain over the downlink to both ESs. The complex baseband signal vector received at the satellite uplink from both users in the first phase is given by

$$y_s = \sum_{i=1}^{2} \mathbf{h}_i^T \mathbf{u}_i s_i + n_s, \tag{18.44}$$

where $(\cdot)^T$ denotes transpose; u_i is the $N_i \times 1$ beamforming vector at ES-i; s_i is the complex-valued symbol with energy E_{s_i}, belonging to an M-PSK constellation, transmitted by ES-i; and n_s denotes the AWGN with zero mean and σ_s^2 variance. In the second phase, the $N_i \times 1$ complex baseband signal vector received at ES-i over the downlink is given by

$$\mathbf{r}_i = a\mathbf{g}_i^T y_s + \mathbf{w}_i, \tag{18.45}$$

where $\mathbf{w}_i$ contains AWGN noise elements with zero mean and σ_i^2 variance.

For removing the self-interference from (18.45), each ES requires the knowledge of its own uplink and downlink channels $\boldsymbol{h}_i$ and $\boldsymbol{g}_i$, respectively. In practice, this knowledge can be generated by using training data. After subtracting the self-interference term from (18.45), we get $\hat{\mathbf{r}}_i = \mathbf{r}_i - a\mathbf{g}_i^T\mathbf{h}_i\mathbf{u}_i s_i$. By multiplying $\tilde{\mathbf{r}}_i$ with $\mathbf{v}_i^H$ (the Hermitian of the combining vector) and from (18.44), the signal received at the ES-i is given by

$$y_i = \mathbf{v}_i^H\tilde{\mathbf{r}}_i = a\mathbf{v}_i^H\mathbf{g}_i^T\mathbf{h}_j\mathbf{u}_j s_j + a\mathbf{v}_i^H\mathbf{g}_i^T n_s + \mathbf{v}_i^H\mathbf{w}_i, \tag{18.46}$$

where $j = 1, 2$ and $i \neq j$.

18.5.1 Beamforming and combining using local channel information

In this scheme, each ES performs beamforming by using its own channel information [10]. In practice, both ESs use their downlink channel estimates.

The transmit weight vector $\mathbf{u}_j$ at ES-j is chosen as

$$\mathbf{u}_j = \frac{\boldsymbol{h}_j^H}{\|\boldsymbol{h}_j^H\|}, \tag{18.47}$$

where $\|\cdot\|$ denotes the Euclidean norm. The combining of the received signal at each ES is also performed by utilizing only their local channel information. The combining vector $\mathbf{v}_i$ at ES-i is given by

$$\mathbf{v}_i = \frac{\boldsymbol{g}_i^T}{\|\boldsymbol{g}_i^T\|}. \tag{18.48}$$

The *instantaneous received SNR at ES-i* is obtained from (18.46) as

$$\gamma_i = \frac{a^2\left|\boldsymbol{v}_i^H\boldsymbol{g}_i^T\boldsymbol{h}_j\boldsymbol{u}_j\right|^2 E(|s_j|^2)}{a^2\sigma_s^2\left|\boldsymbol{g}_i^*\boldsymbol{v}_i\right|^2 + \sigma_i^2\|\boldsymbol{v}_i\|^2}, \tag{18.49}$$

where E(·) is the expectation operator. By substituting the value of $\mathbf{u}_j$ from (18.47) and $\mathbf{v}_i$ from (18.48) in (18.49), and after some algebra, we get

$$\gamma_i = \frac{a^2\|\boldsymbol{g}_i\|^2\|\boldsymbol{h}_j\|^2 E_{s_j}}{a^2\sigma_s^2\|\boldsymbol{g}_i\|^2 + \sigma_i^2}. \tag{18.50}$$

From (18.46), the conditional PDF of y_i is given by

$$f(y_i|\mathbf{g}_i, \mathbf{h}_j, \mathbf{v}_i, \mathbf{u}_j, s_j) = \frac{exp(-(|y_i - a\boldsymbol{v}_i^H\boldsymbol{g}_i^T\boldsymbol{h}_j\boldsymbol{u}_j s_j|^2/a^2\sigma_s^2|\boldsymbol{g}_i^*\boldsymbol{v}_i|^2 + \sigma_i^2\|\boldsymbol{v}_i\|^2))}{\pi(a^2\sigma_s^2|\boldsymbol{g}_i^*\boldsymbol{v}_i|^2 + \sigma_i^2\|\boldsymbol{v}_i\|^2)}. \tag{18.51}$$

The detector of symbol s_j chooses the value of s_j which maximizes this conditional PDF by minimizing $|y_i - a\|\mathbf{g}_i\|\|\mathbf{h}_j\|s_j|^2$.

18.5.2 Received SNR optimal beamforming and combining

From the principle of maximum ratio transmission [26], the transmit weight vector $\mathbf{u}_j$ at ES-j is chosen by

$$\mathbf{u}_j = \frac{(\boldsymbol{g}_i\boldsymbol{h}_j^T)^H\boldsymbol{v}_i}{\|(\boldsymbol{g}_i\boldsymbol{h}_j^T)^H\boldsymbol{v}_i\|}, \tag{18.52}$$

where $\|\cdot\|$ denotes the Euclidean norm. The *instantaneous received SNR at ES-i* is obtained from (18.46) as

$$\gamma_i = \frac{\|(\boldsymbol{g}_i\boldsymbol{h}_j^T)^H\boldsymbol{v}_i\|^2}{\sigma_n^2\|\boldsymbol{g}_i^H\boldsymbol{v}_i\|^2 + \sigma_i^2\|\boldsymbol{v}_i\|^2}. \tag{18.53}$$

From (18.49), we can write

$$\gamma_i = \frac{\|(\boldsymbol{g}_i\boldsymbol{h}_j^T)^H\boldsymbol{v}_i\|^2}{\boldsymbol{v}_i^H(\sigma_n^2\boldsymbol{g}_i\boldsymbol{g}_i^H + \sigma_i^2\boldsymbol{I}_{N_i})\boldsymbol{v}_i}, \tag{18.54}$$

where $\mathbf{I}_{N_i}$ denotes the $N_i \times N_i$ identity matrix. Let us assume $\mathbf{A}_i\mathbf{A}_i^H = \sigma_n^2\mathbf{g}_i\mathbf{g}_i^H + \sigma_i^2\mathbf{I}_{N_i}$ and $\mathbf{z}_i = \mathbf{A}_i^H\mathbf{v}_i$. By substituting these values in (18.49) and after some algebra, we get $\gamma_i = \|(\mathbf{A}_i^{-1}\mathbf{g}_i\mathbf{h}_j^T)^H\mathbf{z}_i\|^2/\|\mathbf{z}_i\|^2$. It can be shown that γ_i is maximized if $\mathbf{z}_i$ is chosen as the eigenvector corresponding to the maximum eigenvalue of $(\mathbf{A}_i^{-1}\mathbf{g}_i\mathbf{h}_j^T)(\mathbf{A}_i^{-1}\mathbf{g}_i\mathbf{h}_j^T)^H$, i.e. $\mathbf{z}_{i,max}$. Hence, the optimal combining vector will be $\mathbf{v}_i = (\mathbf{A}_i^H)^{-1}\mathbf{z}_{i,max}$. By substituting these values of $\mathbf{v}_i$ and $\mathbf{u}_j$ in (18.51), a detector of symbol s_j can be obtained [27].

18.6 Analytical performance of TWSR scheme based on local channel information

First we discuss the analytical performance of the scheme based on local channel information in terms of SER and diversity order. Let $\|\mathbf{h}_j\|^2 = x$, $\|\mathbf{g}_i\|^2 = y$, $\bar{\gamma} = E_{s_j}/\sigma_s^2$, and $C = \sigma_i^2/(a^2\sigma_s^2)$. From (18.50), we get

$$\gamma_i = \frac{\bar{\gamma}xy}{y+C}. \tag{18.55}$$

The ES-i-satellite link is modelled as SR fading channel. An approximate PDF of $\|\mathbf{g}_i\|^2$ is given by [28]

$$\begin{aligned} f_{\|\mathbf{g}_i\|^2}(z) = \alpha_i^{N_i}\sum_{l_i=0}^{c_i}\binom{c_i}{l_i}\beta_i^{c_i-l_i}\Bigg(&\frac{z^{d_i-l_i-1}}{\Gamma(d_i-l_i)} \\ &\times {}_1F_1(d_i; d_i-l_i; -(\beta_i-\delta_i)z) + \frac{\epsilon_i\delta_i z^{d_i-l_i}}{\Gamma(d_i-l_i+1)} \\ &\times {}_1F_1(d_i+1; d_i-l_i+1; -(\beta_i-\delta_i)z)\Bigg). \end{aligned} \tag{18.56}$$

18.6.1 Expression of the SER

From (18.55), MGF of the instantaneous received SNR γ_i can be written as

$$M_{\gamma_i}(s) = \int_0^\infty \int_0^\infty e^{-s\frac{\bar{\gamma}xy}{y+C}} f_{\|\mathbf{h}_j\|^2}(x) f_{\|\mathbf{g}_i\|^2}(y)dxdy. \tag{18.57}$$

After many algebra, it can be shown that the MGF of the scheme is given by [10]

$$\begin{aligned} M_{\gamma_i}(s) = & \frac{\alpha_i^{N_i}\tilde{\alpha}_j^{N_j}\sum_{l_j=0}^{\tilde{c}_j}\binom{\tilde{c}_j}{l_j}\tilde{\beta}_j^{\tilde{c}_j-l_j}}{(\tilde{\beta}_j-\tilde{\delta}_j)^{\tilde{d}_j}} \\ & \times \sum_{l_i=0}^{c_i}\binom{c_i}{l_i}\beta_i^{c_i-l_i}(s\bar{\gamma})^{l_j}(\mathcal{K}(l_i,d_i,l_j,\tilde{d}_j,\bar{\gamma}_j,k) + \epsilon_i\delta_i \times \mathcal{K}(l_i,d_i+1,l_j,\tilde{d}_j,\bar{\gamma}_j,k) \\ & + \frac{\tilde{\epsilon}_j\tilde{\delta}_j}{\tilde{\beta}_i-\tilde{\delta}_j}\mathcal{K}(l_i,d_i,l_j,\tilde{d}_j+1,\bar{\gamma}_j,k) + \frac{\epsilon_i\delta_i\tilde{\epsilon}_j\tilde{\delta}_j}{\tilde{\beta}_j-\tilde{\delta}_j}\mathcal{K}(l_i,d_i+1,l_j,\tilde{d}_j+1,\bar{\gamma}_j,k)), \end{aligned} \tag{18.58}$$

where

$$\begin{aligned} \mathcal{K}(l_i,d_i,l_j,\tilde{d}_j,\bar{\gamma}_j,k) = & \sum_{k=0}^{\tilde{d}_j-l_j}\binom{\tilde{d}_j-l_j}{k}\frac{\bar{\gamma}_j^{-(l_j+k+d_i-l_i)}}{\Gamma(d_i)\Gamma(\tilde{d}_j)C^{l_j+k}} \\ & \times G_{23}^{22}\left(\frac{\beta_i-\delta_i}{\bar{\gamma}_j}\middle| \begin{matrix} 1-d_i, 1-(l_j+k+d_i-l_i) \\ 0, \tilde{d}_j-l_j-k-d_i+l_i, 1-d_i+l_i \end{matrix}\right), \end{aligned} \tag{18.59}$$

where $\bar{\gamma}_j = \frac{\bar{\gamma}s+\tilde{\beta}_j-\tilde{\delta}_j}{(\tilde{\beta}_j-\tilde{\delta}_j)C}$, and $G_{p,q}^{m,n}\left(\cdot\middle|\begin{matrix}\cdots\\ \cdots\end{matrix}\right)$ is the Meijer-G function [16, Eq. (9.301)]. The SER of the considered scheme for M-PSK constellation can be calculated by using the relation given in [10].

18.6.2 Diversity order

Diversity order of the considered system can be derived by using the expression of asymptotic MGF and it can be derived by Slater's theorem [10] and utilizing the fact that for $z \to 0$, ${}_pF_q(a_1, a_2, \ldots, a_p; b_1, b_2, \ldots, b_q; z) \to 1$ [29]. By substituting $l_i = c_l, l_l = \tilde{c}_l, k = k_l = 0$ and using Lemma 1 (assuming very large SNR), the asymptotic MGF (depending only upon the lowest power of the average SNR) can be expressed as follows:

1. For $N_j > N_i$

$$\begin{aligned} M_{\gamma_i}(s) \approx & \frac{\alpha_i^{N_i}\tilde{\alpha}_j^{N_j}(s\bar{\gamma})^{-N_i}}{(\tilde{\beta}_j-\tilde{\delta}_j)^{N_j-N_i}C^{-N_i}}\Bigg(\mathcal{T}(N_i,N_j,\tilde{c}_j,\tilde{d}_j) \\ & + \frac{\tilde{\epsilon}_j\tilde{\delta}_j}{\tilde{\beta}_j-\tilde{\delta}_j}\mathcal{T}(N_i,N_j+1,\tilde{c}_j,\tilde{d}_j+1)\Bigg), \end{aligned} \tag{18.60}$$

where $\mathcal{T}(N_i,N_j,\tilde{c}_j,\tilde{d}_j) = \Gamma(N_j-N_i)\Gamma(\tilde{c}_j+N_i)/(\Gamma(N_i)\Gamma(\tilde{d}_j))$.

2. For $N_i > N_j$

$$M_{\gamma_i}(s) \approx \frac{\alpha_i^{N_i}\tilde{\alpha}_j^{N_j}(s\bar{\gamma})^{-N_j}}{(\beta_i - \delta_i)^{N_i - N_j} C^{-N_j}} \left(\mathcal{T}(N_j, N_i, c_i, d_i) \right.$$

$$\left. + \frac{\epsilon_i \delta_i}{\beta_i - \delta_i} \mathcal{T}(N_j, N_i + 1, c_i, d_i + 1) \right). \tag{18.61}$$

3. For $N_i = N_j$

$$M_{\gamma_i}(s) \approx \frac{\alpha_i^{N_i}\tilde{\alpha}_j^{N_i}(s\bar{\gamma})^{-N_i}}{C^{-N_i}} \left(\frac{2}{\Gamma(N_i)} + \frac{\tilde{\epsilon}_j \tilde{\delta}_j}{\tilde{\beta}_j - \tilde{\delta}_j} + \frac{\epsilon_i \delta_i}{\beta_i - \delta_i} \frac{1}{d_i \Gamma(N_i)} \right). \tag{18.62}$$

It can be seen from (18.60), (18.61), and (18.62) that the diversity of the presented two way relaying scheme is limited by $\min(N_1, N_2)$.

18.6.3 Numerical results and discussion

All links are assumed to be the SR fading LMS links. The analytical and simulation results are plotted for AS ($b = 0.126$, $m = 10.1$, $\Omega = 0.835$) and FHS ($b = 0.063$, $m = 0.739$, $\Omega = 8.97 \times 10^{-4}$).

In Figure 18.6, the SER versus SNR performance of the two-way relaying scheme with $N_1 = N_2 = 2, 3, 4, 5$ and QPSK constellation is shown. For $N_1 = N_2 = 2, 3$ case, it is assumed that the ES-1-to-satellite LMS channel experiences AS, whereas ES-2 faces FHS; this fading scenario is denoted by AS/FHS. For the remaining two cases, i.e. $N_1 = N_2 = 4, 5$, it is assumed that both ESs experience FHS, i.e. FHS/FHS fading scenario exists. A tight matching of the theoretical and simulated SER values is evident from Figure 18.6; further, by adding an additional spatial dimension on the ESs helps significantly in overcoming the severe effects of shadowing. For example, the TWSR system with $N_1 = N_2 = 3$ performs approximately 5.5 dB better than that with $N_1 = N_2 = 2$, at SER $= 10^{-2}$, as seen in Figure 18.6. The aforementioned analytical and simulated SER values are obtained by assuming that perfect knowledge of the CSI, i.e. g_i and h_j, is available at both ESs. We have also shown the SER versus SNR plots for the erroneous CSI case, when there is 20% mean square error in the estimation of the shadowing parts of g_i and h_j with $N_1 = N_2 = 3, 4$ and AS/FHS fading scenario. It can be seen from the figure that for $N_1 = N_2 = 3$ and SER $= 10^{-3}$, there is approximately 1 dB loss of SNR by using the erroneous CSI. However, by increasing the number of antennas to $N_1 = N_2 = 4$, additional performance gain of approximately 2.75 dB can be achieved with the estimated CSI compared to the perfect CSI case with $N_1 = N_2 = 3$, as seen in Figure 18.6.

In Figure 18.7, the analytical SER of the considered TWSR scheme is plotted for large value of the SNR. The analytical values are obtained for QPSK constellation, $N_1 = 2, 5, 7$, and $N_2 = 2, 3, 4$, for FHS/FHS and AS/FHS fading scenarios. We have also plotted the ideal diversity plots by using the relation $\kappa / \bar{\gamma}^{\delta}$, where κ is a positive-valued constant and δ denotes the diversity order, for indicating the slope of the decay of SER versus SNR plot at high SNR values. The diversity order of the system is

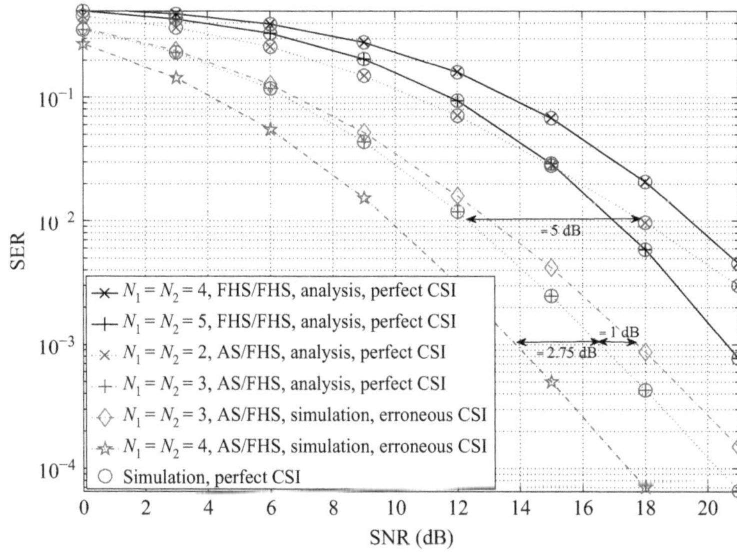

Figure 18.6 Analytical and simulated SER versus SNR performances of the beamforming and combining based two-way satellite cooperative system with QPSK constellation and $\boldsymbol{N_1 = N_2 = 2, 3, 4, 5}$

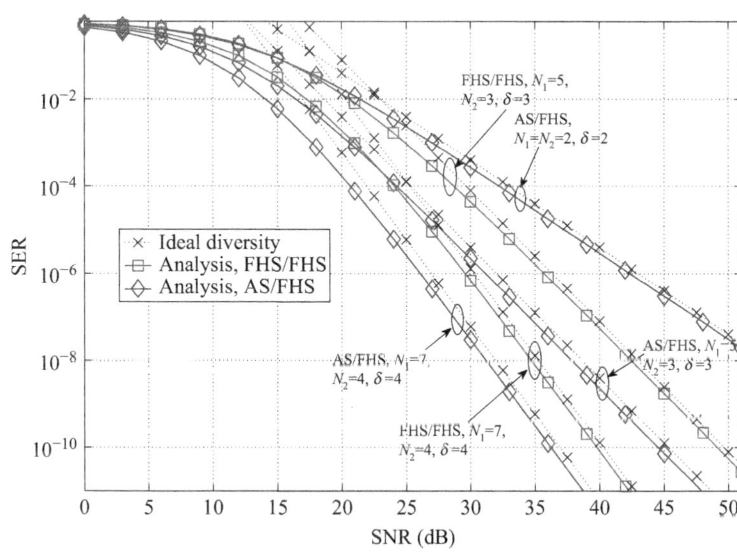

Figure 18.7 Diversity performance of the beamforming and combining based two-way satellite cooperative system with QPSK constellation, $\boldsymbol{N_1 = 2, 5, 7}$*, and* $\boldsymbol{N_2 = 2, 3, 4}$

independent of the fading scenario, as can be seen from the figure. For example, for $\{N_1 = 5, N_2 = 3\}$ and $\{N_1 = 7, N_2 = 4\}$ antenna configurations, the diversity order is three (min(5,3)) and four (min(7,4)), respectively, for FSH/FHS and AS/FHS fading distributions. Further, having better fading in a link does not help in gaining any additional diversity; for $N_1 = N_2 = 2$ and AS/FHS case, the diversity order is two only. Overall, the diversity order of the presented scheme depends upon $\min(N_1, N_2)$ in Figure 18.7.

18.7 Analytical performance of TWSR scheme based on optimal beamforming and combining

In this section, the analytical performance of the system is derived in terms of approximate SER and diversity order. Note that $E\{\mathbf{g}_i\mathbf{g}_i^H\} = \eta_i\mathbf{I}_{N_i}$ and $E\{|g_{ik}|^2\} = \eta_i$ for each $k = 1, 2, \ldots, N_i$; η_i is a function of b_i, m_i and Ω_i (as per [1]). We can write $\gamma_i \approx \bar{\gamma}_i\lambda_{i,j}$ after some algebra in (18.49) by considering high SNR scenario, but it can be shown by simulation that this approximation works well for all considered values of the SNR. Here $\lambda_{i,j}$ is the maximum eigenvalue of $\mathbf{g}_i\mathbf{h}_j^T\mathbf{h}_j^*\mathbf{g}_i^H$ and $\bar{\gamma}_i = 1/(\sigma_i^2 + \eta_i\sigma_n^2)$ denotes the *average SNR at ES-i*, where $(\cdot)^*$ denotes the complex conjugate. As $\mathbf{g}_i\mathbf{h}_j^T\mathbf{h}_j^*\mathbf{g}_i^H = \|\mathbf{h}_j\|^2\mathbf{g}_i\mathbf{g}_i^H$ and $\mathbf{g}_i\mathbf{g}_i^H$ contains only one eigenvalue, i.e. $\|\mathbf{g}_i\|^2$, therefore, $\lambda_{i,j} = \|\mathbf{g}_i\|^2\|\mathbf{h}_j\|^2$. The PDF of $\lambda_{i,j}$ can be derived as

$$f_{\lambda_{i,j}}(y) = \alpha_i^{N_i}\sum_{l_i=0}^{c_i}\binom{c_i}{l_i}\beta_i^{c_i-l_i}\alpha_j^{N_j}\sum_{l_j=0}^{c_j}\binom{c_j}{l_j}\beta_j^{c_j-l_j}\left[\frac{I_1(d_i,d_j,y)}{\Gamma(d_i-l_i)\Gamma(d_j-l_j)}\right.$$
$$+\frac{\epsilon_j\delta_jI_1(d_i,d_j+1,y)}{\Gamma(d_i-l_i)\Gamma(d_j-l_j+1)}+\frac{\epsilon_i\delta_iI_1(d_i+1,d_j,y)}{\Gamma(d_i-l_i+1)\Gamma(d_j-l_j)}$$
$$\left.+\frac{\epsilon_i\delta_i\epsilon_j\delta_jI_1(d_i+1,d_j+1,y)}{\Gamma(d_i-l_i+1)\Gamma(d_j-l_j+1)}\right], \tag{18.63}$$

where

$$I_1(d_i,d_j,y) = \frac{\Gamma(d_i-l_i)\Gamma(d_j-l_j)y^{d_j-l_j-1}}{\Gamma(d_i)\Gamma(d_j)(\beta_i-\delta_i)^{d_i-l_i-d_j+l_j}}$$
$$\times G_{24}^{22}\left((\beta_i-\delta_i)(\beta_j-\delta_j)y\,\middle|\,\begin{matrix}1-d_j,1-l_i-d_j+l_j\\0,d_i-l_i-d_j+l_j,1-d_j+l_j,1-d_j+l_j\end{matrix}\right) \tag{18.64}$$

and $G_{p,q}^{m,n}\left(\cdot\middle|\begin{matrix}\cdots\\\cdots\end{matrix}\right)$ is the Meijer-G function [16, Eq. (9.301)].

18.7.1 Expression of SER

The MGF of γ_i can be derived after many algebra, as

$$M_{\gamma_i}(s) = \alpha_i^{N_i} \sum_{l_i=0}^{c_i} \binom{c_i}{l_i} \beta_i^{c_i-l_i} \alpha_j^{N_j} \sum_{l_j=0}^{c_j} \binom{c_j}{l_j} \beta_j^{c_j-l_j} \left[\frac{M_1(d_i,d_j,s)}{\Gamma(d_i-l_i)\Gamma(d_j-l_j)} \right.$$
$$+ \frac{\epsilon_j\delta_j M_1(d_i,d_j+1,s)}{\Gamma(d_i-l_i)\Gamma(d_j-l_j+1)} + \frac{\epsilon_i\delta_i M_1(d_i+1,d_j,s)}{\Gamma(d_i-l_i+1)\Gamma(d_j-l_j)}$$
$$\left. \times \frac{\epsilon_i\delta_i\epsilon_j\delta_j M_1(d_i+1,d_j+1,s)}{\Gamma(d_i-l_i+1)\Gamma(d_j-l_j+1)} \right], \tag{18.65}$$

where

$$M_1(d_i,d_j,s) = \frac{\Gamma(d_j-l_j)(\beta_i-\delta_i)^{-(d_i-l_i-d_j+l_j)}}{(\Gamma(d_i-l_i))^{-1}(s\bar{\gamma}_i)^{d_j-l_j}\Gamma(d_i)\Gamma(d_j)}$$
$$\times G_{34}^{23}\left(\frac{s^{-1}(\beta_i-\delta_i)}{\bar{\gamma}_i(\beta_j-\delta_j)^{-1}} \middle| \begin{matrix} 1-d_j, 1-l_i-d_j+l_j, 1-d_j+l_j \\ 0, d_i-l_i-d_j+l_j, 1-d_j+l_j, 1-d_j+l_j \end{matrix} \right) \tag{18.66}$$

By using (18.66) and standard relation of SER in terms of MGF [27], the SER of the considered system can be obtained.

18.7.2 Diversity order

The diversity order can be determined by the high SNR rate of decay of $M_{\gamma_i}(s)$. Hence, we can find the diversity order of the studied scheme by using the asymptotic MGF. It is shown in [10] that for $z \to 0$, the Meijer-G function can be approximately written as

$$G_{pq}^{mn}\left(z \middle| \begin{matrix} a_1, a_2, \ldots, a_p \\ b_1, b_2, \ldots, b_q \end{matrix} \right)$$
$$\approx \sum_{h=1}^{m} \frac{\prod_{\substack{j=1 \\ j\neq h}}^{m} \Gamma(b_j-b_h) \prod_{j=1}^{n} \Gamma(1+b_h-a_j) z^{b_h}}{\prod_{j=m+1}^{q} \Gamma(1+b_h-b_j) \prod_{j=n+1}^{p} \Gamma(a_j-b_h)}. \tag{18.67}$$

From (18.65) and (18.67), we can write

$$M_{\gamma_i}(s) \approx \alpha_i^{N_i} \sum\nolimits_{l_i=0}^{c_i} \binom{c_i}{l_i} \beta_i^{c_i-l_i} \alpha_j^{N_j} \sum\nolimits_{l_j=0}^{c_j} \binom{c_j}{l_j} \beta_j^{c_j-l_j} \left[\frac{\epsilon_i\delta_i \tilde{M}_1(d_i+1,d_j,s)}{\Gamma(d_i-l_i+1)\Gamma(d_j-l_j)} \right.$$
$$\left. + \frac{\epsilon_j\delta_j M_1(d_i,d_j+1,s)}{\Gamma(d_i-l_i)\Gamma(d_j-l_j+1)} + \frac{\tilde{M}_1(d_i,d_j,s)}{\Gamma(d_i-l_i)\Gamma(d_j-l_j)} + \frac{\epsilon_i\delta_i\epsilon_j\delta_j \tilde{M}_1(d_i+1,d_j+1,s)}{\Gamma(d_i-l_i+1ght)\Gamma(d_j-l_j+1)} \right], \tag{18.68}$$

where

$$\tilde{M}_1(d_i, d_j, s)$$
$$= \frac{\Gamma(d_i - l_i)\Gamma(d_j - l_j)}{\Gamma(d_i)\Gamma(d_j)(\beta_i - \delta_i)^{d_i - l_i - d_j + l_j}} (s\bar{\gamma}_i)^{-d_j + l_j} \sum_{h=1}^{2} \frac{\prod_{\substack{j=1 \\ j \neq h}}^{2} \Gamma(b_j - b_h)^* \prod_{j=1}^{3} \Gamma(1 + b_h - a_j)}{\prod_{j=3}^{4} \Gamma(1 + b_h - b_j)}$$
$$\times \left(\frac{(\beta_i - \delta_i)(\beta_j - \delta_j)}{s\bar{\gamma}_i} \right)^{b_h}, \qquad (18.69)$$

$a_1 = 1 - d_j$, $a_2 = 1 - l_i - d_j + l_j$, $a_3 = 1 - d_j + l_j$, $b_1 = 0$, $b_2 = d_i - l_i - d_j + l_j$, $b_3 = 1 - d_j + l_j$, and $b_4 = 1 - d_j + l_j$.

For diversity order calculation, we take $l_i = c_i$ and $l_j = c_j$. From (18.69), we can write

$$\tilde{M}_1(d_i, d_j, s) = \frac{\Gamma(d_i - c_i)\Gamma(d_j - c_j)}{\Gamma(d_i)\Gamma(d_j)(\beta_i - \delta_i)^{d_i - c_i - d_j + c_j}} (s\bar{\gamma}_i)^{-d_j + c_j} \left(\frac{\Gamma(b_2)\prod_{j=1}^{3}\Gamma(1 - a_j)}{\prod_{j=3}^{4}\Gamma(1 - b_j)} \right.$$
$$\left. + \frac{\Gamma(b_1 - b_2)\prod_{j=1}^{3}\Gamma(1 + b_2 - a_j)\left(\frac{(\beta_i - \delta_i)(\beta_j - \delta_j)}{s\bar{\gamma}_i}\right)^{b_2}}{\prod_{j=3}^{4}\Gamma(1 + b_2 - b_j)} \right). \qquad (18.70)$$

These are some observations for the diversity order of the studied scheme.

- It can be seen from (18.70) that for $b_2 > 0$ the lowest power of $\bar{\gamma}_i$ is $-d_j + c_j = -N_j$.
- From (18.70), it can be noticed that for $b_2 < 0$ the lowest power of $\bar{\gamma}_i$ is $-d_j + c_j - b_2 = -d_i + c_i = -N_i$.
- Therefore the diversity order of the studied scheme is $\min(N_i, N_j)$.

18.8 Numerical results and discussion

Numerical results are shown for $\bar{\gamma}_1 = \bar{\gamma}_2 = \bar{\gamma}$, which we call the SNR, in Figure 18.8. All results are shown for one of the ESs. It is assumed that all links are i.i.d. SR fading. The simulation and analytical results are plotted for all shadowing scenarios, given in [1]. Simulated SER versus SNR performance is shown in Figure 18.8 in AS, FHS, and ILS environment, for the presented scheme and existing beamforming and combining based scheme [6] with $N_i = N_j = 2$, QPSK constellation, perfect knowledge of CSI at all nodes, and unity transponder gain of the satellite. A close match in between analytical and simulated values verifies the correctness of our analytical results for all fading scenarios. It can be noticed from figure that the studied scheme performs noticeably better than the existing scheme for FHS/FHS (both ESs-satellite links experience FHS), AS/AS (both ESs-satellite link experience AS), and ILS/ILS (both ESs-satellite links experience ILS) environment. For example, an SNR

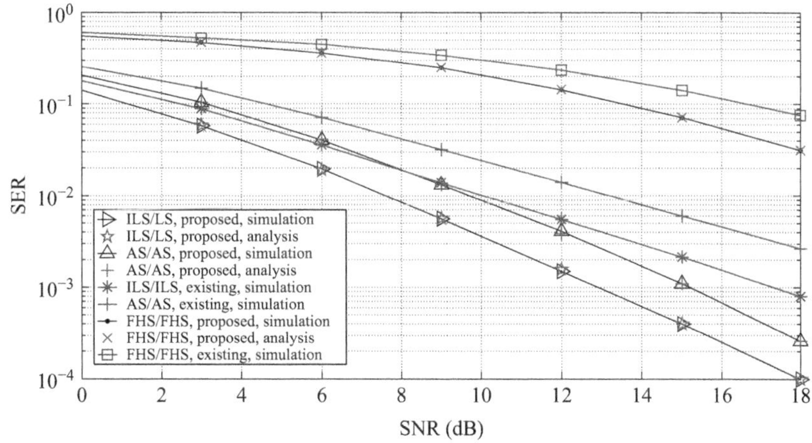

Figure 18.8 SER versus SNR performance of the presented scheme and same rate existing beamforming and combining based scheme [10] with $N_i = N_j = 2$ *over i.i.d. SR fading channels*

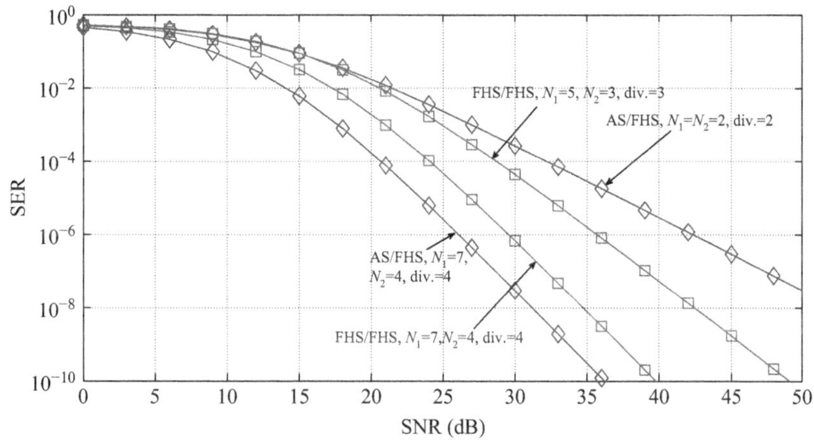

Figure 18.9 Analytical diversity performance of the beamforming and combining based two-way satellite cooperative system with i.i.d. SR fading and with QPSK constellation, $N_1 = 2,5,7$, $N_2 = 2,3,4$

gain of approximately 3.7 dB at an SER of 8×10^{-2}, 4.3 dB at an SER of 2×10^{-3}, and 4.6 dB at an SER of 8×10^{-4} can be obtained by using the studied scheme, in FHS/FHS, AS/AS, and ILS/ILS environment, respectively, as compared to the existing scheme [10]. The SNR gain is very significant for the satellite systems, because in satellite systems each additional dB transmit power significantly increases the weight, size, and cost of the satellite.

Analytical performance of the scheme is plotted by using (18.70) for i.i.d. SR fading with $N_1 = 2, 5, 7$, $N_2 = 2, 3, 4$, and with QPSK constellation in Figure 18.9.

It can be noticed from Figure 18.9 that the diversity order of the scheme is $\min(N_1, N_2)$. For example, $N_1 = 5$ and $N_2 = 3$, the diversity order of the considered scheme is three. Further, the fading scenario does not affect the diversity order of the scheme, as seen from the figure.

18.9 Conclusions

We have discussed the problems associated with the TWSR in this chapter. A training protocol for the TWSR system has been discussed and studied in detail. This training protocol is used to estimate the CSI required for self-interference cancellation and symbol decoding with sufficiently low estimation noise. Performance of this training-based scheme has been analysed in terms of BER and average capacity. Then, differential modulation-based TWSR has been discussed. The use of differential modulation allows for obtaining a differential detector which does not require any channel information in the destination ES. This useful virtue of the proposed differential detector allows for avoiding the difficulty of channel estimation in two-way AF satellite communication. Further, two beamforming and combining schemes for TWSR have been discussed in this chapter. In first scheme, the calculation of beamforming and combining vectors has been performed by utilizing local channels of ESs, whereas second scheme is SNR optimal and the beamforming and combining vectors have been calculated by using maximum eigenvalue criterion. It can be concluded that SNR optimal beamforming and combining outperforms the local channel-based beamforming and combining scheme. All the presented schemes are very useful for practical implementation of TWSR communication systems.

References

[1] Abdi A., Lau W., Alouini M.-S., Kaveh M. 'A new simple model for land mobile satellite channels: First and second order statistics'. *IEEE Trans. Wirel. Commun.* 2003, vol. 2(3), pp. 519–528.

[2] Jo K.Y. *Satellite Communications Network Design and Analysis*. Norwood, MA: Artech House, 2011.

[3] Shankar M.R.B., Zheng G., Maleki S., Ottersten, B. 'Feasibility study of full-duplex relaying in satellite networks'. *Proceedings of the 16th International Workshop on Signal Processing Advances in Wireless Communications (SPAWC)*; Stockholm, 2015, IEEE, pp. 560–564.

[4] Nwankwo C.D., Zhang L., Quddus A., Imran M.A., Tafazolli R. 'A survey of self-interference management techniques for single frequency full duplex systems'. *IEEE Access* 2017, vol. PP(99), pp. 1–25 **DOI:** 10.1109/ACCESS.2017.2774143.

[5] Carrier-in-Carrier (CinC) Technology. Canada: Gilat Satellite Networks Ltd. (NASDAQ:GILT). Available: https://www.gilat.com/ [accessed 14 Jan 2018].

[6] Telesat Carrier-in-Carrier Technology: http:/www.telesat.com/ [accessed 14 Jan 2018].

[7] Rankov B., Wittneben A. 'Achievable rate regions for the two-way relay channel'. *Proceedings of the IEEE International Symposium on Information Theory (ISIT)*; Seattle, WA, July 2006, pp. 1668–1672.

[8] Ji B., Huang Y., Wang H., Yang L. 'Performance analysis of two-way relaying satellite mobile communication'. *Proceedings of the 6th International ICST Conference on Communications and Networking in China (CHINACOM)*; Harbin, China, Aug. 2011, pp. 1099–1103.

[9] Xu C., Wilson S. 'Comparing two-way relay protocols for satellite communication'. [Online]. Available: http://www.cs.virginia.edu/_cx7m/comp.pdf.

[10] Arti M.K., Bhatnagar M.R. 'Two-way mobile satellite relaying: A beamforming and combining based approach'. *IEEE Commun. Lett.* 2014, vol. 18(7), pp. 1187–1190.

[11] Bhatnagar M.R. 'Making two-way satellite relaying feasible: A differential modulation based approach'. *IEEE Trans. Commun.* 2015, vol. 63(8), pp. 2836–2847.

[12] Pratt T., Bostian C., Allnutt J., *Satellite Communications*, 2nd ed. New York: John Wiley & Sons, 2003.

[13] Bhatnagar, M.R. 'Differential decoding of SIM DPSK over FSO MIMO links'. *IEEE Commun. Lett.* 2013, vol. 17(1), pp. 79–82.

[14] Bhatnagar M.R., Hjørungnes A., Song L. 'Differential coding for non-orthogonal space-time block codes with non-unitary constellations over arbitrarily correlated Rayleigh channels. *IEEE Trans. Wirel. Commun.* 2009, vol. 8(8), pp. 3985–3995.

[15] Bhatnagar M.R., Hjørungnes A. 'Differential coding for MAC based two-user MIMO communication systems'. *IEEE Trans. Wireless Commun.* 2012, vol. 11(1), pp. 9–14.

[16] Gradshteyn I.S., Ryzhik I.M., *Table of Integrals, Series, and Products*, 6th ed. San Diego, CA, USA: Academic Press; 2000.

[17] Gao F., Zhang R., Liang Y.-C. 'Optimal channel estimation and training design for two-way relay networks'. *IEEE Trans. Commun.* 2009, vol. 57(10), pp. 3024–3033.

[18] Abdallah S., Psaromiligkos I.N. 'Exact Cramer–Rao bounds for semi-blind channel estimation in amplify-and-forward two-way relay networks employing square QAM'. *IEEE Trans. Wireless Commun.* 2014, vol. 13(12), pp. 6955–6967.

[19] Couvreur C. 'The EM algorithm: A guided tour'. *Proceedings of the 2nd IEEE European Workshop on Computational Intensive Methods Control Signal Process*; Prague, Czech Republic, 1996, pp. 1–6.

[20] Lu J., Letaief K.B., Chuang J.C.-I., Liou M.L. 'M-PSK and M-QAM BER computation using signal-space concepts'. *IEEE Trans. Commun.* 1999, vol. 47(2), pp. 181–184.

[21] Arti M.K. 'Two-way satellite relaying with estimated channel gains'. *IEEE Trans. Commun.* 2016, vol. 64(7), pp. 2808–2820.

[22] Yilmaz F., Alouini M.-S. 'A unified MGF-based capacity analysis of diversity combiners over generalized fading channels'. *IEEE Trans. Commun.* 2012, vol. 60(3), pp. 862–875.

[23] Prudnikov A.P., Brychkov Y.A., Marichev O.I. *Integrals and Series*, 1st ed. New York, USA: Gordon and Breach Science Publishers; 1990.

[24] Proakis J.G. *Digital Communications*, 4th ed. Singapore: McGraw-Hill; 2001.

[25] Bhatnagar M.R., Hjørungnes A. *Cooperative Communications for Improved Wireless Network Transmission: Frameworks for Virtual Antenna Array Applications.* Hershey, PA: IGI Global; 2009, Chapter 12: Single and Double Differential Coding in Cooperative Communications, pp. 321–351.

[26] Lo T.K.Y. 'Maximum ratio transmission'. *IEEE Trans. Commun.* 1999, vol. 47(10), pp. 1458–1461.

[27] Arti M.K. 'A novel beamforming and combining scheme for Two-Way AF satellite systems'. *IEEE Trans. Veh. Technol.* 2017, vol. 66(2), pp. 1248–1256.

[28] Bhatnagar M.R., Arti M.K. 'On the closed-form performance analysis of maximal ratio combining in Shadowed-Rician fading LMS channels'. *IEEE Commun. Lett.* 2014, vol. 18(1), pp. 54–57.

[29] Abramowitz M., Stegun, I.A. *Handbook of Mathematical Functions.* New York, USA: Dover Publications; 1972.

List of Acronyms

3GPP	Third Generation Partnership Project
5G	Fifth Generation
ACI	Adjacent Channel Interference
ACK	Acknowledgement
ACM	Adaptive Coding and Modulation
ACRDA	Asynchronous Contention Resolution Diversity ALOHA
ACTN	Abstraction and Control of Transport Networks
ADC	Analog-to-Digital Converter
AF	Amplify-and-Forward
AM	Amplitude Modulation
AO	Adaptive Optics
APD	Avalanche Photo-Diode
API	Application Programming Interface
APSK	Amplitude Phase Shift Keying
ARQ	Automatic Repeat Request
AWGN	Additive White Gaussian Noise
BC	Broadcast Channel
BDP	Bandwidth Delay Product
BER	Bit Error Rate
BFN	Beamforming Network
BHBTP	Beam-hopping Burst Time Plan
BHC	Beam-hopping Configuration
BN	Backhaul Node
BoD	Bandwidth on Demand
BPF	Band Pass Filter
BPSK	Binary Phase Shift Keying
BS	Broadcast Service
BSM	Broadband Satellite Multimedia
BSTP	Beam-Switching Time Plan
CAPEX	Capital Expenditures
CBRS	Citizens Broadband Radio Service
CC	Convolution Code
CCDF	Complementary Cumulative Distribution Function
CCM	Constant Coding and Modulation
CCSDS	Consultative Committee for Space Data Systems
CDF	Cumulative Distribution Function
CDMA	Code Division Multiple Access
CDN	Content Delivery Network
CED	Conventional Energy Detector
CEPT	European Conference for Postal and Telecommunications
CER	Codeword Error Rate
CoMP	Coordinated Multipoint
COTS	Commercial-Off-The-Shelf
CP	Cyclic Prefix

CPA	Course Pointing Assembly
CR	Cognitive Radio
CRA	Contention Resolution ALOHA
CRDSA	Contention Resolution Diversity Slotted ALOHA
CSA	Coded Slotted ALOHA
CSI	Channel State Information
CSR	Cell Switch Router
CU	Capacity Unit
CWDM	Coarse Wavelength Division Multiplexing
DAA	Digital Antenna Arrays
DAC	Digital-to-Analog Converter
DAMA	Demand Assignment Multiple Access
dB	Decibel
DD	Direct Detection
DF	Decode-and-Forward
DoF	Degree of Freedom
DPC	Dirty Paper Coding
DPSK	Differential PSK
DR	Data Rate
DSL	Digital Subscriber Line
DSP	Digital Signal Processing
DTH	Direct to Home
DTP	Digital Transparent Processor
DVB	Digital Video Broadcasting
DVB-RCS	DVB -Return Channel via Satellite
DVB-S2x	DVB-Second Generation extension
DWDM	Dense Wavelength Division Multiplexing
E2E	End-to-End
ECRA	Enhanced Contention Resolution ALOHA
ED	Energy Detection
EDRS	European Data Relay System
EGC	Equal Gain Combining
EHF	Extremely High Frequency
EIRP	Effective Isotropic Radiated Power
eMBB	Enhanced Mobile Broadband
EMEA	Europe, the Middle East, and Africa
EO	Earth Observation
EPC	Evolved Packet Core
EPS	Evolved Packet System
ES	Earth Station
ESA	European Space Agency
ESIM	Earth Stations in Motion
ESOA	EMEA Satellite Operators Association
ESOMPS	Earth Stations On Moving Platforms
ESSA	Enhanced Spread Spectrum ALOHA
ETSI	European Telecommunications Standards Institute
FCC	Federal Communications Commission
FDM	Frequency Division Multiplexing
FDoA	Frequency Difference of Arrival
FEC	Forward Error Correction
FFR	Full Frequency Reuse
FFT	Fast Fourier Transform
FoA	Frequency of Arrival
FOTA	Firmware Over the Air
FPA	Fine Pointing Assembly
FPGA	Field Programmable Gate-Array

FR4	Frequency Reuse 4
FS	Fixed Service
F-SIM	Fixed Satellite Interactive Multimedia
FSS	Fixed Satellite Service
FTN	Faster-than-Nyquist
FWD	Forward
GaA	Gallium Arsenide
GaN	Gallium Nitride
GAA	General Authorized Access
GBR	Guaranteed Bit Rate
GD	Group Delay
GEO	Geostationary Earth Orbit
GPS	Global Positioning System
GW	Gateway
HD	High Definition
HDFSS	High Density FSS
HDR	High Data rate
HEO	Highly Elliptical Orbit
HNM	Hybrid Network Manager
HPA	High Power Amplifier
HTS	High Throughput Satellites
HTTP	Hypertext Transfer Protocol
IC	Interference Cancellation
ICO	Intermediate Circular Orbit
ID	Iterative Decoding
IETF	Internet Engineering Task Force
IFFT	Inverse Fast Fourier Transform
IM	Information Model
IMD	Intermodulation
IMUX	Input Multiplexer
IoT	Internet of Things
IP	Internet Protocol
IPTV	Internet Protocol Television
IRF	Intelligent Router Facility
IRSA	Irregular repetition Slotted ALOHA
IRT	Index-of-Refraction Turbulence
ISI	Intersymbol Interference
ISNR	interference-to-Signal-plus-Noise Ratio
ISP	Internet Service Provider
ITU-R	International Telecommunication Union -Radiocommunication
JT	Joint Transmission
KPI	Key Performance Indicator
LC-MAMP	Load-Controlled Multiple-Active Multiple-Passive
LC-PAA	Load-Controlled Parasitic Antenna Array
LC-SAMP	Load-Controlled Single-Active Multiple-Passive
LCT	Laser Communication Terminal
LDPC	Low-Density Parity Check
LEO	Low Earth Orbit
LLR	Log-Likelihood Ratio
LMS	Least-Mean Square
LNA	Low Noise Amplifier
LO	Local Oscillator
LOS	Line-of-Sight
LS	Least Square
LSA	Licensed Shared Access
LTE	Long Term Evolution

LTWTA	Linearized TWTA
LUT	Look-Up Table
M&C	Management and Control
M2M	Machine to Machine
MAC	Medium Access Control
MBH	Mobile Backhaul
MCM	Multicarrier Modulation
MCN	Mobile Core Network
MEC	Multi-access Edge Computing
MEO	Medium Earth Orbit
MFTDMA	Multi-Frequency Time Division Multiple Access
MIMO	Multiple Input Multiple Output
MME	Mobility Management Entity
MMSE	Minimum MSE
mMTC	Massive Machine Type Communications
MNO	Mobile Network Operator
ModCod	Modulation and Coding
MPLS	Multi-Protocol Label Switching
MRC	Maximum Ratio Combining
MSE	Mean Square Error
MSS	Mobile Satellite System
MUI	Multiuser Interference
MU-MIMO	Multiuser MIMO
MuSCA	Multi-Slot Coded ALOHA
MVNO	Mobile Virtual Network Operator
NBI	North-Bound Interface
NCC	Network Control Centre
NE	Network Element
NEO	Network Operation
NEP	Noise Equivalent Power
NetIC	Network Information and Control
NFV	Network Function Virtualization
NGSO	Non-Geostationary
N-LOS	Non-Line of Sight
NMC	Network Management Centre
NMSE	Normalized MSE
NNLS	Non-Negative Least Square
NRZ	Non-Return-to-Zero
NS	Network System
OBO	Output Back-Off
OBP	On-Board Processer
OF	Open Flow
OFDM	Orthogonal Frequency Division Multiplexing
OGS	Optical Ground Station
OLEODL	Optical LEO Downlinks
OMUX	Output Multiplexer
ONF	Open Networking Foundation
OOB	Out-of-Band
OOK	On-Off-Keying
OPEX	Operating Expenses
OPLL	Optical Phase-Locked Loop
OSPF	Open Shortest Path First
OSS	Operational Support Systems
OVN	Operator Virtual Network
P2MP	Point-to-Multi-Point

P2P	Point-to-Point
PAA	Point-Ahead Angle
PAPR	Peak-to-Average Power Ratio
PAT	Pointing, Acquisition and Tracking
PCCH	Physical Control Channel
PCE	Path Computation Engine
PD	Predistortion
PDCH	Physical Data Channel
PDF	Probability Density Function
PER	Packet Error Rate
PFD	Power Flux Density
PIN	Positive-Intrinsic-Negative
PKT	Packet
PLH	Physical Layer Header
PLR	Packet Loss Ratio
PM	Phase Modulation
PMP	Point-to-Multipoint
PN	Phase Noise
PPB	Photons Per Bit
PPM	Pulse Position Modulation
PPP	Public Private Partnership
PSD	Power Spectral Density
PU	Primary User
QAM	Quadrature Amplitude Modulation
QID	Queue Identifier
QoE	Quality of Experience
QoS	Quality of Service
QPSK	Quadrature Phase Shift Keying
RA	Random Access
RAB	Radio Access Bearer
RAN	Random Access Network
RAT	Radio Access Technology
RF	Radio Frequency
RFE	Receiver Front End
RRC	Root-Raised Cosine
RRM	Radio Resource Management
RS	Reed–Solomon
RTN	Return
RTT	Round Trip Time
RV	Random Variable
RZ	Return-to-Zero
SA	Slotted ALOHA
SAP	Service Access Point
SAS	Spectrum Access System
SatCom	Satellite Communications
SBI	South-Bound Interface
SC	Selection Combining
SCM	Single-Carrier Modulation
SCPC	Single Channel Per Carrier
SDE	Stochastic Differential Equation
SDN	Software Defined Networking
SDR	Software Defined Radio
SER	Symbol Error Rate
SF	Super-Frame
SFFI	SF format indicator

SFN	Single Frequency Network
SI	Scintillation Index
SIC	Successive Interference Cancellation
SISO	Single-Input-Single-Output
SLA	Service Level Agreement
SLP	Symbol-Level Precoding
S-MIM	S-band Mobile Interactive Multimedia
SMU	Spectrum Monitoring Unit
SNIR	Signal-to-Noise-plus-Interference Ratio
SNL	Shot-Noise-Limited
SNO	Satellite Network Operator
SNR	Signal-to-Noise Ratio
SOHO	Small Office Home Office
SOSF	Start-of-SF
SOTA	Software Over the Air
SR	Sum-Rate
SU	Secondary User
SUCC	Satellite Use Case Category
SU-MIMO	Single-User MIMO
SVNO	Satellite Virtual Network Operator
SwC	Switching Combining
TBS	Transportable Base Station
TCP	Transmission Control Protocol
TD	Total Degradation
TDM	Time Division Multiplexing
TDMA	Time Division Multiple Access
TDOA	Time Difference of Arrival
TE	Traffic Engineering
TS	Time Slot
TSN	Timing Slicing Number
TWSR	Two-Way Satellite Relaying
TWTA	Travelling Wave Tube Amplifier
UAS	Unmanned Aerial System
UAV	Unmanned Aerial Vehicles
UDP	User Datagram Protocol
UE	User Equipment
UHD	Ultra High Definition
URLLC	Ultra-Reliable Low Latency Communications
UT	User Terminal
VCM	Variable Coding and Modulation
VHF	Very High Frequency
VHTS	Very High Throughput Satellite
VLSNR	Very Low SNR
VM	Virtual Machine
VN	Virtual Network
VNF	Virtual Network Function
VNO	Virtual Network Operator
VoD	Video on Demand
VSAT	Very Small Aperture Terminal
WDM	Wavelength Division Multiplexing
WRC	World Radio-communication Conference
WRF	Weather Research Forecasting
ZF	Zero Forcing
ZFBF	Zero Forcing Beamforming

Index